Lecture Notes in Computer Science 8794

Commenced Publication in 1973
Founding and Former Series Editors:
Gerhard Goos, Juris Hartmanis, and Jan van Leeuwen

T0212926

Ying Tan   Yuhui Shi
Carlos A. Coello Coello (Eds.)

Advances in Swarm Intelligence

5th International Conference, ICSI 2014
Hefei, China, October 17-20, 2014
Proceedings, Part I

 Springer

Volume Editors

Ying Tan
Peking University
Key Laboratory of Machine Perception (MOE)
School of Electronics Engineering and Computer Science
Department of Machine Intelligence
Beijing 100871, China
E-mail: ytan@pku.edu.cn

Yuhui Shi
Xi'an Jiaotong-Liverpool University
Department of Electrical and Electronic Engineering
Suzhou 215123, China
E-mail: yuhui.shi@xjtlu.edu.cn

Carlos A. Coello Coello
CINVESTAV-IPN
Investigador Cinvestav 3F, Depto. de Computación
México, D.F. 07300, Mexico
E-mail: ccoello@cs.cinvestav.mx

ISSN 0302-9743   e-ISSN 1611-3349
ISBN 978-3-319-11856-7   e-ISBN 978-3-319-11857-4
DOI 10.1007/978-3-319-11857-4
Springer Cham Heidelberg New York Dordrecht London

Library of Congress Control Number: 2014949260

LNCS Sublibrary: SL 1 – Theoretical Computer Science and General Issues

Typesetting: Camera-ready by author, data conversion by Scientific Publishing Services, Chennai, India

Printed on acid-free paper

Springer is part of Springer Science+Business Media (www.springer.com)

Preface

This book and its companion volume, LNCS vols. 8794 and 8795, constitute the proceedings of the fifth International Conference on Swarm Intelligence (ICSI 2014) held during October 17–20, 2014, in Hefei, China. ICSI 2014 was the fifth international gathering in the world for researchers working on all aspects of swarm intelligence, following the successful and fruitful Harbin event (ICSI 2013), Shenzhen event (ICSI 2012), Chongqing event (ICSI 2011) and Beijing event (ICSI 2010), which provided a high-level academic forum for the participants to disseminate their new research findings and discuss emerging areas of research. It also created a stimulating environment for the participants to interact and exchange information on future challenges and opportunities in the field of swarm intelligence research.

ICSI 2014 received 198 submissions from about 475 authors in 32 countries and regions (Algeria, Australia, Belgium, Brazil, Chile, China, Czech Republic, Finland, Germany, Hong Kong, India, Iran, Ireland, Italy, Japan, Macao, Malaysia, Mexico, New Zealand, Pakistan, Romania, Russia, Singapore, South Africa, Spain, Sweden, Taiwan, Thailand, Tunisia, Turkey, United Kingdom, United States of America) across six continents (Asia, Europe, North America, South America, Africa, and Oceania). Each submission was reviewed by at least two reviewers, and on average 2.7 reviewers. Based on rigorous reviews by the Program Committee members and reviewers, 105 high-quality papers were selected for publication in this proceedings volume with an acceptance rate of 53.03%. The papers are organized in 18 cohesive sections, 3 special sessions and one competitive session, which cover all major topics of swarm intelligence research and development.

As organizers of ICSI 2014, we would like to express sincere thanks to University of Science and Technology of China, Peking University, and Xi'an Jiaotong-Liverpool University for their sponsorship, as well as to the IEEE Computational Intelligence Society, World Federation on Soft Computing, and International Neural Network Society for their technical co-sponsorship. We appreciate the Natural Science Foundation of China for its financial and logistic support. We would also like to thank the members of the Advisory Committee for their guidance, the members of the International Program Committee and additional reviewers for reviewing the papers, and the members of the Publications Committee for checking the accepted papers in a short period of time. Particularly, we are grateful to Springer for publishing the proceedings in the prestigious series of Lecture Notes in Computer Science. Moreover, we wish to express our heartfelt appreciation to the plenary speakers, session chairs, and student helpers. In addition, there are still many more colleagues, associates, friends, and supporters who helped us in immeasurable ways; we express our sincere gratitude to them all. Last but not the least, we would like to thank all the speakers, authors, and participants for their great contributions that made ICSI 2014 successful and all the hard work worthwhile.

July 2014   Ying Tan
                                                                        Yuhui Shi
                                                            Carlos A. Coello Coello

Organization

General Chairs

Russell C. Eberhart   Indiana University-Purdue University, USA
Ying Tan   Peking University, China

Programme Committee Chairs

Yuhui Shi   Xi'an Jiaotong-Liverpool University, China
Carlos A. Coello Coello   CINVESTAV-IPN, Mexico

Advisory Committee Chairs

Gary G. Yen   Oklahoma State University, USA
Hussein Abbass   University of New South Wales, ADFA, Australia
Xingui He   Peking University, China

Technical Committee Chairs

Xiaodong Li   RMIT University, Australia
Andries Engelbrecht   University of Pretoria, South Africa
Ram Akella   University of California, USA
M. Middendorf   University of Leipzig, Germany
Kalyanmoy Deb   Indian Institute of Technology Kanpur, India
Ke Tang   University of Science and Technology of China, China

Special Sessions Chairs

Shi Cheng   The University of Nottingham, Ningbo, China
Meng-Hiot Lim   Nanyang Technological University, Singapore
Benlian Xu   Changshu Institute of Technology, China

Competition Session Chairs

Jane J. Liang   Zhengzhou University, China
Junzhi Li   Peking University, China

Publications Chairs

Radu-Emil Precup   Politehnica University of Timisoara, Romania
Haibin Duan   Beihang University, China

Publicity Chairs

Yew-Soon Ong   Nanyang Technological University, Singapore
Juan Luis Fernandez Martinez   University of Oviedo, Spain
Hideyuki Takagi   Kyushu University, Japan
Qingfu Zhang   University of Essex, UK
Suicheng Gu   University of Pittsburgh, USA
Fernando Buarque   University of Pernambuco, Brazil
Ju Liu   Shandong University, China

Finance and Registration Chairs

Chao Deng   Peking University, China
Andreas Janecek   University of Vienna, Austria

Local Arrangement Chairs

Wenjian Luo   University of Science and Technology of China,
                                       China
Bin Li   University of Science and Technology of China,
                                       China

Program Committee

Kouzou Abdellah   University of Djelfa, Algeria
Ramakrishna Akella   University of California at Santa Cruz, USA
Rafael Alcala   University of Granada, Spain
Peter Andras   Newcastle University, UK
Esther Andrés   INTA, USA
Sabri Arik   Istanbul University, Turkey
Helio Barbosa   Laboratório Nacional de Computação
                                       Científica, Brazil
Carmelo J.A. Bastos Filho   University of Pernambuco, Brazil
Christian Blum   Technical University of Catalonia, Spain
Salim Bouzerdoum   University of Wollongong, Australia
Xinye Cai   Nanhang University, China
David Camacho   Universidad Autonoma de Madrid, Spain
Bin Cao   Tsinghua University, China
Kit Yan Chan   DEBII, Australia

Mu-Song Chen	Da-Yeh University, Taiwan
Walter Chen	National Taipei University of Technology, China
Shi Cheng	The University of Nottingham Ningbo, China
Leandro Coelho	Pontifícia Universidade Católica do Parana, Brazil
Chenggang Cui	Shanghai Advanced Research Institute, Chinese Academy of Sciences, China
Chaohua Dai	Southwest Jiaotong University, China
Arindam K. Das	University of Washington, USA
Prithviraj Dasgupta	University of Nebraska at Omaha, USA
Kusum Deep	Indian Institute of Technology Roorkee, India
Mingcong Deng	Tokyo University of Agriculture and Technology, Japan
Yongsheng Ding	Donghua University, China
Madalina-M. Drugan	Vrije University, The Netherlands
Mark Embrechts	RPI, USA
Andries Engelbrecht	University of Pretoria, South Africa
Fuhua Fan	Electronic Engineering Institute, China
Zhun Fan	Technical University of Denmark, Denmark
Komla Folly	University of Cape Town, South Africa
Shangce Gao	University of Toyama, Japan
Ying Gao	Guangzhou University, China
Shenshen Gu	Shanghai University, China
Suicheng Gu	University of Pittsburgh, USA
Ping Guo	Beijing Normal University, China
Haibo He	University of Rhode Island, USA
Ran He	National Laboratory of Pattern Recognition, China
Marde Helbig	CSIR: Meraka Institute, South Africa
Mo Hongwei	Harbin Engineering University, China
Jun Hu	Chinese Academy of Sciences, China
Xiaohui Hu	Indiana University Purdue University Indianapolis, USA
Guangbin Huang	Nanyang Technological University, Singapore
Amir Hussain	University of Stirling, UK
Hisao Ishibuchi	Osaka Prefecture University, Japan
Andreas Janecek	University of Vienna, Austra
Changan Jiang	RIKEN-TRI Collaboration Center for Human-Interactive Robot Research, Japan
Mingyan Jiang	Shandong University, China
Liu Jianhua	Fujian University of Technology, China
Colin Johnson	University of Kent, USA
Farrukh Khan	FAST-NUCES Islamabad, Pakistan
Arun Khosla	National Institute of Technology, Jalandhar, India

Franziska Klügl   Örebro University, Sweden
Thanatchai
   Kulworawanichpong   Suranaree University of Technology, Thailand
Germano Lambert-Torres   Itajuba Federal University, Brazil
Xiujuan Lei   Shaanxi Normal University, China
Bin Li   University of Science and Technology of China,
                                   China
Xiaodong Li   RMIT University, Australia
Xuelong Li   Chinese Academy of Sciences, China
Yangmin Li   University of Macau, China
Jane-J. Liang   Zhengzhou University, China
Andrei Lihu   Politehnica University of Timisoara, Romania
Fernando B. De Lima Neto   University of Pernambuco, Brazil
Ju Liu   Shandong University, China
Wenlian Lu   Fudan University, China
Wenjian Luo   University of Science and Technology of China,
                                   China
Jinwen Ma   Peking University, China
Chengying Mao   Jiangxi University of Finance and Economics,
                                   China
Michalis Mavrovouniotis   De Montfort University, UK
Bernd Meyer   Monash University, Australia
Martin Middendorf   University of Leipzig, Germany
Sanaz Mostaghim   Institute IWS, Germany
Jonathan Mwaura   University of Pretoria, South Africa
Pietro S. Oliveto   University of Sheffield, UK
Feng Pan   Beijing Institute of Technology, China
Bijaya Ketan Panigrahi   IIT Delhi, India
Sergey Polyakovskiy   Ufa State Aviation Technical University, USA
Thomas Potok   ORNL, USA
Radu-Emil Precup   Politehnica University of Timisoara, Romania
Kai Qin   RMIT University, Australia
Quande Qin   Shenzhen University, China
Boyang Qu   Zhongyuan University of Technology, China
Robert Reynolds   Wayne State University, USA
Guangchen Ruan   Indiana University Bloomington, USA
Eugene Santos   Dartmouth College, USA
Gerald Schaefer   Loughborough University, USA
Kevin Seppi   Brigham Young University, USA
Zhongzhi Shi   Institute of Computing Technology, CAS,
                                   China
Pramod Kumar Singh   ABV-IIITM Gwalior, India
Ponnuthurai Suganthan   Nanyang Technological University, Singapore
Mohammad Taherdangkoo   Shiraz University, Iran
Hideyuki Takagi   Kyushu University, Japan
Ying Tan   Peking University, China

Additional Reviewers

Alves, Felipe
Bi, Shuhui
Chalegre, Marlon
Cheng, Shi
Dong, Xianguang
Gonzalez-Pardo, Antonio
Haifeng, Sima
Hu, Weiwei
Jun, Bo
Lacerda, Marcelo
Lee, Jie
Li, Yexing
Ling, Haifeng
Menéndez, Héctor
Pei, Yan

Rakitianskaia, Anna
Singh, Garima
Singh, Pramod
Wang, Aihui
Wen, Shengjun
Wenbo, Wan
Wu, Peng
Wu, Zhigang
Xiao, Xiao
Xin, Cheng
Yang, Wankou
Yang, Zhixiang
Yu, Czyujian
Zheng, Zhongyang

Table of Contents – Part I

Particle Swarm Optimization

Ant Colony Optimization for Travelling Salesman Problem

Artificial Bee Colony Algorithms

Artificial Immune System

Evolutionary Algorithms

Neural Networks and Fuzzy Methods

Hybrid Methods

Multi-objective Optimization

Multi-agent Systems

Evolutionary Clustering Algorithms

Table of Contents – Part II

Classification Methods

GPU-Based Methods

Scheduling and Path Planning

Wireless Sensor Network

Power System Optimization

Other Applications

Special Session on Swarm Intelligence in Image and Video Processing

Special Session on Applications of Swarm Intelligence to Management Problems

Special Session on Swarm Intelligence for Real-World Application

Special Session on ICSI 2014 Competition on Single Objective Optimization

Comparison of Different Cue-Based Swarm Aggregation Strategies

Farshad Arvin[1], Ali Emre Turgut[2], Nicola Bellotto[1], and Shigang Yue[1]

[1] Computational Intelligence Lab (CIL), School of Computer Science, University of Lincoln, Lincoln, LN6 7TS, UK
[2] Laboratory of Socioecology and Social Evolution, KU Leuven, B-3000 Leuven, Belgium
{farvin,syue}@lincoln.ac.uk

Abstract. In this paper, we compare different aggregation strategies for cue-based aggregation with a mobile robot swarm. We used a sound source as the cue in the environment and performed real robot and simulation based experiments. We compared the performance of two proposed aggregation algorithms we called as the *vector averaging* and *naïve* with the state-of-the-art cue-based aggregation strategy BEECLUST. We showed that the proposed strategies outperform BEECLUST method. We also illustrated the feasibility of the method in the presence of noise. The results showed that the vector averaging algorithm is more robust to noise when compared to the naïve method.

Keywords: swarm robotics, collective behavior, cue-based aggregation.

1 Introduction

Aggregation is a widely observed phenomenon in social insects and animals such as cockroaches, honeybees and birds [1]. It provides additional capabilities to animals such as forming a spore-bearing structure by slime mold or building a nest by termites [2]. In general, two types of aggregation mechanisms are observed in nature: cue-based or clueless. In cue-based aggregation, animals follow external cues to identify optimal zones such as a humid location for sow bugs and then they aggregate on these zones. Whereas in clueless aggregation, animals aggregate at random locations in an environment such as aggregation of cockroaches [3]. From swarm robotics perspective [4], aggregation can be defined as gathering randomly distributed robots to form an aggregate. Due to limited sensing capabilities of the robots, aggregation turns out be one of the challenging tasks in swarm robotics.

Many different studies have been performed in cue-based and clueless aggregation in swarm robotics. We first discuss cue-based aggregation. In one of the earliest studies on cue-based aggregation, Kube and Zhang [5] proposed an aggregation algorithm in which robots are required to aggregate around a light box and then push it. Melhuish et al. [6] proposed an algorithm for aggregation of robots around an infrared (IR) transmitter. The robots after reaching the IR transmitter start to emit sound resembling the vocalization of frogs and birds in order to help the other robots to estimate the size of the aggregate they are in. Honeybee aggregation is another example of cue-based aggregation method that was studied in [7,8]. In these studies, micro robots were deployed in a gradually lighted environment to mimic the behavior of honeybees which

Y. Tan et al. (Eds.): ICSI 2014, Part I, LNCS 8794, pp. 1–8, 2014.

aggregate at a zone that has the optimal temperature. An aggregation algorithm called BEECLUST that relies on inter-robot collisions was proposed [9]. The aggregation method has been used and evaluated in several researches [10,11,12]. In another study [13], two modifications on BEECLUST – *dynamic velocity* and *comparative waiting time* – were applied to increase the performance of aggregation. In addition, a fuzzy-based reasoning method has been proposed in [14,15] which increases the performance of the system significantly.

Clueless aggregation mechanism was employed in various studies. Trianni et al. [16] proposed an aggregation behavior of mobile robots using artificial evolution with static and dynamic strategies. In static strategy, when robots create an aggregate, they are not allowed to leave it. However, in the dynamic strategy, robots are allowed to leave an aggregate with a certain probability depending on the size of the aggregate. Static strategy resulted in many small compact aggregates, whereas with the dynamic one robots are able to form a few larger aggregates. Soysal and Şahin [17] proposed a probabilistic aggregation method using a combination of basic behaviors: obstacle avoidance, approach, repel, and wait. They studied the effects of various parameters such as control mechanisms, time, and arena size in performance of the system. This study was continued by using an evolutionary approach in order to investigate the various effective parameters in aggregation, such as the number of generations, the number of simulation steps which are used for fitness evaluations, population size, and size of arena [18]. Bayrindir and Şahin [19] proposed a macroscopic model to study the effects of population size and probability of leaving an aggregate on aggregation performance in a clueless aggregation scenario.

In this paper, different than the previous studies, we compare three different aggregation mechanisms in a cue-based aggregation scenario. Specifically, we extend the BEECLUST algorithm and propose two new algorithms which we call as *vector averaging* and *naïve* aggregation algorithms. Through systematic real-robot and simulation-based experiments, we analyze and compare the performance differences between those two algorithms and the BEECLUST algorithm.

2   Aggregation Methods

2.1   BEECLUST

BEECLUST aggregation [9] follows a simple algorithm as shown in the Fig. 1. When a robot detects another robot in the environment, it stops and measures the magnitude of the ambient audio signal and waits based on this magnitude. The higher the magnitude is the longer the waiting time (w) becomes. The waiting time is estimated by the following formula assuming that four microphones are used to detect the ambient audio signal:

$$w(t) = w_{max} \cdot \frac{M_a(t)^2}{M_a(t)^2 + \mu}, \tag{1}$$

where $M_a(t) = \frac{1}{4} \sum_{i=1}^{4} M_i(t)$ is the average magnitude of the four microphones, M_i is the magnitude of signal from the ith microphone ranging from 0 and 255, w_{max} is the maximum waiting time, and μ is a parameter which changes the steepness of the waiting curve. w_{max} and μ are determined empirically: $w_{max} = 65$ sec and $\mu = 5500$.

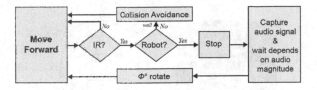

Fig. 1. Control diagram of the aggregation

When the waiting time is over, the robot rotates ϕ degree, which is a random variable drawn uniformly within $[-180°, 180°]$.

2.2 Naïve Method

In the naïve aggregation method, we employ a deterministic decision making mechanism based on both the intensity and the direction of the sound signal. The waiting time is still calculated using (1) based on the average intensity (M_a), but in addition to that, we estimate the direction of the sound source by setting it to the angle of the sound sensor that has the highest reading. $\phi = \theta_i$, $i \in \{1, 2, 3, 4\}$ and θ_i is the angle of the sensor $\{45°, 135°, 225°, 315°\}$ with respect to the frontal axis of the robot having the highest reading.

2.3 Vector Averaging Method

In vector averaging method, the direction information and intensity of the sound source are utilized. We employ an averaging calculation based on both intensity and direction of sound signal to estimate ϕ.

$$\phi = \text{atan2}\left(\frac{\sum_{i=1}^{4} \hat{M}_i \sin(\beta_i)}{\sum_{i=1}^{4} \hat{M}_i \cos(\beta_i)}\right) \tag{2}$$

ϕ is the estimated angular position of the source speaker, β_i is the angular distance between ith microphone and the robot's head. $\hat{M}_i, i \in \{1, 2, 3, 4\}$ is the captured audio signal's intensity levels from microphone i.

3 Experimental Setup

3.1 Robot Platform

AMiR (Autonomous Miniature Robot) as shown in Fig. 2 (a) is an open hardware mobile robot platform [20]. AMiR is specifically developed for swarm robotics studies. The robot has two small geared DC motors that make it move with a maximum speed of $8.6\frac{cm}{s}$ [21]. Six IR proximity sensors using $60°$ topology is used, which allows AMiR to scan its surrounding area without turning [22]. AMiR also has an audio extension module which is composed of four condenser microphones ($45°$, $135°$, $225°$, and $315°$) as shown in Fig. 2 (b).

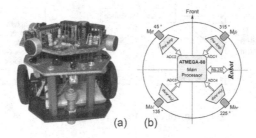

(a)   (b)

Fig. 2. (a) Autonomous Miniature Robot (AMiR) is equipped with an audio extension module and (b) architecture of developed audio signal processor module

3.2   Simulation Software

In order to test the proposed algorithm in large scale, Player/Stage simulation software is utilized. We modeled AMiR and its sensors in Player and used Stage as the simulation platform.

3.3   Experiment Configuration

Real Robot Experiments. In real robot experiments, due to laboratory limitations, a rectangular arena with a size of 120x80 cm^2 is utilized. When compared to the sensing range of the robots, the arena is approximately two times larger than the total sensing area of 6 robots. The experiments are performed with different number of robots $N \in \{3, 4, 5, 6\}$. We placed a sound source at one of the edges of the arena, that plays a single tone of frequency of approximately 1050 Hz. The sound source serves as the environmental cue in the experiment. With the current configuration of the arena, the intensity and waiting times are shown in Fig. 3. Each experiment is repeated 10 times and at the start of each run, the robots are placed in the arena with random positions and orientations.

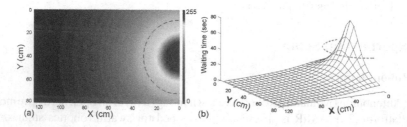

(a)   (b)

Fig. 3. (a) Intensity of audio signals in the arena and (b) relative waiting time in different positions of the arena. Dashed arc shows the predefined optimal aggregation zone around the sound source.

Simulation-Based Experiments. In simulation-based experiments, in order to use large number of robots $N \in \{5, 10, 15, 20, 25\}$, we used an arena with a size of 240×160 cm^2. Similar to the real-robot experiments, the experiments are repeated 10 times and position and orientation of the robots are set randomly at the beginning of each experiment. We perform two sets of experiments: One set without sensing noise, the other with sensing noise. In the latter, we add noise to the sound measurements, which is modeled as a uniformly distributed random variable. For the ith microphone, the noise is added as $\overline{M_i} = |M_i + \sigma\rho|$ where M_i is sensor reading of ith microphone, $\sigma \in \{0.1, 0.3, 0.5\}$ is the noise scaling factor which determines the amount of noise to be added to the reading and ρ is a random value within $[-255, +255]$.

3.4 Metrics

In this study, we are interested in having fast aggregation around the sound source. Therefore, we use aggregation time as one of our metrics. In order to calculate aggregation time, T, we define an aggregation area which is shown with a dashed arc in Fig. 3. The aggregation time is the duration of an experiment in which the number of the robots aggregated in the aggregation area reaches 80% of the total robots.

4 Results and Discussion

4.1 Real Robot Experiments

We first performed experiments with real robots. In general, when the number of robots increases, the aggregation time reduces significantly as shown in Fig. 4a, since increasing the population size increases the number of collisions eventually causing faster aggregation. It should be noted that, the reduction rate in the aggregation time is not the same in the three algorithms. Results show that, vector averaging aggregation is faster than naïve and BEECLUST.

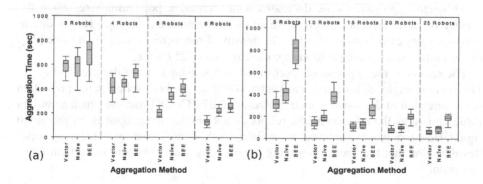

Fig. 4. Aggregation time as a function of population size for vector averaging, naïve, and BEECLUST aggregation methods with (a) real robot and (b) simulated robot experiments

4.2   Simulation-Based Experiments

In these experiments, the aggregation methods are implemented using the simulator. Fig. 4b shows the aggregation time as a function of population size in vector averaging, naïve, and BEECLUST methods. Results show that, population size has a direct impact on the aggregation time. Simulation results also showed that the vector averaging method performs faster aggregation in comparison with naïve and BEECLUST owing to more precise estimates ϕ values after each collision, which increases the performance of the aggregation. In case of BEECLUST that relies on random rotation, the aggregation time is longer than the others. Due to random rotations, robots occasionally move in opposite direction of the sound source, that results in robots leaving the aggregation zone. Therefore, it increases the aggregation time and number of collisions.

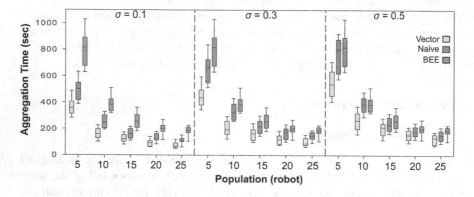

Fig. 5. Effects of different noise values in different population sizes for vector averaging, naïve and BEECLUST methods

Although aggregation time decreases with increasing population size, when the density of swarm reaches a certain value the performance decreases. This is due to over-crowding effect observed, when the density of robots reaches a certain value [23]. In our current setting, the performance increases up to 25 robots.

The results of the experiments performed with sensing noise is depicted in Fig. 5 . The rotation angle, ϕ, for naïve and vector averaging methods relies on the prediction of the direction of the sound source. However, in BEECLUST method which a random value is used as the rotation angle, robots do not employ the magnitude of the audio signal. Hence, only naïve and vector averaging methods are tested against sensing noise. Results show that the vector averaging algorithm is more robust to noise than naïve algorithm.

4.3   Statistical Analysis

We statistically analyzed the results of simulated robot experiments with and without noise using analysis of variance (ANOVA), the F-test method (see Table 1).

Population size has the highest influence on aggregation time for the vector averaging method ($F_\rho = 116.60$), which means that increasing the swarm density improves the performance of the vector averaging method the most. Noise has the least influence on the vector averaging method (7.327, and 12.39 for the vector averaging and naïve methods, respectively).

Table 1. Results of F-test in analysis of variance (ANOVA)

Experimental Setup	Parameters	Aggregation Method		
		Averaging	Naïve	BEECLUST
Without Noise	Population	116.60	85.51	76.75
With Noise	Noise	7.32	12.39	–
	Population	34.28	28.43	–

5 Conclusion

In this paper we evaluate three cue-based aggregation methods, namely, vector averaging, naïve and BEECLUST. The aggregation methods were implemented using real- and simulated-robots with different population sizes. The results showed that the proposed vector averaging method improves the performance of the aggregation significantly. The swarm density has direct impact on the performance of the aggregation. Hence, an increase in population improves the performance of aggregation in all three algorithms. In addition to these, results revealed that the additional noise has less impact on vector averaging method in comparison to the naïve method.

Acknowledgments. This work was supported by EU FP7-IRSES projects EYE2E (269118), LIVCODE (295151) and HAZCEPT (318907). The second author would thank TUBITAK-2219-Program.

References

1. Camazine, S., Franks, N., Sneyd, J., Bonabeau, E., Deneubourg, J.L., Theraulaz, G.: Self-organization in biological systems. Princeton University Press (2003)
2. Parrish, J., Edelstein-Keshet, L.: Complexity, pattern, and evolutionary trade-offs in animal aggregation. Science 284(5411), 99–101 (1999)
3. Jeanson, R., Rivault, C., Deneubourg, J.L., Blanco, S., Fournier, R., Jost, C., Theraulaz, G.: Self-organized aggregation in cockroaches. Animal Behaviour 69(1), 169–180 (2005)
4. Şahin, E., Girgin, S., Bayındır, L., Turgut, A.E.: Swarm robotics. In: Blum, C., Merkle, D. (eds.) Swarm Intelligence, vol. 1, pp. 87–100. Springer, Heidelberg (2008)
5. Kube, C., Zhang, H.: Collective robotics: From social insects to robots. Adaptive Behavior 2(2), 189–219 (1993)
6. Melhuish, C., Holland, O., Hoddell, S.: Convoying: Using chorusing to form travelling groups of minimal agents. Robotics and Autonomous Systems 28(2), 207–216 (1999)
7. Schmickl, T., Thenius, R., Moeslinger, C., Radspieler, G., Kernbach, S., Szymanski, M., Crailsheim, K.: Get in touch: Cooperative decision making based on robot-to-robot collisions. Autonomous Agents and Multi-Agent Systems 18(1), 133–155 (2009)

8. Kernbach, S., Thenius, R., Kernbach, O., Schmickl, T.: Re-embodiment of Honeybee Aggregation Behavior in an Artificial Micro-Robotic System. Adaptive Behavior 17(3), 237–259 (2009)
9. Schmickl, T., Hamann, H.: Beeclust: A swarm algorithm derived from honeybees. In: Xiao, Y., Hu, F. (eds.) Bio-inspired Computing and Communication Networks (2010)
10. Bodi, M., Thenius, R., Szopek, M., Schmickl, T., Crailsheim, K.: Interaction of robot swarms using the honeybee-inspired control algorithm beeclust. Mathematical and Computer Modelling of Dynamical Systems 18(1), 87–100 (2012)
11. Kengyel, D., Thenius, R., Crailsheim, K., Schmick, T.: Influence of a social gradient on a swarm of agents controlled by the beeclust algorithm. In: European Conference on Artificial Life, pp. 1041–1048 (2013)
12. Hereford, J.: Beeclust swarm algorithm: Analysis and implementation using a markov chain model. International Journal of Innovative Computing and Applications 5(2), 115–124 (2013)
13. Arvin, F., Samsudin, K., Ramli, A.R., Bekravi, M.: Imitation of honeybee aggregation with collective behavior of swarm robots. International Journal of Computational Intelligence Systems 4(4), 739–748 (2011)
14. Arvin, F., Turgut, A.E., Yue, S.: Fuzzy-based aggregation with a mobile robot swarm. In: Dorigo, M., Birattari, M., Blum, C., Christensen, A.L., Engelbrecht, A.P., Groß, R., Stützle, T. (eds.) ANTS 2012. LNCS, vol. 7461, pp. 346–347. Springer, Heidelberg (2012)
15. Arvin, F., Turgut, A.E., Bazyari, F., Arikan, K.B., Bellotto, N., Yue, S.: Cue-based aggregation with a mobile robot swarm: A novel fuzzy-based method. Adaptive Behavior 22(3), 189–206 (2014)
16. Trianni, V., Groß, R., Labella, T.H., Şahin, E., Dorigo, M.: Evolving aggregation behaviors in a swarm of robots. In: Banzhaf, W., Ziegler, J., Christaller, T., Dittrich, P., Kim, J.T. (eds.) ECAL 2003. LNCS (LNAI), vol. 2801, pp. 865–874. Springer, Heidelberg (2003)
17. Soysal, O., Şahin, E.: Probabilistic aggregation strategies in swarm robotic systems. In: Swarm Intelligence Symposium, pp. 325–332. IEEE (2005)
18. Soysal, O., Bahçeci, E., Şahin, E.: Aggregation in swarm robotic systems: Evolution and probabilistic control. Turkish Journal of Electrical Engineering & Computer Sciences 15(2), 199–225 (2007)
19. Bayindir, L., Şahin, E.: Modeling self-organized aggregation in swarm robotic systems. In: Swarm Intelligence Symposium, pp. 88–95. IEEE (2009)
20. Arvin, F., Samsudin, K., Ramli, A.R.: Development of a Miniature Robot for Swarm Robotic Application. International Journal of Computer and Electrical Engineering 1, 436–442 (2009)
21. Arvin, F., Bekravi, M.: Encoderless position estimation and error correction techniques for miniature mobile robots. Turkish Journal of Electrical Engineering & Computer Sciences 21, 1631–1645 (2013)
22. Arvin, F., Samsudin, K., Ramli, A.R.: Development of IR-Based Short-Range Communication Techniques for Swarm Robot Applications. Advances in Electrical and Computer Engineering 10(4), 61–68 (2010)
23. Hamann, H.: Towards swarm calculus: Universal properties of swarm performance and collective decisions. In: Dorigo, M., Birattari, M., Blum, C., Christensen, A.L., Engelbrecht, A.P., Groß, R., Stützle, T. (eds.) ANTS 2012. LNCS, vol. 7461, pp. 168–179. Springer, Heidelberg (2012)

PHuNAC Model: Emergence
of Crowd's Swarm Behavior

Olfa Beltaief[1], Sameh El Hadouaj[2], and Khaled Ghedira[3]

[1] Laboratory *SOIE*, ISG of Tunis, Tunisia
beltaief.olfa@gmail.com
[2] Laboratory *SOIE*, FSEG Nabeul, Tunisia
hadouaj@yahoo.fr
[3] Laboratory *SOIE*, ISG of Tunis, Tunisia
khaled.ghedira@isg.rnu.tn

Abstract. The swarm behavior of pedestrians in a crowd, generally, causes a global pattern to emerge. A pedestrian crowd simulation system must have this emergence in order to prove its effectiveness. For this reason, the aim of our work is to demonstrate the effectiveness of our model PHuNAC (Personalities' Human's Nature of Autonomous Crowds) and also prove that the swarm behavior of pedestrians' agents in our model allows the emergence of these global patterns. In order to validate our approach, we compared our system with real data. The conducted experiments show that the model is consistent with the various emergent behaviors and thus it provides realistic simulated pedestrian's behavior.

Keywords: Multi-agent systems, autonomous crowd model, swarm behavior.

1 Introduction

Pedestrian crowd is a set of people with different personalities who are gathered in the same location. These different personalities shape the swarm behavior of the pedestrian crowd. Indeed, personality is a pattern of behavioral and mental traits for each pedestrian [1]. The simulation of these complex phenomena has attracted considerable attention. However, a great number of these models do not cover all the psychological factors necessary for a pedestrian located in a crowd (see [2] and [3]). Moreover, many of these models simulate only homogenous personalities of pedestrians. In the reality, there are different personalities. Experimentations' results of these models show that they lack realism. For this purpose, the goal of our work is to reproduce realistic collective behavior simulated situations by simulating a realistic pedestrian behavior and personality. In this context, we opted for the PHuNAC (Personalities' Human's Nature of Autonomous Crowds) model [4, 5] which includes the necessary psychological factors for a pedestrian located in a crowd and which integrates heterogeneous crowd.

The plan of our paper is as follows: section 2 describes related works. Section 3 describes our model. Section 4 presents our experimentations and a discussion of this work. Finally, in section 5 we conclude and we give perspectives for this work.

Y. Tan et al. (Eds.): ICSI 2014, Part I, LNCS 8794, pp. 9–18, 2014.

2   Related Works

Several pedestrians' crowd simulation models were proposed. These models could be classified into two families: macroscopic models and microscopic models. Macroscopic approaches include regression (e.g., [6]) and fluid dynamic models (e.g., [7], [8], etc.). Microscopic approaches include rule-based models (e.g., [9], [10], etc.), cellular automata models (e.g., [11], [12], etc.), agent-based (e.g., [1], [3], [13], [14], etc.) and social force models (e.g., [15], etc.). The macroscopic approaches simulate the behavior of the crowd as a whole. In fact, macroscopic models don't consider individual features such as physical abilities, direction of movement, and individual positioning [16]. This causes a lack of realism [17]. On the other hand, the microscopic approaches are interested in the behavior, actions and decisions of each pedestrian and his interaction with others [17]. Therefore, the microscopic models allow us to obtain more realistic simulations. For this reason, in our work, we adopt a microscopic approach and more specifically an agent-based model. In fact, the agent-based models are able, over others approaches, to be flexible, to provide a natural description of the system and to capture emergent phenomena and complex human behaviors [4, 18]. In other words, they may reflect more the reality. Several models based on multi-agents and psychological theories have been proposed. We cite namely [3, 13, 14], etc. However, a great number of these models do not cover all the psychological factors necessary for a pedestrian located in a crowd (see [2] and [3]). Therefore the produced pedestrian crowd situations lack realism. Indeed, many of these models simulate only homogenous pedestrians. In the reality, there are different personalities. The goal of our work is to reproduce realistic pedestrian crowd simulated situations by simulating a realistic pedestrian behavior. In this context, we proposed, firstly, our HuNAC (Human's Nature of Autonomous Crowds) model [4]. The design of the pedestrian behavior is based on psychosocial and psychophysical studies of normative personalities. In a second time, in order to have more realistic simulation, we opted for the PHuNAC [5] (Personalities' Human's Nature of Autonomous Crowds) version which integrates heterogeneous crowd.

3   PHuNAC Model

In our PHuNAC model ([4, 5]), each pedestrian is represented by an agent. Each agent has its own autonomous behavior and personality. In the following sections, we present firstly the personality of pedestrians in our model and secondly the behavior of each pedestrian.

3.1   The Personality of Pedestrians

There are several theories that classify personality types such as: MBTI, Big Five, NPA, etc. It must be noted that MBTI and Big Five theory are the widely accepted theories by psychologists [19-22]. However, the Big Five theory is not used in the organizational behavior area in contrast to the MBTI theory [23]. Since the pedestrian crowd behavior is an organizational behavior [24], the MBTI theory is more linked to the crowd pedestrian simulation field. For this reason, we choose the MBTI theory.

In order to integrate personality in our model, we are based on the MBTI theory [25]. In fact, there are 16 personality types (according to MBTI theory). It is the combination of the eight preferences and each combination presents a personality:

- Extraversion (E) or Introversion (I)
- Sensing (S) or iNtuition (N)
- Thinking (T) or Feeling (F)
- Judging (J) or Perceiving (P)

For example, an ESFJ pedestrian is an extraversion, sensing, feeling and judging person.

The dichotomies of the MBTI theory are still qualifying values that we cannot implement and integrate into our PHuNAC model. For this reason, we had to find a method that quantifies the MBTI theory. In this context, we used the method of 2-qubits [26, 27]. The 2-qubit model can generate artificial personalities, which can be compared to the real world. In this method, we need only three parameters which describe the complete distribution without the need for additional parameters. The three variables are namely τ, θ and α with τ is the entanglement factor and θ and α are the parameters of superposition (for more details see [26], [27]).We can produce various artificial personalities through the following equations [26, 27]:

$$\bar{E} = (2\,\tau^2 - 1)\cos\alpha - 2\tau\sqrt{1 - \iota^2}\sin\alpha \tag{1}$$

$$\bar{T} = (2\,\tau^2 - 1)\cos\theta - 2\tau\sqrt{1 - \tau^2}\sin\theta \tag{2}$$

$$\bar{S} = (2\,\tau^2 - 1)\sin\theta - 2\tau\sqrt{1 - \tau^2}\cos\theta \tag{3}$$

The choice of personality is through formulas which allow us to correctly determine a personality with consistent manner (see equations 1, 2 and 3). Each personality characteristic is represented by the letter $\bar{C}$ with $\bar{C} \in \{value(\bar{E}), value(\bar{T}), value(\bar{S})\}$. It is important to note that the characteristic value in psychology varies over time with a constant probability. Therefore, we must determine this probability. The likelihood of a feature of personality pedestrian i is represented by $\Psi_i(\bar{C})$ (see equation 4).

$$\Psi_i(\bar{C}) = \frac{\bar{C} * 100}{\max \bar{C}} \tag{4}$$

For reasons of simplification, we have decoupled the characteristics $\bar{T}$ and $\bar{S}$. In order to decouple these characteristics, we use the platonic concept which is called" ideal" [28]. In the present application, Plato's concept of "ideal" rigorously suggests matching any maximum possible 100% decoupled attitude score with equivalent 100% scores of the associated original MBTI attitudes. Geometrically speaking, this generates a new coordinate system with axes rotated 45° from the original ones. The combination of two approaches gives the following equation (see equation 5).

$$\Psi_i(\overline{TS}) = \frac{\Psi_i(\overline{T}) + \Psi_i(\overline{S})}{2} \tag{5}$$

3.2   Pedestrian Behavior

The behavior of a pedestrian agent is divided into three phases: strategic phase, tactical Phase and operational phase. The strategic phase defines the global plan of a pedestrian agent. The tactical phase represents seeking information and taking decisions. These decisions allow the pedestrian agent to avoid people and obstacles. Each pedestrian agent has its own perception of the environment. Researchers [29], [30] found that pedestrians predict the "cost" of each sidewalk facility in terms of the convenience and   speed   to reach a destination and that the cost is based on their personal expectations. For this purpose, the pedestrian agent in our model divides the corridor in which it is situated in a set of lanes (see Fig.1). The width of each lane is equal to 60 cm (60 cm is the width of the body ellipse of a normal person [31]).

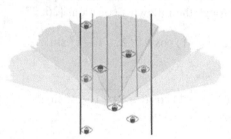

Fig. 1. Lanes' construction

Being situated in a corridor (namely C_i), each time t, the pedestrian agent has to choose the adequate lane.   Indeed, the agent has to choose the closest, the fastest, the least dense lane, having its same direction, and allowing it to avoid the collisions. In other words, the pedestrian agent chooses the easiest lane for its movement. To achieve this goal, the pedestrian agent P determines for each lane a score function (namely S_L). It is very important to note that if we take into account, in our model, only the normative personality (A normal person always chooses the optimal path), the function of the score will be based only on the optimal choice of the way. For this reason, we need to integrate into the score function of our model the "personality" factor. The choice of path should be sometimes optimal and sometimes not optimal this will depend on the agent pedestrian satisfaction and specifically on its personality.   The score function is based on the following attributes (namely A_j):

- Speed: the speed of a lane is the speed of the slower pedestrian in this lane.
- Density: the density of a lane is equal to the number of pedestrians situated in the lane divided by its surface.

- Direction: the direction of a lane is the direction of the majority of pedestrians present on this lane.
- Proportion of pedestrians with the same direction: it is a measure that allows determining the direction of the lane compared of the pedestrian direction.
- Distance of a lane from a pedestrian: it is the Euclidian distance which separates the actual lane of a pedestrian from the lane to be reached.
- Distance of an obstacle, situated in the lane to reach, from the pedestrian: it is the Euclidian distance which separates the pedestrian from the first obstacle which can cross it, placed in the lane to be reached.
- Extraversion: determines how much an agent pedestrian attached to its group.

Each attribute A_j is balanced by a weight W_j which gives us the following evaluation function:

$$S_L = (\sum_i f(A_i) * W_i * F_i) * \Psi_i(\overline{TS}) + f(\Psi(\overline{E})) * W_{\overline{E}} \qquad (6)$$

Note that $f(A_i)$ is the function of the attribute A_i. F_i is the favoring lane variable. F_i is equal to -1 in the case of a penalty and equal to +1 in the case of recompense. For example, if the attribute favors the choice of the lane, this attracts pedestrian to choose the lane in consideration. So the lane score must increase. Therefore F_i must be equal to +1. The idea behind the assignment of weights is to give each attribute a priority. To determine the different weight values, we are based on the normative goals priorities of a normative pedestrian. Indeed, the goal with a highest priority has the highest weight (see [4]). After determining for each lane L the score S_L the pedestrian agent chooses the lane with the highest score.

Finally, in the operational phase, the pedestrian agent determines the direction and the speed suited to reach the chosen lane in the previous phase. In order to resume the behavior of a pedestrian agent in our model, we give the following algorithm.

```
For each L_j in C_i
  S_Lj • Calculate_function_score ()
End for each
Adequate_lane • Max_score (S_L)
If Current_lane = Adequate_lane then
  Move_in_the_same_direction ()
Else
  Change_direction ()
  Move_to_ adequate_lane ()
End if
```

At each time t, the agent pedestrian must observe its environment (the current corridor C_i) and determines for each lane L_j a score S_{Lj}. After that, the agent pedestrian determines the adequate lane for its movement (the lane with the highest score). If the adequate lane is equals to the current lane of the pedestrian agent then it must still move in the same direction. In the other case, it must change its direction and move to the adequate lane.

4   Experiments and Discussions

To implement our model we used NetLogo. Our experiments consist in comparing real and simulated crowds. Firstly, we present qualitative experiments and then we present quantitative experiments.

4.1   Qualitative Experiments

The most common form of experimental results of crowd simulation systems are animation screen-shots. They are used by ([32], [33], [34], etc.) to show scenarios where agents perform a believable swarm behavior. If the visual output seems to exhibit real situations, it is assumed that the system is successful in simulating crowds [35]. In the following we compare the PHuNAC model results with real emergent swarm behaviors of pedestrian's crowd in urban areas and in buildings.

It is important to note that if two different flows are directed in opposite directions inside a single corridor, pedestrians tend to emerge in lanes when the density is high enough. Indeed, the number and dimension of lanes depend on dimension of the corridor [36].

The first experiment evaluates whether our crowd model can reproduce this phenomena. Firstly, we create a street with 9 m width. Subsequently, we defined the density of the crowd in 1.6 p/m². The result of our experiment shows that pedestrian agents emerge in four lanes (see Fig.2). Then, we create another street with 4.2 m width and we defined the same density of the crowd (1.6 p/m²). The result of our experiment shows that pedestrian agents emerge in two lanes (see Fig.2). This result shows that, for a constant density of pedestrians, the number of lanes depend on the dimension of the corridor.

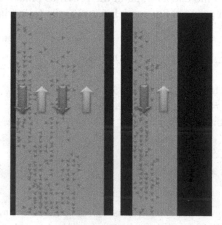

Fig. 2. Lane formation

It is well known that a self-organization occurs in crossing pedestrian flows. When flows cross vertically, diagonal stripe pattern emerges and the congestion degree of each flows varies in the crossing area [22].

In the second experiment, we simulate the crossing flows. Fig. 3 shows a screen-shot of the simulation. We observed that a diagonal stripe pattern emerged after two flows have crossed. This result is consistent with the phenomenon mentioned in [22].

Fig. 3. Diagonal stripe pattern emerges.

4.2   Quantitative Experiments

We present in this section quantitative experiments. We follow the literature in the measurement of two main features of pedestrian's movements: the average velocity and traffic flow. For this purpose, we have been based on the works achieved by Fruin (see [31]). On the basis of real data, the latter has pointed out laws that describe the walking Level of service. These laws represent the distribution of the average velocity as a function of the crowd density. He has also pointed out laws describing the rela-tion between crowd flow and crowd density.

In order to obtain a curve of average velocity as a function of density, at time t, we created an initial population density of 0.01 pedestrians/ m². Subsequently, we meas-ured the average velocity of the crowd. At time t +1, we added another population of the same density which gives us a density of 0.02 pedestrians /m². Similarly, we measured the average velocity of the crowd. We repeated this experiment several times. At the end of the experiment, we obtained a curve of average velocity as a function of density. Fig.4 compares the curves of the average velocity of the PHu-NAC model with curve representing the real data (Fruin [31]). We observe that the velocity curve of our model is very close to the velocity curve representing real data.

It is very important to note that the personality's variation affects the flow charac-teristics [30]. In fact, the flow characteristics include pedestrian's velocity and pede-strian's flow rates. For this purpose, we must take into consideration not only the velocity but also the flow rates. We repeated the same previous experiment, but this time we measured the flow of pedestrian traffic rather than the average velocity. Fig.5 shows the comparison of the flow rates between PHuNAC model and real data. We observe that the flow rate curve of our model is very close to the flow rates curve representing real data.

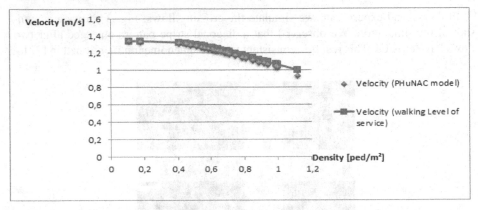

Fig. 4. Comparison of average velocity curves

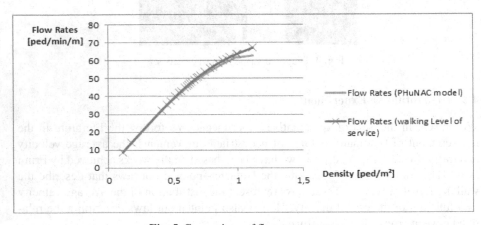

Fig. 5. Comparison of flow rates curves

5   Conclusion

We present in this work the PHuNAC model which is based on multi-agent systems and that includes the major psychological factors. The PHuNAC model is based on the MBTI theory. The MBTI theory is a psychological theory which classifies personality types. The aim of our work is to demonstrate the effectiveness of our model PHuNAC and also prove that the swarm behavior of pedestrians' agents in our model allows the emergence of these global patterns. In order to validate our approach, we compared our system with emergent behaviors. The conducted experiments show that the model is consistent with the various emergent behaviors and thus it provides realistic simulated pedestrian's behavior. It is very important to note that our PHuNAC model doesn't consider emergency situations. We can improve PHuNAC model by considering the evacuation movements.

References

1. Durupinar, F., Allbeck, J.M., Pelechano, N., Badler, N.I.: Creating crowd variation with the ocean personality model. In: Padgham, L., Parkes, D.C., Mller, J. (eds.) AAMAS (3), pp. 1217–1220. IFAAMAS (2008)
2. Moussaïd, M., Perozo, N., Garnier, S., Helbing, D., Theraulaz, G.: The Walking Behaviour of Pedestrian Social Groups and Its Impact on Crowd Dynamics. PLoS ONE 5(4), e10047+ (2010)
3. Fridman, N., Kaminka, G.A.: Towards a cognitive model of crowd behavior based on social comparison theory. In: AAAI, pp. 731–737 (2007)
4. Beltaief, O., Hadouaj, S., Ghedira, K.: Multi-agent simulation model of pedestrians crowd based on psychological theories. In: 4th International Conference on Logistics, LOGISTIQUA (2011)
5. Beltaief, O., et al.: PHuNAC Model: Creating Crowd Variation with the MBTI Personality Model. In: Corchado, J.M., et al. (eds.) PAAMS 2014. CCIS, vol. 430, pp. 179–190. Springer, Heidelberg (2014)
6. Helbing, D.: A fluid-dynamic model for the movement of pedestrians. Complex Systems 6, 391–415 (1992)
7. Smith, R.A.: Density, velocity and flow relationships for closely packed crowds. Safety Science 18, 321–327 (1995)
8. Cusack, M.A.: Modelling aggregate-level entities as a continuum. Mathematical and Computer Modelling of Dynamical Systems 8, 33–48 (2002)
9. Reynolds, C.W.: Flocks, herds, and schools: A distributed behavioral model. In: SIGGRAPH 1987: Proceedings of the 14th Annual Conference on Computer Graphics and Interactive Techniques, New York, NY, USA, pp. 25–34 (1987)
10. Xiong, M., Lees, M., Cai, W., Zhou, S., Low, M.Y.-H.: A Rule-Based Motion Planning for Crowd Simulation. In: Ugail, H., Qahwaji, R., Earnshaw, R.A., Willis, P.J. (eds.) 2009 International Conference on Cyber Worlds, Bradford, West Yorkshire, UK, September 7-11, pp. 88–95. IEEE Computer Society (2009)
11. Klupfel, H.: The simulation of crowds at very large events. In: Schadschneider, A., Khne, R., Pschel, T., Schreckenberg, M., Wolf, D.E. (eds.) Trafic and Granular Flow 2005, Berlin (2006)
12. Weifeng, Y., Hai, T.K.: A novel algorithm of simulating multi-velocity evacuation based on cellular automata modeling and tenability condition. Physica A: Statistical Mechanics and its Applications 379(1), 250–262 (2007)
13. Cherif, F., Chighoub, R.: Crowd simulation influenced by agent's socio-psychological state (2010)
14. Musse, S.R., Thalmann, D.: A model of human crowd behavior: Group interrelationship and collision detection analysis. In: Workshop Computer Animation and Simulation of Eurographics, pp. 39–52 (1997)
15. Helbing, D., Farkas, I., Moln'ar, P., Vicsek, T.: Simulation of pedestrian crowds in normal and evacuation situations. In: Schreckenberg, M., Sharma, S.D. (eds.) Pedestrian and Evacuation Dynamics, Berlin, pp. 21–58 (2002)
16. Still, G.: Crowd Dynamics, pp. 640–655 (2000)
17. Challenger, R., Clegg, C.W., Robinson, M.A.: Understanding Crowd Behaviours, Cabinet Office Emergency Planning College (2009)
18. Bonabeau, E.: Agent-based modeling: Methods and techniques for simulating human systems. Proceedings of the National Academy of Sciences of the United States of America 99(Suppl. 3), 7280–7287 (2002)

19. Furnham, A.: The big five versus the big four: The relationship between the Myers-Briggs Type Indicator (MBTI) and NEO-PI five factor model of personality. Personality and Individual Differences (1996)
20. Johnsson, F.: Personality measures under focus: The NEO-PI-R and the MBTI. Griffith University Undergraduate Student Journal 1, 32–42 (2009)
21. Wurster, C.D. : Myers-Briggs Type Indicator: A cultural and ethical evaluation. Executive Research Project. National Defense University, Washington D.C. (1993) (retrieved May 5, 2008)
22. Ando, K., Oto, H., Aoki, T.: Forecasting the Flow of People. Railway Research Review 45(8), 8–13 (1988) (in Japanese)
23. Howard, P.J.: The Mbti And The Five Factor Model. Center for Applied Cognitive Studies (CentACS), Charlotte (2006)
24. Helbing, D., Molnar, P., Farkas, I.J., Bolay, K.: Self-organizing pedestrian movement. Environment and Planning B: Planning and Design 28(3), 361–383 (2001)
25. JMH Consultancy, Introducing the Myers Briggs Type Indicator (MBTI) (2006)
26. Oosterhout, D.: Automatic Modeling of Personality Types. at the Universiteit van Amsterdam (2009)
27. Zevenbergen, R.: Automatic modeling of personality types: the 2-qubit model. Universiteit van Amsterdam (2010)
28. Wilde, D.J.: Jung's Personality Theory Quantified (2011), ISBN: 978-0-85729-099-1 (Print) 978-0-85729-100-4
29. Hoogendoorn, S.P.: Pedestrian Flow Modeling by Adaptive Control. In: Transportation Research Board Annual Meeting (2004a)
30. Burden, A.M.: New york city pedestrian level of service study phase i. Technical report, City of New York (2006)
31. Fruin, J.J.: Pedestrian Planning and Design. Metropolitan association of urban designers and environmental, New York (1971)
32. Shao, W., Terzopoulos, D.: Autonomous pedestrians. Graphical Models 69(5-6), 246–274 (2007)
33. Treuille, A., Cooper, S., Popovi´c, Z.: Continuum crowds. In: SIG-GRAPH 2006: ACM SIGGRAPH 2006 Papers, New York, NY, USA, pp. 1160–1168 (2006)
34. Pelechano, N., Allbeck, J.M., Badler, N.I.: Controlling individual agents in high-density crowd simulation. In: SCA 2007: Proceedings of the 2007 ACM SIG-GRAPH/Eurographics Symposium on Computer Animation, vol. 16, pp. 99–108. Eurographics Association, Aire-la-Ville (2007)
35. Tasse, F.P.: Crowd simulation of pedestrians in a virtual city. Technical Report Honours Project Report, Virtual Reality Special Interest Group, Computer Science Department, Rhodes University, Grahamstown, South Africa (November 2008)
36. Zanlungo, F.: Microscopic dynamics of articial life systems. Springer (2007)

A Unique Search Model for Optimization

A. S Xie

School of Public Policy and Management, Tsinghua University, Beijing 100084, China
shermanxas@163.com

Abstract. According to benchmark learning theory in business management, a kind of competitive learning mechanism based on dynamic niche was set up. First, by right of imitation and learning, all the individuals within population were able to approach to the high yielding regions in the solution space, and seek out the optimal solutions quickly. Secondly, the premature convergence problem got completely overcame through new optimal solution policy. Finally, the algorithm proposed here is naturally adaptable for the dynamic optimization problems. The unique search model was analyzed and revealed in detail.

Keywords: Benchmark learning, search model, evolutionary algorithm, swarm intelligence.

1   Introduction

Intelligent computation, also known as natural computation, is kind of optimization model, which was inspired by the principles of natural world, especially the biological world. Many optimization algorithms are included in the field of intelligent computation, which mainly consists of evolutionary algorithms (EAs) and swarm intelligences (SIs) et al. These intelligent optimization algorithms have special distinguishing features from each other, but they all have some common deficiencies. First of all, they all try to search the optimal solution by the individual's random drift in the solution space, yet the search direction and search purpose of the random drift are indeterminate and uncertain. Furthermore, they are all population convergence-oriented. Benchmarking [1] originally means that a surveyor's mark on a permanent object of predetermined position and elevation used as a reference point. As a kind of management idea and management method, benchmarking learning originated from enterprise management domain, and it means that some outstanding enterprise can be set as a standard, by which other companies can be measured or judged, and improved consequently.

2   The Benchmarking Learning Algorithm (BLA)

Benchmarking learning, in short, is to find the best case and learn something from it via imitate, others will improve themselves and even beyond the opponent. In the process of searching and learning, the self-organization learning in each niche

Y. Tan et al. (Eds.): ICSI 2014, Part I, LNCS 8794, pp. 19–26, 2014.

population goes like this: each individual will conduct external benchmarking learning first, that is to say, it will adjust its search direction and search step according to the best individual within the whole ecological system. Namely, it will decrease the distance with the external benchmarking. If its evaluation function value does not be improved, the individual will pass into the internal benchmarking learning stage. It will adjust its search direction and search step according to the best individual within the niche population to which it belongs, namely, it will reduce the distance with the internal benchmarking. If its evaluation function value does not be improved yet, the individual will carry out self-learning. It means that the individual will get its dual individual via dual operation. What is more, the best individual in each niche population will be fixed with the program process goes, because every niche population will exchange its best individual with other niche populations. So the internal benchmarking in each population, namely, the learning object of the staff inside each population will be replaced by other internal benchmarking. The three learning operations mentioned above will not be executed according to the order but executed selectively, only when its evaluation function value does not be improved through one learning operation, the individual will conduct another one.

The idea of benchmarking learning for optimization problems is distinctive, so it is very important that how to use the usual encoding methods--float-point encoding method and binary encoding method, to give expression to this idea. In BLA described here, the three learning operation, is extremely helpful to express this idea. The details about the three learning operations can be given as follow.

2.1   External Benchmarking Learning

Let x_E^{best} be the best individual, whose evaluation function value is the maximal or the minimal according to the optimization purpose, in the whole ecological system, that is, the global optimal individual and the external benchmarking, let G_E^{best} be its corresponding gene expression, let G_K^i be the gene expression of x_K^i, which is the i-st individual within niche population P_K, then, the external learning rate of x_K^i can be given as rule (1):

$$\begin{cases} \max f(x): & Grate_K^i = Grate' + f_K^i / \tilde{f}_K - 1 \\ \min f(x): & Grate_K^i = Grate' + \tilde{f}_K / f_K^i - 1 \end{cases} \tag{1}$$

Wherein, $Grate'$ stands for the initial value of the external learning rate, f_K^i stands for the value of the evaluation function of x_K^i, $\tilde{f}_K$ stands for the average value of P_K.

If binary encoding method was put to use, external benchmarking learning was carried out by x_K^i means that the gene-bits in G_K^i, which are different from that in G_E^{best}, would be replaced by the gene-bits in G_E^{best} with a probability of $Grate_K^i$. That is to say, x_K^i took the initiative to narrow the Hamming distance with x_E^{best}.

If float-point encoding method was involved, external benchmarking learning was conducted by x_K^i means that with a probability of $Grate_K^i$, G_K^i would be updated according to rule (2) as below. That is to say, x_K^i took the initiative to reduce the Euclidean distance with x_E^{best}.

$$G_K^i = G_K^i + \lambda (G_E^{best} - G_K^i)\qquad(2)$$

Wherein, $\lambda \in [0,1]$, stands for the shift step length of individual x_k^i. Experiments show that the optimization effect will be better if λ is proportional to the search space, or fixed dynamically according to the evaluation function value in the process of learning. But this is not the focal point of this paper, so it will not be took into further discussion.

2.2   Internal Benchmarking Learning

Let x_K^{best} be the best individual, whose evaluation function value is the maximal or the minimal according to the optimization purpose, in niche population P_K, namely, the local optimal individual and the internal benchmarking; let G_K^{best} be its corresponding gene expression; let G_K^i be the gene expression of x_K^i, which is the i-st individual in niche population P_K.Then, the internal learning rate of x_K^i can be given as rule (3):

$$\begin{cases} Binary\text{:} & Brate_K^i = Brate' - HD_{k,h}/Length + 1 \\ Float\text{:} & Brate_K^i = Brate' - ED_{k,h}/Radius + 1 \end{cases}\qquad(3)$$

Wherein, $Brate'$ stands for the initial value of the internal learning rate; $HD_{K,h}$ stands for the Hamming distance between x_k^i and x_k^{best}; $Length$ stands for the length of the gene expression encoding; $ED_{K,h}$ stands for the Euclidean distance between x_k^i and x_K^{best}, namely, $ED_{k,h} = \sqrt{\sum_1^n (x_i^{best} - x_i)^2}$. $Radius$ stands for the diameter of the search space, that is, $Radius = \sqrt{\sum_1^n (b_i - a_i)^2}$. Here, x_i is the i-st dimension of the gene expression, and $x_i \in [a_i, b_i]$.

Similar to external benchmarking learning, if binary encoding method was put to use, internal benchmarking learning was carried out by x_k^i means that the gene-bits in G_k^i, which are different from that in G_K^{best}, would be replaced by the gene-bits in G_K^{best} with a probability of $Brate_k^i$. That is to say, x_K^i took the initiative to narrow the Hamming distance with x_K^{best}.

If float-point encoding method was adopted, internal benchmarking learning was carried out by x_k^i means that with a probability of $Brate_K^i$, G_K^i would be updated according to rule (4) as below. That is to say, x_k^i took the initiative to diminish the Euclidean distance with x_K^{best}.

$$G_K^i = G_K^i + \lambda (G_K^{best} - G_K^i)\qquad(4)$$

Wherein, $\lambda \in [0,1]$, stands for the shift step length of individual x_k^i.

It seems that there is no difference between external benchmarking learning and internal benchmarking learning, because they are all apt to narrow the Hamming (or Euclidean) distance between an individual and the best one. As a matter of fact, there is great difference between external and internal benchmarking learning. It is of importance to reduce the Hamming (or Euclidean) distance. For one thing, it is helpful for the populations to carry out intensive search, which contributes to forming cluster effect and seeking out the global optimal solution quickly. For another, it is very helpful to maintain the population diversity, because the learning object of each individual within the population is changing constantly and dynamically, therefore, the clustering hierarchy in the whole ecological system is changing dynamically as well.

2.3   Internal Benchmarking Learning

Let x_K^i be the i-st individual in niche population P_K; let f_K^i be the evaluation function value of x_K^i; let $\tilde{f}_K$ be the average value of the niche population P_K; let G_K^i be the gene expression of x_K^i. Then, the self-learning rate of x_K^i can be given as below.

$$\begin{cases} \max f(x): & Srate_K^i = Srate' * \tilde{f}_K / f_K^i \\ \min f(x): & Srate_K^i = Srate' * f_K^i / \tilde{f}_K \end{cases} \quad (5)$$

Wherein, $Srate'$ stands for the initial value of the self-learning rate.

If binary encoding method was put to use, self-learning was carried out by x_K^i means that each gene-bit in G_K^i will conduct dual mapping [2] with a probability of $Srate_K^i$ shown as below.

```
primary gene: 1  0  1  0  0  1  0  1  1  0  0  0  1  0  0
dual gene:    0  1  0  1  1  0  1  0  0  1  1  1  0  1  1
```

Fig. 1. Binary dual mapping

If float-point encoding method was put to use, x_K^i carried out self-learning means that x_K^i would make use of logistic chaos mapping to help itself to jump out of the current region. Let $G_K^i = [x_1, x_2, \cdots, x_{n-1}, x_n]$, $x_i \in [a_i, b_i]$. Then G_K^i would be updated according to rule (6) as below:

$$\begin{cases} \lambda_i(0) = \dfrac{x_i - a_i}{b_i - a_i} \\ \lambda_i(t+1) = \delta \lambda_i(t)(1 - \lambda_i(t)) \\ x_i(t) = a_i + \lambda_i(t)(b_i - a_i) \\ \delta \in [2, 4], i = 1, 2, 3, \cdots, n \end{cases} \quad (6)$$

From rule (5), it is easy to see that when the optimization purpose is to obtain the maximum value of the evaluation function, if the evaluation function value of some individual is smaller than the average value of the niche population it belongs, its self-learning desire will be heightened quickly and its self-learning rate will increase to a larger number, and so it is more likely to obtain its dual individual to enhance the evaluation function value. But if its evaluation function value is bigger than the average value of the niche population it belongs, its self-learning desire will fade away quickly and its self-learning rate will decrease to a smaller number, and so it is great helpful to protect against the ascendant genes to be destroyed. In like manner, when the optimization purpose is to obtain the minimum value of the evaluation function, the self-learning desire will adjust accordingly.

2.4   Pseudocode

Let $E = \{P_1, P_2 \ldots P_{np}\}$ be the whole ecological system consists of np niche populations, N_i be the number of individuals in P_i, P_i^j be the j-th individual in P_i, P_i^{best} be the best individual in P_i. Let f_i^j be the evaluation function value of P_i^j, $\bar{f}_i$ be the average value of P_i at current generation, $\bar{f}_E$ be the average value of E at current generation, P_{best} be the best individual in E. Let $Grate'$ be the external learning rate, $Brate'$ be the internal learning rate, $Srate'$ be the self-learning rate. Let max_gen be the maximum iteration times. Then the pseudocode for BLA can be given as below.

(1) Initialize the np, N_i, $Grate'$, $Brate'$, $Srate'$ and other parameters if necessary.

(2) *for gen=1:max_gen, do*
    (a) *for i=1:np, do*
        i .Evaluate f_i^j
        ii .Evaluate $\bar{f}_i$
        iii.Find and record P_i^{best} **(Setting the internal benchmarking)**
    (b) Find out and record P_{best} **(Setting the external benchmarking)**
    (c) Find out, record and update the best individual so far in E. **(The global optimal solution)**
    (d) evaluate $\bar{f}_E = \left(\sum \bar{f}_i\right)/np$
    (e) *for i=1:np, do*
        i . P_i^j conduct external benchmarking learning
        ii .if f_i^j does not be improved, then, P_i^j will conduct internal benchmarking learning
        iii. if f_i^j does not be improved yet, then, P_i^j will carry out self-learning
    (f) if $\bar{f}_E$ does not be improved or the best individual in E does not be replaced *do*
        P_i will exchange its best individual with other niche populations **(Namely, each niche populations set a new external benchmarking)**

(3) Output the global optimal solution

3   Main Features

The main features and advantages of BLA, which are different from and superior to other optimization methods, were described, analyzed and given as below.

3.1   The Unique Search Model

The general framework of evolutionary algorithms (EAs) represented by genetic algorithm (GA) is that all the individuals in ecological system are carrying out genetic operations including selection, crossover and mutation at random. Though the crossover rate and mutation rate can be fixed adaptively to reduce the randomness of the genetic operation, EAs appeared more stochastic and less intelligent. In general, the search strategy of EAs is still passively adaptive. In the particle swarm optimization (PSO), particles are apt to follow both the best position they had drifted and the position of local optimal particle, so the search behavior of particles within PSO is intelligent in a certain extent. However, PSO can't maintain the population diversity natively and can't get over the premature convergence problem. What is more, no other encoding methods, but only the float-point encoding method can be used in PSO. A kind of PSO based on binary encoding method was proposed [3] by Kennedy and Eberhart, but it run at the sacrifice of PSO's limited intelligence and became purely a kind of random search algorithm. The artificial fish swarm algorithm (AFSA) was designed through simulating the behavior patterns of fish swarm, such as prey, swarm and follow. Like PSO, AFSA is not able to keep the population diversity and overcome the premature convergence problem as well. The ants in the ant colony optimization (ACO) would find out the best route according to the principle—the thicker the pheromones, the closer the route. ACO takes advantage of positive feedback mechanism, but it can't get over the distraction of the local extremum. Besides that, ACO is suitable for discrete problem like traveling salesperson problem (TSP), and it is hard to convergence to the global optimal solution when the number of cities is too great. The simulated annealing (SA) makes use of the metropolis acceptance criteria to help to jump out of the local extremum regions, but it includes too many repetitive iterations and this requirement is difficult to be met in the practical applications. Therefore, SA is just a heuristic random process based on Monte Carlo method. The taboo search (TS) and the predatory search (PS) are more likely to act as a kind of unique search strategy and search pattern. The former would flag the region which had been searched to reduce the repetitive iterations, and the later has no detailed computing method, it is just a strategy to balance the global search and local search, and just a way to keep the algorithm better both in exploration and exploitation.

The general framework of BLA proposed in this paper is that all the individuals in ecological system are selectively executing learn-actions by themselves and their learning objects are constantly changing. The purpose and direction of the learn-actions are very definite and clear, namely, every individual in ecological system wants to grow up and become the learning object of other individuals by simulating and learning from the best individual. When some niche population find out a new global or local optimal solution, that is, setting a new external or internal benchmarking, it will attract lots of individuals, who belong to other niche populations, to join in and help to search a better global or local optimal solution. But if the niche population can't seek out a new global or local optimal solution all the time, its members will be all attracted into other niche populations until its extinction. In a similar way, if some individual can chase down a better global optimal solution, it will attract a lot of individuals into its local region to create and form a new niche population. This is the dynamic niche technique proposed in this paper. Therefore, BLA is a type of learning-competition optimization model, and also a competition-learning optimization model. All in all, BLA described here, whose framework is brief and clear, is easy to be programmed. It is a learning-search strategy as well as a competitive optimizing method. Obviously, BLA, a new kind of search model, which was designed based on management theory and method in business word, is different from all the existing optimization methods, which were designed based on the principles of the natural world, especially the biological world.

3.2   Less Repetitive Operations in Search Process

One of the remarkable characteristics of BLA is less repetitive operations in the search process. A great many of repetitive genetic manipulations would be conducted in the process of evolutionary algorithms (EAs) represented by genetic algorithms (GAs). For example, an individual may carry out crossover operation with one more individuals in the search process, yet its corresponding objective function value did not return to the program immediately. And due to the impact of random selection, after cross-operation was completed, some individuals within the population may be not involved in the crossover operation and their gene structures did not change, while some individuals within the population may conduct cross-operation for many times and their good gene structure may have been damaged. In the standard genetic algorithm(SGA) and the vast majority of its improved versions, each individual's objective function value was evaluated after the three core genetic operations, namely selection, crossover and mutation, was carried out in sequence, which in fact is equal to that each individual conducted a series of repetitive genetic manipulations before its objective function value was evaluated. Additionally, the individual in simulated annealing (SA) will conduct a large number of exploratory movements in the thermal equilibrium phase, which makes no contribution to find out the global optimal solution in most of the time. Though the particles in particle swarm optimization(PSO) have no repetitive movements in the search process, if the objective function value was not improved after once drifting, only at the next iteration were the search direction and drift step of particles corrected. In the search process of BLA, individual's learning behavior is conducted selectively, namely on condition that its objective function value was not be improved after carrying out previous learning strategy, the individual would conduct next learning strategy. Which not only contributes to create no useless and repetitive actions, but also helps to amend the gene structure before once search process came to an end.

3.3   No Useless Operations in Search Process

Useless operations, here, refers in particular to some operations in algorithm, which make no contribution to the optimization problems in question. Take the function optimization problems with real type of decision variables as an example, if the float-point encoding method was adopted, then a great deal of useless operations would appeared in the search process of the most of popular optimization methods, such as PSO, SA and AFSA. The search operations in these methods often make the individuals fly out of solution space, namely the decision variables are apt to beyond the scope of the variable range and become illegal solutions after a series of search operations. To solve this problem, the general approach is that if some decision variable is over the boundary, then it would be replaced by the boundary value or modulus of itself, or it would be assigned a new value within the variable range. But as a matter of fact, these solutions are not only another form of repetitive operations, in essence, but also a kind of useless operation, which completely destroyed the intelligence of the methods and was not conducive to solving the problems in question. In the search process of BLA, every individual in the niche population is carrying out benchmarking-oriented search actions which would narrow the Euclidean distance with the benchmarking. Because the learning object is legitimate, the individual would not fly out of the solution space. Therefore, there are no useless operations in BLA as PSO, SA and AFSA did, this will be verified by the function optimization experiments in the later section.

4   Conclusions

In this paper, a competitive learning mechanism based on dynamic niches was set up according to the core values of benchmarking. BLA, which originated from benchmarking theory of business management, is different from the EIOMs, which stem from the biological activities of nature. So BLA is brand new and it is a newborn member of the family comprising the modern intelligent optimization methods. However, as other optimization methods, BLA also involved a number of controls parameters, such as learning rates, etc. and how to set these controls parameters to optimize BLA to achieve the best effect, which itself is also a combinatorial optimization problem, is one of our next research topics.

References

1. Dong-long, Y.: Benchmarking: How to learn from benchmark enterprises. China Social Sciences Press, Beijing (2004) (in Chinese)
2. Yang, S.: Non-stationary problem optimization using the primal-dual genetic algorithm. Evolutionary Computation 3, 2246–2253 (2003)
3. Kennedy, J., Eberhart, R.C.: A discrete binary of the particle swarm optimization. In: Proc. IEEE International Conference on Systems, Man and Cybernetic, pp. 4104–4209. IEEE Service Center, Piscataway (1997)

Improve the 3-flip Neighborhood Local Search by Random Flat Move for the Set Covering Problem

Chao Gao, Thomas Weise, and Jinlong Li

School of Computer Science and Technology,
University of Science and Technology of China (USTC),
Hefei, China
chao.gao.ustc@gmail.com,
{tweise,jlli}@ustc.edu.cn

Abstract. The 3-flip neighborhood local search (3FNLS) is an excellent heuristic algorithm for the set covering problem which has dominating performance on the most challenging crew scheduling instances from Italy railways. We introduce a method to further improve the effectiveness of 3FNLS by incorporating random flat move to its search process. Empirical studies show that this can obviously improve the solution qualities of 3FNLS on the benchmark instances. Moreover, it updates two best known solutions within reasonable time.

Keywords: Set Covering Problem, 3-flip Local Search, Random Flat Move.

1 Introduction

The set covering problem (SCP) is a prominent combinatorial optimization task which asks to find a collection of subsets to cover all the elements at the minimal cost. Formally, it is defined as: Given a universal set E which contains m elements, n subsets which are $S_1 \cup S_2 \cup ... \cup S_n = E$ and each subset has a cost, find a set of subsets F at the minimal total cost but still cover all elements in X, i.e., $\bigcup_{s \in F} s = E$. In literature, the SCP is generally described as integer linear programming form, as follows:

$$\min \sum_{j=1}^{n} c_j \cdot x_j \tag{1}$$

s.t.

$$\sum_{j=1}^{n} a_{ij} \cdot x_j \geq 1, \quad i \in M, \text{ where } M = \{1, 2, \ldots, m\}, \tag{2}$$

$$x_j \in \{0, 1\}, \quad j \in N, \text{ where } N = \{1, 2, \ldots, n\} \tag{3}$$

Y. Tan et al. (Eds.): ICSI 2014, Part I, LNCS 8794, pp. 27–35, 2014.
© Springer International Publishing Switzerland 2014

The zero-one matrix $A = \{a_{ij}\}_{m \times n}$ represents the problem instance, and $a_{ij} = 1$ means subset S_j is able to cover the element i. For each variable $x_j = 1$ indicates that S_j is selected, and 0 otherwise. In literature, the SCP is also viewed as to find a set of columns to cover all the rows at the minimal total cost, where N is the set of all columns that each has a cost, M is the set of rows that need to be covered.

The SCP is NP-hard in the strong sense [10], thus no complete algorithms with polynomial time complexity are known for SCP. It has many real-world applications, such as crew scheduling from bus, railway and airline systems[1,6,15]. A number of algorithms have been proposed for SCPs, among them the exact algorithms are shown not suitable to tackle large-scale problems because of their untolerable time consumption [5]. Therefore, approximation and heuristic algorithms are also widely studied by researchers both from the operations research and artificial intelligence communities. A variety of approximation or heuristic algorithms have been proposed, including the greedy algorithm [7], randomized greedy procedures [17,8], simulated annealing [11], genetic algorithm [3], ant colony optimization algorithm [13,14], artificial bee colony algorithm [16] and the Meta-RaPS approach by Lan et al. [12].

However, among these heuristics, only a few of them are able to tackle the very large-scale instances from the Italy railways [6,4], which contain to up millions of columns and thousands of rows. These instances, firstly distributed by a FASTER competition in 1994, are generally referred as the most challenging SCP instances from the OR-Library [2]. Caprara et al. proposed a Lagrangian-based heuristic (CFT) and a greedy procedure and obtained very impressive results on this set of instances and other random generated benchmark instances [4]. They won the first prize of the competition. Later, the 3-flip neighborhood local search (3FNLS) proposed by Yagiura et al. [18] is able to surpass CFT on these crew scheduling instances. We emphasize that CFT is essential to the success of 3FNLS, for 3FNLS also uses the same subgradient method implementation proposed in CFT to solve the Lagrangian relaxation of SCP.

In this paper, we introduce a search strategy named random flat move to further improve the effectiveness of 3FNLS. The experimental results show that it is effective, especially on the very large-scale instances, for it generally produces better solution qualities than the original 3FNLS within the same time limits. Further, it discovered two new best known solutions for the largest two instances within reasonable time.

The rest of this paper is organized as follows: In Section 2, we give description of the 3FNLS algorithm. Then our improvement strategy is described in Section 3. Section 4 is the computational results and comparisons. Conclusion and future work are finally presented in Section 5.

2   3FNLS Review

To make this article self-contained, it is necessary to introduce the basic concepts of 3NFLS before presenting our improvement method. However, because the ideas in 3FNLS are complicated, we suggest the readers to refer [18] for the details. In this section, we only provide brief introduction to the key procedures.

The overall procedure of 3FNLS is shown as Algorithm 1, in which the *SUB-GRADIENT* method is used to solve the Lagragian dual relaxation of SCP to obtain a Lagrangian multiplier vector u, and then for each column j a reduced cost is calculated as $c_j(u) = c_j - \sum_{i \in M} a_{ij} \cdot u_i$. Because of the *integrality property*, an optimal solution u^* to the dual of LP relaxation of SCP is also the optimal solution to the Lagrangian dual problem [9]. For a good Lagrangian multiplier vector u, the reduced cost $c_j(u)$ can give reliable information for the goodness of column j, because each column j with $x_j = 1$ in an optimal solution tends to have small $c_j(u)$ value.

From Algorithm 1 we can see that in 3FNLS, initially, the candidate solution x is obtained greedily, and UB is set to *cost(x)*. 3FNLS calls the subgradient method many times. At the first time, $u_i^0 = \min\{c_j/I_j | i \in I_j\}$ ($\forall j \in N$ and I_j is the set of rows j covers), otherwise, u^0 is set to u^+, where u^+ is the Lagrangian multiplier vector obtained by the first call. Let x^* be the stored best solution during the search, then UB is always maintained as *cost(x*)*.

An essential feature of 3FNLS is its problem size reduction heuristic, which is indispensable when facing the very large-scale instances, because directly local search on the whole columns is quite expensive. At first, the selected columns are determined by columns with the first $\alpha \cdot min\_free \cdot$ small $c_j(u)$, where min_free and α are program parameters which are set to 100 and 3, respectively. This is corresponding to Line 4 in Algorithm 1.

Whenever some iterations of local search are finished, the selected columns are adjusted by randomly fixing some 'good columns' to 1 to call the subgradiment method, and then some new columns with $c_j(u') < 0$ are added to the selected column set, where u' is the new Lagrangian multiplier. Let $N_{selected}$ be the selected columns which the local search is conducted on, N_1 be the set of columns in x fixed to 1 (not permitted to flip to 0 during the next period of local search). This is shown as Line 31 in Algorithm 1.

In Algorithm 1, *r-flip is possible*($r \leq 3$) means that there is at least one *r-flip* to decrease a penalty function defined in 3FNLS. At most 3-flip is permitted by 3FNLS and when there are no flips within 3-flip to find to reduce the penalty function value, the penalty weights of rows are updated. The details of how the penalty weights are updated is complicated. Usually, the penalty weights are updated by an increasing manner with the information provided by the last flip. Only when a certain rule is violated, then the penalty weights of rows will be decreased to make sure that there are *possible flips* again. The interested readers are suggested to refer [18] for details.

3   Random Flat Move for 3FNLS

From Algorithm 1, we can see that 3FNLS can be viewed as a multi-start algorithm. Each period of local search starts from fixing some variables in $N_{selected}$ to 1 and calling the subgradient method, and then based on the information from

---

Algorithm 1. The 3FNLS algorithm

Input: A SCP instance
Output: Best found solution x^*
1 Initiate candidate solution x greedily and $UB \leftarrow cost(x)$;
2 Initiate u^0, intiate penalty weights for all rows in M ;
3 $u^+ \leftarrow SUBGRADIENT(UB, u^0)$;
4 Select a subset of columns to $N_{selected}$ based on their reduced cost;
5 $trial \leftarrow 1$;
6 **while** *Time not exceeded* **do**
7   **while** *Time not exceeded* **do**
8   **if** *one flip is possible* **then** process one flip;
9   **if** *better solution is detected* **then**
10 | update x^* and UB;
11   **end**
12   *continue*;
13   **if** *two flip is possible* **then** process two flip;
14   **if** *better solution is detected* **then**
15 | update x^* and UB;
16   **end**
17   *continue*;
18   **if** *three flip is possilbe* **then**
19   **if** *better solution is detected* **then**
20 | update x^* and UB;
21   **end**
22   process three flip;
23   *continue*;
24   **else**
25 | *break*;
26   **end**
27   **end**
28   update penalty weights of rows;
29   **if** *penalty weigts is updated by decrease* **then**
30   **if** *x^* has not been updated for at least $mintr\_lsl$ iterations* **then**
31 | modify variable fixing;
32   **end**
33   **end**
34    $trial \leftarrow trial + 1$;
35 **end**

---

solving Lagrangian relaxation, it continues to flip variables to decrease the *pcost* of the candidate solution until the stop condition is reached. Because the N_1 is determined by the stored best solution and current candidate solution, thus the quality of the stored best solution can directly influence the performance of the following period of local search.

The problem size reduction heuristic (or variable fixing) divides all the columns into two parts, in which the $N_{selected}$ represents the columns selected into the search.

Then, at the beginning of each period of local search, it further reduces the search size by fixing some columns in $N_{selected}$ as 1, which means they are not permitted to be flipped to 0 the following period of local search. It is easy to see that the correctness of selecting columns to fix to 1 is crucial to the search, because wrong fixing could drastically mislead the search. As the columns in N_1 is also selected heuristically, it is obvious that the correctness of the selection of columns to fix to 1 can not be guaranteed.

However, we observe that 3FNLS only updates the stored best solution when another better solution is detected; i.e., only when $cost(x) < UB$ and x is feasible. In the variable fixing modification algorithm, the N_1 is chosen from the intersection between the current candidate solution x and the stored best solution x^* which could be a no-promising local optimum. Therefore we propose a simple search strategy to 3FNLS, which is randomly update the x^* when x becomes feasible and $cost(x) = UB$ by a probability. The idea of random flat move is that the stored best solution could be on a local optima plateau, neutral walking at a probability may lead to a better chance for finding a portal. The modification of the local search of 3FNLS is as below:

Algorithm 2. Random flat move for 3FNLS

```
1  if r-flip is possible then
2      if x is feasible then
3          if UB > cost(x) then
4              x* ← x;
5              UB ← cost(x);
6          end
7          if UB = cost(x) and rand(0, 1) > prfm then
8              x* ← x;
9          end
10     end
11 end
```

In Algorithm 2, *r-flip* refers to one, two or three flip in Algorithm 1, and *prfm* is the probability of the flat move. We set *prfm* to 0.5 in our implementation.

4   Experimental Results

In order to show the effectiveness of our search strategy to 3FNLS, we test the modified algorithm on instances from the OR-Library [2], which contains the randomly generated instances as well as the very large-scale crew scheduling instances from Italy railways.

4.1   The Benchmark Instances

We test 3FNLS on 4 type random instances and 7 challenging instances from Italy railways, shown in Table 1. For type NRE to NRH, each type contains 5 instances. The density is the number of non-zero entries in the problem instance matrix. The optima of these instances are all unknown.

Table 1. Details of the test instances

Instance type	m	n	Range of cost	Density(%)	Number of instances
NRE	500	5000	1–100	10	5
NRF	500	5000	1–100	20	5
NRG	1000	10000	1–100	2	5
NRH	1000	10000	1–100	5	5
RAIL507	507	63009	1– 2	1.2	1
RAIL516	516	47311	1– 2	1.3	1
RAIL582	582	55515	1– 2	1.2	1
RAIL2536	2536	1081841	1– 2	0.4	1
RAIL2586	2586	920683	1– 2	0.4	1
RAIL4284	4284	1092610	1– 2	0.2	1
RAIL4872	4872	968672	1– 2	0.2	1

For Table 1, we can see that one obvious characteristic of these instances is that they all have many more columns (n) than rows (m), especially for the crew scheduling instances from railways, with up to 1 million columns, whereas no more than 5 thousands rows.

4.2   Comparison Results

The source code of 3FNLS is provide by the author Mutsunori Yagiura, written in C. Our improvement algorithm with random flat move (3FNLS-rfm) is directly modified upon the source of 3FNLS. Both the two algorithms are compiled in g++ with -O2 option, run on the same Intel(R) Xeon(R) E5450 3.00 GHz CPU machine with 16 GB RAM, under 64-bit Linux system. Due to the randomness of the algorithms, for each instance, 10 independent runs are performed with random seeds from 11 to 20. All times are measured in CPU seconds in our experiments.

The results are reported as the best solutions (*best*) obtain from the 10 runs, the average solution of the 10 runs (*mean*), number of runs that the best is detected (*#best*), and the average times over the runs that detecting the best (*Avg Time*). The time limit for instances from NRE to NRH is set to 20 seconds, RAIL507, RAIL516 and RAIL582 is set to 200 seconds, and RAIL2536, RAIL2586, RAIL4284 and RAIL4872 is set to 2000 seconds. The comparison results are shown in Table 2.

From Table 2, we can see that for instances from NRE to NRH, both 3FNLS and 3FNLS can achieve the best known solution (BKS) within short times. The only instance they have not all success is NRE2, where the solution quality of 3FNLS-rfm is still better than 3FNLS. For the NRG type instances, the average times of 3FNLS are only slightly smaller than that of 3FNLS-rfm.

Table 2. Comparison results of 3FNLS and 3FNLS-rfm on benchmark instances

Instance	BKS	3FNLS				3FNLS-rfm			
		best	mean	#best	Avg Time	best	mean	#best	Avg Time
NRE1	29	29	29	10	0.24	29	29	10	0.24
NRE2	30	30	30.7	3	6.74	30	30.2	8	12.89
NRE3	27	27	27	10	0.30	30	30	10	0.31
NRE4	28	28	28	10	0.24	28	28	10	0.23
NRE5	28	28	28	10	0.25	28	28	10	0.25
NRF1	14	14	14	10	0.36	14	14	10	0.36
NRF2	15	15	15	10	0.31	15	15	10	0.31
NRF3	14	14	14	10	0.30	14	14	10	0.31
NRF4	14	14	14	10	0.28	14	14	10	0.27
NRF5	13	13	13	10	0.54	13	13	10	0.27
NRG1	176	176	176	10	0.65	176	176	10	0.62
NRG2	154	154	154	10	0.86	154	154	10	1.25
NRG3	166	166	166	10	3.38	166	166	10	5.94
NRG4	168	168	168	10	0.99	168	168	10	1.27
NRG5	168	168	168	10	0.86	168	168	10	1.29
NRH1	63	63	63	10	1.82	63	63	10	2.47
NRH2	63	63	63	10	1.94	63	63	10	2.04
NRH3	59	59	59	10	1.17	59	59	10	0.63
NRH4	58	58	58	10	0.07	58	58	10	0.89
NRH5	55	55	55	10	0.60	55	55	10	0.59
RAIL507	174	174	174	10	26.45	174	174	10	9.70
RAIL516	182	182	182	10	1.67	182	182	10	1.64
RAIL582	211	211	211	10	2.83	211	211	10	2.31
RAIL2536	690[a]	691	691.3	7	709.83	**690**	690.2	9	910.80
RAIL2586	945[a]	946	947.0	2	1269.67	**945**	946.7	1	1002.79
RAIL4284	1064[a]	1063	1064.1	3	1350.27	**1062***	1063.3	2	1826.34
RAIL4872	1528[a]	1528	1530.0	2	1214.32	**1527***	1529.0	1	910.40
SUM	6136	6137	6143.1	237	4529.61	6133	6138.4	241	4596.41

[a] The BKSs of RAIL2536, RAIL2586, RAIL4284 and RAIL4872 are all previously found by 3FNLS, reported in [18].

[+] The better solution of 3FNLS-rfm is highlighted by boldface.

[+] The updated BKS of 3FNLS-rfm is with a following asterisk.

In Table 2, we highlight the better solutions of 3FNLS-rfm than 3FNLS in boldface, and the updated best known solutions with a following asterisk. It is easy to see that 3FNLS-rfm has 4 better best solutions than 3FNLS on the 7 railway instances. Moreover, 3FNLS-rfm also has better solution qualities than 3FNLS on the instance RAIL507, RAIL516 and RAIL582, because its average times are generally smaller than 3FNLS. For RAIL2536 and RAIL2586, the origianl 3FNLS fails to achieve the BKS of these two instances within 2000 seconds. For the RAIL4284 and RAIL4872, 3FNLS-rfm has updated the best known solutions of these two instances. The results in Table 2 show that 3FNLS-rfm can obviously improve the performance of 3FNLS, especially on the challenging large-scale railway problems, for it is always achieves better solution qualities, given the same time limits on the same machine.

5 Conclusion and Future Work

In this paper, we have reviewed the state-of-the-art 3-flip neighborhood local search (3FNLS) algorithm for the set covering problem. Through the analysis of the process of 3FNLS, we notice that it can be regarded as a multi-start algorithm, which starts each period of local search by solving the Lagrangian and fixing some variables to 1. However, 3FNLS does not allow the *stored best solution* to move on the other 'equal-best 'solutions which may be portal to better solutions. Therefore, we propose a random flat move strategy to 3FNLS to make sure that the *stored best solution* can be updated by a probability on the possible plateau.

The proposed strategy has been tested on instances from the OR-Library, and the computational comparison results show that our strategy can obviously improve the performance of 3FNLS on the very large-scale instances. Moreover, it has updated the best known solutions of the last two instances. Observing that the very large-scale railway instances are commonly regarded as most challenging in the SCP benchmark instances, we believe our improvement method should be worth of existing.

In this paper, the probability of our flat move is intuitively set to 0.5, which may not be the ideal value for this algorithm. In the future, further empirical studies will be conducted to explain the behaviors of 3FNLS with different flat move probabilities. The relationship between the effectiveness of 3FNLS and instance features is also worth for further study.

Acknowledgments. The authors are grateful for M. Yagiura for providing their source of 3FNLS. This research work was partially supported by an EPSRC grant (No. EP/I010297/1), the Fundamental Research Funds for the Central Universities (WK0110000023), the National Natural Science Foundation of China under Grants 61150110488, the Special Financial Grant 201104329 from the China Postdoctoral Science Foundation, and the Chinese Academy of Sciences (CAS) Fellowship for Young International Scientists 2011Y1GB01.

References

1. Balas, E., Carrera, M.C.: A dynamic subgradient-based branch-and-bound procedure for set covering. Operations Research 44(6), 875–890 (1996)
2. Beasley, J.E.: Or-library: distributing test problems by electronic mail. Journal of the Operational Research Society, 1069–1072 (1990)
3. Beasley, J.E., Chu, P.C.: A genetic algorithm for the set covering problem. European Journal of Operational Research 94(2), 392–404 (1996)
4. Caprara, A., Fischetti, M., Toth, P.: A heuristic method for the set covering problem. Operations Research 47(5), 730–743 (1999)

5. Caprara, A., Toth, P., Fischetti, M.: Algorithms for the set covering problem. Annals of Operations Research 98(1-4), 353–371 (2000)
6. Ceria, S., Nobili, P., Sassano, A.: A lagrangian-based heuristic for large-scale set covering problems. Mathematical Programming 81(2), 215–228 (1998)
7. Chvatal, V.: A greedy heuristic for the set-covering problem. Mathematics of Operations Research 4(3), 233–235 (1979)
8. Feo, T.A., Resende, M.G.: A probabilistic heuristic for a computationally difficult set covering problem. Operations Research Letters 8(2), 67–71 (1989)
9. Fisher, M.L.: The lagrangian relaxation method for solving integer programming problems. Management Science 27(1), 1–18 (1981)
10. Gary, M.R., Johnson, D.S.: Computers and Intractability: A Guide to the Theory of NP-completeness (1979)
11. Jacobs, L.W., Brusco, M.J.: Note: A local-search heuristic for large set-covering problems. Naval Research Logistics 42(7), 1129–1140 (1995)
12. Lan, G., De Puy, G.W., Whitehouse, G.E.: An effective and simple heuristic for the set covering problem. European Journal of Operational Research 176(3), 1387–1403 (2007)
13. Lessing, L., Dumitrescu, I., Stützle, T.: A comparison between aco algorithms for the set covering problem. In: Dorigo, M., Birattari, M., Blum, C., Gambardella, L.M., Mondada, F., Stützle, T. (eds.) ANTS 2004. LNCS, vol. 3172, pp. 1–12. Springer, Heidelberg (2004)
14. Ren, Z.G., Feng, Z.R., Ke, L.J., Zhang, Z.J.: New ideas for applying ant colony optimization to the set covering problem. Computers & Industrial Engineering 58(4), 774–784 (2010)
15. Smith, B.M., Wren, A.: A bus crew scheduling system using a set covering formulation. Transportation Research Part A: General 22(2), 97–108 (1988)
16. Sundar, S., Singh, A.: A hybrid heuristic for the set covering problem. Operational Research 12(3), 345–365 (2012)
17. Vasko, F.J.: An efficient heuristic for large set covering problems. Naval Research Logistics Quarterly 31(1), 163–171 (1984)
18. Yagiura, M., Kishida, M., Ibaraki, T.: A 3-flip neighborhood local search for the set covering problem. European Journal of Operational Research 172(2), 472–499 (2006)

The Threat-Evading Actions of Animal Swarms without Active Defense Abilities

Qiang Sun[1], XiaoLong Liang[2], ZhongHai Yin[1], and YaLi Wang[1]

[1] Science College, Air Force Engineering University, Xi'an, China
s126email@126.com
[2] Air Traffic Control and Navigation College, Air Force Engineering University,
Xi'an, China

Abstract. For hunting foods, migrating and breeding, some small animals tend towards flocking. Once encountering predators, these swarms usually present some special threat-evading behaviors and show incredible regularity in movement by coordinating with each other. In this paper, the actions of a small fish swarm evades a blacktip reef shark are taken as an example, we put forward the concept of predator threat-field and present a mathematical model to describe the threat-field intensity, demonstrate the decision-making process of preys escaping the predator's threat by revising R-A model. The validity of the model is shown by simulations.

Keywords: Swarm, decision-making, threat-field, threat-evading, movement.

1   Introduction

While hunting food or moving, an animal swarm is likely to encounter a predator's attack. The responses of different kinds of animals to predator threat are quite different. The responses to attacks roughly fall into two categories: with and without active defense abilities. The paper focus on the phenomena of none active defense abilities swarms evade predator threat.

A typical scene was recorded by videos and photos[1, 2], as illustrated in Figure 1.

Fig. 1. Fish swarm avoiding a blacktip reef shark

Y. Tan et al. (Eds.): ICSI 2014, Part I, LNCS 8794, pp. 36–43, 2014.
© Springer International Publishing Switzerland 2014

According to these photos and videos, when there is no predator attack, individuals in the fish shoal are closely related to one another and the behaviors of fish are obviously regular. When encountering predator attack, the behaviors of individuals evading dangers present obvious regularity. Then, what is the law of the predator threat? How does an individual make the decision to evade the threat? How does the threat information transfer among individuals without any special signal sent by individuals in the shoal or even without the knowledge of the threat location? These problems will be dealt with in this paper.

2   The Previous Relational Works

In 1987, Reynolds put forward the Boids model to study the bird flying behavior[3]. He gave three simple rules each individual (agent) must abide, that is, *collision avoidance*, *speed matching* and *flock centering*. Iain D. Couzin et al presented the R-A model[4,5] which can be used to describe quite well the close relations among swarm members, reflect social interaction behaviors among the individuals in a swarm.

Suppose the swarm consists of N agents in the R-A model. Respectively, denote the location, speed and acceleration of individual i by vectors x_i, v_i, a_i $(x_i, v_i, a_i \in R^n)$. The movement of individuals is controlled by the following dynamics equation:

$$\begin{cases} \dot{x}_i = v_i \\ \dot{v}_i = a_i \end{cases}. \tag{1}$$

where $\dot{x}_i$ and $\dot{v}_i$ are the derivatives of x_i and v_i, respectively. The acceleration a_i of individual i can be represented as follows[6-7]

$$a_i = c_1 V_{ri}^0 + c_2 V_{mi}^0 + c_3 V_{ai}^0 + c_4 V_{Ri}^0. \tag{2}$$

where the vectors V_{ri}^0, V_{mi}^0, V_{ai}^0 and V_{Ri}^0 are, respectively, the corresponding unit vectors of vector V_{ri}, V_{mi}, V_{ai} and V_{Ri}. V_{ri} is repulsion-action direction of individuals in region R_r. V_{mi} is the sum speed vector of individuals in region R_m; Vector V_{ai} is the direction of individuals' attract-action to individual i in region R_a; V_{Ri} is a random vector which gives the preference movement direction of individual i.

$$V_{ri} = \sum_{j \in R_r} \frac{x_i - x_j}{\| x_i - x_j \|}, \quad V_{mi} = \sum_{j \in R_m} v_j, \quad V_{ai} = \sum_{j \in R_a} \frac{x_j - x_i}{\| x_j - x_i \|}. \tag{3}$$

The constants $c_1 \sim c_4$ are, respectively, the weight coefficient of each term and they give the individual's decision-makings in each direction in the movement process. The acceleration and speed values of each individual are subject to the following constraints

$$a_i = \begin{cases} a_i & \| a_i \| \leq A_{max} \\ A_{max} \dfrac{a_i}{\| a_i \|} & \| a_i \| > A_{max} \end{cases}, \quad v_i = \begin{cases} v_i & \| v_i \| \leq V_{max} \\ V_{max} \dfrac{v_i}{\| v_i \|} & \| v_i \| > V_{max} \end{cases}. \quad (4)$$

where A_{max} and V_{max} are, respectively, the reachable maximum acceleration and reachable maximum speed value for an individual.

3   Threat-Evading Mechanism

3.1   The Threat-Field Concept and Model

In the course of predator attack prey swarm, we can observe and understand the behavior is that the predators follow the prey-swarm when they found them, then rapidly chase and try to rush the prey-swarm, so they can acquire hunting opportunity. When the distance between predator and prey-swarm reduces to a certain degree, the preys begin to show stress-response to evade the menace, and quickly run away along a direction to reduce the menace. The shorter the distance is, the greater menace to be perceived by preys, the more intense the threat-evading reactions are.

For a simple analysis and expression, we suppose the prey-swarm consists of N individuals. At the moment t, the position vector of individual i is $x_i(t)$, the speed vector is $v_i(t)$ and the acceleration vector is $a_i(t)$. The corresponding position vector of predator is $h(t)$ and the speed vector of predator is $s(t)$.

Obviously, there are varying degree threats to the preys around the predator. The threat has its source direction and corresponding intensity. Therefore, there is a threat-field around the predator. It can be described as a vector field. See Figure 2, the triangle represents the predator H and the circularity represents the prey Q_i. Within the predator's threat-field, prey Q_i can sense the threat of the predator H. The threat can be described as a vector $w_i(t)$ and the direction is from H to Q_i.

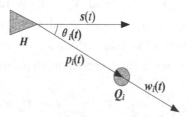

Fig. 2. The threat vector

The relative position between predator H and prey Q_i can be expressed as a vector $p_i(t) = x_i(t) - h(t)$. The distance between them is $\rho_i(t) = \| p_i(t) \|$ and the angle between the relative position direction $p_i(t)/\| p_i(t) \|$ and predator movement direction $s(t)/\| s(t) \|$ is $\theta_i(t)$ ($\theta_i(t) \in [0, \pi]$).

In terms of only consider single factor, apparently, the longer the distance $\rho_i(t)$ is, the smaller the threat to prey becomes; the lager the angle $\theta_i(t)$ is, the smaller the threat to prey is. This means that the threat intensity is related to two factors: $\rho_i(t)$ and $\theta_i(t)$. If the distance $\rho_i(t)$ is only take into consideration, for $\rho_i(t) = 0$, the threat-field intensity will be the maximum, and for $\rho_i(t) = \infty$, the threat-field intensity will be the minimum 0. Moreover, the threat-field intensity will decrease with the decrease of distance $\rho_i(t)$. According simulation experiment, we found the exponential function $e^{-\rho_i^2(t)}$ can describe this feature. If the angle $\theta_i(t)$ is only considered, for $\theta_i(t) = 0$, the threat level will be maximal, and for $\theta_i(t) = \pi$, the threat intensity will be minimal (0). The threat-field intensity should decrease with the decrease of angle $\theta_i(t)$. According simulation experiment, we found the function $\frac{1}{2}(\cos\theta_i(t) + 1)$ is able to meet this requirement. Furthermore, the orders of magnitude of the two functions are consistent with each other. So the arithmetic mean expression of the two functions is defined as threat-field intensity, that is

$$w(\rho_i(t), \theta_i(t)) = \frac{1}{2}[e^{-\rho_i^2(t)} + \frac{1}{2}(\cos\theta_i(t) + 1)] \ . \tag{5}$$

where $\rho_i(t) = \| p_i(t) \|$, $\cos\theta_i(t) = \dfrac{p_i(t) \cdot s(t)}{\| p_i(t) \| \cdot \| s(t) \|}$. The maximal value by this threat-field intensity definition is 1 and the minimal value is 0.

3.2   Threat-Evading Decision-Making

In our works, the threat-evading movement model is based on the R-A model (see Eq. (2)). In R-A model, V_{Ri} is a random vector used to express the movement preference of an individual. When an individual falls into predator's threat-field, its movement preference ought to be threat-evading. Therefore, the individual random preference vector V_{Ri} is considered to be replaced by the threat-evading direction vector to characterize an individual's movement actions. By adjusting of the weight coefficients $c_1 \sim c_4$ according to the threat-field intensity, an individual's threat-evading desire can be expressed. For example, when the threat intensity is stronger, the value of $c_1 \sim c_3$ will be reduced and the value of the c_4 will be increased, so as to represent the increase of an individual's threat-evading desire degree.

Apparently, predator threat intensity is decrease from predator to beyond. The shorter the distance between the prey and the predator is, the greater threat intensity the prey senses. Obviously, there exists a distance d_0. It is that a prey can bear the greatest threat intensity. That is, d_0 is a critical distance value, if the predator-prey distance is less than d_0, the prey will sense that threat is too violent to bear. Thus, the prey will show the most intense threat-evading response, escape from threat desperately. In addition, it can be sure that there is a maximal sense distance, denoted by d_1, it is the distance that a prey rely on their own sensing organs to perceive nearby predators. We call d_1 as the prey's maximal appreciable threat distance.

From the perspective that prey evade a predator's threat, the individuals within the prey-swarm can be divided into three layers by two concentric circles, the radii respectively is d_0 and d_1 and the centre of a circle is predator H. In the inner-most layer, that is, the neighborhood $U(H, d_0)$, the prey individuals will perceive the strongest threat, so their threat-evading desire is the biggest. Obviously, the most reasonable threat-evading direction is the one that can decrease the threat intensity most quickly. Suppose $T_i(t)$ is the gradient vector of the threat-field intensity function $w(\rho_i(t), \theta_i(t))$, that is

$$T_i(t) = grad \ w = \left(\frac{\partial w}{\partial x_i}, \frac{\partial w}{\partial y_i} \right)_t . \tag{6}$$

where x_i, y_i are, respectively, the two coordinate components of the individual position vector $x_i = (x_i, y_i)^T$. According to the calculus theory, vector $T_i(t)$ is the direction that threat intensity function's value increase fastest. Therefore, vector $-T_i(t)/\|T_i(t)\|$ represents the direction in which the threat intensity decrease most quickly, so it is the best evading direction, that is, in R-A model,

$$V_{Ri} = -\frac{T_i(t)}{\|T_i(t)\|} . \tag{7}$$

In addition, if a individual in the inner-most layer (that is, in neighborhood $U(H, d_0)$), its threat-evading desire is the strongest, so its movement decision is evading the threat desperately. Therefore, in R-A model, take the weight factors as $c_1 = c_2 = c_3 = 0$, $c_4 > 0$, so as to indicate that the individuals' strongest desire in the process of evading threat.

For individuals in the middle layers, that is outside the neighborhood $U(H, d_0)$ but within the neighborhood $U(H, d_1)$, from the inside to the outside, the threat intensity decreases. When individuals make movement direction decision, they will consider both the threat-evading direction and the other factors, such as inter-individual distances, matching speed and group-keeping. That is, in the R-A model, the preference vector is still taken V_{Ri} as eq.(7). In order to express the variations of considered factors, in an individual's decision-making process, as the prey-predator distance varies from d_0 to d_1, the weight factors $c_1 \sim c_3$ increase linearly from 0 to a normal value (that is, the value in the state without threat) and the weight factor c_4 descend linearly form maximum to a normal value.

The individuals in the outermost layer (i.e., outside the neighborhood $U(H, d_1)$) cannot directly sense the threat of predators, so they determine their movement direction based on the surrounding individuals' locations, movement status as well as their own preferences. That is, the individual movement decision-making is consistent with the motion equations (see Eq. (1)~(4)) of R-A model. Still, V_{Ri} is a random vector representing individual movement preference.

3.3   Threat Information Transfer

Although the above analysis doesn't mention the problems of information transfer, it can be ascertained that there are specific information transfer methods in the process of threat-evasion. The information transfer process can be understood as follows: first of all, a part of individuals within the threat-field (the distance between the two sides is less than maximal appreciable threat distance d_1) can sense the predator's threat rely on their own sensing organs. They show unusual threat-evading stress-response actions. For those individuals beyond the maximal appreciable threat distance d_1, even if those individuals who have sensed the threat don't send any clear alarm signal such as wow or ultrasonic waves, they can still sense the threat through the near threatened individuals' abnormal threat-evading actions. For example, some other individuals around an individual suddenly get very close to it and even knocked into it; the neighboring individuals move suddenly to a particular direction with strong desire. Through these abnormal behaviors, the individual will be able to realize the threat, and thus knows the threat. Because of social interactions among the individuals in a swarm, the individual realizes threat-evading movement implicitly.

4   Simulations

In order to observe the threat-field effects on the prey individuals' escape direction, numerical simulations are carried out specially. The simulation method is as follows: fix a moving predator to a certain fixed position and put some prey individuals around the predator. The distance between the prey and predator is d_0. The prey individuals' initial speed is 0. In the case of only taking the threat-field action into account and without considering the individuals' interaction, simulate these individuals movement process. The specific implement is, in R-A model, take $c_1 = c_2 = c_3 = 0$, $c_4 = 1$ and V_{Ri} is the negative gradient vector of the threat-field.

The simulation result is shown in Figure 3(a). It can be found that, within the range of maximal threat distance d_1, the prey individuals escape along the direction of negative gradient vector of threat-field. When they go beyond the maximal threat distance d_1, their movement direction is no longer subject to threat-field.

Actually, the prey individual's threat-evading movement direction is affected not only by threat-field, but also by the neighboring individuals' movement states, furthermore, the predator is still in the state of constant movement. So the prey's threat-evading movement process is more complex. Now the prey evading threat simulation in the state of moving predator is presented in the following part.

The purpose of the predator is to approach the prey-swarm as close as possible. Suppose the predator is to determine its moving direction based on the swarm center and the average speed of the all of the preys in the swarm. So in the process of simulation calculation, the acceleration of the predator is taken as the direction vector pointing to the prey swarm center and all prey's average speed vectors, i.e.

$$a_H(t + \Delta t) = \overline{x}(t) - h(t) + \overline{v}(t) \ . \tag{8}$$

where $h(t)$ is the current position vector of predator, $\overline{x}(t)$ is the geometry center of prey-swarm. $\overline{v}(t)$ is the average speed of the prey-swarm. Considering the predator would have the greatest acceleration limit, if $\| x(t) - h(t) + \overline{v}(t) \| > a_{H\,max}$, then

$$a_H(t + \Delta t) = a_{H\,max} \frac{\overline{x}(t) - h(t) + \overline{v}(t)}{\| \overline{x}(t) - h(t) + \overline{v}(t) \|} \ . \tag{9}$$

where $a_{H\,max}$ is the maximal acceleration that predators can reach.

According to the acceleration of the predator and the current speed, the movement speed of the predator can be calculated

$$v_H(t + \Delta t) = a_H(t)\Delta t + v_H(t) \ . \tag{10}$$

Considering the maximum speed limit, if $\| a_H(t)\Delta t + v_H(t) \| > v_{H\,max}$, then

$$v_H(t + \Delta t) = v_{H\,max} \frac{a_H(t)\Delta t + v_H(t)}{\| a_H(t)\Delta t + v_H(t) \|} \ . \tag{11}$$

where $v_{H\,max}$ is the maximum speed that predators can reach.

According to the predator speed and current position, the new position of the predator is

$$x_H(t + \Delta t) = v_H(t)\Delta t + x_H(t) \ . \tag{12}$$

The movement of the prey can be controlled by Eq. (1)~(4). In R-A model, the weight factors $c_1 \sim c_4$ and vector V_{Ri} is determined according the methods of sections 3 (Threat-evading mechanism) in this paper. During the simulation, some data need to be given, including the prey's body length, maximum speed and acceleration, maximal threat-evading distance, maximal sensing range and the inter-individual repulsion distance and social interaction neighborhood radius. In addition, the predator's body length, maximal speed and maximal acceleration are also needed. In the process of simulation, a simulation result video file is recorded. Here the simulated screenshot is presented, as shown in Figure 3(b).

From the simulation results, individuals in the swarm are closely-related and the social interaction s very clear. Encountering a predator's attack, the threat-evading actions of individuals in the swarm are obviously regular, the same as show in the photographs and videos.

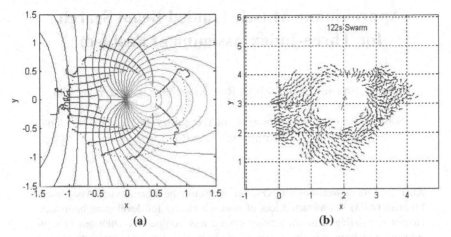

(a)   (b)

Fig. 3. Simulation screenshot of preys evading the predator

5   Conclusion

In this paper, a mathematical model is presented to demonstrate the factors and laws of a predator giving threat to the preys in a dense swarm. The decision-making rules of prey individuals' threat-evading actions are studied and the threat information transfer methods among the prey individuals are analyzed. It is very important for us to understand the swarm threat-evading action mechanism, the threat-field formation mechanism and the decision-making rules of prey evading the threat, as well as the threat information transfer law. The simulation results are consistent with the actual situation, which shows the validity of the models. The models and methods presented in this paper can be used in the research such as robots swarm and aircraft cluster etc. artificial agents system's threat-evading and obstacle-avoiding action control, providing the bionics principle for automatic control method.

References

1. Wilkison, P.: Blacktip Reef Shark, Maldives. National Geographic (July 2011)
2. http://blog.renren.com/share/224777634/7486626661
3. http://v.youku.com/v_playlist/f5749147o1p12.html
4. Reynolds, C., Flocks, H., et al.: A distributed behavioral model. Computer Graphics 21(1), 25–34 (1987)
5. Iain, D.C., Jens, K., Richard, J., Graeme, D.R., Nigel, R.F.: Collective Memory and Spatial Sorting in Animal Groups. Journal of Theory Biology 218(1), 1–11 (2002)
6. Ialn, D.C., Jens, K., Nigel, R.F., Slmon, A.L.: Effective leadership and decision-making in animal groups on the move. Nature 433(3), 513–516 (2005)
7. Olfati-Saber, R.: Flocking for multi-agent dynamic system: algorithms and theory. IEEE Transaction on Automatic Control, 401–415 (2006)
8. Pedrami, R., Gordon, B.W.: Control and analysis of energetic swarm systems. In: Proceedings of the 2007 American Control Conference, pp. 1894–1899 (2007)

Approximate Muscle Guided Beam Search
for Three-Index Assignment Problem

He Jiang, Shuwei Zhang, Zhilei Ren, Xiaochen Lai, and Yong Piao

Software School, Dalian University of Technology, Dalian, 116621, China
jianghe@dlut.edu.cn

Abstract. As a well-known NP-hard problem, the Three-Index Assignment Problem (AP3) has attracted lots of research efforts for developing heuristics. However, existing heuristics either obtain less competitive solutions or consume too much time. In this paper, a new heuristic named Approximate Muscle guided Beam Search (AMBS) is developed to achieve a good trade-off between solution quality and running time. By combining the approximate muscle with beam search, the solution space size can be significantly decreased, thus the time for searching the solution can be sharply reduced. Extensive experimental results on the benchmark indicate that the new algorithm is able to obtain solutions with competitive quality and it can be employed on instances with large-scale. Work of this paper not only proposes a new efficient heuristic, but also provides a promising method to improve the efficiency of beam search.

Keywords: Combinatorial Optimization, Heuristic, Muscle, Beam Search.

1   Introduction

The Three-Index Assignment Problem (AP3) was first introduced by Pierskalla [1, 2]. It is a NP-hard problem with wide applications, including addressing a rolling mill, scheduling capital investments, military troop assignment, satellite coverage optimization [1, 2], scheduling teaching practice[3], and production of printed circuit boards [4]. It can be viewed as an optimization problem on a 0-1 programming model:

$$\min \sum_{i \in I} \sum_{j \in J} \sum_{k \in K} c_{ijk} x_{ijk} \quad . \tag{1}$$

subject to

$$\sum_{j \in J} \sum_{k \in K} x_{ijk} = 1 , \ \forall i \in I \quad . \tag{2}$$

$$\sum_{i \in I} \sum_{k \in K} x_{ijk} = 1 , \ \forall j \in J \quad . \tag{3}$$

$$\sum_{i \in I} \sum_{j \in J} x_{ijk} = 1 , \ \forall k \in K \quad . \tag{4}$$

$$x_{ijk} \in \{0,1\} , \ \forall i \in I, j \in J, k \in K \quad . \tag{5}$$

where $I = J = K = \{1,2,3,...,n\}$.
The solution of AP3 can be presented by two permutations:

Y. Tan et al. (Eds.): ICSI 2014, Part I, LNCS 8794, pp. 44–52, 2014.
© Springer International Publishing Switzerland 2014

$$\min \sum_i^n c_{i,p(i),q(i)} \ , \quad p(i), q(i) \in \pi_N \ .\tag{6}$$

where π_N presents the set of all permutations on the integer set $N=\{1,2,\dots,N\}$. Here c_{ijk} represents the cost of a triple $(i,j,k)\in I\times J\times K$.

Due to its intractability, lots of exact and heuristic algorithms are proposed to solve it, including Balas and Saltzman [5], Crama and Spieksma[6], Burkard and Rudolf [7], Pardalos and Pitsoulis[8], Voss [9], Aiex, Resende, Pardalos, and Toraldo[10], Huang and Lim [11], Jiang, Xuan, and Zhang [12]. Among these algorithms, LSGA proposed by Huang and Lim [11], and AMGO proposed by Jiang, Xuan, and Zhang [12] perform better than the other heuristics. LSGA can obtain a solution within quite a short time, but on difficult instances LSGA might not perform well in terms of the solution quality, while AMGO can obtain better solutions with high quality, but on large instances the running time is intolerable. It would be ideal to achieve a good trade-off between solution quality and running time.

To tackle the challenges in balancing solution quality and running time, we propose a new heuristic named Approximate Muscle guided Beam Search (AMBS). It combines two phases. In the first phase, a multi-restart local search algorithm is used to generate a smaller search space, which we call "approximate muscle". In the latter phase, beam search is employed to obtain a high quality solution. By combining the approximate muscle and beam search, we can obtain solutions with relatively high quality in a short time. Experimental results on the standard AP3 benchmark indicate that in terms of solution quality, the solutions obtained by AMBS are better than LSGA and not worse than the pure beam search, while in terms of running time, AMBS can deal with large instances that AMGO and the pure beam search cannot.

The rest of this paper is organized as follow. In Section 2, a review of the muscle and beam search is given. In section 3, the framework of AMBS is proposed. Experiment results are reported in Section 4. In Section 5, the conclusion is presented.

2   Muscle and Beam Search

In this section, we present the two concepts the muscle and beam search. For each concept, we first briefly review its related work and then present its details.

2.1   Muscle

The proposition of the concept muscle is inspired by the backbone. The backbone means the shared common parts of optimal solutions for an instance. It is an important tool for NP-hard problem. In contrast to the backbone, the muscle is the union of optimal solutions. It was first proposed by Jiang, Xuan, and Zhang in 2008 [12]. Some efficient algorithms have been proposed using the muscle. For example, Jiang and Chen developed an algorithm for solving the Generalized Minimum Spanning Tree problem with the muscle [13]. Obviously, if the muscle could be obtained, the search space for an instance would be decreased sharply. However, Jiang has proved that there is no polynomial time algorithm to obtain the muscle for AP3 problem [12].

Now that the muscle cannot be obtained directly, there are some other ways to approximate it. Experiments conducted by Jiang indicate that the probability that the union of local optima contains the optimal solution increases with the growth of the number of local optimum, while the size increases slower [12]. Hence, the union of local optima can approximate the muscle, it can be named the approximate muscle.

2.2   Beam Search

Beam search is a widely-used heuristic algorithm. For example, Cazenave combined Nested Monte-Carlo Search with beam search to enhance Nested Monte-Carlo Search [14], López-Ibáñez and Blum combined beam search with ant colony optimization to solve the travelling salesman problem with time windows [15].

Beam search can be viewed as an adaptation of branch-and-bound search. The standard version of beam search builds its search tree using breadth-first search. At each level of the search tree, a heuristic algorithm is employed to estimate all the successors, and then a predetermined number of best nodes are stored, while the others are pruned off permanently. This number is called the beam width. By varying the beam width, beam search varies from greedy search (the beam width equals to 1) to a complete breadth-first search (no limit to the beam width). By limiting the beam width, the complexity of the search becomes polynomial. In this way, beam search can find a solution with relatively high quality within practical time. We call the standard version of beam search as the pure beam search, to distinguish it with AMBS.

3   Approximate Muscle Guided Beam Search for AP3

In this section, we introduce the detail of AMBS. We will first present the framework of our algorithm, then we will show the details in the following subsections.

3.1   AMBS for AP3

The algorithm is shown in Algorithm 1. The instance $AP3(I,J,K,c)$, the number of sampling k, and the beam width $width$ are the inputs. The output is the solution s^*. A instance is stored in a 3-dimesional array, and the solutions is stored in two arrays.

Two phases are in the algorithm. In the beginning of the search phase, the order of search level is sorted ascending by the number of triples. When calculating the lower bound of each branch, more time will be consumed when the branch is at the higher level. Thus, after the sorting, fewer nodes are in the higher level, and searching time is reduced. More details about building the search tree is introduced in section 3.3.

In the following subsections, we will discuss the details of two phases, respectively.

3.2   Approximate Muscle for AP3

In the first phase, we use the union of local optima to approximate the muscle. The detail is shown in Algorithm 2. The inputs are an AP3 instance and the number of sampling.

The output includes the approximate muscle a_muscle, and the best solution s is obtained. The approximate muscle is stored in a 3-dimensional array, where the cost is the same value as the instance if it is sampled, or infinite if not.

The approximate muscle is initialized as an empty set first. Then k local optima are obtained to make up the approximate muscle. A random solution is generated, then a local search algorithm is applied to obtain a local optimum. The local search algorithm we use here is the Hungarian local search proposed by Huang and Lim [11]. The best local optimum is recorded.

3.3   Beam Search for AP3

In the second phase, we use beam search to find a better solution. Before the introduction of beam search for AP3, we will first present how we build the breadth-first search tree. An instance of AP3 can be represented as a three-dimensional matrix. First, the matrix is divided into n layers. For example, an instance with the size of 4 is divided into 4 layers. Select a layer to be level 1 of the search tree. Then another layer is selected to build next level. Since one triple has been determined in level 1, there are 9 successors, and 144 nodes in all in level 2. In this way, the tree is built. If using the muscle to build the tree, only the triples in the muscle are considered.

Algorithm 1. AMBS for AP3
Input: AP3instance $AP3(I,J,K,c)$, k, $width$
Output: solution s^*
Begin
//the sampling phase
(1)   obtain the approximate muscle a_muscle and a solution s' as the upper bound
with $GenerateAM\ (AP3(I,J,K,c),k)$;
//the search phase
(2)   sort the search order of the approximate muscle and get the $order$;
(3)   obtain the solution s^* with $BS(a\_muscle,width,s',order)$;
End

Algorithm 2. GenerateAM (Generate Approximate Muscle)
Input: AP3instance $AP3(I,J,K,c)$, k
Output: a_muscle, solution s'
Begin
(1) $a\_muscle = \varnothing$
(2)   for $counter = 1$ to k do
(3)   for $i = 1$ to n do $p[i] = i$, $q[i] = i$;
(4)   for $i = 1$ to n do
(5)   let j_1, j_2 be two random integers between 1 and n;
(6)   swap $p[i]$ and $p[j_1]$; swap $q[i]$ and $q[j_2]$;
(7)   let $s = \{(i, p[i], q[i]) \mid 1 \le i \le n\}$;
(8)   obtain a local optimum s_{local} by applying the local search to s;
(9) $a\_muscle = a\_mucle \bigcup s_{local}$;
(10)   if $c(s_{local}) < c(s')$ then $s' = s_{local}$
End

The detail of beam search for AP3 is presented in Algorithm 3. The inputs are the approximate muscle *a_muscle*, the beam width *width*, the best local solution *s'*, and the search order *order*. The output is the solution s^* of this AP3 instance.

In the algorithm, a candidate represents a branch to be searched. When the search comes to a certain level, the lower bounds of all the successors are generated. The arrays *fp* and *fq* are used to record the determined triples to guarantee the constraints. Then the lower bounds are calculated. The lower bound includes three parts: the sum of the triples' cost in the candidate, the cost of the triple relevant to the successor, and lower bound of the sub-problem. The sub-problem is the approximate muscle without the layers containing the determined triples. The lower bound calculating method is proposed by Kim et al. [16]. In the end, at most *width* successors with smaller lower bound than the cost of *s'* are kept to be the new candidates. After searching, a local search algorithm is employed to the remaining candidates. Then the best (including *s'*) is chosen to be the solution s^*. Since the approximate muscle is stored in the same way as an instance, beam search algorithm can be used to solve AP3 problem independently.

Algorithm 3. BS (Beam Search)
Input: a_muscle , $width$, solution s' , $order$
Output: solution s^*
Begin
(1)   for every *level* based on *order* in the search tree do
(2)   for every *candidate* do
(3)   for every triple $(i, j, k) \in candidate$ do $fp[j] = true, fq[k] = true$;
(4)   for every triple $(order[level], j, k) \in a\_muscle$ do
(5)   if $fp[j] = false$ and $fq[k] = false$ then
(6)   generate the sub-problem; calculate the lower bounds;
(7)   sort the bounds of all *candidate*;
(8)   for $i = 1$ to *width* do
(9)   if lower bound of the branch $< c(s')$ then
(10)   this branch belongs to the new *candidates*;
(11)   else break;
(12)   employ the local search algorithm on every *candidate* and choose the best to be the solution s^*
End

---

4   Experimental Result

In this section, we first show the parameter tuning result. Then we present the results of our algorithm on the benchmark. The codes are implemented with C++ under windows 7 using visual studio 2010 on a computer with Intel Core i3-M330 2.13G. The time in the tables is measured in seconds.

4.1   Parameter Tuning

Two parameters are used in AMBS, the number of sampling and the beam width. We determine the number of sampling as 1000, the same value in AMGO [12]. As for the beam width, we test different beam widths {100, 200, 300, 400} on 4*5 instances from Balas and Saltzman Dataset (see Section 4.2) and 6 instances from Crama and Spieksma Dataset (see Section 4.3).

Table 1 shows the result of our parameter tuning. We run the algorithm 10 times on each instances of the Balas and Saltzman Dataset with each beam width, while run it once on Crama and Spieksma Dataset since the result varies little. The value of Balas and Saltzman Dataset is the average value of each size. The result indicates that the solution quality and the running time rise with the increase of the beam width. Note that the running time of 3DA99N1, 3DA198N1 and 3D1299N1 vary little in different beam widths. This is because the approximate muscle space of each of the instance is so small. The running time of 3DI198N1 is the longest. When the beam width is 300, the running time is about 20 minutes. In order to balance the quality of the solution and the running time, we determine the beam width as 300 in the rest of experiments.

Table 1. Beam Width Tuning

Instance Id	Width=100		Width=200		Width=300		Width=400	
	Cost	Time	Cost	Time	Cost	Time	Cost	Time
BS_14_x	10	0.74	10	1.07	10	1.31	10	1.72
BS_18_x	6.86	2.48	6.66	4.25	6.48	5.97	6.46	7.82
BS_22_x	4.86	6.75	4.62	12.00	4.34	17.13	4.34	22.45
BS_26_x	2.74	15.01	2.34	27.05	2.1	39.16	2.08	51.01
3DA99N1	1608	7.01	1608	7.39	1608	7.24	1608	7.16
3DA198N1	2662	62.05	2662	65.00	2662	63.15	2662	64.34
3DIJ99N1	4797	16.59	4797	18.53	4797	20.33	4797	22.11
3DI198N1	9685	479.66	9684	863.94	9684	1219.82	9684	1566.40
3D1299N1	133	1.70	133	1.75	133	1.73	133	1.73
3D1198N1	286	169.37	286	276.14	286	383.95	286	573.86

4.2   Balas and Saltzman Dataset

This dataset is generated by Balas and Saltzman[5]. It contains 60 instances with size of 4, 6, 8, ..., 26. For each size, five instances are generated randomly.

Table 2. Balas and Saltzman Dataset (12*5 instances)

n	Opt.	LSGA		AMGO		Beam Search		AMBS	
		Cost	Time	Cost	Time	Cost	Time	Cost	Time
4	42.2	42.2	0	42.2	0.01	42.2	0.01	42.2	0.01
6	40.2	40.2	0.01	40.2	0.03	40.2	0.03	40.2	0.03
8	23.8	23.8	0.03	23.8	0.06	23.8	0.06	23.8	0.06
10	19	19	0.37	19	0.11	19	0.14	19	0.11
12	15.6	15.6	0.87	15.6	0.18	15.6	0.62	15.6	0.26
14	10	10	1.73	10	0.26	10	7.62	10	1.31
16	10	10	1.89	10.16	0.52	10	22.74	10	3.32
18	6.4	7.2	2.95	6.4	0.97	6.4	49.65	6.48	5.97
20	4.8	5.2	4.01	4.8	1.67	4.8	98.18	4.88	10.43
22	4	5.6	4.54	4	6.26	4.24	185.70	4.34	17.13
24	1.8	3.2	5.66	1.96	12.16	2.38	313.60	2.28	26.95
26	1.3	3.6	10.78	1	6.62	2.38	526.59	2.1	39.16

Table 2 shows the result on this dataset. The column "Opt." is the optimal solution reported by Balas and Saltzman[5]. "LSGA" is the result reported in Huang's paper [11] using a PIII 800MHz PC. "AMGO" is the results of the program implemented according to Jiang's paper [12]. "Beam Search" is the result of the pure beam search. "AMBS" is the result of our algorithm. The results of AMGO, the pure beam search and AMBS are the average cost after running 10 times on each instance. The solutions of size 26 found by AMGO are 0,0,2,1,2 respectively, the average cost is 1. Because the average of any five integers cannot be 1.3, it may be a typo in Balas's paper.

The result indicates that AMBS can get solutions with higher quality than LSGA. AMGO generates the best solutions, and the running time is quite short, because it employs a global search on the approximate muscle and the search space is quite small. AMBS uses an incomplete search and needs to estimate the lower bound of each branch, thus, the quality of solutions is a little worse and the running time is longer than AMGO. Compared with the pure beam search, the running time of AMBS is about one-tenth of the pure beam search, but the quality is comparable. This is because that there are much fewer successors in the approximate muscle.

4.3   Crama and Spieksma Dataset

This dataset is generated by Crama and Spieksma[6]. Three types are in the dataset. In each type, three instances have the size of 33, and three have the size of 66.

Table 2 shows the result on this dataset. AMGO, the pure beam search and AMBS are executed once since the result varies little. There are some cells with no value, because the running time is longer than 30 minutes, and we regard it unacceptable.

Table 3. Crama and Spieksma Dataset. (18 instances)

n	Instance Id	LSGA		AMGO		Beam Search		AMBS	
		Cost	Time	Cost	Time	Cost	Time	Cost	Time
33	3DA99N1	1608	0.03	1608	7.60	1608	649.74	1608	7.24
33	3DA99N2	1401	0.11	1401	7.11	1401	1733.90	1401	6.52
33	3DA99N3	1604	0.11	1586	7.61	1604	1606.99	1604	7.30
66	3DA198N1	2662	0.55	2662	71.22	-	-	2662	63.15
66	3DA198N2	2449	0.27	-	-	-	-	2449	74.20
66	3DA198N3	2758	0.58	-	-	-	-	2758	82.07
33	3DIJ99N1	4797	0.11	-	-	-	-	4797	20.33
33	3DIJ99N2	5067	0.26	-	-	-	-	5067	35.95
33	3DIJ99N3	4287	0.26	-	-	-	-	4287	26.07
66	3DI198N1	9684	4.86	-	-	-	-	9684	1219.82
66	3DI198N2	8944	3.35	-	-	-	-	8944	929.51
66	3DI198N3	9745	3.09	-	-	-	-	9745	767.66
33	3D1299N1	133	0.01	-	-	133	3.50	133	1.73
33	3D1299N2	131	0.03	-	-	131	1128.17	131	3.94
33	3D1299N3	131	0.02	131	1.98	131	580.97	131	3.31
66	3D1198N1	286	0.15	-	-	-	-	286	383.95
66	3D1198N2	286	0.16	-	-	-	-	286	341.05
66	3D1198N3	282	0.23	-	-	-	-	282	329.67

This result indicates that AMBS is able to run on every instance, and obtain a solution with high quality, while AMGO and the pure beam search are not able to deal with a number of instances. The running time of LSGA is quite short with high quality solution. This is because LSGA is an iterative algorithm and the instances in this dataset are easy to solve. If the instance is hard to solve, like the large instance in Balas and Saltzman Dataset, the quality of solutions of LSGA might not be that high.

5   Conclusion

In this paper, we propose a new heuristic named Approximate Muscle guided Beam Search (AMBS) for AP3 problem. This algorithm combines the approximate muscle and beam search. In this way, AMBS can achieve a good trade-off between the solution quality and the running time. Experimental results indicate that the new algorithm is able to obtain solutions with competitive quality, even on large instances.

Acknowledgement. This work was supported in part by the Fundamental Research Funds for the Central Universities under Grant DUT13RC(3)53, in part by the New Century Excellent Talents in University under Grant NCET-13-0073, in part by China Postdoctoral Science Foundation under Grant 2014M551083, and in part by the National Natural Science Foundation of China under Grant 61175062 and Grant 61370144.

References

1. Pierskalla, W.P.: The tri-substitution method for the three-dimensional assignment problem. CORS Journal 5, 71–81 (1967)
2. Pierskalla, W.P.: Letter to the Editor—The Multidimensional Assignment Problem. Operations Research 16, 422–431 (1968)
3. Frieze, A.M., Yadegar, J.: An Algorithm for Solving 3-Dimensional Assignment Problems with Application to Scheduling a Teaching Practice. The Journal of the Operational Research Society 32, 989–995 (1981)
4. Crama, Y., Kolen, A.W.J., Oerlemans, A.G., Spieksma, F.C.R.: Throughput rate optimization in the automated assembly of printed circuit boards. Ann. Oper. Res. 26, 455–480 (1991)
5. Balas, E., Saltzman, M.J.: An Algorithm for the Three-Index Assignment Problem. Operations Research 39, 150–161 (1991)
6. Crama, Y., Spieksma, F.C.R.: Approximation algorithms for three-dimensional assignment problems with triangle inequalities. European Journal of Operational Research 60, 273–279 (1992)
7. Burkard, R.E., Rudolf, R., Woeginger, G.J.: Three-dimensional axial assignment problems with decomposable cost coefficients. Discrete Applied Mathematics 65, 123–139 (1996)
8. Pardalos, P.M., Pitsoulis, L.S.: Nonlinear assignment problems: Algorithms and applications. Springer (2000)
9. Voss, S.: Heuristics for Nonlinear Assignment Problems. In: Pardalos, P., Pitsoulis, L. (eds.) Nonlinear Assignment Problems, vol. 7, pp. 175–215. Springer, US (2000)

10. Aiex, R.M., Resende, M.G.C., Pardalos, P.M., Toraldo, G.: GRASP with Path Relinking for Three-Index Assignment. INFORMS J. on Computing 17, 224–247 (2005)
11. Huang, G., Lim, A.: A hybrid genetic algorithm for the Three-Index Assignment Problem. European Journal of Operational Research 172, 249–257 (2006)
12. Jiang, H., Xuan, J., Zhang, X.: An approximate muscle guided global optimization algorithm for the Three-Index Assignment Problem. In: IEEE Congress on Evolutionary Computation, CEC 2008 (IEEE World Congress on Computational Intelligence), pp. 2404–2410 (2008)
13. Jiang, H., Chen, Y.: An efficient algorithm for generalized minimum spanning tree problem. In: Proceedings of the 12th Annual Conference on Genetic and Evolutionary Computation, pp. 217–224. ACM, Portland (2010)
14. Cazenave, T.: Monte Carlo Beam Search. IEEE Transactions on Computational Intelligence and AI in Games 4, 68–72 (2012)
15. López-Ibáñez, M., Blum, C.: Beam-ACO for the travelling salesman problem with time windows. Computers & Operations Research 37, 1570–1583 (2010)
16. Kim, B.-J., Hightower, W.L., Hahn, P.M., Zhu, Y.-R., Sun, L.: Lower bounds for the axial three-index assignment problem. European Journal of Operational Research 202, 654–668 (2010)

Improving Enhanced Fireworks Algorithm with New Gaussian Explosion and Population Selection Strategies*

Bei Zhang, Minxia Zhang, and Yu-Jun Zheng

College of Computer Science and Technology, Zhejiang University of Technology,
Hangzhou 310023, China
zhangbei-zjut@outlook.com,zmx@zjut.edu.cn,yujun.zheng@computer.org

Abstract. Fireworks algorithm (FWA) is a relatively new metaheuristic in swarm intelligence and EFWA is an enhanced version of FWA. This paper presents a new improved method, named IEFWA, which modifies EFWA in two aspects: a new Gaussian explosion operator that enables new sparks to learn from more exemplars in the population and thus improves solution diversity and avoids being trapped in local optima, and a new population selection strategy that enables high-quality solutions to have high probabilities of entering the next generation without incurring high computational cost. Numerical experiments show that the IEFWA algorithm outperforms EFWA on a set of benchmark function optimization problems.

Keywords: global optimization, fireworks algorithm (FWA), Gaussian explosion, population selection.

1 Introduction

Initially proposed by Tan and Zhu [1], fireworks algorithm (FWA) is a relatively new nature-inspired optimization method mimicking the explosion process of fireworks for optimization problems. In FWA, a solution to the problem is analogous to a firework or a spark, and an explosion is analogous to a stochastic search in the solution space around the firework. By explosion, fireworks with better fitness tend to generate more sparks within smaller explosion ranges in order to intensify local search, while fireworks with worse fitness generate fewer sparks within larger explosion ranges to facilitate global search, as illustrated in Fig. 1 [1]. Numerical experiments on a set of benchmark functions show that, FWA has more rapid convergence speed than some typical particle swarm optimization (PSO) algorithms such as [2] and [3].

Since its proposal, FWA has attracted much attention. Zheng et al. [4] developed a new hybrid FWA by combining it with differential evolution (DE) [5], which selects new fireworks for the next generation from highly ranked solutions

---

* Supported by grants from National Natural Science Foundation (No. 61105073) and Zhejiang Provincial Natural Science Foundation (No. LY14F030011) of China.

Y. Tan et al. (Eds.): ICSI 2014, Part I, LNCS 8794, pp. 53–63, 2014.

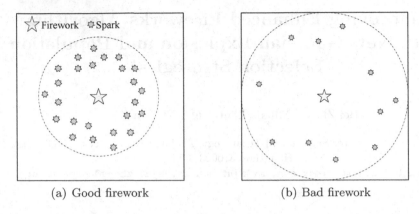

(a) Good firework   (b) Bad firework

Fig. 1. Illustration of fireworks explosions in FWA

and updates these newly generated solutions with the DE mutation, crossover, and selection operators. Pei et al. [6] studied the influence of approximation model, sampling method, and sampling number on the acceleration performance of FWA, and improved the algorithm by using an elite strategy for enhancing the search capability. The hybrid algorithm proposed by Zhang et al. [7] introduces the migration operator of biogeography-based optimization (BBO) [8] to FWA, which can effectively enhance information sharing among the population, and thus improves solution diversity and avoids premature convergence.

In [9] Ding et al. studied the parallel implementation of FWA on GPUs. In [10] Zheng et al. presented a multiobjective version of FWA, which has shown the great success for variable-rate fertilization in oil crop production. FWA has also been successfully applied to many other practical problems [11,12,13,14].

An important improvement to FWA is the enhanced FWA (EFWA) proposed by Zheng et al. [15], which tackles the limitations of the original FWA by developing new explosion operators, new strategy for selecting population for the next generation, new mapping strategy for sparks out of the search space, and new parameter mechanisms. Consequently, EFWA outperforms FWA in terms of convergence capabilities, meanwhile reducing the runtime significantly.

In this paper, we further improve EFWA in two aspects, i.e., the Gaussian explosion operator and the population selection strategy. We develop a new Gaussian explosion operator that enables new sparks to learn from more exemplars in the population, and thus improves solution diversity and avoids being trapped in local optima. Moreover, we propose a new population selection strategy that enables high-quality solutions to have high probabilities of entering the next generation, without incurring high computational cost. Numerical experiments show that the proposed algorithm, named IEFWA, outperforms EFWA on a set of benchmark function optimization problems.

In the rest of the paper, Section 2 introduces FWA and EFWA, Section 3 describes IEFWA in detail, Section 4 presents the experiments, and Section 5 concludes.

2   FWA and EFWA

FWA is a metaheuristic optimization method inspired by the phenomenon of fireworks explosion, where a solution to the problem is analogous to a firework or a spark. The key principle of FWA is that fitter fireworks can explode more sparks within a smaller area, while worse fireworks generate fewer sparks within a larger amplitude. The basic framework of FWA is as follows:

1. Randomly initialize a certain number of locations to set off fireworks.
2. For each firework, perform a *regular explosion* operation to generate a set of sparks.
3. Select a small number of fireworks, on each of which perform a *Gaussian explosion* operation to generate a few number of sparks.
4. Choose individuals from the current generation of fireworks and sparks to enter into the next generation.
5. Repeat Step 2-4 until the termination condition is satisfied.

For regular explosion of FWA, the number of sparks s_i and the explosion amplitude A_i of the ith firework X_i are respectively calculated as follows:

$$s_i = M_e \cdot \frac{f_{\max} - f(X_i) + \epsilon}{\sum_{j=1}^{n}(f_{\max} - f(X_j)) + \epsilon} \tag{1}$$

$$A_i = \widehat{A} \cdot \frac{f(X_i) - f_{\min} + \epsilon}{\sum_{j=1}^{n}(f(X_j) - f_{\min}) + \epsilon} \tag{2}$$

where n is the size of population, $f(X_i)$ is the fitness value of X_i, $f_{\max}$ and $f_{\min}$ are respectively the maximum and minimum fitness values among the n fireworks, M_e and $\widehat{A}$ are two parameters for respectively controlling the total number of sparks and the maximum explosion amplitude, and ϵ is a small constant to avoid zero-division-error.

To avoid overwhelming effects of splendid fireworks, the number of sparks is further bounded as follows (where $s_{\min}$ and $s_{\max}$ are control parameters and *round* rounds a number to its closest integer):

$$s_i = \begin{cases} s_{\min} & \text{if } s_i < s_{\min} \\ s_{\max} & \text{if } s_i > s_{\max} \\ round(s_i) & \text{else} \end{cases} \tag{3}$$

According to Eq. (2), the explosion amplitude may be too small for very good fireworks. EFWA tackles this issue by setting at each dimension d a lower limit of explosion amplitude $A_{\min}^d$, which decreases with the number of generations (or function evaluations) t as follows:

$$A_{\min}^d = A_{init} - \frac{A_{init} - A_{final}}{t_{\max}}\sqrt{(2t_{\max} - t)t} \tag{4}$$

where A_{init} and A_{final} are respectively the initial and final minimum explosion amplitude, and $t_{\max}$ is the maximum number of generations (or function evaluations).

When performing a regular explosion on X_i, FWA computes an offset displacement as $rand(-1, 1) \cdot A_i$, which is added to z random dimensions of X_i. EFWA modifies the explosion operator by computing a different displacement value for each dimension d:

$$X_j^d = X_i^d + rand(-1, 1) \cdot A_i^d \tag{5}$$

Gaussian explosion of FWA is mainly used for improving solution diversity. At each generation, FWA randomly chooses a small number M_g of fireworks, and for each firework obtains a Gaussian spark by computing its location as:

$$X_j^d = X_i^d \cdot N(1, 1) \tag{6}$$

where $N(1, 1)$ is a Gaussian random number with mean 1 and standard deviation 1. However, such a mechanism often causes many sparks to be very close to the origin of the search space, and fireworks already close to the origin cannot escape from this location. To tackle this, EFWA uses the following Gaussian explosion that makes the new spark learn from the best individual X_{best} found so far:

$$X_j^d = X_i^d + (X_{\text{best}}^d - X_i^d) \cdot N(0, 1) \tag{7}$$

Algorithm 1 and Algorithm 2 respectively describes the regular explosion and Gaussian explosion procedures used in EFWA, where X_i denotes the firework to be exploded, and D is the dimension of the problem.

Algorithm 1. The regular explosion used in EFWA.

---
1   Calculate s_i and A_i for firework X_i;
2   **for** $k = 1$ **to** s_i **do**
3   Initialize a spark $X_j = X_i$;
4   **for** $d = 1$ **to** D **do**
5   **if** $rand(0, 1) < 0.5$ **then**
6             $X_j^d = X_i^d + rand(-1, 1) \cdot A_i^d$;
7   **if** X_j^d is out of the search range **then**
8   Randomly set X_j^d in the search range;
9   Add X_j as a new spark.

Algorithm 2. The Gaussian explosion in EFWA.

---
1   Initialize a spark $X_j = X_i$;
2   **for** $d = 1$ **to** D **do**
3   **if** $rand(0, 1) < 0.5$ **then**
4          $X_j^d = X_i^d + (X_{\text{best}}^d - X_i^d) \cdot N(0, 1)$;
5   **if** X_j^d is out of the search range **then**
6   Randomly set X_j^d in the search range;
7   Return X_j as a new Gaussian spark.

At each generation, the best individual among all the sparks and fireworks is always chosen to the next generation. In FWA, the other $(n-1)$ fireworks are selected according to the probabilities proportional to their distances to other individuals, which is computationally expensive. Therefore, EFWA employs a very simple strategy that randomly selects the remaining $(n-1)$ fireworks.

3   An Improved EFWA

The proposed IEFWA intends to improve EFWA in two aspects: the Gaussian explosion operator and the population selection strategy.

3.1   A New Gaussian Explosion Operator

As shown in Eq. (6), the Gaussian explosion of EFWA makes the new spark learn from the best individual X_{best} found so far. This is similar to the mechanism of learning from the global best in PSO [2]. Thus EFWA also partly suffers the problem of premature convergence as PSO: If the X_{best} is just a local optimum or very close to it, the Gaussian sparks will be heavily attracted by the local optimum; if there is no new best solution found for a certain number of generations, more and more sparks will converge to the local optimum, and the algorithm is easily trapped.

To tackle this issue, we develop a new Gaussian explosion operator that enables new Gaussian sparks to learn from not only the current best but also other exemplars in the population. Our tactic is very simple: When exploding a firework X_i, at each dimension d, we first randomly choose two individuals from the current population, and then select the one with higher fitness value, denoted by X_{lbest}, as the exemplar that replaces X_{best} in Eq. (7):

$$X_j^d = X_i^d + (X_{\text{lbest}}^d - X_i^d) \cdot N(0,1) \tag{8}$$

In this way, every Gaussian spark has a chance to learn from different exemplars at different dimensions, and thus the solution diversity can be increased greatly. The combination of the regular explosion of EFWA and the new Gaussian explosion of IEFWA can balance the exploration and exploitation much more effectively. Algorithm 3 presents the new Gaussian explosion procedure.

Algorithm 3. The Gaussian explosion in IEFWA.

---

1   Initialize a spark $X_j = X_i$;
2   for $d = 1$ to D do
3   if $rand(0,1) < 0.5$ then
4   Randomly choose two individuals from the population;
5   Set X_{lbest} to the better one between them;
6            $X_j^d = X_i^d + (X_{\text{lbest}}^d - X_i^d) \cdot N(0,1)$;
7   if X_j^d is out of the search range then
8   Randomly set X_j^d in the search range;
9   Return X_j as a new Gaussian spark.

3.2   A New Population Selection Strategy

When selecting individuals to the next generation, the original FWA uses a distance-based selection strategy, which is effective in terms of population diversity but incurs high computational cost. EFWA employs a random selection strategy that is computationally efficient but may lose some high-quality individuals which may ultimately lead to the global optimum.

In general, we want that high-fitness individuals have high selection probability, while low-fitness individuals still have chances of entering into the next generation. Here we employ the pairwise comparison method used in evolutionary programming (EP) [16] to determine whether an individual should survive to the next generation. That is, for each individual, we randomly choose q opponents from the current generation, and conduct q pairwise comparisons between the test individual and the opponents. If the individual is fitter, it receives a "win". Finally, among all the individuals in the current generation, n individuals that have the most wins enter into the next generation.

EP has a fixed population size and uses a fixed q value of about 10~20. However, in FWA the total number of fireworks and sparks often varies from generation to generation, and thus we set q to a random value in the range $[1, np]$, where np is the total number of individuals in the current generation. Algorithm 4 presents the population selection procedure used in IEFWA.

Algorithm 4. The population selection in IEFWA.

1   Let $q = rand(1, np)$;
2   **for each** X_i in the current generation **do**
3   Let $Wins(X_i) = 0$;
4   Randomly choose q opponents from the population;
5   **for each** opponent X_i' **do**
6   **if** $f(X_i) > f(X_i')$ **then**
7               $Wins(X_i) \leftarrow Wins(X_i) + 1$;
8   Sort np individuals in decreasing order of $Wins(X_i)$;
9   Return the first n individuals.

---

4   Computational Experiment

4.1   Experimental Settings

We test the performance of the proposed IEFWA on a set of 18 benchmark functions denoted as f_1–f_{18}, which are summarized in Table 1. Here f_1–f_{13} include unimodal and simple multimodal functions taking from [17], and f_{14}–f_{18} are shifted and rotated (SR) functions taking from [18]. All the functions are high-dimensional problems, and in this paper we use 30-D problems.

To evaluate our new strategies, besides the proposed IEFWA that uses both the new Gaussian explosion operator and the new population selection strategy, we also implement another version that uses only the new Gaussian explosion

Table 1. A summary of the benchmark functions used in the paper (for f_{14}–f_{18}, $\boldsymbol{M}$ is the rotation matrix and $\boldsymbol{o}_i$ is the shifted global optimum [18])

Name	Function	Range
Sphere	$f_1(x) = \sum\limits_{i=1}^{D} x_i^2$	$[-100, 100]^D$
Schwefel 2.22	$f_2(x) = \sum\limits_{i=1}^{D} |x_i| + \prod\limits_{i=1}^{D} |x_i|$	$[-10, 10]^D$
Schwefel 1.2	$f_3(x) = \sum\limits_{i=1}^{D} \left(\sum\limits_{j=1}^{i} x_j \right)^2$	$[-100, 100]^D$
Schwefel 2.21	$f_4(x) = \max\limits_{i}\{|x_i|, 1 \leq i \leq D\}$	$[-100, 100]^D$
Rosenbrock	$f_5(x) = \sum\limits_{i=2}^{D-1} (100(x_i^2 - x_{i-1})^2 + (x_i - 1)^2)$	$[-30, 30]^D$
Step	$f_6(x) = \sum\limits_{i=1}^{D} \left(\lfloor x_i + 0.5 \rfloor \right)^2$	$[-100, 100]^D$
Quartic	$f_7(x) = \sum\limits_{i=1}^{D} i x_i^4 + rand[0, 1)$	$[-1.28, 1.28]^D$
Schwefel	$f_8(x) = 418.9829 \times D - \sum\limits_{i=1}^{D} x_i \sin(|x_i|^{\frac{1}{2}})$	$[-500, 500]^D$
Rastrigin	$f_9(x) = \sum\limits_{i=1}^{D} (x_i^2 - 10\cos(2\pi x_i) + 10)$	$[-5.12, 5.12]^D$
Ackley	$f_{10}(x) = -20\exp\left(-0.2\sqrt{\frac{1}{D}\sum\limits_{i=1}^{D} x_i^2} \right)$ $-\exp\left(\frac{1}{D}\sum\limits_{i=1}^{D} \cos(2\pi x_i)\right) + 20 + e$	$[-32, 32]^D$
Griewank	$f_{11}(x) = \frac{1}{4000}\sum\limits_{i=1}^{D} x_i^2 - \prod\limits_{i=1}^{D} \cos(\frac{x_i}{\sqrt{i}}) + 1$	$[-600, 600]^D$
Penalized1	$f_{12}(x) = \frac{\pi}{30}\Big(10\sin^2(\pi y_1) + \sum\limits_{i=1}^{D-1} (y_i - 1)^2(1 + 10\sin^2($ $\pi y_{i+1})) + (y_D - 1)^2\Big) + \sum\limits_{i=1}^{D-1} u(x_i, 10, 100, 4)$	$[-50, 50]^D$
Penalized2	$f_{13}(x) = 0.1\Big(\sin^2(3\pi x_1) + \sum\limits_{i=1}^{D-1} (x_i - 1)^2(1 + \sin^2(3\pi x_{i+1}))$ $+ (x_D - 1)^2(1 + \sin^2(2\pi x_D)) \Big) + \sum\limits_{i=1}^{D-1} u(x_i, 5, 100, 4)$ where $u(x_i, a, k, m) = \begin{cases} k(x_i - a)^m, & x_i > a \\ 0, & -a \leq x_i \leq a \\ k(-x_i - a)^m, & x_i < -a \end{cases}$	$[-50, 50]^D$
SR Bent Cigar	$f_{14}(\boldsymbol{x}) = 200 + y_1^2 + 10^6 \sum_{i=2}^{D} y_i^2,\quad \boldsymbol{y} = \boldsymbol{M}(\boldsymbol{x} - \boldsymbol{o}_1)$	$[-100, 100]^D$
SR Discus	$f_{15}(\boldsymbol{x}) = 300 + 10^6 y_1^2 + \sum_{i=2}^{D} y_i^2,\quad \boldsymbol{y} = \boldsymbol{M}(\boldsymbol{x} - \boldsymbol{o}_2)$	$[-100, 100]^D$
SR Ackley	$f_{16}(\boldsymbol{x}) = 500 + f_{10}(\boldsymbol{y}),\quad \boldsymbol{y} = \boldsymbol{M}(\boldsymbol{x} - \boldsymbol{o}_3)$	$[-100, 100]^D$
SR Griewank	$f_{17}(\boldsymbol{x}) = 700 + f_{11}(\boldsymbol{y}),\quad \boldsymbol{y} = \boldsymbol{M}\left(\frac{600(\boldsymbol{x} - \boldsymbol{o}_4)}{100}\right)$	$[-100, 100]^D$
SR Rastrigin	$f_{18}(\boldsymbol{x}) = 900 + f_9(\boldsymbol{y}),\quad \boldsymbol{y} = \boldsymbol{M}\left(\frac{5.12(\boldsymbol{x} - \boldsymbol{o}_5)}{100}\right)$	$[-100, 100]^D$

operator, denoted as IEFWA1. For the sake of fair comparison, the maximum number of function evaluations (NFE) is set to 200,000 for every problem.

We have compared EFWA, IEFWA1, and IEFWA on the 18 test problems, using $n = 5$, $M_e = 50$, $M_g = 5$, $\widehat{A} = 40$, $s_{max} = 40$, $s_{min} = 2$, $A_{init} = 0.02(X_{max} - X_{min})$ and $A_{final} = 0.001(X_{max} - X_{min})$, as suggested in [1] and [15]. The experiments are conducted on a computer of Intel Core i5-2520M processor and 4GB DDR3 memory. Each algorithm has been run 60 times (with different random seeds) on each problem.

4.2 Experimental Results

Table 2 presents the mean and standard deviation of the best fitness values obtained by the three algorithms averaged over 60 runs. The bold values indicate the best results among the three algorithms. We have also conducted paired t-tests between IEFWA and the other two algorithms, and mark $^+$ before the mean values in columns 2 and 4 if IEFWA has statistically significant performance improvement over the corresponding algorithms (at 95% confidence level).

Table 2. The experimental results of the three algorithms

ID	FWA		IEFWA1		IEFWA	
	mean	std	mean	std	mean	std
f_1	$^+$5.01E+00	(9.01E−01)	$^+$7.48E−04	(3.51E−04)	**5.32E−05**	(5.10E−05)
f_2	$^+$9.17E−01	(1.49E−01)	$^+$1.18E−02	(2.35E−03)	**1.25E−03**	(8.58E−04)
f_3	$^+$6.23E+01	(1.45E+01)	$^+$3.79E−02	(1.23E−02)	**1.98E−03**	(1.63E−03)
f_4	$^+$9.55E−01	(6.79E−02)	**2.01E−01**	(4.29E−02)	3.29E−01	(3.30E−01)
f_5	$^+$2.18E+02	(2.57E+02)	$^+$1.00E+02	(1.57E+02)	**5.34E+01**	(3.29E+01)
f_6	$^+$3.87E+00	(9.99E−01)	**0.00E+00**	(0.00E+00)	**0.00E+00**	(0.00E+00)
f_7	$^+$1.83E−02	(1.36E−02)	$^+$1.75E−02	(1.18E−02)	**8.22E−03**	(5.68E−03)
f_8	5.30E+03	(8.56E+02)	**5.12E+03**	(7.47E+02)	5.20E+03	(4.37E+02)
f_9	$^+$1.27E+02	(2.02E+01)	**1.83E+01**	(5.95E+00)	9.55E+01	(1.90E+01)
f_{10}	$^+$1.16E+00	(1.72E−01)	$^+$1.88E−02	(5.88E−01)	**1.80E−03**	(1.21E−03)
f_{11}	$^+$1.03E+00	(2.21E−02)	$^+$2.48E−02	(2.20E−02)	**9.90E−03**	(8.87E−03)
f_{12}	$^+$1.12E+01	(2.77E+00)	$^+$1.42E+00	(1.59E+00)	**9.35E−01**	(9.43E−01)
f_{13}	$^+$8.18E−01	(2.39E−01)	$^+$9.67E−05	(1.33E−04)	**2.76E−05**	(5.98E−05)
f_{14}	$^+$6.86E+06	(1.27E+06)	$^+$8.87E+03	(1.09E+04)	**5.90E+03**	(1.25E+04)
f_{15}	$^+$3.99E+02	(2.76E+01)	$^+$3.53E+02	(2.25E+01)	**3.35E+02**	(1.77E+01)
f_{16}	$^+$5.21E+02	(7.56E−02)	$^+$5.20E+02	(4.78E−03)	**5.20E+02**	(1.82E−03)
f_{17}	$^+$9.58E+06	(1.60E+06)	$^+$1.02E+04	(3.64E+03)	**3.07E+03**	(2.05E+03)
f_{18}	$^+$1.06E+03	(4.80E+01)	1.02E+03	(5.00E+01)	**1.01E+03**	(3.12E+01)

As we can see from the results, EFWA never obtains the best mean value on any of the 18 problems, and the two IEFWA versions both achieve considerable performance improvement over EFWA. In terms of statistical tests, except that on f_8 there is no significant difference between EFWA and IEFWA, on the

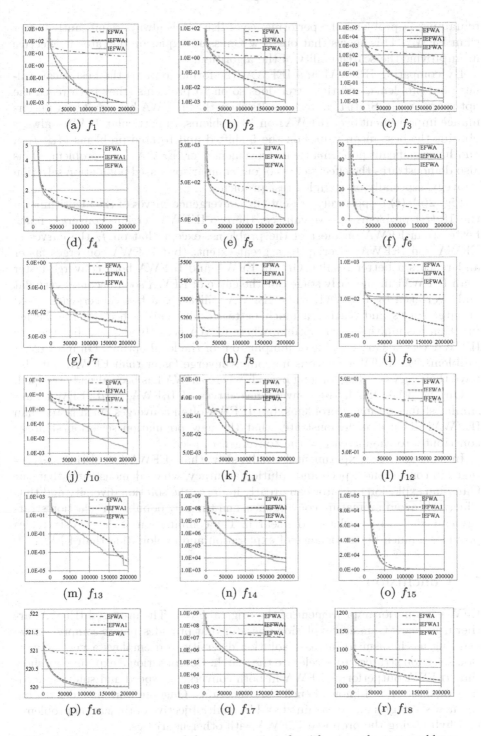

Fig. 2. Convergence curves of the comparative algorithms on the test problems

remaining 17 problems the performance of IEFWA is always significantly better than EFWA. This shows that our new Gaussian explosion operator is effective in improving the search ability of the algorithm.

By comparing IEFWA1 and IEFWA, the former obtains the best mean values on 4 problems, and the latter does so on 15 problems (they both reach the optimum on f_6). Statistical test results show that IEFWA has significant performance improvement over IEFWA1 on 13 problems. In particular, IEFWA always obtains the best mean values on the 5 shifted and rotated problems (f_{14}–f_{18}), and has significant performance improvement over IEFWA1 on 4 problems. This also demonstrates the effectiveness of our comparison-based population selection strategy, especially on complex test problems.

Fig. 2(a)–(t) respectively present the convergence curves of the algorithms on the 18 test problems. As we can see, the two IEFWA versions converges much better than EFWA on most of the problems, except that on f_7 the curves of EFWA and IEFWA1 overlap to a great extent, (but IEFWA converges faster and reaches a better result). On f_{10} EFWA and IEFWA both converge faster than IEFWA1 in the early stage but the curve of EFWA soon becomes very flat and overtaken by IEFWA1, while IEFWA still keeps a fast convergence speed at later stages and reaches a much better result. This is because the Ackley function has multiple local optima, where EFWA is easily trapped, but the two IEFWA versions are capable of jumping out of the local optima. On most other problems, the IEFWA versions not only converge faster than EFWA, but also have their curves falling for long periods where EFWA has been already trapped.

On the other hand, the convergence curves of IEFWA1 and IEFWA have similar shapes on many problems, but IEFWA often converges much faster than IEFWA1. This also demonstrates that the new population selection strategy contributes to the increase of the convergence speed.

In summary, the experimental results show that IEFWA has obvious advantages in convergence speed and solution accuracy, which demonstrates that our Gaussian explosion operator can greatly improve the solution diversity and thus effectively avoid premature convergence, and the new population selection strategy can efficiently accelerate the search. The combination of the two strategies provides a much better balance of exploration and exploitation than EFWA.

5   Conclusion

EFWA is a major improvement of the original FWA. The proposed IEFWA further uses a new Gaussian explosion operator that provides a more comprehensive learning mechanism to increase solution diversity, and employs a new population selection strategy to accelerate the search. Computational experiments show that IEFWA outperforms EFWA in both convergence speed and solution accuracy on a set of well-known benchmark functions. Ongoing work includes testing the new strategies for constrained and/or multiobjective optimization problems, and hybridizing the proposed IEFWA with other heuristics.

References

1. Tan, Y., Zhu, Y.: Fireworks algorithm for optimization. In: Tan, Y., Shi, Y., Tan, K.C. (eds.) ICSI 2010, Part I. LNCS, vol. 6145, pp. 355–364. Springer, Heidelberg (2010)
2. Kennedy, J., Eberhart, R.: Particle swarm optimization. In: Proceeding of the IEEE International Conference on Neural Networks, vol. 4, pp. 1942–1948 (1995)
3. Tan, Y., Xiao, Z.: Clonal particle swarm optimization and its applications. In: Proceeding of the IEEE Congress on Evolutionary Computation, pp. 2303–2309 (2007)
4. Zheng, Y.J., Xu, X.L., Ling, H.F., Chen, S.Y.: A hybrid fireworks optimization method with differential evolution operators. Neurocomputing (2014), doi:10.1016/j.neucom.2012.08.075
5. Storn, R., Price, K.: Differential evolution: A simple and efficient heuristic for global optimization over continuous spaces. J. Global Optim. 11(4), 341–369 (1997)
6. Pei, Y., Zheng, S., Tan, Y., et al.: An empirical study on influence of approximation approaches on enhancing fireworks algorithm. In: Proceeding of the 2012 IEEE International Conference on Systems, Man, and Cybernetics, pp. 1322–1327 (2012)
7. Zhang, B., Zhang, M.X., Zheng, Y.J.: A hybrid biogeography-based optimization and fireworks algorithm. In: Proceeding of the IEEE Congress on Evolutionary Computation (2014)
8. Simon, D.: Biogeography-based optimization. IEEE Trans. Evol. Comput. 12(6), 702–713 (2008)
9. Ding, K., Zheng, S., Tan, Y.: A GPU-based parallel fireworks algorithm for optimization. In: Proceeding of the 15th Annual Conference on Genetic and Evolutionary Computation Conference, pp. 9–16 (2013)
10. Zheng, Y.J., Song, Q., Chen, S.Y.: Multiobjective fireworks optimization for variable-rate fertilization in oil crop production. Applied Soft Computing 13(11), 4253–4263 (2013)
11. Gao, H., Diao, M.: Cultural firework algorithm and its application for digital filters design. International Journal of Modelling, Identification and Control 14(4), 324–331 (2011)
12. Janecek, A., Tan, Y.: Iterative improvement of the multiplicative update nmf algorithm using nature-inspired optimization. In: Proceeding of the 7th International Conference on Natural Computation, vol. 3, pp. 1668–1672 (2011)
13. Janecek, A., Tan, Y.: Using population based algorithms for initializing nonnegative matrix factorization. In: Tan, Y., Shi, Y., Chai, Y., Wang, G. (eds.) ICSI 2011, Part II. LNCS, vol. 6729, pp. 307–316. Springer, Heidelberg (2011)
14. He, W., Mi, G., Tan, Y.: Parameter optimization of localconcentration model for spam detection by using fireworks algorithm. In: Tan, Y., Shi, Y., Mo, H. (eds.) ICSI 2013, Part I. LNCS, vol. 7928, pp. 439–450. Springer, Heidelberg (2013)
15. Zheng, S., Janecek, A., Tan, Y.: Enhanced fireworks algorithm. In: Proceeding of the IEEE Congress on Evolutionary Computation, pp. 2069–2077 (2013)
16. Back, T., Schwefel, H.P.: An overview of evolutionary algorithms for parameter optimization. Evolutionary Computation 1(1), 1–23 (1993)
17. Yao, X., Liu, Y., Lin, G.: Evolutionary programming made faster. IEEE Trans. Evol. Comput. 3(2), 82–102 (1999)
18. Liang, J.J., Qu, B.Y., Suganthan, P.N.: Problem definitions and evaluation criteria for the CEC 2014 special session and competition on single objective real-parameter numerical optimization. Tech. Rep. 201311, Computational Intelligence Laboratory, Zhengzhou University, Zhengzhou, China (2014)

A Unified Matrix-Based Stochastic Optimization Algorithm

Xinchao Zhao[1] and Junling Hao[2]

[1] School of Science,
Beijing University of Posts and Telecommunications, Beijing 100876, China
zhaoxc@bupt.edu.cn
[2] School of Statistics,
University of International Business and Economics, Beijing 100029, China
haojunling@uibe.edu.cn

Abstract. Various metaheuristics have been proposed recently and each of them has its inherent evolutionary, physical-based, and/or swarm intelligent mechanisms. This paper does not focus on any subbranch, but on the metaheuristics research from a unified view. The population of decision vectors is looked on as an abstract matrix and three novel basic solution generation operations, $E[p(i,j)]$, $E[p(c \cdot i)]$ and $E[i, p(c \cdot i + j)]$, are proposed in this paper. They are inspired by the elementary matrix transformations, all of which have none latent meanings. Experiments with real-coded genetic algorithm, particle swarm optimization and differential evolution illustrate its promising performance and potential.

Keywords: Matrix-Based optimization, MaOA, metaheuristics, stochastic optimization.

1 Introduction

A metaheuristic [9] is a higher-level procedure or heuristic designed to find, generate, or select a lower-level procedure or heuristic that may provide a sufficiently good solution to an optimization problem, especially with incomplete or imperfect information or limited computation capacity. Compared to optimization algorithms and iterative methods, metaheuristics [1] do not guarantee that a globally optimal solution can be found on some class of problems. By searching over a large set of feasible solutions, metaheuristics can often find good solutions with less computational efforts than the iterative methods or simple heuristics.

There are a wide variety of metaheuristics, which are implemented as a form of stochastic optimization, for example, simulated annealing (SA) [8], tabu search (TS) [4], ant colony optimization (ACO) [3], artificial bee colony (ABC) [6], genetic algorithms (GA) [5], particle swarm optimization (PSO) [7], and differential evolution (DE) [13] etc. All metaheuristics have their inherent evolutionary, physical-based, and/or swarm intelligence mechanisms. They are problem-independent strategies to guide the search process. Metaheuristics maintain a population of decision vectors (SA excluded as a single solution) and the corresponding generation operations for new solutions. The population of decision

Y. Tan et al. (Eds.): ICSI 2014, Part I, LNCS 8794, pp. 64–73, 2014.

vectors of all metaheuristics can be mathematically modeled as a matrix, which is independent of any biological, physical or swarm intelligence mechanisms.

From a different and unified view, a *matrix optimization algorithm* (MaOA) is proposed in this paper. The operation object of MaOA is a matrix which is composed of a population solutions. Three types of elementary matrix transformations in linear algebra [11], E[i,j], E[c·i] and E[i, c·i+j], are modeled into its basic new solution generation strategies, E[p(i,j)], E[p(c·i)] and E[i, p(c·i+j)], as algorithmic operations of MaOA.

The rest of this paper is organized as follows. Fundamental knowledge are presented in Section 2. Algorithm MaOA is proposed in Section 3. Experimental comparisons are given in Section 4. This paper is concluded in Section 5.

2 Fundamental Knowledge

2.1 Genetic Algorithm

Over the last few decades, evolutionary algorithms (EAs) have shown tremendous success in solving complex optimization problems. EA family contains a number of different algorithms, however, genetic algorithm (GA) is the most popular and widely used in practice. It mimics the process of natural selection which generates solutions to optimization problems using such bio-inspired techniques as crossover, mutation and selection operations [5].

For GA, especially for binary GA, crossover is a key operator while mutation is usually a background operator. However, mutation operation is also very important to maintain population diversity and to enhance GA's performance. Traditional GA uses binary coded strings which are called chromosomes. But other coding methods, such as gray code, integer code, and real code, are also usually used in practice. Mutation operation is the main/only generation strategy for evolution strategy and evolutionary programming.

2.2 Particle Swarm Optimization

PSO is a stochastic global optimization algorithm which emulates swarm behaviors of birds flocking. It was introduced by Kennedy and Eberhart in 1995 [7], which is a population-based iterative learning algorithm that shares some common characteristics with other EAs [5]. However, PSO searches for an optimum through each particle flying in the search space and adjusting its flying trajectory according to its personal best experience and its neighborhood's best experience. Owing to its simple concept and high efficiency, PSO has become a widely adopted optimization technique and has been successfully applied to many real-world problems

2.3 Differential Evolution

Differential evolution [13] is one of the best general purpose evolutionary optimization methods available. It is known as an efficient global optimization method for continuous problems. Similar with EAs, it starts with an initial population vectors, which are randomly generated when none preliminary knowledge about the solution space is available. In DE, there exist many trial vector generation strategies, followed by crossover and selection operations. Moreover, three crucial control parameters involved in DE, i.e., population size, scaling factor, and crossover rate, may also have significant influence on its performance.

2.4   Three Types of Elementary Matrices

A matrix [11] is a rectangular array of numbers - in other words, numbers are grouped into rows and columns. We use matrix to represent and to solve system of linear equations. Elementary operations for matrix play a crucial role in finding the inverse, solving linear systems, finding the eigenvalues and eigenvectors and so on. As we know, the transpose operation interchanges the rows and the columns of a matrix easily. Thus, we will only discuss elementary row operations in the following.

An elementary matrix differs from the identity matrix by one single elementary row operation. Left multiplication (pre-multiplication) by an elementary matrix represents an elementary row operation for a given matrix. There are three types of elementary matrices [11] as follows, which correspond to three types of row operations

- Interchanging two rows, $E[i, j]$;
- Multiplying a row by a nonzero constant, $E[c \cdot i]$;
- Adding a row to another one multiplied by a nonzero constant, $E[i, c \cdot i + j]$.

3   Matrix Optimization Algorithm (MaOA)

3.1   Motivation to Propose MaOA

A lot of metaheuristics [10] have been proposed in recent several decades. Each of them has its inherent biological, physical, and/or swarm intelligent mechanisms and its characteristic generation operations. It is possible to confuse the users how to choose the fittest algorithm for their problems, especially for practitioners. Even though solving methods have been chosen, the next puzzle is how to modify the present problems to adapt the initial algorithmic operations. Are there the common features of them? Can a unified stochastic optimization algorithm be proposed? Can it be modeled with an abstract mathematical concept? It is also hoped to be free from any latent meaning.

This paper will propose a novel population-based matrix optimization algorithm (MaOA). It mathematically models the operation object of population algorithm as an abstract matrix, which has nothing latent biological, physical or swarm inspiring mechanisms. Its new solution generation strategies are modeled with three most fundamental matrices, i.e., elementary matrices/transformations. They includes interchanging two rows of a matrix, $E[i, j]$, multiplying a row by a nonzero constant, $E[c \cdot i]$ and adding a row to another one multiplied by a nonzero constant, $E[i, c \cdot i+j]$. Of course, these three types of elementary operations need some adaption to make the algorithmic operations meaningful.

The proposed MaOA aims at

- being faithful to the meaning of operation matrix as far as possible;
- conforming to the inherent search mechanism of iteration algorithm as far as possible;
- less parameters as far as possible.

3.2   Three Elementary Matrix Operations

This section will model the algorithmic operations in terms of mathematics with three types of elementary matrices/transformations.

Probabilistic Dimensionwise Interchanging Two Rows. As we know, simple interchange two rows for a solution matrix is meaningless. Therefore, the elementary matrix operation of interchanging two rows, $E[i, j]$, is modeled as the probabilistic dimensionwise interchanging two rows, $E[p(i, j)]$.

- Randomly select two rows from solution matrix R_i and R_j;
- if *rand* < *pSwap*
-      for each dimension of R_i and R_j
-      if *rand* > *pSwap*
-         interchange this component of R_i and R_j;

where *rand* is a random number in $[0, 1]$, *pSwap* is the interchanging probability. The sum of the probability of dimensionwise component interchanging and the probability of interchanging operation executed to the chosen two rows is kept to be 1 in order to keep algorithm as simple as possible.

Probabilistic Dimensionwise Multiplying a Row by Constant. The elementary matrix operation of multiplying a row by a nonzero constant, $E[c \cdot i]$, is modeled as the probabilistic dimensionwise multiplying a row by a random number, $E[p(c \cdot i)]$.

- for each row, R_i, of solution matrix;
-      for each dimension of R_i
-         if *rand* < *pMult*
-            this component of R_i is multiplied by a *Gaussian* random constant;

where *rand* is a random number in $[0, 1]$, *pMult* is the probability of being multiplied by a constant of a dimension.

Probabilistic Dimensionwise Adding a Row to Another One Multiplied by Constant. The elementary matrix operation of adding a row to another one multiplied by a constant, $E[i, c \cdot i+j]$, is modeled as the probabilistic dimensionwise adding a row to another one multiplied by constant, $E[i, p(c \cdot i+j)]$.

– for each row, R_i, of solution matrix;
– if $rand > pSum$
–   select another row, R_j, of solution matrix;
–   for each dimension of R_i and R_j
–   if $rand < pSum$
–   this component of R_j is added to R_i before multiplying a *Gaussian* random constant;

where *rand* is a random number in $[0, 1]$, *pSum* is the probability of dimension-wise adding a row to another one before multiplying a constant. The sum of the probability of adding the other row to the current one and the probability of adding operation executed to each dimension of the corresponding row is kept to be 1 in order to keep algorithm as simple as possible.

3.3   Selection Operation of MaOA

All the above three matrix-inspired operations are exploration-oriented generation strategies. It is natural that selection pressure of solution matrix is lacking. Therefore its performance is possible to be dissatisfied although it has good population diversity.

Based on these observations, selection operation, $(\mu + \lambda)$, with high selection pressure is adopted. μ is the solution number of parental population and λ is the population size of the offspring. μ and λ are factually the same in this algorithm. Elitism strategy is also utilized in MaOA besides $(\mu + \lambda)$ selection.

3.4   Repair Operator

The new solutions are possible to be out of search range after some generation operations. A simple and popular repair operator is adopted for this case, which works as follows: if the d-th element, $x_{i,d}^k$, of the solution vector $x_i^k = (x_{i,1}^k, \cdots, x_{i,d}^k, \cdots, x_{i,D}^k)$ is out of search range $[LB_d, UB_d]$, then $x_{i,d}^k$ is reset as follows:

$$
x_{i,d}^k = \begin{cases} \min\{UB_d, 2LB_d - x_{i,d}^k\}, & \text{if } x_{i,d}^k < LB_d \\ \max\{LB_d, 2UB_d - x_{i,d}^k\}, & \text{if } x_{i,d}^k > UB_d \end{cases} . \tag{1}
$$

The repair operator is adopted by all the algorithms in this paper in order to keep the difference among algorithms minimum.

4   Experimental Comparisons

In this section, how MaOA performs is compared with other metaheuristic search algorithms. The main operations and parameters of GA, PSO and DE algorithms are reviewed firstly.

4.1   Benchmarks

Thirteen benchmarks, the first 13 benchmarks from reference [14], are adopted to validate the performance of MaOA These benchmark functions are the classical functions which are widely utilized by many researchers [10,15,16].

4.2   Comparison Opponents and Parameters Settings

Three competitors, i.e., PSO, real-coded GA (RGA), and DE (DE/rand/1) are used in this paper to empirically verify the new proposed MaOA. Repair operator and elitism strategy are all used in four algorithms. For the comparison of four competitors PSO, RGA, DE and MaOA, maximal function evaluation number is 60000; population size is 30; dimension of benchmark is 30; all the results are obtained from 50 independent runs.

PSO. PSO approach updates the velocity and positions of particle population with the fitness information obtained from the evolving environment. Individuals of population can be expected to move towards better positions. The velocity and position updating equations are given as Eqs.(2, 3).

$$v_{i,d}^{k+1} = \omega v_{i,d}^{k} + c_1 r_1^k (p_{i,d}^k - x_{i,d}^k) + c_2 r_2^k (p_{g,d}^k - x_{i,d}^k) \ . \tag{2}$$

$$x_{i,d}^{k+1} = x_{i,d}^{k} + v_{i,d}^{k+1} \ . \tag{3}$$

where $v_{i,d}^k / x_{i,d}^k / p_{i,d}^k$ are the d-th dimensional velocity/position/personal best of particle i in cycle k; $p_{g,d}^k$ is the d-th dimension of the gbest in cycle k; ω is the inertia weight; c_1 and c_2 are positive constants; r_1 and r_2 are two random numbers in $[0, 1]$. Parameter $\omega = 1/(2 * log(2))$, $c_1 = 0.5 + log(2)$ and $c_2 = c_1$.

RGA. In RGA, arithmetic crossover, *Gaussian* mutation and $(\mu + \lambda)$ selection are used as described in [12]. Repair operator and elitism strategy are also used. Crossover and mutation probabilities are 0.8 and 0.1. Selection $(\mu + \lambda)$ is used to keep the difference minimum to other algorithms.

DE. DE has many generation strategies [2], among which DE/rand/1/binomial is the most popular one. It is given as Eq.(4).

$$v_i^k = x_{r_1}^k + F(x_{r_2}^k - x_{r_3}^k) \ . \tag{4}$$

where $r_1, r_2, r_3 \in [1, PS]$ are random and mutually distinct integers, and they are also different with the vector index i. PS is population size.

Then a binomial crossover operator on x_i^k and v_i^k to generate a trial vector u_i^k. Selection $(\mu + \lambda)$ is used.

Constant parameters F and CR are not utilized in this paper which vary with different solutions. For solution i, its F is decided with a normal random number with 0.5 mean and 0.3 variance.

$$F_i = normrnd(0.5, 0.3) .\tag{5}$$

For each solution, its CR is set 0.9 or 0.2 with 50% probability individually.

4.3 Numerical Comparison

The final results of PSO, RGA, DE and MaOA in 50 independent runs are presented in Table 1.

Generally speaking, MaOA performs best from nine of thirteen functions and RGA and DE performs best from three of thirteen functions which can be observed from Table 1. MaOA, RGA and DE all obtained the same and the best results on function f_6. MaOA not only performs best on nine functions, but also obtains significantly better results over its competitors on functions f_1, f_2, f_4, f_5, f_9-f_{11}. All these observations can also be verified by the t-test values.

4.4 Evolving Behaviors Comparison

In order to further verify the evolutionary performance of MaOA, the average best function values found-until-now of PSO, RGA, DE and MaOA at every iteration over 50 runs versus the function evaluations are plotted in Fig. 1. Due to the space limit, ten Benchmarks, including five unimodal functions f_2, f_4-f_7 and five multimodal functions f_9-f_{13}, are selected to illustrate their evolutionary performance.

Fig. 1 illustrates the excellent convergence properties and fast convergence speeds of MaOA for most of the functions. As shown in Table 1 that RGA, DE and MaOA all obtained the true optima of function f_6, however, its evolving performance tells us that MaOA has fastest convergence speed when comparing with RGA and DE.

5   Conclusions, Discussions and Further Questions

Independent from various biological, physical and swarm mechanisms, an abstract mathematical concept, matrix-based stochastic optimization algorithm (MaOA) is proposed in this paper. Three elementary matrix-based transformations are modeled as its generation strategies.

Only preliminary algorithm model and framework are proposed in this paper. As we know, matrix is an inherent transformation in linear algebra. Therefore, MaOA is possible to lead a wide and further development in the areas of evolutionary computation and swarm intelligence.

– various matrix-based transformations are possible to modify and/or improve the current new solution generation strategies;

Table 1. Result comparisons between PSO, RGA, DE and MaOA. Data are the statistical results over 50 independent runs, where "Mean" and "STD" are the average results and the standard deviation of the final results in 50 runs. † is the two-tailed t-test value with 49 degrees of freedom, which is significant at $\alpha = 0.05$. "–" for t-test means that algorithms obtained the same results and no t-test values were provided.

Function	Items	PSO	RGA	DE	MaOA
f_1	Mean	6.78e+1	9.69e-3	2.34e-2	**1.31e-31**
	STD	1.19e+2	2.24e-3	1.65e-1	3.99e-31
†	t-test	-30.52	-4.02	-7.63	
f_2	Mean	3.78e+1	2.57e-1	1.86e-12	**7.34e-24**
	STD	2.04e+1	4.38e-2	9.11e-12	1.44e-23
†	t-test	-41.47	-13.05	-2.44	
f_3	Mean	8.24e+3	**1.66e+1**	1.06e+4	2.35e+2
	STD	4.65e+3	6.71	4.28e+3	5.23e+2
†	t-test	-12.08	2.95	-17.05	
f_4	Mean	3.37e+2	1.15e-1	5.94	**1.23e-4**
	STD	7.14	1.56e-2	6.22	3.95e-4
†	t-test	-52.21	-6.75	-33.35	
f_5	Mean	1.32e+6	3.48e+2	2.72e+5	**2.67e+1**
	STD	3.26e+6	6.51e+2	1.88e+6	3.33e-1
†	t-test	-5.49	-2.02	-4.85	
f_6	Mean	7.75e+2	**0**	**0**	**0**
	STD	1.03e+3	0	0	0
†	t-test	-5.30	–	–	
f_7	Mean	1.09	4.61e-2	1.27e-2	**6.43e-3**
	STD	5.04e-1	1.44e-2	4.22e-3	2.46e-3
†	t-test	-19.23	-15.18	-9.11	
f_8	Mean	4.28e+3	6.08e+3	**2.67e+2**	5.50e+3
	STD	8.65e+2	9.68e+2	1.90e+2	3.50e+2
†	t-test	9.22	-4.02	19.84	
f_9	Mean	1.23e+2	1.75e+1	1.28e+1	**0**
	STD	2.39e+1	4.09	1.11e+1	0
†	t-test	-36.55	-30.42	-7.67	
f_{10}	Mean	1.10e+1	7.43e-2	5.44e-2	**6.71e-15**
	STD	3.73	1.26e-2	2.22e-1	1.72e-15
†	t-test	-41.61	-20.86	-17.34	
f_{11}	Mean	1.59	6.54e-1	4.76e-3	**0**
	STD	9.84e-1	8.14e-1	9.90e-3	0
†	t-test	-11.43	-5.68	-3.39	
f_{12}	Mean	1.28e+1	6.88e-1	7.73e-1	9.05e-2
	STD	4.95	1.04	4.57	1.82e-2
†	t-test	-20.07	-3.48	1.93	
f_{13}	Mean	4.66e+1	**1.42e-3**	1.61e+2	1.00
	STD	5.72e+1	1.59e-3	1.12e+3	1.16e-1
†	t-test	-5.63	61.05	-10.01	

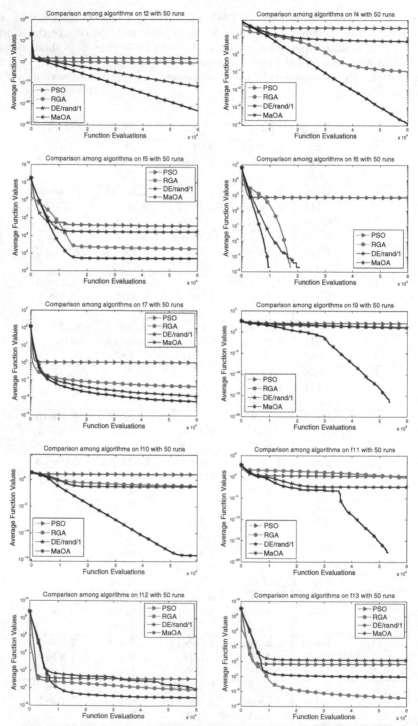

Fig. 1. Average Best Evolving Performance Comparisons among PSO, RGA, DE and MaOA for unimodal functions f_2, f_4-f_7 and multimodal functions f_9-f_{13}.

- the eigenvectors of a matrix deeply characterize the properties of the matrix, so they are possible to play some special roles for algorithm design;
- the principal components of a set of vectors [16] have abundant information of the vectors set, so they also have promising potentials.

Acknowledgement. This research is supported by National Natural Science Foundation of China (61105127, 61375066). A part of this work is supported by Chinese Scholarship Council for the academic visiting.

References

1. Blum, C., Roli, A.: Metaheuristics in combinatorial optimization: Overview and conceptual comparison. ACM Computing Surveys 35(3), 268–308 (2003)
2. Das, S., Suganthan, P.N.: Differential Evolution: A Survey of the State-of-the-Art. IEEE Transactions on Evolutionary Computation 15(1), 4–31 (2011)
3. Dorigo, M., Stutzle, T.: Ant Colony Optimization. The MIT Press, Cambridge (2004)
4. Glover, F., Laguna, M.: Tabu Search. Kluwer Academic Publishers, Norwell (1997)
5. Goldberg, D.E.: Genetic Algorithms in Search, Optimization and Machine Learning. Addison-Wesley, Reading (1989)
6. Karaboga, D., Basturk, B.: A powerful and efficient algorithm for numerical function optimization: Artificial bee colony (ABC) algorithm. Journal of Global Optimization 39(3), 459–471 (2007)
7. Kennedy, J., Eberhart, R.C.: Particle swarm optimization. In: Proc. of IEEE Int. Conf. on Neural Networks IV, pp. 1942–1948. IEEE Press, Piscataway (1995)
8. Kirkpatrick, S., Gelatt, C.D., Vecchi, M.P.: Optimization by simulated annealing. Science 220(4598), 671–680 (1983)
9. Leonora, B., Dorigo, M., Gambardella, L.M., Gutjahr, W.J.: A survey on metaheuristics for stochastic combinatorial optimization. Natural Computing 8(2), 239–287 (2009)
10. Mirjalili, S., Mirjalili, S.M., Lewis, A.: Grey Wolf Optimizer. Advances in Engineering Software 69, 46–61 (2014)
11. Ju, Y.M.: Linear algebra, 2nd edn. Tsinghua University Press, Beijing (2002)
12. Rashedi, E., Nezamabadi-Pour, H., Saryazdi, S.: GSA: A gravitational search algorithm. Information Sciences 179(13), 2232–2248 (2009)
13. Storn, R., Price, K.: Differential evolution: A simple and efficient heuristic for global optimization over continuous spaces. J. Global Optimization 11(4), 341–359 (1997)
14. Yao, X., Liu, Y., Lin, G.M.: Evolutionary Programming Made Faster. IEEE Trans. Evol. Comput. 3(2), 82–102 (1999)
15. Zhan, Z.H., Zhang, J., Li, Y., et al.: Orthogonal learning particle swarm optimization. IEEE Trans. Evolut. Comput. 15(6), 832–847 (2011)
16. Zhao, X.C., Lin, W.Q., Zhang, Q.F.: Enhanced Particle Swarm Optimization based on Principal Component Analysis and Line Search. Applied Mathematics and Computation 229(25), 440–456 (2014)

Chaotic Fruit Fly Optimization Algorithm

Xiujuan Lei[1,2], Mingyu Du[1], Jin Xu[2], and Ying Tan[2]

[1] School of Computer Science, Shaanxi Normal University, Xi'an, 710062, China
xjlei168@163.com
[2] School of Electronics Engineering and Computer Science, Peking University,
Beijing,100871, China (Visiting Scholar)

Abstract. Fruit fly optimization algorithm (FOA) was a novel swarm intelligent algorithm inspired by the food finding behavior of fruit flies. Due to the deficiency of trapping into the local optimum of FOA, a new fruit fly optimization integrated with chaos operation (named CFOA) was proposed in this paper, in which logistic chaos mapping was introduced into the movement of the fruit flies, the optimum was generated by both the best fruit fly and the best fruit fly in chaos. Experiments on single-mode and multi-mode functions show CFOA not only outperforms the basic FOA and other swarm intelligence optimization algorithms in both precision and efficiency, but also has the superb searching ability.

Keywords: Fruit fly optimization, logistic chaos, function optimization.

1 Introduction

Bio-inspired algorithms provide a new perspective for solving complex problems by mimicking the biological behaviors and nature phenomenon, with the characteristics of high robust, low complexities, excellent efficiency and superb performance, and also overcoming the weakness in searching and calculation for finite solutions and high complexity in traditional algorithms. As a significant branch of bio-heuristic research, swarm intelligence is inspired by the behavior of birds, fish, ants and bee colonies and so on in order to search global optima. Besides the characteristics of the meta-heuristic algorithms, swarm intelligent algorithms have the advantages of easy operating and having good parallel architecture. In recent years, novel swarm intelligent optimization algorithms spring up continually and have driven many researches. For example, particle swarm optimization algorithm (PSO) [1], proposed in 1995, imitated the behavior of birds; Bacterial foraging optimization algorithm (BFO) [2], introduced in 2002, simulated the foraging of bacteria; Glowworm swarm optimization algorithm (GSO) [3], developed in 2009, inspired by the glowworms for searching the light. Artificial bee colony algorithm (ABC) has two different mechanisms consisting of foraging behavior [4] and propagating behavior [5]. Swarm intelligent algorithms have been applied in many fields such as function optimization [6, 7], traveling salesman problem [8], path planning [9], image segmentation[10], spam detection [11], data clustering [12], and functional modules detection in protein-protein interaction network [13, 14] *etc.*.

Y. Tan et al. (Eds.): ICSI 2014, Part I, LNCS 8794, pp. 74–85, 2014.

Fruit fly optimization algorithm (FOA) [15] is a novel swarm intelligent algorithm proposed by Pan in 2011, mimicking the foraging behavior of fruit flies for searching global optimum. With the outstanding olfactory, fruit flies can perceive the smell in the air even the food source beyond 40 meters and fly toward it. Then, after it gets close to the food location, it can also use its sensitive vision to find food and the company's flocking location, and also fly towards that direction.

FOA has been applied in many field such as neural network parameters optimization [16], [17], financial distress [18], PID controller [19], scheduling [20], and knapsack [21] and so on. Because FOA is a novel algorithm, its application in scientific fields is not very extensive, what's more, its weakness avoid it using for many fields. In order to overcome the weakness, we adopt chaos to the basic FOA.

Chaos [22] is a stochastic phenomenon created in nonlinear and ensured system with the characteristics of randomness, regularity, ergodicity, and sensitive to the initial values, which makes it applied to many scientific research field such as image processing [23], signal processing [24], electric power system [25], optical assessment [26], and neural networks [27] and so on. Due to the features of chaos are corresponding to the features in swarm intelligence, chaos is combined with swarm intelligent algorithms for optimization problems to strengthen the performance of the swarm intelligent algorithms, such as PSO [28, 29], ABC [30], and bat algorithm [31] and so on.

In this paper, logistic chaos operation is introduced into FOA in the movement of the fruit flies, in which the optimum was generated by both the best fruit fly and the best fruit fly in chaos. Besides, the calculation of high dimension distance was adopted to basic FOA to overcome the drawback that it is only used for one dimensional problems.

The rest of this paper is organized as follows. Section 2 introduces the basic concepts and principles including the chaotic mapping, distance metric, and basic FOA. Section 3 provides the chaotic fruit fly optimization algorithm. Results from experiments are described in Section 4. Finally, in section 5, conclusions about the paper and future research are shown.

2   Basic Concepts and Principles

2.1   The Chaotic Mapping

Some statistic distribution is used for enhancing the randomness of algorithms, such as uniform and Gaussian distribution. With the randomness properties, chaos is a superb choice to generate random data. Because of the chaotic characteristics of ergodicity and mixing of chaos, algorithms can potentially carry out iterative search steps at higher speeds than standard stochastic search with standard probability distributions [32]. As a typical chaotic system, logistic mapping is the most representative chaotic mapping with simple operation and well dynamic randomness introduced by May [33] in 1976. Logistic mapping is defined as:

$$z(t+1) = \mu z(t)(1 - z(t)) \quad z \in (0,1) \quad 0 < \mu \le 4 \ . \tag{1}$$

in which c is a control parameter and determines whether chaotic variable z stabilizes at a constant value and t denotes the iteration number. Variable z cannot be assigned to 0, 0.25, 0.75, 0.5 and 1. When $\mu = 4$, the sequence of the logistic mapping is chaotic. In later experiments, $\mu = 4$ is adopted.

2.2   Basic Fruit Fly Optimization Algorithm

Fruit fly optimization algorithm is a novel swarm intelligent optimization algorithm with the property of simple operation. Figure 1 shows the fruit fly and group iterative food searching process of fruit fly [15].

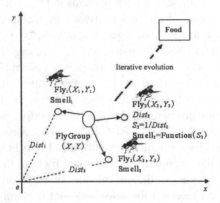

Fig. 1. Illustration of the group iterative food searching of fruit fly

According to the basic FOA [15], several steps are involved as below:

Step 1. Randomly initialize fruit fly swarm location which is shown in Fig.1. The initial location is marked as (*InitX_axis, InitY_axis*).

Step 2. Give the random direction and distance for the search of food using osphresis by an individual fruit fly. New location can be calculated using:

$$x(t + 1) = x(t) + randomvalue_x$$
$$y(t + 1) = y(t) + randomvalue_y \qquad (2)$$

where *randomvalue* is the movment value in each coordinate. As shown in Fig.1, Fly group move to the new locations like Fly1, Fly2, Fly3, the new locations compose the new fly group and new locations take place of the former fly group locations for calculation.

Step 3. Due to the food location cannot be known, the distance to the origin is thus estimated first, marked as *Dist* calculated by:

$$Dist_i = \sqrt{x_i^2 + y_i^2} \qquad (3)$$

The smell concentration judgment value (S) is calculated, and this value is the reciprocal of *Dist*.

Step 4. Substitute smell concentration judgment value (*S*) into smell concentration judgment function (or called Fitness function) so as to find the smell concentration (*Smell$_i$*) of the individual location of the fruit fly.

$$Smell_i = Function(S_i) . \tag{4}$$

Step 5. Find out the fruit fly with minimal smell concentration (finding the maximal value marked as [*bestSmell bestIndex*]) among the fruit fly swarm.

Step 6. Keep the best smell concentration value (marked as *Smellbest*) and *x*, *y* coordinates, and at this moment, the fruit fly swarm will use vision to fly towards that location.

Step 7. Enter iterative optimization to repeat the implementation of *Steps* 2-5, then judge if the smell concentration is superior to the previous iterative smell concentration, if so, implement *Step* 6.

3   Chaotic Fruit Fly Optimization Algorithm

3.1   Principle of Chaotic Fruit Fly Optimization Algorithm

As we know above, only one variable is referred in the basic FOA, we tried to seek for the algorithm for multiple variables, so comes to chance the distance metrics in tradition. In general, basic FOA is a powerful algorithm in swarm intelligent algorithms with the features of simple calculation and high efficiency.

As a consequence of basic FOA, several local optima are achieved instead of global optima. Aimed at this deficiency, chaotic mapping is adopted to improve the performance of basic FOA escaping from the local optima in this paper. The modified FOA proposed in this paper is marked as chaotic fruit fly optimization algorithm (CFOA, for short).

3.2   Distance Metric

Distance is the metric for two variables in similarity, the larger distance, the more difference is. Several distance metrics [34] are used frequently such as Euclidean distance, Mahattan distance, Minkowski distance and Mahalanobis distance. Euclidean distance is taken advantage to calculate the distance resulting in variable one dimension, while experiments show Mahalanobis distance performs well in high dimension as a consequence of vector variable. As can be seen above, high dimension problems are not involved in the basic FOA. It becomes obviously that Euclidean distance is not appropriate for the high dimension problems; meanwhile the complexity for calculating is a very time-consuming process. Hence, distance metric is redesigned to make the algorithm apply to problems in high dimension and reduce the computational complexity. Due to the unknown location of the food source, we assume that it locates in zero in coordinates, then absolute distance is adopted in each dimension for lessening the calculation complexity and insuring the vector result required. That is to say, smell concentration judgment value (*S*) is a multidimensional variable for high dimension problems in CFOA instead of single dimension smell concentration judgment value in FOA.

3.3   New Location Update

The new location of the fruit fly group is combined the best location in the basic movement (x_b) with the best chaotic location (x_c) in logistic mapping. The new location is defined as:

$$x(t + 1) = x(t) + ax_b(t) + (1 - a)x_c(t)r \ . \tag{5}$$

where t stands for the iteration, r is a random number, a denotes the balance parameter ranging from 0 to 1. If $a = 1$, new location depends on the movement of the fruit group independently; if $a = 0$, new location only depends on the chaotic mapping. In order to acquire the outperformance, random number r is introduced to avoid the absoluteness and increase the possibility of seeking for the global optimum.

3.4   CFOA Process

CFOA includes several steps as below:

Step 1. Initialization. Initialize the locations of the first fruit fly group, where uniform distribution is used for experiments to generate the random locations between the *max* and *min* values in the real models. The maximum iteration t_{max}, group size n, problem dimension d, and the bound values should be given at the beginning.

Step 2. Fly group movement. According to the new location calculation method, use Eq. (5) to get the new location. Let the best location in the basic movement be equal to the best chaotic location ($x_b = x_c$) in the initial stage of the algorithm.

Step 3. Calculation for smell concentration. As discussed above, absolute distance is introduced to calculate the smell concentration judgment value (S). After that, perform the 4th step of basic FOA in section 2.3, using Eq. (4) to achieve the value of smell concentration. Smell is the objective function value as well.

Step 4. Frist selection. Find out the best location (x_b) in fruit group with minimal smell concentration, mark the value of smell ($smell_1$) as the same operation of 5th step of basic FOA in section 2.2

Step 5. Chaotic operation. Let the whole fruit group in logistic mapping. On account of data in logistic mapping ranges from 0 to 1, variables in fruit group should be standardized in order to match the variable z in logistic mapping. Assume variables of fruit group (x) ranging from the low bound (*low*) to up bound (*up*), standardized variable (z') defines as:

$$z' = \frac{x - low}{up - low} \qquad x \in \left[low, up\right] \ . \tag{6}$$

z' is matching the variable z in logistic mapping, operate the Eq. (1) to transform $z'(n)$ to $z'(n+1)$, where n denoting the iteration n in searching space.

After the chaotic operation, the variable $z'(n+1)$ ranges from 0 to 1, therefore, inverse substitution should be taken to transform $z'(n+1)$ in logistic mapping to data in fruit group. Corresponding to Eq. (6), the substitution is presented below:

$$x' = z'(up - low) + low \quad z' \in (0,1) \ . \tag{7}$$

Step 6. Best selection. Following the chaotic operation, Find out the best *x'* in fruit group with the minimal smell concentration, mark the best chaotic location (x_c) and value of smell (*smell₂*) similar to the 3th step above. Compare the value of best smell in the basic movement and chaotic movement, mark the smaller as the best smell (*bestsmell*).

Step 7. Enter iterative optimization to judge whether the iteration achieve n_{max} or not, if archives, end up the optimization get rid of loops and output global optima. Otherwise, go to the *Step 2*.

4   Evaluation and Analysis of Experimental Results

Algorithms are tested in Matlab 7.13 and experiments are executed on Pentium dual-core processor 3.10 GHz PC with 4G RAM. 6 Benchmark functions are experimented to testifying the CFOA algorithm compared with PSO [1], BFO [2], GSO [3] and the basic FOA algorithms.

4.1   Benchmark Functions

In the experiments, benchmark functions are used to demonstrate the performance of the algorithm shown in Table 1. Among which, the former 3 functions (*f1-f3*) are unimodal and the others (*f4-f6*) are multimodal functions.

Table 1. Benchmark functions.

Id	Name	Equation	Domain
$f1$	Sphere	$\sum\limits_{i=1}^{d} x_i^2$	±5.12
$f2$	Tablet	$10^6 x_1^2 + \sum\limits_{i=2}^{d} x_i^2$	±100
$f3$	Quadric	$\sum\limits_{i=1}^{d} (\sum\limits_{j=1}^{i} x_j)^2$	±100
$f4$	Rastrigin	$\sum\limits_{i=1}^{d} (x_i^2 - 10\cos(2\pi x_i) + 10)$	±5.12
$f5$	Ackley	$20 + e - 20e^{-0.2\sqrt{\frac{1}{d}\sum\limits_{i=1}^{d} x_i^2}} - e^{\frac{1}{d}\sum\limits_{i=1}^{d}\cos(2\pi x_i)}$	±32
$f6$	Schaffer	$\dfrac{\sin^2 \sqrt{x_1^2 + x_2^2}}{(1 + 0.001(x_1^2 + x_2^2))^2} + 0.5$	±100

4.2   Parameters Setting

It can be seen from swarm intelligent algorithms such as PSO and GSO, the group size assignment is 50 in general. Here, the group size ranging from 10 to 60 are tested in search of the most suitable value, taking Sphere function in 30 dimensions as an example, which is shown in Table 2. "Convergence" is defined as the iteration where the value of objective function becomes changeless. "Time" denotes the running time.

Table 2. Influences of group size in CFOA

Size	Convergence	Time(s)
10	31	0.0921
20	20	0.1909
30	20	0.3081
40	21	0.3890
50	19	0.3720
60	18	0.4827

From Table 2, we can see that running time increase along with group size. In pace with group size augments, convergent iteration varies slightly. When group size ranged from 10 to 20, running time double increased, but from 20 to 30, running time changes little than double. As noted above, group size should be assigned to 20.

Maximal iteration is set to 1000 in comparison between CFOA and FOA algorithm due to the curves converge approximately about 600 iteration; when comparing with other algorithms in Section 4.4, the maximal iteration is set to 200; other parameters are set in Table 3.

Table 3. Parameters of algorithms

Algorithm	Parameters
PSO	$n = 50, w = 0.8, \ c_1 = 2, \ c_2 = 2$
BFO	$n = 20, \ n_c = 10, \ n_s = 5, \ n_r = 2, \ c_r = 0.025$
GSO	$n = 20, \ \rho = 0.4, \ \gamma = 0.6, \ l_i(0) = 4, \ n_t = 4, \ r_d = 50, \ r_s = 50$

4.3   Comparison between CFOA and FOA

By virtue of FOA without the ability to handle high dimensional problems, the distance metric mentioned in section 3.2 is adopted for high dimensional problems here. We compare the best, mean and worst values of the 150 times running for the benchmark functions in 30 dimensions, along with the convergent searching curves gained by FOA and CFOA in Table 4 and Fig. 2, respectively.

Table 4. Comparison of performances between CFOA and FOA

Function	FOA			CFOA		
	Best	Mean	Worst	Best	Mean	Worst
Sphere	1.0588e-04	1.1214e-04	1.1370e-04	0	0	0
Tablet	2.4690	2.7316	2.9027	0	0	0
Quadric	0.0016	0.0017	0.0018	0	0	0
Rastrigin	0.0211	0.2606	0.0219	0	0	0
Ackley	0.0077	0.0080	0.0082	-8.8818e-16	-8.8818e-16	-8.8818e-16
Schaffer	1.0700e-04	1.1320e-04	1.1535e-04	0	0	0

As can be seen from Table 4, CFOA outperforms the basic FOA in aspects of best, mean and worst values, the results win a perfect victory to the problems, reaching the precise value in real problems model mostly, expect for the Ackley problem. On the other hand, FOA is a stable algorithm which can be referred from the little deviation among the best, mean and worst values in Table 4. However, CFOA is more stable than FOA from the data in Table 4.

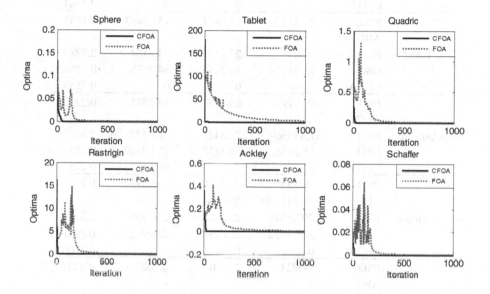

Fig. 2. Searching curves between FOA and CFOA

Basic FOA is not convergent commendably in the problem of Tablet, which can be brought out from Fig. 2. We can see the curves of basic FOA changes sharply and go oscillating in the former iterations, most get smooth in about 300 iterations, and even some diverges in the last such as Tablet. However CFOA converges faster evidently and its curves are smoother than FOA.

4.4 Comparison among CFOA and other Swarm Intelligent Algorithms

As noted above, some typical swarm intelligent optimization algorithms were emerged, in which PSO a well-known algorithm is applied in various fields, after that came BFO, then GSO in recent years. CFOA with PSO, BFO, and GSO are compared to reveal its excellent performance. Maximal iteration is assigned to 200 and $d = 30$. In addition, ABC cited in [35] is used for comparison. The best, mean and worst values are shown in Table 5. Searching curves of CFOA, PSO, BFO, and GSO are compared show in Fig. 3.

Table 5. Comparison of the best value among CFOA and other algorithms

Function	Algorithm	Best value	Mean value	Worst value	Time(s)
Sphere	PSO	0.7082	4.2435	16.3361	0.1536
	ABC	6.9216e-06	9.71e-06	1.7306e-05	-
	BFO	0.0569	27.9043	34.4376	2.2938
	GSO	15.1463	24.1407	37.9462	0.8207
	CFOA	0	0	0	0.1584
Tablet	PSO	9.2533	47.5864	151.2483	0.1633
	ABC	-	-	-	-
	BFO	0.8019	27.8247	54.2484	2.8789
	GSO	21.5472	26.2276	84.2679	0.7496
	CFOA	0	0	0	0.1858
Quadric	PSO	0.1471	6.8473	18.2545	0.2270
	ABC	-	-	-	-
	BFO	1.4393e-07	35.8423	62.8472	5.4017
	GSO	55.2578	89.1475	107.3562	0.8838
	CFOA	0	0	0	0.2755
Rastrigin	PSO	20.1543	70.5196	135.2792	0.1681
	ABC	9.6741e-04	0.0024	0.0054	-
	BFO	207.5546	222.6783	236.8989	2.7732
	GSO	203.8972	227.2889	278.0422	0.8361
	CFOA	0	0	0	0.1700
Ackley	PSO	0.0924	2.3001	4.1087	0.2427
	ABC	-	-	-	-
	BFO	0.0940	1.4919	5.3649	2.8435
	GSO	4.3467	4.6689	5.5228	0.8678
	CFOA	-8.8818e-16	-8.8818e-16	-8.8818e-16	0.2138
Schaffer	PSO	0.0372	0.0622	0.0782	0.2109
	ABC	0.8701	1.0657	1.2542	-
	BFO	0.0165	0.0372	0.0412	2.4955
	GSO	0.0372	0.0372	0.0372	0.9135
	CFOA	0	0	0	0.1952

From Table 5 we can see that CFOA reaches the best value of problems mostly to be the best in actual, values among the best, mean and worst are equivalent. Other algorithms cannot gain the accurate value in actual and even traps into local optima. Furthermore, although PSO runs faster than the other algorithms, CFOA has the advantage in running time, better than PSO. Values obtained by CFOA, which exceeded far from BFO and GSO, better than PSO as well. As a consequence, CFOA is a superb algorithm with outstanding robustness and wonderful accuracy for the functions above.

We tested PSO, BFO and GSO along with CFOA of performance in convergence and searching abilities. Searching curves of algorithms show in Fig.3, we can see that the convergent iteration begin to converge and the values changing in the iteration period. CFOA reaches the smallest values in Fig.3, what's more, the CFOA curves

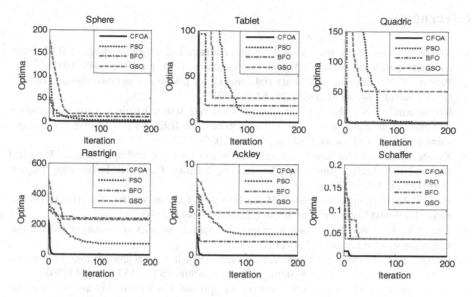

Fig. 3. Searching curves of CFOA and other swarm intelligent algorithms

converges better than the other three algorithms, especially for the functions such as Tablet, Quadric, Rastrigin and Ackley. In addition, the approximately equivalent values are apparently for the Schaffer function.

5   Conclusion and Discussion

Aiming at the deficiencies of trapping into local optimum, converging slowly as well as not suitable for high dimension problems in the basic fruit fly optimization algorithm, chaotic fruit fly optimization algorithm is presented in this paper. In the first place, we modified the distance metric to suit the high dimension problems, absolute distance is adopted here in each dimension to transform the distance into a vector. Secondly, we introduced logistic mapping, the famous typical chaotic mapping to the new algorithm to expand the searching space. Last but not least, new location update is designed to improve the optimum gained by the group. As superior results gained above, CFOA algorithm performs outstanding in both optima searching and running time, not only outperforms the basic FOA, but also other swarm intelligent algorithms. We are intending to apply it to other fields for scientific research in the near future to testify whether it works well.

Acknowledgement. This paper is supported by the National Natural Science Foundation of China (61100164, 61173190), Scientific Research Start-up Foundation for Returned Scholars, Ministry of Education of China ([2012]1707) and the Fundamental Research Funds for the Central Universities, Shaanxi Normal University (GK201402035, GK201302025).

References

1. Kennedy, J., Eberhard, R.C.: Particle swarm optimization. In: Proceedings of IEEE International Conference on Neural Networks, Piscataway, USA, pp. 1942–1948 (1995)
2. Passino, K.M.: Biomimicry of bacterial foraging for distributed optimization and control. IEEE Control Systems Magazine 22, 52–67 (2002)
3. Krishnanand, K.N., Ghose, D.: Detection of multiple source locations using a glowworm metaphor with applications to collective robotics. In: IEEE Swarm Intelligence Symposium, Pasadena, California, USA, pp. 84–91 (2005)
4. Karaboga, D.: An idea based on honey bee swarm for numerical optimization. Technical Report-TR06, Erciyes University, Engineering Faculty, Computer Engineering Department 129(2), 2865–2874 (2005)
5. Hadded, O.B., Afshar, A., Marino, M.A.: Honey-bees mating optimization (HBMO) algorithm. Earth and Environmental Science 20(5), 661–680 (2006)
6. Sun, S.Y., Li, J.W.: A two-swarm cooperative particle swarms optimization. Swarm and Evolutionary Computation 15, 1–18 (2014)
7. Chatzis, S.P., Koukas, S.: Numerical optimization using synergetic swarms of foraging bacterial populations. Expert Systems with Applications 38(12), 15332–15343 (2011)
8. Mavrovouniotis, M., Yang, S.X.: Ant colony optimization with immigrants schemes for the dynamic travelling salesman problem with traffic factors. Applied Soft Computing 13(10), 4023–4037 (2013)
9. Ma, Q.Z., Lei, X.J., Zhang, Q.: Mobile Robot Path Planning with Complex Constraints Based on the Second-order Oscillating Particle Swarm Optimization Algorithm. In: 2009 World Congress on Computer Science and Information Engineering, Los Angeles, USA, vol. 5, pp. 244–248 (2009)
10. Lei, X.J., Fu, A.L.: Two-Dimensional Maximum Entropy Image Segmentation Method Based on Quantum-behaved Particle Swarm Optimization Algorithm. In: Proceedings of the 4rd International Conference on Natural Computation, Jinan, China, vol. 3, pp. 692–696 (2008)
11. Tan, Y.: Particle Swarm Optimizer Algorithms Inspired by Immunity-Clonal Mechanism and Their Application to Spam Detection. International Journal of Swarm Intelligence Research 1(1), 64–86 (2010)
12. Kuo, R.J., Syu, Y.J., Chen, Z.Y., Tien, F.C.: Integration of particle swarm optimization and genetic algorithm for dynamic clustering Original Research Article. Information Sciences 195, 124–140 (2012)
13. Lei, X.J., Tian, J.F., Ge, L., Zhang, A.D.: Clustering and Overlapping Modules Detection in PPI Network Based on IBFO. Proteomics 13(2), 278–290 (2013)
14. Lei, X.J., Wu, S., Ge, L., Zhang, A.D.: The Clustering Model and Algorithm of PPI Network Based on Propagating Mechanism of Artificial Bee Colony. Information Sciences 247, 21–39 (2013)
15. Pan, W.T.: A new evolutionary computation approach: Fruit fly optimization algorithm. In: 2011 Conference of Digital Technology and Innovation Management, Taipei (2011)
16. Li, H.Z., Guo, S., Li, C.J., et al.: A hybrid annual power load forecasting model based on generalized regression neural network with fruit fly optimization algorithm. Knowledge-Based Systems 37, 378–387 (2013)
17. Lin, S.M.: Analysis of service satisfaction in web auction logistics service using a combination of fruit fly optimization algorithm and general regression neural network. Neural Comput. Appl. 22(3-4), 783–791 (2013)

18. Pan, W.T.: A new Fruit Fly Optimization Algorithm: Taking the financial distress model as an example. Knowledge-Based Systems 26, 69–74 (2012)
19. Wang, S., Yan, B.: Fruit fly optimization algorithm based fractional order fuzzy-PID controller for electronic throttle. Nonlinear Dynamics 73(1-2), 611–619 (2013)
20. Zheng, X.L., Wang, L., Wang, S.Y.: A novel fruit fly optimization algorithm for the semiconductor final testing scheduling problem. Knowledge-Based Systems 57, 95–103 (2014)
21. Wang, L., Zheng, X.L., Wang, S.Y.: A novel binary fruit fly optimization algorithm for solving the multidimensional knapsack problem. Knowledge-Based System 48, 17–23 (2013)
22. Pecora, L., Carroll, T.L.: Synchronization in chaotic system. Phy. Rev. Lett. 64(8), 821–824 (1990)
23. Zhou, Y.C., Bao, L., Chen, C.L.P.: A new 1D chaotic system for image encryption. Signal Processing 97, 172–182 (2014)
24. Kassem, A., Hassan, H.A.H., Harkouss, Y., et al.: Efficient neural chaotic generator for image encryption. Digital Signal Processing 25, 266–274 (2014)
25. Ugur, M., Cekli, S., Uzunoglu, C.P.: Amplitude and frequency detection of power system signals with chaotic distortions using independent component analysis. Electric Power Systems Research 108, 43–49 (2014)
26. Petrauskiene, V., Survila, A., Fedaravicius, A., et al.: Dynamic visual cryptography for optical assessment of chaotic oscillations. Optics & Laser Technology 57, 129–135 (2014)
27. Yang, G., Yi, J.Y.: Delayed chaotic neural network with annealing controlling for maximum clique problem. Neurocomputing 127(15), 114–123 (2014)
28. Wang, J.Z., Zhu, S.L., Zhao, W.G., et al.: Optimal parameters estimation and input subset for grey model based on chaotic particle swarm optimization algorithm. Expert Systems with Applications 38(7), 8151–8158 (2011)
29. Lei, X. J., Sun, J.-J., Ma, Q.-Z.: Multiple Sequence Alignment Based on Chaotic PSO. In: Cai, Z., Li, Z., Kang, Z., Liu, Y. (eds.) ISICA 2009. CCIS, vol. 51, pp. 351–360. Springer, Heidelberg (2009)
30. Gao, W.F., Liu, S.Y., Jiang, F.: An improved artificial bee colony algorithm for directing orbits of chaotic systems. Applied Mathematics and Computation 218, 3868–3879 (2011)
31. Gandomi, A.H., Yang, X.S.: Chaotic bat algorithm. Journal of Computational Science 5(2), 224–232 (2014)
32. Coelho, L., Mariani, V.C.: Use of chaotic sequences in biologically inspired algorithm for engineering design optimization. Expert Systems with Applications 34, 1905–1913 (2008)
33. May, R.: Simple mathematical models with very complicated dynamics. Nature 261, 459–467 (1976)
34. Zerzucha, P., Walczak, B.: Concept of (dis)similarity in data analysis. TrAC Trends in Analytical Chemistry (38), 116–128 (2012)
35. Lei, X.J., Huang, X., Zhang, A.D.: Improved Artificial Bee Colony Algorithm and Its Application in Data Clustering. In: The IEEE Fifth International Conference on Bio-Inspired Computing, Theories and Applications (BIC-TA 2010), Changsha, China, pp. 514–521 (2010)

A New Bio-inspired Algorithm:
Chicken Swarm Optimization

Xianbing Meng[1,2], Yu Liu[2], Xiaozhi Gao[1,3], and Hengzhen Zhang[1]

[1] College of Information Engineering, Shanghai Maritime University,
1550 Haigang Avenue, Shanghai, 201306, P.R. China
[2] Chengdu Green Energy and Green Manufacturing R&D Center,
355 Tengfei Road No. 2, Chengdu, 610200, P.R. China
[3] Department of Electrical Engineering and Automation, Aalto University School
of Electrical Engineering, Otaniementie 17, FI-00076 Aalto, Finland
x.b.meng12@gmail.com, yu.liu@vip.163.com

Abstract. A new bio-inspired algorithm, Chicken Swarm Optimization (CSO),
is proposed for optimization applications. Mimicking the hierarchal order in the
chicken swarm and the behaviors of the chicken swarm, including roosters,
hens and chicks, CSO can efficiently extract the chickens' swarm intelligence
to optimize problems. Experiments on twelve benchmark problems and a speed
reducer design were conducted to compare the performance of CSO with that of
other algorithms. The results show that CSO can achieve good optimization results in terms of both optimization accuracy and robustness. Future researches
about CSO are finally suggested.

Keywords: Hierarchal order, Chickens' behaviors, Swarm intelligence, Chick-
en Swarm Optimization, Optimization applications.

1   Introduction

Bio-inspired meta-heuristic algorithms have shown proficiency of solving a great
many optimization applications [1, 2]. They exploit the tolerance for imprecision and
uncertainty of the optimization problems and can achieve acceptable solutions using
low computing cost. Thus the mate-heuristic algorithms, like Particle Swarm Optimization (PSO) [3], Differential Evolution (DE) [2], Bat Algorithm (BA) [1], have attracted great research interest for dealing with optimization applications.

New algorithms are still emerging, including krill herb algorithm [4], and social
spider optimization [5] et al. All these algorithms extract the swarm intelligence from
the laws of biological systems in nature. However, to learn from the nature for developing a better algorithm is still in progress.

In this paper, a new bio-inspired optimization algorithm, namely Chicken Swarm
Optimization (CSO) is proposed. It mimics the hierarchal order in the chicken swarm
and the behaviors of the chicken swarm. The chicken swarm can be divided into
several groups, each of which consists of one rooster and many hens and chicks.
Different chickens follow different laws of motions. There exist competitions between
different chickens under specific hierarchal order.

Y. Tan et al. (Eds.): ICSI 2014, Part I, LNCS 8794, pp. 86–94, 2014.

The rest of paper is organized as follows. Section 2 introduces the general biology of the chicken. The details about the CSO are discussed in Section 3. The simulations and comparative studies are presented in section 4. Section 5 summaries this paper with some conclusions and discussions.

2   General Biology

As one of the most widespread domestic animals, the chickens themselves and their eggs are primarily kept as a source of food. Domestic chickens are gregarious birds and live together in flocks. They are cognitively sophisticated and can recognize over 100 individuals even after several months of separation. There are over 30 distinct sounds for their communication, which range from clucks, cackles, chirps and cries, including a lot of information related to nesting, food discovery, mating and danger. Besides learning through trial and error, the chickens would also learn from their previous experience and others' for making decisions [6].

A hierarchal order plays a significant role in the social lives of chickens. The preponderant chickens in a flock will dominate the weak. There exist the more dominant hens that remain near to the head roosters as well as the more submissive hens and roosters who stand at the periphery of the group. Removing or adding chickens from an existing group would causes a temporary disruption to the social order until a specific hierarchal order is established [7].

The dominant individuals have priority for food access, while the roosters may call their group-mates to eat first when they find food. The gracious behavior also exists in the hens when they raise their children. However, this is not the case existing for individuals from different groups. Roosters would emit a loud call when other chickens from a different group invade their territory [8].

In general, the chicken's behaviors vary with gender. The head rooster would positively search for food, and fight with chickens who invade the territory the group inhabits. The dominant chickens would be nearly consistent with the head roosters to forage for food. The submissive ones, however, would reluctantly stand at the periphery of the group to search for food. There exist competitions between different chickens. As for the chicks, they search for the food around their mother.

Each chicken is too simple to cooperate with each other. Taken as a swarm, however, they may coordinate themselves as a team to search for food under specific hierarchal order. This swarm intelligence can be associated with the objective problem to be optimized, and inspired us to design a new algorithm.

3   Chicken Swarm Optimization

Given the aforementioned descriptions, we can develop CSO mathematically. For simplicity, we idealized the chickens' behaviors by the following rules.

(1) In the chicken swarm, there exist several groups. Each group comprises a dominant rooster, a couple of hens, and chicks.

(2) How to divide the chicken swarm into several groups and determine the identity of the chickens (roosters, hens and chicks) all depend on the fitness values of the chickens themselves. The chickens with best several fitness values would be acted as roosters, each of which would be the head rooster in a group. The chickens with worst several fitness values would be designated as chicks. The others would be the hens. The hens randomly choose which group to live in. The mother-child relationship between the hens and the chicks is also randomly established.

(3) The hierarchal order, dominance relationship and mother-child relationship in a group will remain unchanged. These statuses only update every several (G) time steps.

(4) Chickens follow their group-mate rooster to search for food, while they may prevent the ones from eating their own food. Assume chickens would randomly steal the good food already found by others. The chicks search for food around their mother (hen). The dominant individuals have advantage in competition for food.

Assume RN, HN, CN and MN indicate the number of the roosters, the hens, the chicks and the mother hens, respectively. The best RN chickens would be assumed to be roosters, while the worst CN ones would be regarded as chicks. The rest are treated as hens. All N virtual chickens, depicted by their positions $x_{i,j}^t$ ($i \in [1, \cdots, N], j \in [1, \cdots, D]$) at time step t, search for food in a D-dimensional space. In this work, the optimization problems are the minimal ones. Thus the best RN chickens correspond to the ones with RN minimal fitness values.

3.1   Movement of the Chickens

The roosters with better fitness values have priority for food access than the ones with worse fitness values. For simplicity, this case can be simulated by the situation that the roosters with better fitness values can search for food in a wider range of places than that of the roosters with worse fitness values. This can be formulated below.

$$x_{i,j}^{t+1} = x_{i,j}^t * (1 + Randn(0, \sigma^2)) . \tag{1}$$

$$\sigma^2 = \begin{cases} 1, if\ f_i \le f_k , \\ \exp\left(\frac{(f_k - f_i)}{|f_i| + \varepsilon}\right), \ otherwise, \end{cases} k \in [1, N], k \ne i . \tag{2}$$

Where $Randn$ (0, σ^2) is a Gaussian distribution with mean 0 and standard deviation σ^2. ε, which is used to avoid zero-division-error, is the smallest constant in the computer. k, a rooster's index, is randomly selected from the roosters group, f is the fitness value of the corresponding x.

As for the hens, they can follow their group-mate roosters to search for food. Moreover, they would also randomly steal the good food found by other chickens, though they would be repressed by the other chickens. The more dominant hens would have advantage in competing for food than the more submissive ones. These phenomena can be formulated mathematically as follows.

$$x_{i,j}^{t+1} = x_{i,j}^t + S1 * Rand * \left(x_{r1,j}^t - x_{i,j}^t\right) + S2 * Rand * \left(x_{r2,j}^t - x_{i,j}^t\right) . \tag{3}$$

$$S1 = \exp\left((f_i - f_{r1}) / (abs(f_i) + \varepsilon) \right) . \tag{4}$$

$$S2 = \exp((f_{r2} - f_i)) \ . \tag{5}$$

Where *Rand* is a uniform random number over [0, 1]. $r1 \in [1, \cdots, N]$ is an index of the rooster, which is the *i*th hen's group-mate, while $r2 \in [1, \cdots, N]$ is an index of the chicken (rooster or hen), which is randomly chosen from the swarm. $r1 \neq r2$.

Obviously, $f_i > f_{r1}, f_i > f_{r2}$, thus $S2 < 1 < S1$. Assume $S1=0$, then the *i*th hen would forage for food just followed by other chickens. The bigger the difference of the two chickens' fitness values, the smaller $S2$ and the bigger the gap between the two chickens' positions is. Thus the hens would not easily steal the food found by other chickens. The reason that the formula form of $S1$ differs from that of $S2$ is that there exist competitions in a group. For simplicity, the fitness values of the chickens relative to the fitness value of the rooster are simulated as the competitions between chickens in a group. Suppose $S2=0$, then the *i*th hen would search for food in their own territory. For the specific group, the rooster's fitness value is unique. Thus the smaller the *i*th hen's fitness value, the nearer $S1$ approximates to 1 and the smaller the gap between the positions of the *i*th hen and its group-mate rooster is. Hence the more dominant hens would be more likely than the more submissive ones to eat the food.

The chicks move around their mother to forage for food. This is formulated below.

$$x_{i,j}^{t+1} = x_{i,j}^t + FL * \left(x_{m,j}^t - x_{i,j}^t\right) \ . \tag{6}$$

Where $x_{m,j}^t$ stands for the position of the *i*th chick's mother $(m \in [1, N])$. FL $(FL \in (0,2))$ is a parameter, which means that the chick would follow its mother to forage for food. Consider the individual differences, the FL of each chick would randomly choose between 0 and 2.

Chicken Swarm Optimization. Framework of the CSO
Initialize a population of N chickens and define the related parameters;
Evaluate the N chickens' fitness values, $t=0$;
While ($t <$ Max_Generation)
If ($t \% G == 0$)
Rank the chickens' fitness values and establish a hierarchal order in the swarm;
Divide the swarm into different groups, and determine the relationship between the chicks and mother hens in a group; End if
For $i = 1 : N$
If $i ==$ rooster   Update its solution/location using equation (1); End if
If $i ==$ hen   Update its solution/location using equation (3); End if
If $i ==$ chick   Update its solution/location using equation (6); End if
Evaluate the new solution;
If the new solution is better than its previous one, update it;
End for
End while

3.2   Parametric Analysis

There exist six parameters in CSO. Humans keep chickens primarily as a source of food. As the food themselves, only hens can lay eggs, which can also be the source of food. Hence keeping hens is more beneficial for human than keeping roosters. Thus *HN* would be bigger than *RN*. Given the individual differences, not all hens would hatch their eggs simultaneously. Thus *HN* is also bigger than *MN*. Though each hen can raise more than one chick, we assume the population of adult chickens would surpass that of the chicks, *CN*. As for *G*, it should be set at an appropriate value, which is problem-based. If the value of *G* is very big, it's not conducive for the algorithm to converge to the global optimal quickly. While if the value of *G* is very small, the algorithm may trap into local optimal. After the preliminary test, $G \in$ [2,20] may achieve good results for most problems.

Furthermore, the formula of the chick's movement can be associated with the corresponding part in DE. If we set *RN* and *MN* at 0, thus CSO essentially becomes the basic mutation scheme of DE. Hence the partial conclusions from the DE [2] can be used. In practice, $FL \in$ [0.4, 1] usually perform well.

4   Validation and Comparison

4.1   Benchmark Problems Optimization

Twelve popular benchmark problems [9, 10] (shown in Table 1) are used to verify the performance of the CSO compared with that of PSO, DE and BA. The statistical results have been obtained, based on 100 independent trials, in all the case studies. The number of iterations is 1,000 in each trial. For a fair comparison, all of the common parameters of these methods, such as the population size, dimensions and maximum number of generations, are set to be the same. The related parameters of these algorithms are showed at Table 2.

Table 1. Twelve benchmark problems

Problem name	ID	Dimension	Bounds	Optimum
High Conditioned Elliptic	F1	20	[-100,100]	0
Bent Cigar	F2	20	[-100,100]	0
Discus	F3	20	[-100,100]	0
Ackley	F4	20	[-32,32]	0
Griewank	F5	20	[-600,600]	0
Sphere	F6	20	[-100,100]	0
Step	F7	20	[-100,100]	0
Powell Sum	F8	20	[-1,1]	0
Rastrigin	F9	20	[-5,10]	0
Axis parallel hyper-ellipsoid	F10	20	[-5.12,5.12]	0
Brown	F11	20	[-1,4]	0
Exponential	F12	20	[-1,1]	-1

Table 2. The related parameter values

Algorithm	Parameters
PSO	$c1=c2=1.49445$, $w = 0.729$
DE	$CR = 0.9$, $F = 0.6$
BA	$\alpha = \gamma = 0.9$, $f_{min} = 0$, $f_{max} = 2$, $A_0 \in [0,2], r_0 \in [0,1]$
CSO	$RN=0.2*N$, $HN=0.6*N$, $CN=N-RN-HN$, $MN=0.1*N$, $G = 10$, $FL \in [0.5, 0.9]$

There are many variants of PSO, DE and BA. In this work, the basic BA and the standard PSO are chosen. As for DE, the DE/rand/1/bin scheme is selected. Table 3 displays the statistical comparison of the four algorithms on twelve benchmark problems. It clearly shows that CSO is superior to PSO, DE and BA on all these problems in terms of accuracy, efficiency and robustness.

The superiority of CSO over PSO, BA and DE should be the case. If we set $RN = CN = 0$, and let $S1$, $S2$ be the parameters like $c1$ and $c2$ in PSO, thus CSO will be similar to the standard PSO. Hence CSO can inherit many advantages of PSO and DE. Moreover, the chickens' swarm intelligence can be efficiently extracted in CSO. Given the diverse laws of the chickens' motions and cooperation between the multi-groups, the search space can be efficiently explored. Under the specific hierarchal order, the whole chicken swarm may behave like a team to forage for food, which can be associated with the objective problems to be optimized. All of these merits enhance the performance of CSO.

Table 3. Statistical comparison of CSO with PSO, DE and BA

Problem	Algorithm	Best	Mean	Worst	Std.
	PSO	12600.53084	49808.05813	101417.29666	2124.09
	CSO	0	0	0	1.89943e-60
F1	BA	3782.10211	30996.60571	84110.41779	30996.6057
	DE	0	0	0	5.99605e-12
	PSO	0	1300	10000	339.7
	CSO	0	0	0	1.35815e-62
F2	BA	1780594.354	2636386.576	3804547.807	36993.4
	DE	0	0.00001	0.00001	1.88465e-6
	PSO	0	0	0	4.66505e-33
	CSO	0	0	0	9.37294e-66
F3	BA	101.92698	2256.99578	6307.26214	132.731
	DE	0	0	0	3.22023e-12
	PSO	0	0	0	1.31168e-16
	CSO	0	0	0	6.12169e-17
F4	BA	1.48288	2.59402	3.07403	0.02297
	DE	0	0.35702	11.64977	0.17096
	PSO	0	0	0	5.36538e-8
	CSO	0	0	0	0
F5	BA	0.004	2.82906	15.42094	0.296984
	DE	0	0	0	1.05399e-12
	PSO	0	0	0	1.8988e-35
	CSO	0	0	0	4.10796e-70
F6	BA	1.867408	2.94197	4.18701	0.0491192
	DE	0	0	0	1.94304e-12
	PSO	0	0	0	0
	CSO	0	0	0	0
F7	BA	1	3.41	6	0.103105
	DE	0	0	0	0
	PSO	0	0	0	4.78569e-60
	CSO	0	0	0	0
F8	BA	0	0	0	5.02596e-8
	DE	0	0	0	0

Table 3. (*Continued*)

Problem	Algorithm	Best	Mean	Worst	Std.
	PSO	10.94454	21.26284	41.78822	0.60048
	CSO	0	0	0	0
F9	BA	88.44729	121.99296	167.60654	1.57913
	DE	8.41884	22.70527	43.9751	0.706825
	PSO	0	0	0	2.13096e-37
	CSO	0	0	0	1.59801e-71
F10	BA	24.34652	39.73613	63.15	0.749752
	DE	0	0	0	4.15726e-14
	PSO	0	1.29	8	0.171595
	CSO	0	0	0	5.71381e-72
F11	BA	4.22819	5.92480	7.52686	0.0726235
	DE	0	0	0	2.37888e-15
	PSO	-1	-1	-1	9.58157e-18
	CSO	-1	-1	-1	0
F12	BA	-0.41494	-0.20415	-0.12952	0.0052408
	DE	-1	-1	-1	1.75511e-16

4.2   Speed Reducer Design

Design of the speed reducer [11] (as shown in Fig. 1) is to design a gearbox, which can be rotated at its most efficient speed. The gearbox is described by the face width $b(x_1)$, module of teeth $m(x_2)$, number of teeth in the pinion $z(x_3)$, length of the first shaft between bearings $h1(x_4)$, length of the second shaft between bearings $h2(x_5)$, diameter of the first shaft $d1(x_6)$, and diameter of the first shaft $d2(x_7)$. The optimization in the design of the speed reducer is to minimize its total weight, subject to constraints on bending stress of the gear teeth, surface stress, transverse deflections of the shafts, and stresses in the shafts. This problem can be formulated as follows.

Minimize $f(\vec{x}) = 0.7854x_1x_2^2(3.3333x_3^2 + 14.9334x_3 - 43.0934) -$
$$1.508x_1(x_6^2 + x_7^2) + 7.4777(x_6^3 + x_7^3) + 0.7854(x_4x_6^2 + x_5x_7^2)$$

Subject to $g_1(\vec{x}) = \frac{27}{x_1x_2^2x_3}$ $g_2(\vec{x}) = \frac{397.5}{x_1x_2^2x_3^2}$ $g_3(\vec{x}) = \frac{1.93x_4^3}{x_2x_6^4x_3}$ $g_4(\vec{x}) = \frac{1.93x_5^3}{x_2x_7^4x_3}$

$g_5(\vec{x}) = \frac{1}{110x_6^3}\sqrt{(\frac{745x_4}{x_2x_3})^2 + 16.9 \times 10^6}$ $g_6(\vec{x}) = \frac{1}{85x_7^3}\sqrt{(\frac{745x_5}{x_2x_3})^2 + 157.5 \times 10^6}$

$g_7(\vec{x}) = \frac{x_2x_3}{40}$ $g_8(\vec{x}) = \frac{5x_2}{x_1}$ $g_9(\vec{x}) = \frac{x_1}{12x_2}$ $g_{10}(\vec{x}) = \frac{1.5x_6+1.9}{x_4}$ $g_{11}(\vec{x}) = \frac{1.1x_7+1.9}{x_5}$

Where $2.6 \leq x_1 \leq 3.6, 0.7 \leq x_2 \leq 0.8, 17 \leq x_3 \leq 28, 7.3 \leq x_4 \leq 8.3, 7.8 \leq x_5 \leq 8.3, 2.9 \leq x_6 \leq 3.9, 5 \leq x_7 \leq 5.5$, $g_i(\vec{x}) \leq 1(i = 1, 2, 3, \cdots 11)$.

Table 4 summarizes a comparison of the results achieved by CSO and other algorithms. It clearly shows that CSO's results outperform all the results achieved by the six methods in terms of both optimization accuracy and robustness. The best solution achieved by CSO is $\vec{x} = (3.5, 0.7, 17, 7.308, 7.802, 3.35, 5.287)$ with $f(\vec{x}) = 2996.60481329$. The constraint values are $\vec{g} = (-0.07, -0.2, -0.5, -0.9, -2.33e-6, -1.06e-5, -0.7, -5.06e-5, -0.58, -0.05, -0.01)$, which indicates that the solution is feasible.

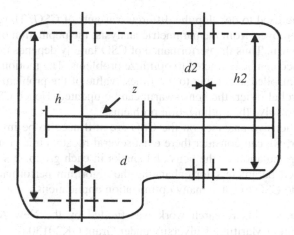

Fig. 1. Speed reducer

Table 4. Optimization results of the speed reducer design

	Robert, et al [11]			Mezura,	Akay, et al	Gandomi	CSO
	SF1(lBest)	SFI(square)	SFI(gBest)	et al [12]	[13]	et al [14]	
Best	3000.737	2996.974	2998.991	2999.264	2997.05841	3000.981	2996.605
Mean	3015.026	3000.278	3011.062	3014.759	2997.05841	3007.2	2997.764
Worst	3044.332	3007.301	3034.973	N/A	N/A	3009	3007.258
Std.	9.95	2.804	9.105	11.0	0	4.963	0.165

5   Discussions

Mimicking the chickens' behaviors, a new bio-inspired algorithm, namely Chicken Swarm Optimization was proposed for optimization problems. The performance of CSO is compared with that of the PSO, DE and BA on twelve benchmark problems. Experiments show that CSO outperforms the PSO, DE and BA in terms of both optimization accuracy and robustness. Moreover, CSO can efficiently solve the speed reducer design, which endues the CSO with a promising prospect of further studying.

One of the reasons that CSO has very promising performance is that CSO inherits major advantages of many algorithms. PSO and the mutation scheme of DE are the special cases of the CSO under appropriate simplifications. What is more significant for the superiority of the CSO is that the chickens' swarm intelligence can be efficiently extracted to optimize problems. The chickens' diverse movements can be conducive for the algorithm to strike a good balance between the randomness and determinacy for finding the optima. The whole chicken swarm consists of several groups, namely multi-swarm. Through integration of the hierarchal order, chickens of the different groups may behave as a team and coordinate themselves to forage for food. Thus CSO can behave intelligently to optimize problems efficiently.

The innovation in this paper not only lies in efficiently extracting the chickens' swarm intelligence to optimize problems, but also making CSO innate multi-swarm method. Multi-swarm technique is usually used to enhance performance of the population-based algorithm. As an innate multi-swarm algorithm, various multi-swarm techniques can be used to develop the different variants of CSO. Thus CSO has good extensibility. Moreover, from the parametric analysis, the population of the hens is the biggest in the swarm. Thus the performance of CSO largely depends on how the hens' swarm intelligence can be extracted to optimize problems. The motion of the hens can be adaptively controlled according to the fitness value of the problem itself. With the dynamical hierarchal order, the hens swarm can be updated. Hence CSO has the self-adaptive ability to solve the optimization problems.

More comprehensive analyses on the CSO are still need to be investigated in the future. Moreover, we can consider there exist several roosters in a group and dynamically adjust the population of the hens and chicks in each group. It's also significant to tune the related parameters for enhancing the algorithm performance, and design the variants of the CSO to solve many optimization applications.

Acknowledgments. This research work was funded by the New Academic Staffs Program of Shanghai Maritime University under Grant GK2013089.

References

1. Yang, X.S.: Bat algorithm: literature review and applications. International Journal of Bio-inspired Computation 5(3), 141–149 (2013)
2. Das, S., Suganthan, P.N.: Differential evolution: A survey of the state-of-the-art. IEEE Transactions on Evolutionary Computation 15(1), 4–31 (2011)
3. Jordehi, A.R., Jasni, J.: Parameter selection in particle swarm optimization: A survey. Journal of Experimental & Theoretical Artificial Intelligence 25(4), 527–542 (2013)
4. Gandomi, A.H., Alavi, A.H.: Krill herd: A new bio-inspired optimization algorithm. Communications in Nonlinear Science and Numerical Simulation 17, 4831–4845 (2012)
5. Cuevas, E., Cienfuegos, M., Zaldivar, D., Cisneros, M.: A swarm optimization algorithm inspired in the behavior of the social-spider. Expert Systems with Applications 40, 6374–6384 (2013)
6. Smith, C.L., Zielinski, S.L.: The Startling Intelligence of the Common Chicken. Scientific American 310(2) (2014)
7. Grillo, R.: Chicken Behavior: An Overview of Recent Science, http://freefromharm.org/chicken-behavior-an-overview-of-recent-science
8. Chicken, http://en.wikipedia.org/wiki/Chicken
9. Tan, Y., Li, J.Z., Zheng, Z.Y.: ICSI, Competition on Single Objective Optimization (2014), http://www.ic-si.org/competition/ICSI.pdf
10. Yang, X.S.: Nature-inspired optimization algorithm. Elsevier (2014)
11. Robert, R., Mostafa, A.: Embedding a social fabric component into cultural algorithms toolkit for an enhanced knowledge-driven engineering optimization. International Journal of Intelligent Computing and Cybernetic 1(4), 563–597 (2008)
12. Mezura, M.E., Hernandez, O.B.: Modified bacterial foraging optimization for engineering design. In: Proceedings of the Artificial Neural Networks in Engineering Conference, vol. 19, pp. 357–364. Intelligent Engineering Systems Through Artificial Neural Networks (2009)
13. Akay, B., Karaboga, D.: Artificial bee colony algorithm for large-scale problems and engineering design optimization. Journal of Intelligent Manufacturing 23(4), 1001–1014 (2012)
14. Gandomi, A.H., Yang, X.S., Alavi, A.H.: Cuckoo search algorithm: A metaheuristic approach to solve structural optimization problems. Engineering with Computers 29, 17–35 (2013)

A Population-Based Extremal Optimization Algorithm with Knowledge-Based Mutation

Junfeng Chen[1], Yingjuan Xie[1], and Hua Chen[2]

[1] College of IOT Engineering, Hohai University, 213022, Changzhou, China
chen-1997@163.com, xieyj@hhuc.edu.cn
[2] Mathematics and Physics Department, Hohai University, 213022, Changzhou, China
chenhua112@163.com

Abstract. Extremal optimization is a dynamic, heuristic intelligent algorithm. It evolves a single solution and makes local modifications to the worst components. In this paper, a knowledge-base mutation operator is presented based on the distribution knowledge of candidate solutions. And then a population-based extremal optimization with knowledge-based mutation is proposed by introducing the idea of swarm evolution. Finally, the proposed method is applied to PID parameter tuning. The simulation results show that the proposed algorithm is characterized by high response speed, small overshoot and steady-state error, and obtains satisfactory control effect.

Keywords: Extremal optimization, Knowledge-based mutation, PID controller.

1 Introduction

Extremal optimization (EO) is a phenomenon-mimicking algorithm inspired by the Bak-Sneppen model of self-organized criticality from the field of statistical physics [1-2]. Unlike Genetic Algorithm (GA) favoring the good genes of the individuals, EO prefers to vary the worst component, together with its nearest neighbors, of the present individual, and the whole system can adaptively evolve into the self-organized critical state.

The basic EO makes the search to approach near-ground state energy quickly but gets stuck in a meta-stable state easily. To overcome this problem, an adjustable parameter τ was introduced into basic EO [3]. Yao et al. [4] proposed Lévy mutation to produce alterable step sizes by means of adjusting the control parameter and Chen et al. [5] adopted the Lévy mutation for EO. Menai et al. [6] introduced an improved EO algorithm based on the Bose-Einstein distribution in quantum physics. Sousa et al. [7] proposed a generalized extremal optimization method. Zeng et al. [8] demonstrated that EO with heuristic initial information is more effective than the one with random initial solution. Moreover, hybrid algorithms [9] were proposed by combining the exploitation ability of EO with the exploration ability of other optimization methods.

However, the above-mentioned EO algorithms cannot guarantee to work well for other optimization problems, and further discussions are needed for their conclusions. In this paper, a novel mutation strategy is proposed by analyzing the distribution of

Y. Tan et al. (Eds.): ICSI 2014, Part I, LNCS 8794, pp. 95–102, 2014.

candidate solutions, and the uncertain information of optimization problem is extracted for guiding the evolution of EO.

The rest of this paper is organized as follows. In Section 2, a new mutation operator is presented based on the distribution knowledge of candidate solutions. In Section 3, a population-based extremal optimization with knowledge-based mutation is proposed by introducing the idea of swarm evolution. In Section 4, the proposed method is applied to PID parameter tuning. Our concluding remarks are contained in Section 5.

2   Knowledge-Based Mutation Strategy

In this section, a knowledge-based mutation operator is proposed to acquire information about the optimization problem. Generally, the landscape information can be approximately described as the distribution of candidate solutions in learning algorithms. Here, we take a population of candidate solutions to illustrate the uncertain information about the problems to be solved. Figure 1 shows the distribution of candidate solutions in EO when dealing with Schwefel's function. The black dots represent the candidate solutions, plotted by the first component X_1 on the horizontal axis and the second component X_2 on the vertical. The contours describe the same evaluation function values. The color is deeper and the evaluation value is smaller.

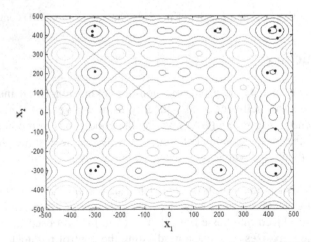

Fig. 1. The distribution of candidate solutions for solving Schwefel's function

From Fig. 1, the candidate solutions are widely dispersed at the level of groups and concentrate at four local areas at the level of individuals. That is, the Schwefel's function has more than one extreme value point and sub-optimal solutions are far from the global optimal solution. Therefore, this uncertain information of the problem being optimized can be extracted from the distribution of candidate solutions.

Cloud model is an effective tool for studying and analyzing the uncertainty information [10]. It represents the fuzziness and randomness and their correlation.

Meanwhile, it converts between qualitative knowledge and its quantitative expression. In this paper, the overall property of candidate solutions is represented by three numerical characteristics: Expected value (Ex), Entropy (En) and Hyper-Entropy (He), which are dependent upon the idea of cloud model. The calculation process of numerical characteristics is as follows.

Algorithm 1. Calculate the numerical characteristics of the candidate solutions

---

Input: candidate solution x_i and its evaluation value $f(x_i)$, $i = 1, 2, \cdots, n$.

Output: the numerical characteristics of the candidate solutions $[Ex, En, He]$

---

(1) Select the candidate solution with the best evaluation value as elite candidate solution x^*, and the expected value $Ex = x^*$.

(2) Calculate the standard variance of x_i for En, i.e., $En = \dfrac{1}{n-1} \sum_{i=1}^{n} (x_i - Ex)^2$.

(3) For each couple of $(x_i, f(x_i))$, calculate $En_i' = \sqrt{\dfrac{-(x_i - \hat{x})^2}{2 \ln f(x_i)}}$.

(4) Calculate the mean of En_i' for En', i.e., $\overline{En'} = \dfrac{1}{n} \sum_{i=1}^{n} En_i'$.

(5) Calculate the hyper-entropy $He = \dfrac{1}{n-1} \sum_{i=1}^{n} (En_i' - \overline{En'})$.

Algorithm 2. Knowledge-based Mutation strategy

Input: the numerical characteristics $[Ex, En, He]$, the size of solution set n.

Output: candidate solution and its evaluation value $\{x_i, f(x_i)\}$, $i = 1, 2, \cdots, n$.

---

(1) Generate a normally distributed random number En_i' with expectation En and variance He, i.e., $En_i' = \text{NormRand}(En, He)$.

(2) Generate a normally distributed random number x_i with expectation Ex and variance En_i', i.e., $x_i = \text{NormRand}(Ex, En_i')$.

(3) Calculate the evaluation value of candidate solution $f(x_i)$. And x_i with evaluation value $f(x_i)$ is a candidate solution in the domain.

(4) Repeat steps 1 to 3 until all the n candidate solutions are generated.

The expected value Ex is the position corresponding to the center of the cloud gravity, whose elements are fully compatible with the knowledge. And the elite candidate solution is the most classical sample while solving optimization problems. So the elite candidate solution is related to the expected value of cloud model. The entropy En reflects the dispersing extent of the candidate solutions, so the entropy can indicate the search scope of EO. The hyper-entropy He indirectly shows dispersion degree and thickness of candidate solutions, which can be used as the stability degree of candidate solutions in the search process of EO.

The knowledge-based mutation is proposed by analyzing the distribution of candidate solutions in the search space, and the uncertain information of optimization problem is extracted according to *Algorithm 1*. The new candidate solutions are further generated on the basis of the numerical characteristics. The detailed process is as *Algorithm 2*.

3   Population-Based EO with Knowledge-Based Mutation

3.1   Population Representation

The basic EO works on a single solution at any time, namely, the point-to-point search. Population-based methods such as GA and Particle Swarm Optimization (PSO), on the contrary, perform search processes which describe the evolution of a set of points in the search space. Consequently, population-based algorithms provide a parallel, intrinsic way for the exploration of the search space. In order to enhance global searching ability of EO, we develop a population-based EO in the light of the warm search strategy.

Consider a set of candidate solutions (hereafter referred to as solution set) X composed by n candidate solutions. Each candidate solution is composed of m components. A population of candidate solutions generated by EO can be written in the following vector and matrix forms:

$$X = \begin{bmatrix} x_1 \\ x_2 \\ \vdots \\ x_n \end{bmatrix} = \begin{bmatrix} x_{11} & x_{12} & \cdots & x_{1m} \\ x_{21} & x_{22} & \cdots & x_{2m} \\ \vdots & \vdots & \vdots & \vdots \\ x_{n1} & x_{n2} & \cdots & x_{nm} \end{bmatrix}. \tag{1}$$

where $x_i = [x_{i1}, x_{i2}, \cdots, x_{im}]$ is the i-th candidate solution, and x_{ij} is the j-th solution component of the i-th candidate solution, $i = 1, 2, \cdots, n$, $j = 1, 2, \cdots, m$.

3.2   Population-Based Extremal Optimization

A population-based EO (PEO) with knowledge-based mutation is proposed in this section. PEO deals with a set of solutions rather than a single solution in every iteration, and each individual is modified by the knowledge-based mutation operator. To be specific, distribution information of excellent individuals is extracted and used to construct the new solutions in the search process. Heuristic information helps contribute to the optimal performance of the proposed EO algorithm. The structure diagram of PEO with uncertainty knowledge is shown in Fig.2.

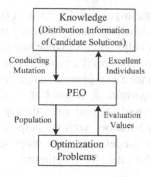

Fig. 2. Structure diagram of PEO with uncertainty knowledge

Algorithm 3. PEO with knowledge-based mutation

Input: initial solution set $X(0) = [x_1(0), x_2(0), \cdots, x_n(0)]$.

Output: the optimal solution and its evaluation value $\{x^*(t), f(x^*(t))\}$.

---

(1) Initialize the solution set $X(0)$ randomly.

(2) Calculate the evaluation value of each candidate solution $f(x_i(t))$ and the mean evaluation value of solution set, i.e., $\overline{f}(X(t))$, $i = 1, 2, \cdots n$. Select the candidate solution with the best evaluation value as $x^*(t)$.

(3) Call **Algorithm** 1 and take the candidate solution satisfying $f(x_i(t)) > \overline{f}(X(t))$ as input. And then, we have numerical characteristics of solutions distribution $[Ex(t), En(t), He(t)]$.

(4) Employ Cauchy and Gauss mutation to generate temporary population $X_{tem1}(t+1)$ and $X_{tem2}(t+1)$, respectively. Calculate the fitness function value for each individual, and sort the solutions according to their fitness values. Denote the best candidate solution as $x^*_{tem}(t+1)$.

(5) If inequality $f(x^*_{tem}(t+1)) > f(x^*(t))$ is met, call **Algorithm** 2 taking the numerical characteristics $[Ex(t), En(t), He(t)]$ as input and generate temporary population $X_{tem3}(t+1)$. Select the top n solutions to form a new solution set $X(t+1)$. Otherwise, the new solution set is equal to $X_{tem3}(t+1)$, i.e., $X(t+1) = X_{tem3}(t+1)$.

(6) Output the best candidate solution $x^*(t)$ and its evaluation value $f(x^*(t))$ if maximum iterations or solution precision criteria is met, otherwise go to Step 2.

As shown in Fig. 2, the candidate solutions with good evaluation function values are first selected in every iteration. The uncertain knowledge is extracted as three numerical characteristics according to the distribution of candidate solutions. After that, heuristic knowledge is used in the mutation operator for constructing the new candidate solutions in the new generation. In general, PEO with uncertainty knowledge bridges the gap between uncertainty knowledge and practical problems to be solved. Meanwhile, uncertainty information helps to concentrate search on those regions in which better solutions may likely exist. The specific process is as *Algorithm 3*.

Mutation strategy in *Algorithm 3* can generate large variations and small variations simultaneously. Therefore, *Algorithm 3* is good at both coarse-grained search and fine-grained search. The proposed method can quickly identify the regions in the search space with high quality solutions with the help of the uncertainty knowledge. None of search parameters are predefined in the proposed algorithm, and the whole system can adaptively approach to the global optimum.

4   PEO for PID Parameter Tuning

A PID controller is a control loop feedback mechanism widely used in industrial control systems, which involves three separate constant parameters, i.e., the proportional, the integral and derivative values, denoted by k_P, k_I, k_D. By tuning the three parameters, the controller can provide control action designed for specific process requirements. PID parameter tuning can be seen as an optimization problem in EO algorithm which aims to seek an optimal or near-optimal set of proportional, integral and derivative values [11]. A second-order system with time-delay is taken as the simulation object, and its transfer function is described as follow.

$$G(s) = \frac{5}{(2s+1)(0.5s+1)} e^{-0.2s}.$$

(2)

In this paper, a candidate solution is composed of three components which represent the proportional, integral and derivative parameters, respectively. According to the control requirements of the system, Integral of Time Multiplied by Absolute Error (ITAE) is used to evaluate performance of the PID controller. The reciprocal of ITAE is selected as the evaluation function in EO.

$$\min_{k_i} f(k_i) = \frac{1}{ITAE} = \frac{1}{\int_0^\infty t|e(t)|dt}.$$

(3)

where $k_i = [k_{iD}, k_{iP}, k_{iI}]$, and $e(t)$ is an error signal in time domain.

To verify the effectiveness and feasibility of the proposed EO algorithm, it is applied to the PID parameter tuning, and PSO, GA and EO are used as comparisons. In the following experiment, the population size of GA, PSO and PEO is set as 40. The same maximum iteration $itermax = 50$ is applied to avoid falling into an infinite loop. The parameters of PID controller range from 0 to 10. For the standard PSO, the cognitive and social scaling parameters are both equal to 2.0. The inertia weight ω decreases linearly from 0.9 to 0.2 with the increase of the iterations. The basic GA adopts non-linear ranking selection, arithmetic crossover and uniform mutation operators, where the selection probability P_s is 0.2, crossover probability P_c is 0.75, and mutation probability P_m is 0.1. The basic EO adopts adaptive Lévy Mutation with the scaling factor $\gamma = 1$.

The algorithms were coded in Matlab 8.0 and the simulations were run on an Intel Core i3 2350M 2.3GHz with 2 GB memory capacity. As these algorithms are nondeterministic algorithms, we set a small positive value 1.0×10^{-6} for each test function. In other words, if a solution falls between this value and that of the actual global optimum, the solution is judged to be acceptable. Moreover, each algorithm is tested 30 times independently to obtain reasonable statistical results.

The performances of these algorithms are evaluated by the indicators of control systems mainly include overshoot ($m_p\%$), the rise time (t_r), setting time (t_s) and steady-state error ($e_{ss}\%$). Figure 3 shows the comparative performance of PEO with GA, PSO and EO. Iterative trajectories of PEO-PID parameters are shown in Fig. 3(a) and the step response under different algorithms-based PID controllers is shown in Fig. 3(b).

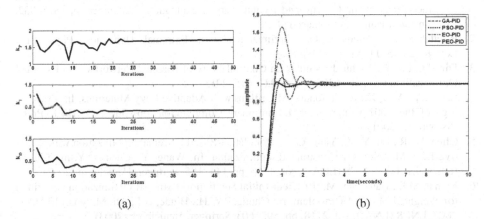

(a)   (b)

Fig. 3. The comparative performance of PEO-PID with GA-PID, PSO-PID and EO-PID

As shown in Fig. 3(a), PEO approaches to an optimal or near-optimal set of proportional, integral and derivative values after 25 times of iterative computation. From Fig. 3(b), the overshoot, setting time and steady-state error obtained by PEO are all smaller than those by GA, PSO and EO. The rise time obtained by PEO is equivalent to that by EO and GA, and smaller than that by PSO. The EO-PID has the largest overshoot, and GA-PID has the longest setting time. Moreover, the PSO-PID obtains well control effect except for the steady-state error.

5   Conclusions

The basic EO deals with a single solution and merely has a mutation operator. In this paper, adaptive knowledge-based mutation is proposed based on the cloud model.

It can extract useful knowledge from the distribution of candidate solutions in the search process and take full advantage of the information to generate the new solution candidates. A population-based EO with knowledge-based mutation is further developed in the light of the warm idea. It combines both attributes in coarse-grained search and fine-grained search. Finally, the proposed PEO is applied to the optimization problem of PID parameter tuning, with PSO and GA as comparisons. The simulation results show that PEO-PID outperforms GA-PID, PSO-PID and EO-PID significantly in the overshoot, setting time and steady-state error, and obtains a satisfactory control effect.

Acknowledgments. The research work reported in this paper is supported by the Fundamental Research Funds for the Central Universities (2013B18614, 2013B09014).

References

1. Boettcher, S., Percus, A.G.: Extremal Optimization: Methods Derived from Co-evolution. In: Proceedings of the Genetic and Evolutionary Computation Conference, pp. 825–832. Morgan Kaufmann, San Francisco (1999)
2. Boettcher, S., Percus, A.G.: Optimization with Extremal Dynamics. Phys. Rev. Lett. 86(23), 5211–5214 (2001)
3. Ding, J., Lu, Y.Z., Chu, J.: Studies on Controllability of Directed Networks with Extremal Optimization. Physica A 392(24), 6603–6615 (2013)
4. Lee, C.Y., Yao, X.: Evolutionary Algorithms with Adaptive Lévy Mutations. In: Proceedings of the 2001 Congress on Evolutionary Computation, pp. 568–575. IEEE Press, Piscataway (2001)
5. Chen, M.-R., Lu, Y.-Z., Yang, G.-k.: Population-Based Extremal Optimization with Adaptive Lévy Mutation for Constrained Optimization. In: Wang, Y., Cheung, Y.-m., Liu, H. (eds.) CIS 2006. LNCS (LNAI), vol. 4456, pp. 144–155. Springer, Heidelberg (2007)
6. Menai, M.E., Batouche, M.: Efficient Initial Solution to Extremal Optimization Algorithm for Weighted MAXSAT Problem. In: Chung, P.W.H., Hinde, C.J., Ali, M. (eds.) IEA/AIE 2003. LNCS (LNAI), vol. 2718, pp. 592–603. Springer, Heidelberg (2003)
7. Sousa, F.L., Vlassov, V., Ramos, F.M.: Generalized Extremal Optimization: An Application in Heat Pipe Design. Appl. Math. Model. 28(10), 911–931 (2004)
8. Zeng, G.Q., Lu, Y.Z., Mao, W.J., Chu, J.: Study on Probability Distributions for Evolution in Modified Extremal Optimization. Physica A 389(9), 1922–1930 (2010)
9. Li, X., Luo, J., Chen, M.R., Wang, N.: An Improved Shuffled Frog-leaping Algorithm with Extremal Optimisation for Continuous Optimisation. Inform. Sciences 192, 143–151 (2012)
10. Li, D.Y., Liu, C.Y., Du, Y., Han, X.: Artificial Intelligence with Uncertainty. Journal of Software 15(11), 1583–1594 (2004)
11. Li, J., Chai, T.Y., Gong, J.K.: Design of PID controller using cross entropy method. Control and Decision 26(5), 794–796 (2011)

A New Magnetotactic Bacteria Optimization Algorithm Based on Moment Migration

Hongwei Mo, Lili Liu, and Mengjiao Geng

Automation College, Harbin Engineering University,
150001 Harbin, China
honwei2004@126.com, {liulilikaoyan,mengjiaogood}@163.com

Abstract. Magnetotactic bacteria is a kind of polyphyletic group of prokaryotes with the characteristics of magnetotaxis that make them orient and swim along geomagnetic field lines. Its distinct biology characteristics are useful to design new optimization technology. In this paper, a new bionic optimization algorithm named Magnetotactic Bacteria Moment Migration Algorithm(MBMMA) is proposed. In the proposed algorithm, the moments of a chain of magnetosomes are considered as solutions. The moments of relative good solutions can migrate each other to enhance the diversity of the MBMMA. It is compared with Genetic Algorithm, Differential Evolution and CLPSO on standard functions problems. The experiment results show that the MBMMA is effective in solving optimization problems. It shows good and competitive performance compared with the compared algorithms.

Keywords: Magnetotactic bacteria, nature inspired computing, moment migration.

1 Introduction

Optimization design problems in engineering solved by the inspiration of biologic systems can be date back to 1940s. In nature, many kinds of animals and insects show the amazing abilities of solving complex problems. Today, a lots of biology inspired algorithms(BIAs) have been proposed to apply for different engineering problems. Swarm Intelligence (SI) is one kind of important BIAs. Some well known SI algorithms include Ant Colony Optimization(ACO)[1], Particle Swarm Optimization (PSO)[2], Artificial Bee Colony (ABC)[3], Artificial Fish Swarm (AFS) [4], Bacterial Foraging Optimization algorithm(BFOA)[5], which mimics the ants, birds, bees, fish and bacteria behaviors, respectively. Among them, PSO had been paid more attention in the past two decades. Although different BIAs had shown different performance in solving optimization problem, 'No free lunch theorem' had told us that there is no universal algorithm which can be better over all possible problems [6]. So it is always necessary for us to develop new algorithm for problem solving.

In nature, magnetotactic bacteria (MTBs) is a special kind of bacteria which have many micro magnetic particles named magnetosome in their bodies. These magnetic particles can generate moments to guide the bacteria to swim along geomagnetic field lines of the earth[7]. Thus, most bacteria can find optimal localities in their

Y. Tan et al. (Eds.): ICSI 2014, Part I, LNCS 8794, pp. 103–114, 2014.

environment to maximize their substrate or energy uptake[8]. Mo had proposed an optimization algorithm named Magnetotactic Bacteria Optimization Algorithm(MBMMA) inspired by the magnetotactic bacteria[9]. But in the MBMMA, the quality and diversity of solutions are mainly subjected to the step of random replacement. The moments don't play the role of regulating the quality of the solutions.

In this paper, a new magnetotactic bacteria moment migration algorithm(MBMMA) is proposed. The moments of magnetosomes in MTBs are considered as the feature values of solutions in the MBMMA. The moments of relative good solutions can migrate to the other solutions. Such a migration strategy can enhance the diversity of solutions in the algorithm and make the algorithm be effective in solving optimization problems.

2   Magnetotactic Bacteria Moment Migration Algorithm

For the MTBs, the efficiency of the magnetotactic response greatly depends on the coordinated movement of each cell's flagellae and the total magnetic dipole moment, which in turn depends on the relative orientation of the magnetosome chains to each other and their polarity distribution. Each cell carries a remanent magnetic moment, the direction of which is given by the orientation of the magnetosome-chain axis and its magnetic polarity[10]. If each cell is to align its magnetosome chain parallel to the other ones, with the same polarity would yield the most efficient swimming way for living. For the algorithm, we consider this state as finding the optimal solution. The interaction energy between different chains in different cells make MTBs strive for better living. The basic optimization process inspired by magnetotactic bacteria can be seen in[9]. In the following we briefly describe he basic operators and the main steps of MBMMA. MBMMA mainly has three steps and three main operators including moment generation, moment migration, moment replacement.

2.1   Interaction Distance

In the algorithm, each solution is looked as a cell containing a magnetosome chain. Before obtaining the interaction energy of cells, the distance d_{ir} of two cells x_i and x_r calculated as follows:

$$d_{i,r} = x_i - x_r \tag{1}$$

Thus, we can get a distance matrix $D = [d_{1,r}, d_{2,r}, ... d_{i,r}, ..., d_{N,r}]'$, where r is a randomly selected integer in $[1, N]$. N is the size of cell population.

2.2   Moments Generation

Based on the distances among cells, the interaction energy e_i between two cells is defined as

$$e_i(t) = \left(\frac{d_{ij}(t)}{1 + c_1 * D_{i,r} + c_2 * d_{pq}(t)} \right)^3 \qquad (2)$$

where t is the current generation number, c_1 and c_2 are constants, d_{ij} is one element of distance matrix D. d_{pq} is a randomly selected element from D. P and r are randomly integers in [1,N]. $q \in [1, J]$ stands for one randomly selected dimension. J is the dimension of a cell. $D_{i,r}$ stands for the Euclidean distance between two cells x_i, x_r.

After obtaining interaction energy, the moments m_i are generated as follows:

$$m_i(t) = \frac{e_i(t)}{B} \qquad (3)$$

where B is a constant .
Then the total moments of a cell is regulated as follows:

$$x_{i,j}(t) = x_{i,j}(t) + m_{r,q}(t) * rand \qquad (4)$$

where $m_{r,q}$ is randomly selected element from m_i. $rand$ is a random number in interval (0,1).

2.3   Moments Migration

After moments generation, the moments migration is realized as follows.
    If $rand > 0.5$, the moments in the cell migrate as follows:

$$x_{ij}(t+1) = x_{rj}(t) \qquad (5)$$

Otherwise,

$$x_{ij}(t+1) = x_{ij}(t) + (x_{cbest,q}(t) - x_{i,q}(t)) \cdot rand \qquad (6)$$

where $x_{cbest,q}$ is the qth dimension of the best individual in the current generation.

2.4   Moments Replacement

After the moments migration, some worse moments are replaced by the following way:

$$x_i(t+1) = m_{r,q}(t) * ((rand(1, J) - 1) * rand(1, J)) \qquad (7)$$

where $m_{r,q}$ is the qth dimension of m_r. r is a randomly integer in [1,N]. $q \in [1, J]$ stands for one randomly selected dimension. $rand(1, J)$ is a random vector with J dimensions.

Generally, A pseudo code of MBMMA is as follows:

    I. Data Structures: Define the simple bounds, determination of algorithm parameters.

      II. Initialization: Randomly create the initial population in the search space.

      III. While stop criteria is not met

        for $i = 1: N$

          interaction distance according to(1)

        end

        for $i = 1: N$

          moments generation according to (2),(3) and (4)

        End

        sort the population according to fitness

        for $i = 1: N$

          moments migration according to (5) and (6)

        end

        sort the population according to fitness

        for $i = 1: N/2$

          moments MTS replacement to (7)

        end

      VI. End while

In the algorithm, the value of test benchmark function is used as the fitness.

3 Simulation Results

To analyze the performance of MBMMA, the experiments are carried out on 10 benchmark functions. These benchmark functions are widely used in evaluating global numerical optimization algorithms. In this section, the benchmark functions are presented firstly. Secondly, the parameter settings of MBMMA and the algorithms chosen for comparison are presented. Finally, the simulation results obtained from different experimental studies are analyzed and discussed.

3.1 Benchmark Functions

A short description of 10 benchmark functions is shown in Tables 1. These test functions, can be classified into two groups. The first six functions $f_1 - f_6$ are unimodal functions. The unimodal functions here are used to test if MBMMA can maintain the fast-converging feature compared with the other methods. The next four functions $f_7 - f_{10}$ are multimodal functions with many local optima. These functions can be used to test the global search ability of the algorithm in avoiding premature convergence.

Initial range, formulation, characteristics, the dimensions and parameters setting of these problems are listed in Tables 1. In Tables 1, characteristics of each function are given under the column titled C. In this column, M means that the function is multi-modal, while U means that the function is unimodal. If the function is separable, abbreviation S is used to indicate this specification. Letter N refers that the function is non-separable. Dimensions of the problems we used can be found in Tables 1 under the column titled D.

Table 1. Classical test functions used in experiments

Function	Range	D	C	Formulation
f_1 : Sphere	[-100, 100]	30	US	$f(x)=\sum_{i=1}^{n}x_i^2$
f_2 : Schwefel2.22	[-10, 10]	30	UN	$f(x)=\sum_{i=1}^{n}\lvert x_i\rvert+\prod_{i=1}^{n}\lvert x_i\rvert$
f_3 : Schwefel1.2	[-100, 100]	30	UN	$f(x)=\sum_{i=1}^{n}(\sum_{j=1}^{i}x_j)^2$
f_4 : Step	[-100, 100]	30	US	$f(x)=\sum_{i=1}^{n}(\lfloor x_i+0.5\rfloor)^2$
f_5 : Quartic	[-1.28, 1.28]	30	US	$f(x)=\sum_{i=1}^{n}ix_i^4+random[0,1)$
f_6 : Rosenbrock	[-30, 30]	30	UN	$f(x)=\sum_{i=1}^{n-1}[100(x_{i+1}-x_i^2)^2+(x_i-1)^2]$
f_7 : Rastrigin	[5.12, 5,12]	30	MS	$f(x)=\sum_{i=1}^{n}[x_i^2-10\cos(2\pi x_i)+10]$
f_8 : Generalized Schwefel	[-500, 500]	30	MS	$f(x)=\sum_{i=1}^{n}-x_i\sin(\sqrt{\lvert x_i\rvert})$
f_9 : Griewank	[-600, 600]	30	MN	$f(x)=\frac{1}{4000}\sum_{i=1}^{n}x_i^2-\prod_{i=1}^{n}\cos\left(\frac{x_i}{\sqrt{i}}\right)+1$
f_{10} : Ackley	[-32, 32]	30	MN	$f(x)=-20\exp\left(-0.2\sqrt{\frac{1}{n}\sum_{i=1}^{n}x_i^2}\right)$ $-\exp\left(\frac{1}{n}\sum_{i=1}^{n}\cos(2\pi x_i)\right)+20+e$

3.2   Experiments Settings

In all experiments, during each run, a maximum number of 3000 generations is used. To reduce statistical errors, each test is repeated 30 times independently and the mean results are used in the comparisons. In order to make a fair comparison, the population size for the algorithms is uniformly set to 40.

The other specific parameters of the MBMMA and the other compared algorithms are given below:

GA Settings[11]: In our experiments, we employ a real number coded standard GA having evaluation, fitness scaling, seeded selection, random selection, crossover, mutation and elite units. Single point crossover operation with the rate of 0.8 is employed. Mutation operation restores genetic diversity lost during the application of reproduction and crossover. Mutation rate in our experiments is 0.01.

DE Settings[12]: In DE, F is a real constant which affects the differential variation between two solutions and set to 0.5 in our experiments. Value of crossover rate, which controls the change of the diversity of the population, is chosen to be 0.9.

CLPSO Settings[13]: In our experiments, cognitive and social components are both set to1.49445. Inertia weight, which determines how the previous velocity of the particle influences the velocity in the next iteration, is linearly from 0.9 to 0.2.

MBMMA setting: In the MBMMA, only the magnetic field B needs to be set up as a parameter, B =0.5.

3.3   Experimental Results and Discussions

The compared results on test functions are listed in Tables 2-3, which are in terms of the mean, best and standard deviation of the solutions obtained in the 30 independent runs by each algorithm.

In order to determine whether the results obtained by MBMMA are statistically different from the results generated by other algorithms, the nonparametric Wilcoxon rank sum tests[14][15][16]are conducted between the MBMMA results and the best result achieved by the other three algorithms for each problem. The h values presented in the Tables 2-3 are the results of Wilcoxon rank sum tests. An h value of one indicates that the performances of the two algorithms are statistically different with 95% certainty, whereas h value of zero implies that the performances are not statistically different.

"+", "=", "-" mean that MBMMA is significantly better, equal and significantly worse, respectively, when compared with other algorithms.

sig–better: The number of test functions on which MBMMA obtains significantly better results.

sig–worse: The number of test functions on which MBMMA obtains significantly worse results.On unimodal functions $f_1 - f_6$, it is relatively easy to locate the global optimum. Therefore, we focus on comparing the performance of the algorithms in terms of solution accuracy.Table 2 presents the mean and the best fitness values yielded by MBMMA and the compared methods after 30 runs. From the results, we observe that for unimodal functions, the proposed MBMMA achieves the highest accuracy or has equal performace with the compared methods. In general, it clearly demonstrates the superior performance of MBMMA to the compared methods. According to the results of Wilcoxon rank sum tests shown in Table 2, the differences between the results obtained by the MBMMA and the other algorithms are statistically significant. MBMMA significantly outperforms three other algorithms on f_1, f_3 , f_5 and f_6.

For f_2 , MBMMA and GA both can obtain the best results and significantly outperform DE and CLPSO. MBMMA has better performance on f_4 compared with DE and CLPSO, and the differences between the results obtained by the MBMMA and GA are not statistically significant. Overall, MBMMA performs better than the compared methods on unimodal functions.On multimodal functions, the global optimum is more difficult to locate. Therefore, in the comparison, we study the accuracy and reliability of MBMMA and the other compared methods.

Table 2. Statistical results on Unimodal Functions obtained by GA, DE, CLPSO and MBMMA

Func.		GA	DE	CLPSO	MBMMA
f_1	Mean	2.3437	1.3287e-45	242.3895	0
	Dev	2.8695	1.8614e-45	110.8980	0
	median	1.8846	5.0650e-46	218.3507	0
	best	0.1391	5.2834e-49	136.6229	0
	worst	9.2630	5.4661e-45	474.8157	0
	h	1+	1+	1+	\
f_2	Mean	0	7.0104e-28	5.6391	0
	Dev	0	1.0003e-27	1.8057	0
	median	0	2.9829e-28	5.4325	0
	best	0	6.6737e-29	3.4951	0
	worst	0	3.2508e-27	9.2423	0
	h	0=	1+	1+	\
f_3	Mean	7.4996e+03	9.6243e-05	1.2308e+04	0
	Dev	1.9150e+03	9.3179e-05	3.1485e+03	0
	median	7.1364e+03	4.4838e-05	1.2019e+04	0
	best	4.9104e+03	1.6823e-06	7.8782e+03	0
	worst	1.1441e+04	2.5282e-04	1.8806e+04	0
	h	1+	1+	1+	\
f_4	Mean	0.5556	7.2222	322.1111	0
	Dev	0.7265	13.6178	130.5703	0
	median	0	2	291	0
	best	0	0	138	0
	worst	2	41	594	0
	h	0=	1+	1+	\
f_5	Mean	0.2010	0.0091	0.2356	3.5985e-0
	Dev	0.0862	0.0056	0.0547	3.4716e-0
	median	0.1824	0.0077	0.2240	2.2675e-0
	best	0.0960	0.0043	0.1802	6.6784e-0
	worst	0.3576	0.0227	0.3526	9.0295e-0
	h	1+	1+	1+	\
f_6	Mean	98.9783	30.4482	2.2177e+03	28.7057
	Dev	45.0647	23.0222	3.0672e+04	0.0012
	median	107.3990	23.2319	1.2884e+04	28.7064
	best	30.7317	0.0426	3.0484e+03	28.7040
	worst	172.4143	69.8208	1.0211e+05	28.7068
	h	1+	1+	1+	\
sig–better		4	6	6	
sig–worse		0	0	0	

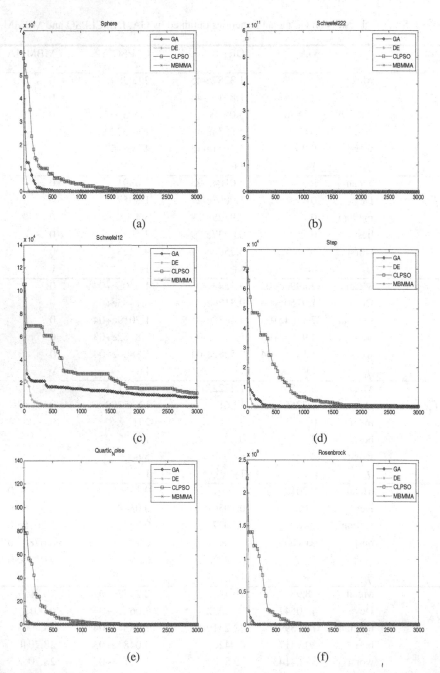

Fig. 1. Convergence curves of six functions (a)Sphere (b) Schwefel2.22 (c) Schwefel1.2 (d) Step (e) Quartic (f) Rosenbrock

Table 3. Statistical results on Multimodal Functions obtained by GA, DE, CLPSO and MBMMA

Func.		GA	DE	CLPSO	MBMMA
f_7	Mean	0	18.2237	46.3823	0
	Dev	0	7.2519	4.8799	0
	median	0	14.9244	43.8456	0
	best	0	8.9546	41.2034	0
	worst	0	32.6665	53.8777	0
	h	0=	1+	1+	\
f_8	Mean	-1.2569e+04	-1.0220e+04	-1.0431e+04	-5.5238e+03
	Dev	0.1030	777.4842	450.1157	31.5695
	median	-1.2569e+04	-1.0545e+04	-1.0310e+04	-5.5359e+03
	best	-1.2569e+04	-1.1329e+04	-1.1195e+04	-5.5376e+03
	worst	-1.2569e+04	-9.1166e+03	-9.8085e+03	-5.4404e+03
	h	1-	1-	1-	\
f_9	Mean	0.0271	1.2724e-47	3.0849	0
	Dev	0.0108	3.6248e-47	1.7968	0
	median	0.0236	4.2497e-49	2.7662	0
	best	0.0149	5.2876e-51	0.9647	0
	worst	0.0504	1.0935e-46	6.6591	0
	h	1+	1+	1+	\
f_{10}	Mean	1.0530	7.0067e-15	5.8445	-8.8818e-16
	Dev	0.4122	2.3685e-15	0.7683	0
	median	0.7878	6.2172e-15	5.7866	-8.8818e-16
	best	0.7171	6.2172e-15	4.7406	-8.8818e-16
	worst	1.7258	1.3323e-14	6.9621	-8.8818e-16
	h	1+	1+	1+	\
sig–better		2	3	3	
sig–worse		1	1	1	

"+", "=", "-" mean that MBMMA is significantly better, equal and significantly worse, respectively, when compared with other algorithms.

sig–better: The number of test functions on which MBMMA obtains significantly better results.

sig–worse: The number of test functions on which MBMMA obtains significantly worse results.

Comparisons of solution accuracy on multimodal functions are given in Table 3. According to the results of Wilcoxon rank sum tests, MBMMA performs significantly better than three compared methods on f_9 and f_{10}. For f_7, MBMMA and GA both can obtain the best results and significantly outperform DE and CLPSO. For f_8, MBMMA shows significantly worse performance compared with the other three methods. Overall, MBMMA performs better than the compared methods on multimodal functions except on f_8.

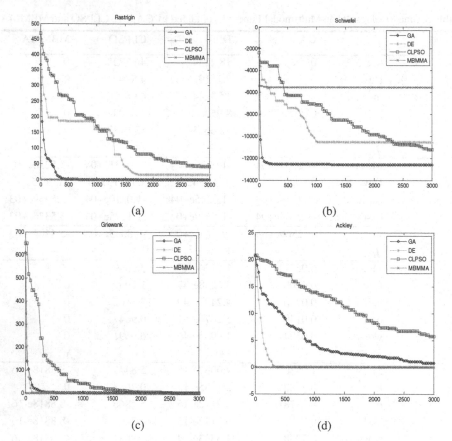

Fig. 2. Convergence curves of six functions (a) Rastrigin (b) Generalized Schwefel (c) Griewank (d) Ackley

In Figure 1 and Figure 2, it can be seen that MBMMA has the fastest convergence speed than all the other three algorithms except on function Generalized Schwefel. CLPSO has the lowest speed.

In total, as seen from the results, MBMMA achieves better performance than compared methods in terms accuracy of global optima for unimodal as well as multi-modal functions except Schwefel function. MBMMA produces better quality of optima and there are significant differences among MBMMA and the compared algorithms.

4   Conclusions

In this paper, we proposes a new Magnetotactic Bacteria Moment Migration Algorithm (MBMMA), which is based on the original idea of magnetotactic bacteria optimization algorithm(MBOA). MBMMA adopts energy function to produce moments. And the obtained moments are used to obtain problem solutions. The moments can migrate among different solutions in each generation. MBMMA has simple procedure and is easy to implement. MBMMA is compared with 3 optimization algorithms including GA, DE and CLPSO. The experimental results show that it is effective in solving optimization problems. In future, MBMMA will be improved to solve more complex problems including constrained optimization, multi-objective optimization and some real engineering problems.

Acknowledgements. This work is partially supported by the National Natural Science Foundation of China under Grant No.61075113, the Excellent Youth Foundation of Heilongjiang Province of China under Grant No. JC201212, the Fundamental Research Funds for the Central Universities No.HEUCFX041306 and Harbin Excellent Discipline Leader, No.2012RFXXG073.

References

1. Dorigo, M., Manianiezzo, V., Colorni, A.: The Ant System: Optimization by a Colony of Cooperating Agents. IEEE Trans. Sys. Man and Cybernetics 26, 1–13 (1996)
2. Kennedy, J., Eberhart, R.: Particle Swarm Optimization. In: IEEE Int. Conf. on Neural Networks, Piscataway, NJ, pp. 1942–1948 (1995)
3. Tereshko, V.: Reaction–diffusion Model of a Honeybee Colony's Foraging Behaviour. In: Deb, K., Rudolph, G., Lutton, E., Merelo, J.J., Schoenauer, M., Schwefel, H.-P., Yao, X. (eds.) PPSN 2000. LNCS, vol. 1917, pp. 807–816. Springer, Heidelberg (2000)
4. Bastos, F., Carmelo, J.A., Lima, N., De Fernando, B.: A Novel Search Algorithm Based on Fish School Behavior. In: IEEE Int. Conf. on Systems, Man, and Cybernetics, Singapore, pp. 32–38 (2002)
5. Müeller, S., Marchetto, J., Airaghi, S., Koumoutsakos, P.: Optimization Based on Bacterial Chemotaxis. IEEE Trans on Evolutionary Computation 6, 16–29 (2002)
6. Wolpert, D.H., Macready, W.G.: No Free Lunch Theorems for Optimization. IEEE Trans. on Evolutionary Computation 1, 67–82 (1997)
7. Faivre, D., Schuler, D.: Magnetotactic Bacteria and Magnetosomes. Chem. Rev. 108, 4875–4898 (2008)
8. Mitchell, J.G., Kogure, K.: Bacterial Motility: Links to the Environment and a Driving Force for Microbial Physics. FEMS Microbiol. Ecol. 55, 3–16 (2006)
9. Mo, H.W., Xu, L.F.: Magnetotactic Bacteria Optimization Algorithm for Multimodal Optimization. In: IEEE Symposium on Swarm Intelligence (SIS), Sinpore (2013)
10. Michael, W., Leida, G.A., Alfonso, F.D., et al.: Barros Magnetic Optimization in a Multicellular Magnetotactic Organism. Biophysical Journal 92, 661–670 (2007)
11. Beyer, H.G.: The Theory of Evolution Strategies. Springer, Heidelberg (2001)
12. Storn, R., Price, K.: Differential Evolution-a Simple and Efficient Heuristic for Global Optimization over Continuous Spaces. Journal of Global Optimization 11, 341–359 (1997)
13. Liang, J.J., Qin, A.K., Suganthan, P.N., Baskar, S.: Comprehensive Learning Particle Swarm Optimizer for Global Optimization of Multimodal Functions. IEEE Trans. Evolut. Comput. 10, 281–295 (2006)
14. García, S., Fernández, A., Luengo, J.: A Study of Statistical Techniques and Performance Measures for Genetics-Based Machine Learning: Accuracy and Interpretability. Soft Comput. Fusion Found. Methodol. Appl. 13, 959–977 (2009)

15. Cai, Y.Q., Wang, J.H., Yin, J.: Learning-enhanced Differential Evolution for Numerical Optimization. Soft Comput. 16, 303–330 (2012)
16. Derrac, J., García, S., Molina, D., Herrera, F.: A Practical Tutorial on the Use of Nonparametric Statistical Tests as a Methodology for Comparing Evolutionary and Swarm Intelligence Algorithms. Swarm Evol. Comput. 1, 3–18 (2011)

A Magnetotactic Bacteria Algorithm
Based on Power Spectrum for Optimization

Hongwei Mo, Lili Liu, and Mengjiao Geng

Automation College, Harbin Engineering University,
150001 Harbin, China
honwei2004@126.com, {liulilikaoyan,mengjiaogood}@163.com

Abstract. Magnetotactic bacteria is one kind of bacteria with magnetic particles called magnetosomes in its body. The magnetotactic bacteria move towards the ideal living conditions under the interaction between magnetic field produced by the magnetic particles chain and that of the earth. In the paper, a new magnetotactic bacteria algorithm based on power spectrum (PSMBA) for optimization is proposed. The candidate solutions are decided by power spectrum in the algorithm. Its performance is tested on 8 standard functions problems and compared with the other two popular optimization algorithms. Experimental results show that the PSMBA is effective in optimization problems and has good and competitive performance.

Keywords: Magnetotactic bacteria algorithm, power spectrum, optimization.

1 Introduction

Optimization is prevalent in almost every field of science and engineering, ranging from profit maximization in economics to signal interference minimization in electrical engineering. Engineering problems with optimization objectives are often difficult and time consuming, and the application of nature or biology-inspired algorithms in combination with the conventional optimization methods has been very successful in the last several decades.

The major goal for the developments of any optimization technique is accurate solution with less complexity. Since 1960s genetic algorithm(GA) has been playing a dominant role in the optimization world [1]. However, the limitation of GA of getting trapped in local minima and need for more computational time forced the researchers to search for more efficient optimization techniques. Since 1980s, more and more Nature inspired Algorithms (NIAs) were developed following GA[2]. The most famous algorithms are ant colony optimization [3] and particle swarm optimization [4]. In recent years, many new bio-inspired computing methods were proposed, such as immune clone selection algorithm [5], artificial bees colony optimization[6], artificial fish

Y. Tan et al. (Eds.): ICSI 2014, Part I, LNCS 8794, pp. 115–125, 2014.

swarm optimization (AFSO) [7], bacterial forging algorithms [8], biogeography-based optimization algorithm [9], firefly algorithm (FA) [10], brain storm optimization (BSO)[11] and so on. Their inspiration sources can be found directly from their names.

The ability of exploration and exploitation and accuracy of global optima are the major criteria for NIAs. Researchers worldwide are striving for improving the existing methods for seeking the best optimal solution.

In this paper, we propose a new optimization technique named as Power Spectrum Based Magnetotactic Bacteria Algorithm which is based on the model of power spectra of the magnetic field noise produced by Nonmotile bacteria Brownian rotation in zero magnetic field. The process of PSMBA follow through power spectrum calculation, bacteria rotation, bacteria swimming and bacteria replacement. It is tested on various ten standard benchmark functions for ensuring the efficiency of the proposed algorithms.

2   Power Spectrum Based Magnettactic Bacteria Algorithm

2.1   Basic Principles of PSMBA

Magnetotactic bacteria occur widely in natural sediments from both marine and freshwater habitats. They produce intracellular, membrane-bounded magnetite (Fe^3O^4.etc.) particles and synthesize a kind of magnetite colloids, a fairly narrow size distribution of particles of colloidal size and specific crystallographic orientations characterize the mineral particles and their enveloping membrane, together called magnetosomes (MTS), which are typically arranged in the form of one or several chains and impart a permanent magnetic dipole moment to the bacterium [12].

Chemla[13] described a study of the dynamics of magnetotactic bacteria in an aqueous medium. They presented a measurement of the magnetic field fluctuations produced by an ensemble of nonmotile bacteria. From this they determined the average rotational drag experienced by the cell and the average magnetic dipole moment. By detecting the fluctuations from an ensemble of motile cells, they illustrated the vastly different dynamics at play.

The power spectrum of the magnetic field noise is generated by the Brownian rotation of nonmotile cells in a zero magnetic field. Translational Brownian motion occurs over very long time scales and does not contribute to the measured noise. To prevent settling of the bacteria, the sample is agitated every 30 min. Calculations show that a single nonmotile bacterium produces a Lorentzian power spectrum

$$S_B(f) \propto \frac{2\tau_0}{1+(2\pi f \tau_0)^2} . \qquad (1)$$

where τ_0 is the characteristic time scale of the Brownian rotation, and $1/2\pi\tau_0$ is the knee frequency. The time τ_0 is $\alpha/2k_BT$, where the rotational drag coefficient α is strongly dependent on the size of the bacterium. Modeling the bacteria crudely as rigid cylinders of length L and diameter d (neglecting the flagellum), one finds

$$\alpha = \frac{\pi \eta L^3}{3}\left[\ln\left(\frac{L}{d}\right) - \gamma\right]^{-1}.$$ (2)

where η is the viscosity of the medium and $\gamma \approx 0.662 - 0.92d\,/\,L$. The average rotational drag coefficient to be $\bar{\alpha} = (3.9 \pm 0.3) \times 10^{-20}\,N \cdot m$ for $T = 290K$. In particular, Chemla observed vibrational and rotational modes of the bacteria associated with the flagellar motor and deduce their frequency and amplitude. So the vibrational and rotational motions of the magnetic bacteria have effect on the measured magnetic field noise power spectrum $S_B(f)$.

From the view of the point of magnetic bacteria dynamics, power spectrum $S_B(f)$ is an important factor which reflects the magnetic bacteria motion states, that is, whether they are straggling for better living conditions. This schema can be used as a way of solving optimization problem.

2.2   Procdures of PSMBA

The PSMBA mainly includes three steps: power spectrum calculation, bacteria rotation, bacteria swimming.

Power Spectrum Calculation. Randomly select two bacteria x_{r1}, x_{r2} $(r1, r2 \in (1, 2, ..., N))$ in the population.

In the algorithm, L is defined as

$$L^t_{r1, r2} = x^t_{r1, j} - x^t_{r2, j}.$$ (3)

Based on(2), for simplifying calculation, $\ln(\frac{L}{d}) \rightarrow \frac{L}{d}$. In each generation, we get

$$\alpha^t_{i, j} = \frac{\pi \eta (L^t_{r1, r2})^3}{3}\left[\frac{L^t_{r1, r2}}{d} - \gamma\right]^{-1}.$$ (4)

where η and γ are constants.

$$\tau^t_{ij} = \frac{\alpha^t_{ij}}{2kT}.$$ (5)

where k and T are contants.

Suppose $S_i = (s_{i1}, s_{i2}, ..., s_{im})$ is the power spectrum of a single bacterium, we have

$$s^t_{ij} = \frac{2\tau^t_{ij}}{1 + (2\pi * t * \tau^t_{ij}))^2}.$$ (6)

Bacteria Rotation. Rotation is an important motion mode for a single bacterium. In the algorithm, a single bacterium rotates as follows:

$$x_{i,j}^{t+1} = x_{p,q}^{t} + s_{p,q}^{t} * rand \quad . \tag{7}$$

p is randomly selected from $[1, N]$. N is the size of bacteria population. $q \in [1, m]$ stands for a randomly selected dimension. m is the dimension of a bacterium. $rand$ is a random number in interval(0,1).

Bacteria Swimming. After rotation, the bacteria swim as a whole as follows: If rand>0.5 , the bacteria swim as follows:

$$x_{i}^{t+1} = x_{i}^{t} + rand * (1 - rand) \quad . \tag{8}$$

Otherwise, they swim as follows :

$$x_{i}^{t+1} = x_{i}^{t} + rand * (rand - 1) \quad . \tag{9}$$

Bacteria Replacement. After rotation and swimming, some bacteria are not fit for living in further. Some worse bacteria are replaced in an ensemble of bacteria. In the algorithm, based on the fitness of each bacterium, half of them are replaced as follows:

$$x_{ij}^{t+1} = s_{p,q}^{t} * ((rand - 1) * rand) \quad . \tag{10}$$

The pseudo random code of PSMBA is described as follows:

Initialization
for t =1 to   NP
        for i =1: N
        calculate power spectrum   according to (3) (4) (5)and
(6)
          bacterium rotation according to (7)
   end
        for i =1: N
        bacteria swimming according to (8) and (9)
        end
      evaluate the quality of X_i according to fitness
       bacteria replacement according to (10)
        end
---

3   Simulation Results

3.1   Parameter Settings and Benchmark Functions

In this section, in order to analyze the performance of PSMBA, the experiments of minimization are carried out on 8 benchmark functions. All of the algorithms used for comparison are the basic version for fair comparison.

In the experiments, the three algorithms are the basic ones without any improvement. In order to make a fair comparison, the values of the common parameters for the algorithms such as population size and generation are chosen to be the same. The population size is set to 30. During each run, a maximum number of 1000 generations is used. The other specific parameters of algorithms are given below:

GA Settings: In our experiments, we employ a real number coded standard GA having evaluation, fitness scaling, seeded selection, random selection, crossover, mutation and elite units. Single point crossover operation with the rate of 0.8 is employed. Mutation operation restores genetic diversity lost during the application of reproduction and crossover. Mutation rate in our experiments is 0.01.

DE Settings[14]: In DE, F is a real constant which affects the differential variation between two solutions and set to 0.5 in our experiments. Value of crossover rate, which controls the change of the diversity of the population, is chosen to be 0.9.

PSMBA setting: The parameter settings of the PSMBA are $\eta = 0.0001$, K=0.0138, T =29, γ −0.762.

Table 1. Benchmark functions used in experiments

Function	Range	D	C	Formulation
f_1: Step	[-100, 100]	30	US	$f(x)=\sum_{i=1}^{n}(\lfloor x_i+0.5\rfloor)^2$
f_2: Sphere	[-100, 100]	30	US	$f(x)=\sum_{i=1}^{n}x_i^2$
f_3: Quartic	[-1.28, 1.28]	30	US	$f(x)=\sum_{i=1}^{n}ix_i^4+random[0,1)$
f_4: Schwefel2.22	[-10, 10]	30	UN	$f(x)=\sum_{i=1}^{n}\lvert x_i\rvert+\prod_{i=1}^{n}\lvert x_i\rvert$
f_5: Rastrigin	[-5.12, 5,12]	30	MS	$f(x)=\sum_{i=1}^{n}[x_i^2-10\cos(2\pi x_i)+10$
f_6: Schaffer	[-100, 100]	2	MN	$f(x)=0.5+\dfrac{\sin^2(\sqrt{x_1^2+x_2^2})-0.5}{(1+0.001(x_1^2+x_2^2))^2}$
f_7: Griewank	[-600, 600]	30	MN	$f(x)=\dfrac{1}{4000}\sum_{i=1}^{n}x_i^2-\prod_{i=1}^{n}\cos\left(\dfrac{x_i}{\sqrt{i}}\right)+1$
f_8: Ackley	[-32, 32]	30	MN	$f(x)=-20\exp\left(-0.2\sqrt{\dfrac{1}{n}\sum_{i=1}^{n}x_i^2}\right)$ $-\exp\left(\dfrac{1}{n}\sum_{i=1}^{n}\cos(2\pi x_i)\right)+20+e$

In order to characterize the type of problems for which the algorithm is suitable and test the performance of PSMBA, we used 10 benchmark problems in order to compare the performance of these algorithms. A short description of 10 benchmark functions is shown in Tables 1. These benchmark functions are widely used in evaluating global numerical optimization algorithms.

Initial range, formulation, characteristics, the dimensions and parameters setting of these problems are listed in Tables 1. In Tables 1, characteristics of each function are given under the column titled C. In this column, M means that the function is multimodal, while U means that the function is unimodal. If the function is separable, abbreviation S is used to indicate this specification. Letter N refers that the function is non-separable. Dimensions of the problems we used can be found in Tables 1 under the column titled D.

Multimodal functions are used to test the ability of algorithms getting rid of local minima. If the exploration process of an algorithm is poor, it cannot search the whole space efficiently and it gets stuck at the local minima. The dimensionality of the search space is an important issue with the problem. Separable function have interrelation among their variables. Therefore, non-separable functions are more difficult than the separable functions.

3.2   Experimental Results

The compared results on 8 functions are listed in Tables 2. The results are shown in Tables 2 in terms of the mean and standard deviation, median, best and worst of the solutions obtained in the 30 independent runs by each algorithm. And the best results of each function are highlighted in boldface.

Table 2. Statistical results of 30 runs obtained byGA, DE, FA and PSMBA algorithms(Mean: Mean of the Best Values, StdDev: Standard Deviation of the Best Values)

Func.	Min			DE	GA	PSMBA
f_1	0		Mean	24.3333	56.4667	**0**
			StdDev	38.8741	21.3101	**0**
			median	10	53.5000	**0**
			best	1	22	**0**
			worst	155	114	**0**
f_2	0		Mean	9.0762e-06	75.5128	**3.4535e-21**
			StdDev	4.1731e-05	34.9841	**5.3211e-22**
			median	4.9260e-08	67.6962	**3.5448e-21**
			best	1.4007e-11	29.6809	**2.3497e-21**
			worst	2.2879e-04	181.2149	**4.3191e-21**
f_3	0		Mean	0.0528	1.5883	**1.1140e-04**
			StdDev	0.0357	2.7530	**7.8158e-05**
			median	0.0500	0.6946	**1.1283e-04**
			best	0.0152	0.1039	**7.2391e-06**
			worst	0.1945	12.9754	**2.6636e-04**
f_4	0		Mean	4.1811e-05	1.5440	**2.3541e-09**
			StdDev	1.9426e-04	1.4077	**1.6878e-10**
			median	4.7228e-07	1.2569	**2.3921e-09**
			best	1.0595e-08	0	**1.9180e-09**
			worst	0.0011	5.5465	**2.6962e-09**

Table 2. (*Continued*)

f_5	0	Mean	66.8015	0	0
		StdDev	40.0307	0	0
		median	61.7680	0	0
		best	17.0802	0	0
		worst	173.7519	0	0
f_6	0	Mean	9.7159e-04	0.0047	0
		StdDev	0.0030	0.0051	0
		median	0	0	0
		best	0	0	0
		worst	0.0097	0.0103	0
f_7	0	Mean	1.7321e-08	0.6632	**2.4429e-26**
		StdDev	4.4353e-08	0.3812	**3.4510e-27**
		median	1.2479e-10	0.6803	**2.4386e-26**
		best	1.0014e-12	0.1138	**1.7831e-26**
		worst	2.0730e-07	1.8282	**3.0857e-26**
f_8	0	Mean	1.8766	4.2105	**1.3520e-10**
		StdDev	1.5001	0.7290	**1.1023e-11**
		median	1.6438	4.1857	**1.3626e-10**
		best	6.8596e-06	2.8520	**1.1096e-10**
		worst	7.7000	6.2631	**1.5192e-10**

As seen from Tables 2, PSMBA is better than DE and GA on 8 test functions, respectively. GA has the same performance with PSMBA on Rastrigin function. In general, the results in Tables 2 demonstrate that PSMBA is better than DE and GA in terms of quality of the final solutions.

To sum up, the results of nonparametric statistical tests consistently demonstrate that PSMBA is significantly better than the competitors.

In order to show the PSMBA's performance in further, we plot the convergence graphs of 8 well known benchmark functions. Fig. 1, 2, 3, 4, 5, 6, 7, 8 show the progress toward optima value in terms of number of generations versus fitness for the 8 benchmark functions, respectively.

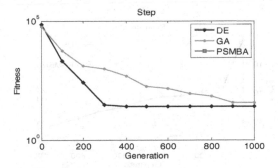

Fig. 1. Experimental results on f1

Based on Fig. 1 and the results presented in Table 2, it is clear that PSMBA has better performance than DE and GA in terms of accuracy and convergence.

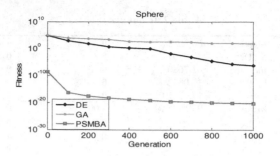

Fig. 2. Experimental results on f2

Based on Fig. 2 and Table 2, PSMBA outperforms all the other algorithms on f2. The comparison results for f2 show that PSMBA has the best performance on convergence and accuracy .

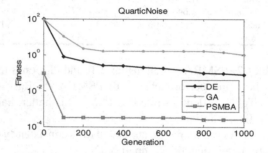

Fig. 3. Experimental results on f3

Based on Fig. 3 and Table 2, the comparison results for f3 show that PSMBA has the best performance on convergence and accuracy. And PSMBA outperforms all the other algorithms on f3.

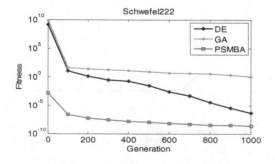

Fig. 4. Experimental results on f4

Based on Fig. 4 and the results presented in Table 2, PSMBA is better than DE and GA on f4.

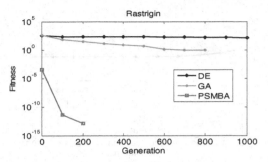

Fig. 5. Experimental results on f5

Based on Fig. 5 and Table 2, PSMBA and GA achieve comparatively better performance in terms of accuracy of global optima and convergence for f5. But PSMBA can achieve the better performance within fewer generations compared with GA.

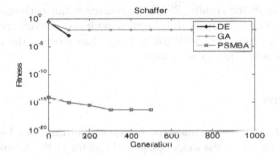

Fig. 6. Experimental results on f6

Based on Fig. 6 and Table 2, PSMBA is better than GA on f6. PSMBA and DE achieve comparatively better performance in terms of accuracy of global optima and convergence for f6.

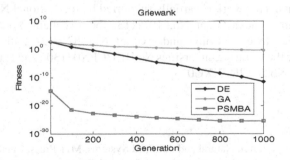

Fig. 7. Experimental results on f7

Based on Fig. 7 and Table 2, the lowest value of fitness obtained by PSMBA for f7 indicates that the PSMBA has better performance than compared techniques in terms of accuracy and convergence.

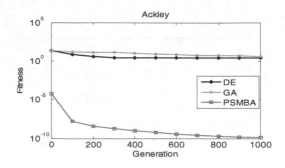

Fig. 8. Experimental results on f8

Based on Fig. 8 and Table 2, PSMBA is better than the compared algorithms on f8. PSMBA is consistent for getting better quality global minima for multimodal function with minimum generations.

In total, as seen from the results, PSMBA achieves better performance than compared methods in terms accuracy of global optima with fast convergence for unimodal as well as multimodal functions. PSMBA is faster than most of the compared algorithms and produces better quality of optima.

4   Conclusions

In this paper, we propose a new magnetic bacteria algorithm based on power spectrum-PSMBA. It is based on the model of power spectra of the magnetic field noise produced by Nonmotile bacteria Brownian rotation in zero magnetic field. It is compared with classical optimization algorithms GA and DE. The experimental results show that it is effective in solving optimization problems. In future, it will be used to solve more complex problems including constrained optimization, multi-objective optimization and some real engineering problems.

Acknowledgements. This work is partially supported by the National Natural Science Foundation of China under Grant No.61075113, the Excellent Youth Foundation of Heilongjiang Province of China under Grant No. JC201212, the Fundamental Research Funds for the Central Universities No.HEUCFX041306 and Harbin Excellent Discipline Leader, No.2012RFXXG073.

References

1. Holland, J.H.: Adaption in Natural and Artificial Systems. MIT Press, Cambridge (1975)
2. Mo, H.W.: Research Development on Nature Inspired Computing. Journal of Intelligence Systems 6, 11–13 (2011) (in Chinese)
3. Dorigo, M., Manianiezzo, V., Colorni, A.: The Ant System: Optimization by A Colony of Cooperating Agents. IEEE Trans. Sys. Man and Cybernetics 26, 1–13 (1996)

4. Eberhart, R.C., Kennedy, J.: A New Optimizer Using Particle Swarm Theory. In: Proceedings of the Sixth International Symposium on Micro Machine and Human Science, Nagoya, Japan, pp. 39–43 (1995)
5. De Castro, L.N., Von Zuben, F.J.: Learning and Optimization Using the Clonal Selection Principle. IEEE Transactions on Evolutionary Computation 6, 239–251 (2002)
6. Tereshko, V.: Reaction–diffusion Model of a Honeybee Colony's Foraging Behaviour. In: Deb, K., Rudolph, G., Lutton, E., Merelo, J.J., Schoenauer, M., Schwefel, H.-P., Yao, X. (eds.) PPSN 2000. LNCS, vol. 1917, pp. 807–816. Springer, Heidelberg (2000)
7. Bastos-Filho, C.J.A., De Lima Neto, F.B.: A Novel Search Algorithm Based on Fish School Behavior. In: IEEE Int. Conf. on Systems, Man, and Cybernetics, Singapore, pp. 32–38 (2002)
8. Passino, K.M.: Biomimicry of Bacterial Foraging for Distributed Optimization and Control. IEEE Control Systems Magazine 22(3), 52–67 (2002)
9. Simon, D.: Biogeography-based Optimization. IEEE Trans. on Evolutionary Computation 12(6), 702–713 (2008)
10. Yang, X.S.: Nature-Inspired Metaheuristic Algorithms. Luniver Press (2008)
11. Shi, Y.: Brain storm optimization algorithm. In: Tan, Y., Shi, Y., Chai, Y., Wang, G. (eds.) ICSI 2011, Part I. LNCS, vol. 6728, pp. 303–309. Springer, Heidelberg (2011)
12. Faivre, D., Schuler, D.: Magnetotactic Bacteria and Magnetosomes. Chem. Rev. 108, 4875–4898 (2008)
13. Chemla, Y.R., Grossman, H.L., Lee, T.S., Clarke, J., Adamkiewicz, M., Buchanan, B.B.: A New Study of Bacterial Motion: Superconducting Quantum Interference Device Microscopy of Magnetotactic Bacteria. Biophysical Journal 76, 3323–3330 (1999)
14. Cai, Y.Q., Wang, J.H., Yin, J.: Learning-enhanced Differential Evolution for Numerical Optimization. Soft Comput. 16, 303–330 (2012)

A Proposal of PSO Particles' Initialization for Costly Unconstrained Optimization Problems: ORTHOinit

Matteo Diez[1], Andrea Serani[1,2], Cecilia Leotardi[1], Emilio F. Campana[1], Daniele Peri[3], Umberto Iemma[2], Giovanni Fasano[4], and Silvio Giove[5]

[1] Natl. Research Council–Marine Tech. Research Inst. (CNR-INSEAN), Italy
{(matteo.diez,emiliofortunato.campana}@cnr.it, c.leotardi@insean.it
[2] University 'Roma Tre', Department of Engineering, Italy
{andrea.serani,umberto.iemma}@uniroma3.it
[3] Natl. Research Council–Ist. Applicazioni del Calcolo 'M. Picone', (CNR-IAC), Italy
d.peri@iac.cnr.it
[4] University Ca'Foscari of Venice, Department of Management, Italy
fasano@unive.it
[5] University Ca'Foscari of Venice, Department of Economics, Italy
sgiove@unive.it

Abstract. A proposal for particles' initialization in PSO is presented and discussed, with focus on costly global unconstrained optimization problems. The standard PSO iteration is reformulated such that the trajectories of the particles are studied in an extended space, combining particles' position and speed. To the aim of exploring effectively and efficiently the optimization search space since the early iterations, the particles are initialized using sets of orthogonal vectors in the extended space (orthogonal initialization, ORTHOinit). Theoretical derivation and application to a simulation-based optimization problem in ship design are presented, showing the potential benefits of the current approach.

Keywords: Global Optimization, Derivative-free Optimization, Deterministic PSO, Particles' Initial Position and Velocity.

1 Introduction

In this paper we consider the solution of the global unconstrained optimization problem

$$\min_{x \in \mathbb{R}^n} f(x), \tag{1}$$

where $f : \mathbb{R}^n \to \mathbb{R}$ is *continuous* and possibly *nondifferentiable*. In particular, we aim at detecting a global minimum x^* of (1), satisfying $f(x^*) \leq f(x)$, for any $x \in \mathbb{R}^n$. Of course we assume that (1) admits solution, which may be guaranteed under mild assumptions on $f(x)$ (e.g., $f(x)$ is coercive with $\lim_{\|x\| \to \infty} f(x) = +\infty$). Furthermore, we also assume that the function $f(x)$ is computationally *expensive*, which possibly discourages the use of asymptotically convergent methods (i.e., iterative methods that only eventually ensure convergence properties to stationary points).

Y. Tan et al. (Eds.): ICSI 2014, Part I, LNCS 8794, pp. 126–133, 2014.
© Springer International Publishing Switzerland 2014

PSO is an iterative method for global optimization, based on updating a population of points (namely *particles*). Preliminary numerical tests performed in [1], for 60 standard problems, suggest that the initial choice of particles' position/velocity may affect significantly the performance of PSO, giving motivation for further investigation of apparently basic and well stated issues for PSO (such as the choice of the initial particles' position/velocity).

Herein, we study the trajectories of particles in an extended space, so that analytical indications will be available in order to suggest the setting of initial particles position and velocity. The current approach starts from considering the results obtained in [2,3,4,5,6], though our partial conclusions in Sections 4-5 are, to the best of our knowledge, novel in the literature.

In Sections 2-3 we first recall the reformulation of PSO, detailed in [7] and [8], while the new proposal of this paper is in Sections 4-5. Conclusions and future work are presented in Section 6. In the following, 'I' indicates the identity matrix and 'e_i' is the i-th unit vector. The Euclidean norm is simply indicated by $\| \cdot \|$.

2   A Reformulation of PSO Iteration

Consider the following standard (and complete) iteration of PSO:

$$\begin{cases} v_j^{k+1} = \chi \left[w^k v_j^k + c_j^k r_j^k (p_j^k - x_j^k) + c_g^k r_g^k (p_g^k - x_j^k) \right], & k \geq 0, \\ x_j^{k+1} = x_j^k + v_j^{k+1}, & k \geq 0, \end{cases} \tag{2}$$

where $j = 1, ..., P$ indicates the j-th particle, P is finite, v_j^k and x_j^k are the *velocity* and the *position* of particle j at step k, and the coefficients $\chi, w^k, c_j^k, r_j^k, c_g^k, r_g^k$ are bounded. Finally, p_j^k and p_g^k satisfy

$$p_j^k = \operatorname*{argmin}_{0 \leq h \leq k} \{f(x_j^h)\}, \quad j = 1, \ldots, P, \qquad p_g^k = \operatorname*{argmin}_{0 \leq h \leq k, \ j=1,\ldots,P} \{f(x_j^h)\}. \tag{3}$$

We can also generalize (2) and the analysis in this paper by assuming that possibly the velocity v_j^{k+1} depends on all the terms $p_h^k - x_j^k$, $h = 1, \ldots, P$, obtaining the so called *fully informed* PSO (FIPS) [9]. This corresponds to allow a more general *social* contribution in PSO iteration. Notwithstanding the latter choice, we prefer to keep the notation as simple as possible, considering the recurrence (2) as is. Without loss of generality at present we focus on the j-th particle and omit the subscript in the recurrence (2), so that $p_j^k = p^k$ and $v_j^k = v^k$.

Assumption 1. *We assume in (2) that* $c_j^k = c$, $r_j^k = r$ *for any* $j = 1, ..., P$, $c_g^k = \bar{c}$, $r_g^k = \bar{r}$ *and* $w^k = w$, *for any* $k \geq 0$.

Using the latter position the iteration (2) is equivalent to the *dynamic, linear* and *stationary system*[1]

$$X(k+1) = \begin{pmatrix} \chi w I & -\chi(cr + \bar{c}\bar{r})I \\ \chi w I & [1 - \chi(cr + \bar{c}\bar{r})]I \end{pmatrix} X(k) + \begin{pmatrix} \chi(crp^k + \bar{c}\bar{r}p_g^k) \\ \chi(crp^k + \bar{c}\bar{r}p_g^k) \end{pmatrix}, \tag{4}$$

---

[1] See also [7,8], whose terminology and symbols are simply reported in this brief section and in the next one. Then, in Section 4 we extend the latter results to our purposes.

where

$$X(k) = \begin{pmatrix} v^k \\ x^k \end{pmatrix} \in \mathbb{R}^{2n}, \qquad k \geq 0.$$

The sequence $\{X(k)\}$ identifies a trajectory in the state space $\mathbb{R}^{2n}$, and since (4) is a linear and stationary system, we may consider the *free response* $X_L(k)$ and the *forced response* $X_F(k)$ of the trajectory $\{X(k)\}$. Then, considering (4) we explicitly obtain, at step $k \geq 0$, $X(k) = X_L(k) + X_F(k)$, where

$$X_L(k) = \Phi(k)X(0), \qquad X_F(k) = \sum_{\tau=0}^{k-1} H(k-\tau)U(\tau), \tag{5}$$

and, after some calculations

$$\Phi(k) = \begin{pmatrix} \chi w I & -\chi(cr + \bar{c}\bar{r})I \\ \chi w I & [1 - \chi(cr + \bar{c}\bar{r})]\, I \end{pmatrix}^k, \quad H(k-\tau) = \begin{pmatrix} \chi w I & -\chi(cr + \bar{c}\bar{r})I \\ \chi w I & [1 - \chi(cr + \bar{c}\bar{r})]\, I \end{pmatrix}^{k-\tau-1}, \tag{6}$$

$$U(\tau) = \begin{pmatrix} \chi(crp^k + \bar{c}\bar{r}p_g^k) \\ \chi(crp^k + \bar{c}\bar{r}p_g^k) \end{pmatrix}. \tag{7}$$

A remarkable observation from the latter formulae is that $X_L(k)$ in (5) uniquely depends on the initial point $X(0)$, and is not affected by the vector p_g^k. On the contrary, $X_F(k)$ in (5) is independent of $X(0)$, being strongly dependent on p_g^k. This implies that the quantities $X_L(k)$ and $X_F(k)$ can be separately computed.

3   Structural Properties of Matrix $\Phi(k)$ and Computation of $X_L(k)$

In order to simplify our analysis, provided that Assumption 1 holds, hereafter we consider the following position in (6)

$$a = \chi w, \qquad \omega = \chi(cr + \bar{c}\bar{r}). \tag{8}$$

Now, we first recall (see [7,2,4]) that in order to ensure necessary conditions which avoid divergence of the trajectories of particles, the relations

$$0 < |a| < 1, \qquad 0 < \omega < 2(a + 1) \tag{9}$$

must hold. Moreover, the only two eigenvalues λ_1 and λ_2 of $\Phi(1)$ coincide if and only if $\omega = (1 \pm \sqrt{a})^2$. Thus, if $\omega \neq (1 \pm \sqrt{a})^2$ then the results in [7] can be applied, so that $X_L(k) = [\Phi(1)]^k X(0)$ can be computed by simply introducing the eigenvalues λ_1 and λ_2 of $\Phi(1)$, yielding the formula

$$[\Phi(1)]^k X(0) = \begin{bmatrix} \gamma_1(k)v^0 - \gamma_2(k)x^0 \\ \gamma_3(k)v^0 - \gamma_4(k)x^0 \end{bmatrix}, \tag{10}$$

where

$$\gamma_1(k) = \frac{\lambda_1^k(a-\lambda_2) - \lambda_2^k(a-\lambda_1)}{\lambda_1 - \lambda_2} \qquad \gamma_2(k) = \frac{\omega(\lambda_1^k - \lambda_2^k)}{\lambda_1 - \lambda_2}$$

$$\gamma_3(k) = \frac{(\lambda_1^k - \lambda_2^k)(a-\lambda_1)(a-\lambda_2)}{\omega(\lambda_1 - \lambda_2)} \qquad \gamma_4(k) = \frac{\lambda_1^k(a-\lambda_1) - \lambda_2^k(a-\lambda_2)}{\lambda_1 - \lambda_2}.$$

4   A Novel Starting Point for Particles in PSO

In this section we study a novel strategy to possibly improve the efficiency of PSO, based on the idea of widely exploring the search space in the early iterations, while maintaining the PSO iteration (2). As stated in the Introduction, our analysis seems more promising when each function evaluation is particularly expensive and time resources are scarce, so that a few iterations of PSO are allowed. We start our analysis using the reformulation in Section 2, in order to impose a novel condition for the choice of initial particles' position/velocity (namely the next relation (14)).

Consider two particles, namely particle j and particle h, such that $1 \leq j \neq h \leq P$; using the theory in Section 2 we can consider their trajectories in the space $\mathbb{R}^{2n}$, so that their initial position and free response are respectively given by (see (10))

Particle j:

$$
X(0)^{(j)} = \begin{pmatrix} v_j^0 \\ x_j^0 \end{pmatrix} \Rightarrow X_L(k)^{(j)} = [\varPhi(1)]^k X(0)^{(j)} = \begin{bmatrix} \xi_1(k)^{(j)} v_j^0 - \xi_2(k)^{(j)} x_j^0 \\ \xi_3(k)^{(j)} v_j^0 - \xi_4(k)^{(j)} x_j^0 \end{bmatrix}, \quad (11)
$$

Particle h:

$$
X(0)^{(h)} = \begin{pmatrix} v_h^0 \\ x_h^0 \end{pmatrix} \Rightarrow X_L(k)^{(h)} = [\varPhi(1)]^k X(0)^{(h)} = \begin{bmatrix} \xi_1(k)^{(h)} v_h^0 - \xi_2(k)^{(h)} x_h^0 \\ \xi_3(k)^{(h)} v_h^0 - \xi_4(k)^{(h)} x_h^0 \end{bmatrix}, \quad (12)
$$

where $\xi_i(k)^{(h)}$, $i = 1, \ldots, 4$ (similarly for $\xi_i(k)^{(j)}$) coincide with $\gamma_i(k)$, $i = 1, \ldots, 4$, if $\omega \neq (1 \pm \sqrt{a})^2$.

Now, observe that at iteration k the velocity v^k of a particle may be regarded as a search direction from the current position x^k. Thus, we can be interested to find out conditions on the initial position of the particles, in order to possibly guarantee the orthogonality of particles' velocity at any iteration $k \geq 0$. The latter fact is expected to possibly favour a better exploration in $\mathbb{R}^{2n}$. However, the latter condition is very tough to impose, without strongly modifying PSO iteration (2). Nonetheless, following the idea in Section 6 of [7], we can attempt for any k to impose the orthogonality of the free responses $\{X_L(k)^{(j)}\}$. In particular, numerical efficiency is ensured by the fact that it is possible to set the initial position and velocity of n particles, in such a way that the corresponding free responses $X_L(k)^{j_1}, \ldots, X_L(k)^{j_n}$ satisfy

$$
\left[X_L(k)^{j_i} \right]^T \left[X_L(k)^{j_h} \right] = 0, \qquad \forall j_i, j_h \in \{j_1, \ldots, j_n\}, \quad i \neq h.
$$

In order to generalize the latter idea we observe here that what really matters is the orthogonality of the search directions of the particles, and possibly not the orthogonality of the entire free responses. On this guideline, here we study the initial position and velocity of $2n$ particles, so that for any k the corresponding free responses $X_L(k)^{j_1}, \ldots, X_L(k)^{j_{2n}}$ satisfy for any $1 \leq j \neq h \leq 2n$ (see (11)-(12))

$$
\left[\xi_1(k)^{(j)} v_j^0 - \xi_2(k)^{(j)} x_j^0 \right]^T \left[\xi_1(k)^{(h)} v_h^0 - \xi_2(k)^{(h)} x_h^0 \right] = 0. \quad (13)
$$

I.e., only the first n entries of the free responses of particles j and h, corresponding to the velocity, are orthogonal. After some computation, the latter relation is equivalent to the conditions

$$0 = \left[\left(\xi_1(k)^{(j)}I \quad -\xi_2(k)^{(j)}I\right)\begin{pmatrix} v_j^0 \\ x_j^0 \end{pmatrix}\right]^T \left[\left(\xi_1(k)^{(h)}I \quad -\xi_2(k)^{(h)}I\right)\begin{pmatrix} v_h^0 \\ x_h^0 \end{pmatrix}\right]$$

$$= \begin{pmatrix} v_j^0 \\ x_j^0 \end{pmatrix}^T \begin{bmatrix} \sigma_1 I & \sigma_2 I \\ \hat{\sigma}_2 I & \sigma_3 I \end{bmatrix} \begin{pmatrix} v_h^0 \\ x_h^0 \end{pmatrix}, \tag{14}$$

where $\sigma_1 = \xi_1(k)^{(j)}\xi_1(k)^{(h)}$, $\sigma_2 = -\xi_1(k)^{(j)}\xi_2(k)^{(h)}$, $\hat{\sigma}_2 = -\xi_2(k)^{(j)}\xi_1(k)^{(h)}$, $\sigma_3 = \xi_2(k)^{(j)}\xi_2(k)^{(h)}$.

Observe that setting the same parameters ω and a in (8) for all the particles (i.e., for any $k \geq 0$ we have $\xi_1(k)^{(j)} = \xi_1(k)^{(h)} = \xi_1(k)$ and $\xi_2(k)^{(j)} = \xi_2(k)^{(h)} = \xi_2(k)$) the matrix

$$\Lambda = \begin{bmatrix} \sigma_1 I & \sigma_2 I \\ \hat{\sigma}_2 I & \sigma_3 I \end{bmatrix} = \begin{bmatrix} \sigma_1 I & \sigma_2 I \\ \sigma_2 I & \sigma_3 I \end{bmatrix} \tag{15}$$

is symmetric and condition (14) indicates that the vectors

$$\begin{pmatrix} v_j^0 \\ x_j^0 \end{pmatrix}, \qquad \begin{pmatrix} v_h^0 \\ x_h^0 \end{pmatrix}. \tag{16}$$

must be *mutually conjugate* (see also [10,11] for a reference). The first relevant property induced by the introduction of conjugacy is that conjugate vectors are linearly independent. This implies that in case the vectors (16) are mutually conjugate, then not only the velocities of the free responses of the particles are orthogonal (as stated in relation (13)), but the vectors (16) will be also *sufficiently well* scattered in $\mathbb{R}^{2n}$. Now, note that if z_i and z_j are distinct eigenvectors of matrix Λ, respectively associated to the eigenvalues λ_i and λ_j, then we simply have $z_i^T \Lambda z_j = z_i^T (\lambda_j z_j) = \lambda_j z_i^T z_j = 0$, where the last equality follows from the fact that distinct eigenvectors of a symmetric matrix are orthogonal. Thus, the eigenvectors of a symmetric matrix are also mutually conjugate directions with respect to that matrix. As a consequence, in order to satisfy condition (14) it suffices to compute the eigenvectors of (15), and set the vectors in (16) as proportional to the latter eigenvectors. After some computation we have for the corresponding $2n$ eigenvectors $u_1^{(i)}$, $u_2^{(i)}$, $i = 1, \ldots, n$, of the matrix in (15) the simple expressions

$$u_1^{(i)} = \begin{pmatrix} -\frac{\sigma_3 - \mu_-}{\sigma_2} e_i \\ e_i \end{pmatrix} \in \mathbb{R}^{2n}, \; u_2^{(i)} = \begin{pmatrix} -\frac{\sigma_3 - \mu_+}{\sigma_2} e_i \\ e_i \end{pmatrix} \in \mathbb{R}^{2n}, \; i = 1, \ldots, n, \tag{17}$$

where $\mu_{\mp} = \left[(\sigma_1 + \sigma_3) \mp \sqrt{(\sigma_1 + \sigma_3)^2 - 4(\sigma_1\sigma_3 - \sigma_2^2)}\right]/2$ are the eigenvalues of matrix Λ.

The last result implies that in order to satisfy the conditions (14), for any $1 \leq j \neq h \leq P \leq 2n$, it suffices to set the initial particle position and velocity (respectively of the i-th and $(n + i)$-th particle) according with the following (ORTHOinit) initialization

$$\begin{pmatrix} v_i^0 \\ x_i^0 \end{pmatrix} = (-1)^{\alpha_{1i}} \rho_i^1 u_1^{(i)}, \qquad \rho_i^1 \in \mathbb{R} \setminus \{0\}, \ \alpha_{1i} \in \{0, 1\}, \ i = 1, \dots, n \qquad (18)$$

$$\begin{pmatrix} v_{n+i}^0 \\ x_{n+i}^0 \end{pmatrix} = (-1)^{\alpha_{2i}} \rho_i^2 u_2^{(i)}, \qquad \rho_i^2 \in \mathbb{R} \setminus \{0\}, \ \alpha_{2i} \in \{0, 1\}, \ i = 1, \dots, n. \qquad (19)$$

Recalling that the choice of the coefficients $\rho_i^1, \rho_i^2, \ i = 1, \dots, n$, in (18)-(19) is arbitrary, we conclude that, in case in (8) $\omega \neq (1 \pm \sqrt{a})^2$, then

- when $P \leq 2n$, the choice (18)-(19) of the particles position and velocity guarantees that *the components of velocity of the free responses* of the particles will be *orthogonal at any iteration $k \geq 0$*, provided that Assumption 1 holds (i.e., no randomness is used in PSO, as in DPSO [12]);
- in case $P > 2n$, the user can adopt the choice (18)-(19) of the particles position and velocity for $2n$ particles, while setting the remaining $(P - 2n)$ particles arbitrarily.

5   Numerical Results

Numerical results for both test functions and a simulation-based design of a high-speed catamaran are performed using DPSO, setting the parameters according to Assumption 1.

Numerical experiments are performed to assess the initialization (18)-(19) on the 60 test functions in [1], varying the initialization of the swarm. Three approaches are used. Specifically, in the first approach the swarm is initialized as shown in [1], using $4n$ particles distributed (following a Hammersley sequence sampling, HSS) over the variables domain and its boundary. The second approach, following the guidelines in the previous sections, consists of using two orthogonal sets of $2n$ particles each (ORTHOinit initialization). For the third approach, an orthogonal set of $2n$ particles is added to the initialization set of the first approach. As shown in Figures 1 and 2, using two ORTHOinit sets of $2n$ particles gives the best performance in terms of evaluation metric Δ (see [1]), for test functions with both $n < 10$ and $n \geq 10$ design variables.

For the catamaran design optimization, the parent hull considered is that of the *Delft* catamaran, a concept ship used for experimental and numerical benchmarks. The optimization problem is taken from [13] and solved by means of stochastic radial-basis functions interpolation [14] of high-fidelity URANS simulations. Six design variables control global shape modifications, based on the Karhunen-Loève expansion of the shape modification vector [15]. The objective is the reduction of the total resistance in calm water at Froude number equal to 0.5. Figure 3 plots the decrease of the objective function in the first twenty DPSO iterations (early iterations), comparing the reference implementation given in

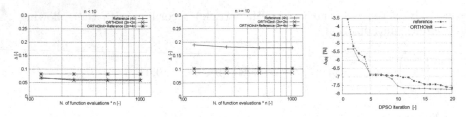

Fig. 1. Test funct., $n < 10$ **Fig. 2.** Test funct., $n \geq 10$ **Fig. 3.** *Delft* catamaran

[1], based on HSS initialization of particles over domain and boundary, with the method in this paper ($4n$ particles are used). ORTHOinit shows a faster progress than the reference implementation, confirming the effectiveness of the present method when a reduced number of iterations is allowed.

6   Conclusions and Future Work

With respect to [7] the theory above yields a guideline for the choice of $2n$ (and not just n) particles' initial position/velocity. This was expected to provide a more powerful tool (as numerical results seem to confirm) for the exploration of the search space. Moreover, the above theory proposes a particles' initialization in PSO which is related to the space dimension n. Though no specific conclusion seems to be drawn by the latter observation, note that most of the exact derivative-free methods for smooth problems, as well as gradient-based methods for continuously differentiable functions, show some analogies. We are persuaded that in our framework an adaptive criterion might be advisable, in order to restart the position and velocity of the particles after a given number of iterations. The latter criterion can indeed monitor the norm $\|X_L(k)^{(j)}\|, j = 1, \ldots, 2n$ (see also Section 5 of [7]), of the free response of particles. When the latter quantity approaches zero, a restart would re-impose orthogonality among the free responses of the particles, using the theory in Section 4.

Acknowledgements. The work of M. Diez is supported by the US Navy Office of Naval Research, NICOP Grant N62909-11-1-7011, under the administration of Dr. Ki-Han Kim and Dr. Woei-Min Lin. The work of A. Serani, C. Leotardi, G. Fasano and E.F. Campana is supported by the Italian Flagship Project RIT-MARE, coordinated by the Italian National Research Council and funded by the Italian Ministry of Education, within the National Research Program 2011-2013.

References

1. Serani, A., Diez, M., Leotardi, C., Peri, D., Fasano, G., Iemma, U., Campana, E.F.: On the use of synchronous and asynchronous single-objective deterministic Particle Swarm Optimization in ship design problems. In: Proceeding of OPT-i, International Conference on Engineering and Applied Sciences Optimization, Kos Island, Greece, June 4-6 (2014)

2. Clerc, M., Kennedy, J.: The Particle Swarm - Explosion, Stability, and Convergence in a Multidimensional Complex Space. IEEE Transactions on Evolutionary Computation 6(1) (2002)
3. Ozcan, E., Mohan, C.K.: Particle Swarm Optimization: Surfing the Waves. In: Proceedings of the 1999 IEEE Congress on Evolutionary Comnputation, pp. 1939–1944. IEEE Service Center, Piscataway (1999)
4. Trelea, I.C.: The particle swarm optimization algorithm: convergence analysis and parameter selection. Information Processing Letters 85, 317–325 (2003)
5. Van den Berg, F., Engelbrecht, F.: A Study of Particle Swarm Optimization Particle Trajectories. Information Sciences Journal (2005)
6. Poli, R.: The Sampling Distribution of Particle Swarm Optimisers and their Stability, Technical Report CSM-465, University of Essex (2007)
7. Campana, E.F., Fasano, G., Pinto, A.: Dynamic analysis for the selection of parameters and initial population, in particle swarm optimization. Journal of Global Optimization 48, 347–397 (2010)
8. Campana, E.F., Diez, M., Fasano, G., Peri, D.: Initial particles position and parameters selection for PSO. In: Tan, Y., Shi, Y., Mo, H. (eds.) ICSI 2013, Part I. LNCS, vol. 7928, pp. 112–119. Springer, Heidelberg (2013)
9. Mendes, R., Kennedy, J., Neves, J.: The fully informed particle swarm: Simpler, maybe better. IEEE Transactions on Evolutionary Computation 8, 204–210 (2004)
10. Hestenes, M.R., Stiefel, E.: Methods of conjugate gradients for solving linear systems. Journal of Research of the National Bureau of Standards 49, 409–435 (1952)
11. Fasano, G., Roma, M.: Iterative Computation of Negative Curvature Directions in Large Scale Optimization. Computational Optimization and Applications 38, 81–104 (2007)
12. Campana, E.F., Liuzzi, G., Lucidi, S., Peri, D., Piccialli, V., Pinto, A.: New Global Optimization Methods for Ship Design Problems. Optimization and Engineering 10, 533–555 (2009)
13. Chen, X., Diez, M., Kandasamy, M., Zhang, Z., Campana, E.F., Stern, F.: High-fidelity global optimization of shape design by dimensionality reduction, metamodels and deterministic particle swarm. Engineering Optimization (in press, 2014), doi: 10.1080/0305215X.2014.895340
14. Volpi, S., Diez, M., Gaul, N.J., Song, H., Iemma, U., Choi, K.K., Campana, E.F., Stern, F.: Development and validation of a dynamic metamodel based on stochastic radial basis functions and uncertainty quantification. Structural Multidisciplinary Optimization (in press, 2014), doi: 10.1007/s00158-014-1128-5
15. Diez, M., He, W., Campana, E.F., Stern, F.: Uncertainty quantification of Delft catamaran resistance, sinkage and trim for variable Froude number and geometry using metamodels, quadrature and Karhunen-Loève expansion. Journal of Marine Science and Technology 19(2), 143–169 (2014), doi:10.1007/s00773-013-0235-0

An Adaptive Particle Swarm Optimization within the Conceptual Framework of Computational Thinking

Bin Li[1], Xiao-lei Liang[2], and Lin Yang[2]

[1] Fujian University of Technology, Department of Transportation,
Minhou University Town. Xueyuan Road No.3, 350118 Fuzhou, China
[2] Wuhan University of Technology, School of Logistics Engineering,
Heping Road. No. 1040, 430063 Wuhan, China
`{mse2007_lb,liangxiaolei,lyang}@whut.edu.cn`

Abstract. The individual learning and team working is the quintessence of particle swarm optimization (PSO). Within the conceptual framework of computational thinking, the every particle is seen as a computing entity and the whole bird community is a generalized distributed, parallel, reconfigurable and heterogeneous computing system. Meanwhile, the small world network provides a favorable tool for the topology structure reconfiguration among birds. So a learning framework of distributed reconfigurable PSO with small world network (DRPSOSW) is proposed, which is supposed to give a systemative approach to improve algorithms. Finally, a series of benchmark functions are tested and contrasted with the former representative algorithms to validate the feasibility and creditability of DRPSOSW.

Keywords: particle swarm optimization, computational thinking, complex adaptive systems, small world network, computational experiments.

1 Introduction

Particle Swarm Optimization (PSO) was initially proposed in 1995 by James Kennedy and Russell Eberhart as a global stochastic search algorithm that is based on swarm intelligence (SI) and inspired by the social behaviors of fish schooling and birds flocking [1]. Seeing that PSO possesses a clear and favorable biology society background as well as a few parameters and simplicity of implementation, it has been widely associated with the various fields whether in the scientific research or engineering practice.

Nevertheless, the original PSO suffers from the chronic illnesses that is trapped into premature convergence and stagnation easily, especially for the complex high-dimension multimodal problems. Consequently, an enormous amount of algorithm variants have been put forward, and has obtained remarkable achievement [2,3]. The self-adaptation and social learning are the primary focus of improvement measures.

Yet the systemative solution to improve PSO is scarce thus far. This paper extends our previous work [4], and concentrates on the above two issues, and proposes a new PSO learning framework based on computational thinking, namely the distributed

Y. Tan et al. (Eds.): ICSI 2014, Part I, LNCS 8794, pp. 134–141, 2014.

reconfigurable PSO with small world network (DRPSOSW), to prevent from premature convergence and keep the biological diversity of PSO as well.

2   Computational Thinking and Complex Adaptive Systems

Computational thinking involves solving problems, designing systems, and understanding human behavior, by drawing on the concepts fundamental to computer science [5]. The nature of computational thinking is abstraction and automation. The essence of computation is a symbol string f is translated into another symbol string g. Seeing that the team working in SI are the evolutionary process in nature, and the PSO searching procedure is in conformity with the essence of computation too. Hence we intend to discuss and improve PSO with computational thinking.

On the other hand, the core idea of complex adaptive system (CAS) theory is adaptability makes complexity, and the adaptability is the focus conception of CAS. The notion of adaptability is no longer just the survival of the fittest and the principle of use and disuse. It is generalized to be a kind of learning process which takes on different temporal scales notwithstanding. Regarding the above nature of computation, we conclude that the adaptability is no other than a generalized computation.

The living prototype of PSO is just CAS beyond question, and the every bird is an agent that is an intelligent entity with sociality, so the bird community is a multi-agent system (MAS) too. The adaptability in PSO is that the birds can interact on each other and their environments with the explicit initiative and purpose. Hence the population has the ability of gathering experiences and continuous learning. The process of self-learning and self-organization is a computation for the whole community, and it can change the individual behavioral style and population social structure as well.

In addition, the bird community is inquired into according to the essence of computation from the following theoretical perspectives. For one thing, the PSO is provided with the inherent parallelism as it is based on population, so the community is considered as a parallel processing system (PPS). For another, although that the birds carry out the flight and foraging is a continuous process, that does not hinder our seeing the population and the concomitant searching behavior as a discrete time dynamic system (DEDS) by the circumscriptions of the multifarious discrete events.

Given all that, the bird community for PSO is the recombination of PPS, DEDS, MAS and CAS from the different visual angles, which all reflect the computational essence ultimately. Meanwhile, the PSO is a singular blend of deterministic and stochastic by the computational thinking. That also gives us a systemative conceptual framework and theoretical principles of the improvement approach to PSO.

3   Compound Computing Architecture for PSO

Computational thinking comes from computer science, and the latter provides the former with adequate nutrition and a place to grow steadily. we can elaborate the computing architecture for the swarm to make a precise definition of computational model of PSO, which is supposed to facilitate the further analysis and improvement on algorithms. Due to its own characteristics of particle swarm and the inherent self-learning pattern, we discuss the computing architecture with different aspects.

Above all, the particle cruise in the problem space independently though it is influenced by the companions, so the whole population performs the parallel computing that is a kind of space and data parallelism. The every bird is just a parallel processing element (PPE). The computing paradigm is the multiple instruction stream multiple data stream (MIMD) according to the classical Flynn's Taxonomy. Furthermore, the computation architecture is massive parallel processing (MPP). It means that each bird possesses its own genetic gene, computing resource, memory space and control logic, and a bird can not access the other bird's inner space, which suggests that the population searching can be also seen as a kind of distributed computation in a sense. The every particle is just a distributed processing unit (DPU), and the information interaction among birds are implemented by the social communication network which is a data redistribution process substantially.

Secondly, seeing that the nonuniform distribution of satisfactory solution and population density, and the dynamic relationship among social communication, the neighborhood of every bird should be always changing throughout the community development. If a bird is regarded as a reconfigurable processing unit (RPU) and the population is executed clustering constantly by the Euclidean distance and/or object function fitness values with the population evolutionary. The every temporary cluster can be seen as an array processor, and the community has the ability to carry out the substantial transformation of data path and control flow by the reconfiguration of the population topology structure. Namely, the searching process can be considered as a reconfigurable computing for CAS.

Lastly, as is known to all, that the bird group possesses and preserves the species diversity is very advantageous to the performance of PSO. As a result, the various measures can be introduced into the initialization and evolution of the particle swarm to achieve the objective. Supposing that the every bird is deemed to be a heterogeneous computing unit (HCU), the whole population searching behavior is no other than a kind of heterogeneous computing.

4   Distributed Reconfigurable PSO with Small World Network

4.1   Artificial Society and Small World Network

On the grounds of the above computational thinking, the every bird in PSO is just about a PPE, DPU, RPU and HCU from a theoretical perspective of high performance computer architecture, and then all of them construct a flexible and efficient distributed, parallel, reconfigurable and heterogeneous computing systems. From another point of view, the each particle is a agent with the very limited computational capabilities and storage capacity with insight into MAS and CAS, and consequently the swarm form a relatively simple artificial society, which implies that the computational experiments (CE) can provide a good road to design and implement DRPSOSW. Whether the former view or the latter, the topology structure and synchronous frequency among birds both are of great significance to the performance.

The small world network (SWN) provides a favorable mathematical model for representing the topology structure and synchronous frequency in the community, and the specific reasons are listed below. First of all, the practical social network is just about SWN, which gives us an inspiration to construct an appropriate topological relationship by the law of learning from nature. Next, the speed of information transmission is rather quick in SWN, and the change on a small quantity of connections can have dramatic effects on the network performance. That is propitious to the fast convergence and achieves the new emerging patterns with keeping diversity of individuals simultaneously. Thirdly, the social communication network of the community should be dynamic and reconfigurable in the each iterative search. The SWN can give security to define the case and reconstruct the topology structure among birds well. Fourthly, as a kind of complex network, the SWN is an intermediate between rule network and stochastic network. The distinguishing features of SWN is incorporating deterministic and stochastic into one, which makes for the balance between exploration and exploitation ability of PSO. Finally, the SWN model can be given a precise definition by a few parameters, furthermore, is easily to design and implement with a low computation complexity. That is beneficial for the permeation and integration between PSO and SWN greatly with the conceptual framework and fundamental principles of CAS and MAS.

4.2 Proposed Learning Framework

With the original intention of the reinforcement of social psychology, we propose a new PSO learning framework that is applying computational thinking into PSO in essence that is DRPSOSW accordingly. The specific definition of is given as follows. The movement patterns of a particle in the D dimension problem space can be represented by a set of difference equations that are defined by (1) and (2) clearly, which both constitute the kernel of self-learning and self-organization mechanism.

$$v_{id}(t+1)=\omega v_{id}(t)+c_1 r_{1d}(t)(p_{id}(t)-x_{id}(t))$$
$$+c_2 r_{2d}(t)(p_{nd}(t)-x_{id}(t))+c_3 r_{3d}(t)(p_{sd}(t)-x_{id}(t)) \tag{1}$$

$$x_{id}(t+1)=x_{id}(t)+v_{id}(t+1) \tag{2}$$

In a parallel manner with the original PSO algorithm, the v_{id} and x_{id} stand for the component of velocity vector and position vector of the first d dimension on the first i particle respectively, and the p_{id} is the current optimal position during the first i particle searching process, and the p_{nd} is the current optimal position of the neighborhood of the particle i, and the p_{sd} is the current optimal position of the whole community experience so far. In addition, the ω indicates the inertia weight, the c_1, c_2 and c_3 are the coefficient of cognitive learning, neighborhood learning and social learning separately. Then again, the r_1, r_2 and r_3 are all the independent random numbers which are located between 0 and 1 with the equiprobability.

For one thing, the formula of (1) makes the community to be the formation of the hybrid and flexible computing architecture from the visual angle of computer architecture. If the every agent is seen as a independent computing unit, the whole community is typical of parallel computing, distributed computation, reconfigurable computing and heterogeneous computing simultaneously. That provides a fundamental framework for DRPSOSW, and gives a computational thinking abstraction in narrow sense too. To be specific, the every component in the formula of (1) can be considered as the effect of different computing architecture. The first item is the embodiment of parallel computing, and the second one is the function of distributed computation, and the third one is the role of reconfigurable computing, and the last one is also the action of parallel computing that is the synchronous communication among PPE. The heterogeneous computing is demonstrated by the scaling diversity of evolutionary population.

For another, the community can be seen as a computational society with the social communication distinguishing features of SWN from the viewpoint of artificial life and complex network. That gives a continuously evolving theoretical principles for DRPSOSW, and provides an automation exploration perspective in a wide sense.

4.3   Computing Paradigm and Parameter Tuning

The DRPSOSW is only a learning framework of SI, and there are a lot of algorithm paradigms and parameters both are still to be determined. The cardinal contents are enumerated as follows. Firstly, the sociality, purpose, adaptability and self-learning are the four fundamental characteristics of SI. Furthermore, the implement of purpose, adaptability and self-learning all depend upon the definition of sociality in the final analysis. The definition of SWN including the relevant technological parameters is of decisive significance for the performance of DRPSOSW. Secondly, the first two components in the equation of (1) are the individual learning section, and the last two one are the team learning section. The proportion of both has vast importance to the behavior of the community. In addition, the definition of inertia weight has a certain effect on the performance of DRPSOSW. Thirdly, the overflow handing is a key point in the computer system. A similar event occurs in the DRPSOSW. The overflow of position and velocity are the inevitable outcome when birds move by the equation of (1) and (2). Nevertheless, the overflow of position and velocity is not always a bad thing. That provide a useful way to escape from local optimums. The computing paradigm of both has the important effect on DRPSOSW for that is a necessary and beneficial complement to the conventional cruise. Finally, the PSO is a kind of intelligent optimization algorithm with the combination of deterministic computing framework and probabilistic adaptive strategy in substance, so the random number generating scheme is vitally important for the approach. However, the probability distribution and the related parameters setting are just about a very large scale, multi-objective, multi-constraint, discrete nonlinear, typical NP complete combinatorial optimization problem. That is a big challenge of DRPSOSW too.

5   Computational Experiments

The computational experiments for DRPSOSW is designed and implemented by AnyLogic 6.9.0 and SQL Server 2012. We establish computational experiment by the comprehensive visual angle of PPS, DEDS, MAS and CAS. The computational experimental machine is the DELL Precision M4800 workstation, which is based on the Intel Core i7-4800MQ CPU and 32G memory. Meanwhile, the operating system (OS) is Win 8 Pro which is a 64 bit OS. In addition, the maximum available memory of AnyLogic is 4G, and the computational accuracy reaches up to 1.0E-38.

Some classical benchmark functions, which exhibit the different characteristics and traps, are selected and tested to judge and evaluate the performance of DRPSOSW. The candidates for testing are Sphere, Ronsenbrock, Griewank, Rastrigrin, Schwefel 2.22 and Ackley, which are called F1, F2, F3, F4, F5 and F6 for short later.

The computing paradigm and parameters settings of DRPSOSW are listed below. The population size is 150, and the iterated searching times is 5000, which is intended to demonstrate the evolutionary process of the community clearly with the benefit of slow motion. The coefficient of inertia is computed according to the user-defined pattern (2) in the literature [6] that be aimed at showing the universality of the self-learning framework. The coefficient of cognition is 0.90, and the coefficient of neighborhood is 0.90, and the coefficient of society is 0.80. The r_i all fulfills the uniform distribution of between 0.0 and 1.0. The SWN adopts the classical WS model, and the connections of every agent is 6, and the neighbor link fraction is 0.6.

We execute the benchmark functions test with the dimensions of 50 and 100, and the each group experiments is carried out by the different random number generation seed from 1 to 50 for the same objective function with the identical dimensions. In other words, 600 times CE are executed, which provides a comprehensive assessment with the premise of having and holding repeatability and randomicity simultaneously. The computational experiments result is showed in Table 1, and the convergence rate (CR) means the average optimization generations for the special problem with the given dimensions. We can conclude that DRPSOSW stays ahead of the many state-of-the-art variants obviously, such as LPSO-TVAC[7] and MPSO[8].

Table 1. DRPSOSW computational experiments result

CE Group	Testing Function	Problem Dimension	Average Fitness	Standard Deviations	Convergence Rate	CR Standard Deviations
1	F1	50	0.0	0.0	2825.0	9.8995
2	F1	100	0.0	0.0	3605.25	71.9554
3	F2	50	0.0	0.0	1580.4	22.7772
4	F2	100	0.0	0.0	1590.6	24.4602
5	F3	50	0.0	0.0	2217.2	12.3774
6	F3	100	0.0	0.0	2628.5	17.6777
7	F4	50	0.0	0.0	323.8	76.6629
8	F4	100	0.0	0.0	328.4	21.5592
9	F5	50	0.0	0.0	3695.0	36.7696
10	F5	100	4.10E-12	3.1617E-12	4208.34	145.2389
11	F6	50	1.33E-14	0.0	2622.40	106.3969
12	F6	100	7.37E-14	7.5364E-15	3113.2	54.8471

Thus the DRPSOSW showed promise even though it is only based upon the fundamental mode of SWN and the inertia weight, as it obtains the theoretical optimal value for F1~F4 and an excellent performance for F5 and F6. Those indicates the stable, efficient and robust searching capability. We take F3 with the dimensions of 50 for example to demonstrate the evolutionary process of DRPSOSW. That we select the random number seed is 5, 16, 27, 38 and 49, the five sets of population current best position fitness (PCBPF) evolutionary curve are demonstrated by Fig. 1.

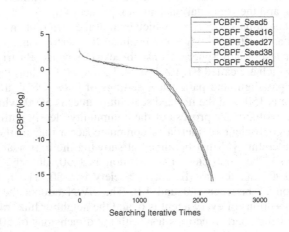

Fig. 1. Population evolutionary process for F3 with the dimensions of 50

Even the DRPSOSW does not achieve the optimal value, it also shows the favorable and stable searching ability whether for the many dimensions or the high dimensions problem space. The distinguishing feature is incarnate by F6 evidently. We select two group experiments for the dimension of 50 and 100 separately, which demonstrate the analogical evolutionary curve. The result is showed by Fig. 2, and the upper part is the CE for the dimension of 50, and the bottom half is the one for the dimension of 100.

6   Conclusions

PSO is a simple and effective SI algorithm, but it is too straight-forward to underline the most arresting features of individual learning and team work that are contained in the social creatures. We treat the living prototype of PSO as a CAS, and then make a precise definition of bird movement patterns by the fusion of computational thinking and SWN to achieve a self-learning and self-adaption framework. That is intended to emphasize the essence of computability, sociality and adaptability of SI to obtain a systemative improving solution to the PSO. However, DRPSOSW is only a conceptual framework, the definition of computing paradigm and the selection of SWN model both are even the more complicated issues than the established work. Much the same thing is found in parameter setting. Those are the further work orientation.

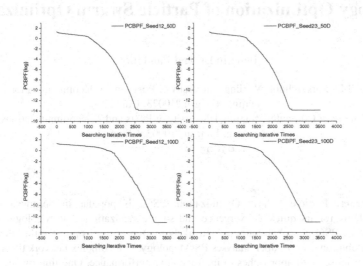

Fig. 2. Population evolutionary process for F6 with the dimensions of 50 and 100

Acknowledgments. This work was partially supported by the National Natural Science Foundation of China Grant No. 61304210, the Founds Project under the Ministry of Education of the PRC for Young people who are devoted to the researches of Humanities & Social Sciences in China Grant No. 11YJC630089, the Natural Science Foundation of Fujian Province in China Grant No. 2012J05108.

References

1. Kennedy, J., Eberhart, R.: Particle swarm optimization. In: IEEE International Conference on Neural Networks, pp. 1942–1948. IEEE Press, New York (1995)
2. Li, X., Yao, X.: Cooperatively coevolving particle swarms for large scale optimization. IEEE Transactions on Evolutionary Computation 16, 210–224 (2012)
3. Chen, W., Zhang, J., Lin, W., et al.: Particle swarm optimization with an aging leader and challengers. IEEE Transactions on Evolutionary Computation 17, 241–258 (2013)
4. Li, B., Li, W.: Simulation based optimization for PSO computational model. Journal of System Simulation 23, 2118–2124 (2011)
5. Wing, J.M.: Computational Thinking. Communications of the ACM 49, 33–35 (2006)
6. Hu, J., Zeng, J.: Selection on inertia weight of particle swarm optimization. Computer Engineering 33, 193–195 (2007)
7. Tripathi, P.K., Bandyopadhyay, S., Pal, S.K.: Multi-objective particle swarm optimization with time variant inertia and acceleration coefficients. Information Sciences 177, 5033–5049 (2007)
8. Ma, G., Zhou, W., Chang, X.: A novel particle swarm optimization algorithm based on particle migration. Applied Mathematics and Computation 218, 6620–6626 (2012)

Topology Optimization of Particle Swarm Optimization

Fenglin Li[1] and Jian Guo[2]

[1] Bell Honors School, Nanjing University of Posts and Telecommunications
Nanjing, Jiangsu 210023, China
[2] College of Computer, Nanjing University of Posts and Telecommunications
Nanjing, Jiangsu 210003, China
guoj@njupt.edu.cn

Abstract. Particle Swarm Optimization (PSO) is popular in optimization problems for its quick convergence and simple realization. The topology of standard PSO is global-coupling and likely to stop at local optima rather than the global one. This paper analyses PSO topology with complex network theory and proposes two approaches to improve PSO performance. One improvement is PSO with regular network structure (RN-PSO) and another is PSO with random network structure (RD-PSO). Experiments and comparisons on various optimization problems show the effectiveness of both methods.

Keywords: particle swarm optimization, topology optimization, complex network.

1 Introduction

Particle swarm optimization (PSO) [1] is one of evolutionary algorithms deriving from observations of bird foraging behavior. Once one bird finds food, an efficient method for other birds to find food is to search in the vicinity of the bird that has found food. This may illustrate an optimizing model: the birds represent a group of searching points of the solution space, the bird that has found food represents the optimum, searching nearby the winner represents a certain updating rule of nodes. Each node is updated according to its location and velocity and that of the current winner. It converges quickly and can be easily realized. Therefore PSO and its improved versions have been applied to various areas to solve practical problems [2] [3] [4].

To improve PSO performance, researchers [5] introduced "inertia weight" to denote the relevance between current status and that after updating. And P. N. Sugant han [6] proposed neighborhood operator approach with linearly-decreasing inertia weight value to adaptively update the group. Asanga et al comes up with a new variant using varying acceleration coefficients [7]. [8] introduces dissipative theory (negative entropy) and develops DPSO which updates velocity and position with two random numbers called "chaos factor".

By analysis of PSO topology, [9] [10] put forward several optimization approaches focusing on specific network structures (e.g. circle, triangle, rectangle, etc.) instead of

Y. Tan et al. (Eds.): ICSI 2014, Part I, LNCS 8794, pp. 142–149, 2014.

the general law of various topologies. Matsushita et al [11] brings up a new method by applying "cooperative coefficient(C)" to help determine neighborhood relations between nodes when updated. It discusses different C values on the performance of the proposed version, but does not give access to general parameter setting rules. Yue-jiao Gong et al has defined a "stagnation coefficient (Sc)" to describe the updating process of nodes that stagnate after a certain number of generations, and applied adaptation mechanism to adjust the randomization and size of neighborhood based on small-world network [12].

As we can easily find that standard PSO has short average path length and simple degree distribution [13]. Meanwhile the topology of standard PSO is fully-connected. This leads to an inevitable problem that particle swarms stop at a local optimum rather than the global one. With some concepts from complex network theory [13], we find that the topology has a large clustering coefficient, which means the swarm is highly converged.

In this paper, we analyze the topology of standard PSO with some complex network parameters and propose two topology variants. One is to introduce a regular network structure (RN-PSO). Each node in RN-PSO is connected to a number of neighbors which are next to the node. The other approach is random topology (RD-PSO), where nodes are connected to a number of random selected neighbors. We apply both approaches to four benchmarks to test their performance.

The remainder of this paper is arranged as follows: Section 2 analyses standard PSO topology with some complex network parameters. We present our two proposed methods in detail in Section 3 and test their performance in Section 4. Section 5 concludes the entire paper.

2   Standard PSO and Its Topology

In standard PSO, each particle has two parameters: location and velocity. The two parameters are updated in every generation according to its own location and velocity and the winner's. We use a vector $X_i=(x_{i1}, x_{i2}, ..., x_{iD})$ to represent the location of particle i, vector $V_i=(v_{i1}, v_{i2}, ..., v_{iD})$ the velocity of the particle, both in D dimensions, and $1 \leq i \leq N$, where N is swarm size. In every generation, each particle updates its location and velocity by the following equations

$$v_{id}(g+1)=wv_{id}(g)+c_1 r_1(l_{id}-x_{id}(g))+c_2 r_2(x_{best}-x_{id}(g)) \qquad (1a)$$
$$x_{id}(g+1)=x_{id}(g)+v_{id}(g+1) \qquad (1b)$$

where w is inertia weight [5] , c_1 and c_2 are acceleration coefficients [7] , and usually $c_1=c_2$. r_1, r_2 are two uniformly distributed random numbers between 0 and 1. l_{id} is the local best position, x_{best} is the global best position so far. $x_{best}-x_{id}(t)$, $l_{id}-x_{id}(t)$ show the distance between current global best and particle i, and the distance between local best and particle x_i, respectively. And g represents the order of generations.

The standard PSO has a topology of centralized pattern, as illustrated in Fig.1. The black circle denotes global best, and the grey circles denote other nodes of the network, the arrows denote how the nodes "learn". We can see that the network structure has pretty short average path length $L=2(N-1)/N$. The global best has direct connection to every node. In this sense, if the network has N nodes, the winner has a degree of $(N-1)$, while the degree of other nodes is one. And clustering coefficient $C=(N-1)/N \rightarrow 1$ [13].

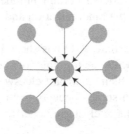

Fig. 1. Topology of standard PSO

3   Optimization of Swarm Structure of PSO

As discussed in Section 2, the topology of standard PSO has inevitable weaknesses in a concentrated network structure, in particular its large clustering coefficient, which constrains strongly a node's searching range. Considering its global-coupling topology, in this section we will propose two optimizing approaches of PSO network structure: RN-PSO (regular network structure) and RD-PSO (random network structure).

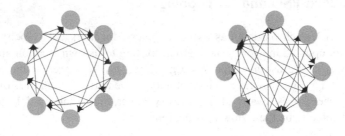

Fig. 2. Topology comparison of RN-PSO and RD-PSO

3.1   PSO with Regular Network Structure: RN-PSO

In this part we will discuss regular network structure and we call it RN-PSO. Each node will "learn" from a number of (i.e., R) orderly selected nodes (i.e., neighbors) to update local best position. The topology might be illustrated by the left graph in Fig.2, where the swarm has 8 particles and arrows from each node denote that it learns from its 3 contiguous neighbors.

In this approach, the network parameters of each node are the same. Nodes are equivalent in the whole network, and degree of each node increases linearly with R. And we can easily know that average path length L decreases and clustering coefficient grows non-linearly. In particular, when $R=1$, the network is a ring, $C=0$. When R comes close to N, the structure becomes tightly-coupled, and we can imagine its performance will go bad.

3.2   PSO with Random Network Structure: RD-PSO

We present another improving method in this subsection: random network structure. Instead of "learning" from a fixed number of contiguous neighbors, each node compares with R randomly selected nodes to update local best position. A simplified topology model with 8 nodes is shown by the right graph in Fig.2, where arrows from each node represent learning from 3 random neighbors in every generation.

As we expect RD-PSO to have better performance, its network topology has advantages that a regular structure does not acquire. Nodes are more flexible and have more chance of a global searching range.

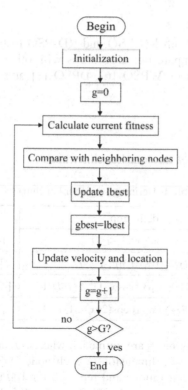

Fig. 3. Flowchart of the proposed algorithms

3.3   Algorithm Description

The algorithms can be realized by the following process.

Step1: Initialize a group of particles. Set this is the first generation of the iterative process, that is, g=0. The position vector of all nodes is $X_i=(x_{i1}, x_{i2}, \ldots, x_{iD})$, velocity vector is $V_i=(v_{i1}, v_{i2}, \ldots, v_{iD})$.

Step2: Calculate the fitness of test function, *currentfit=f(X)*. Let the initializing position and current fitness be the current local best position (*lbest*) and local best fitness (*lbestfit*), respectively. For each node *i*, find the fitness of *R* neighbors (for RN-PSO, the *R* neighbors are orderly following node *i*; for RD-PSO, the *R* neighbors are random selected) and compare.

Step3: Update *lbest*.

Step4: Let global best position (*gbest*) be the best of *lbest*, global best fitness (*gbestfit*) be the corresponding fitness value.

Step5: Update the velocities and locations of all nodes according to Eq. (1).

Step6: Begin the next generation and go to Step2.

The flowchart is shown in Fig.3.

4   Simulation

In this section, we apply both RN-PSO and RD-PSO to four benchmarks to test their performance. To compare our algorithms with other PSO variants, we also present simulation results of WPSO [6], DPSO [8] and IPSO [11] along with standard PSO.

4.1   Settings

Table 1 shows definitions of the four benchmarks.

Table 1. Definitions of four benchmarks

name	definition	domain	optima
Sphere	$f_1(x)=\sum x_i^2, 1 \leq i \leq D$	[-5.12,5.12]	0
Step	$f_2(x)=\sum (\lfloor x_i \rfloor +0.5\rfloor)^2, 1 \leq i \leq D$	[-100,100]	0
Rosenbrock	$f_3(x)=\sum 100(x_{i+1}-x_i^2)^2+(x_i-1)^2, 1 \leq i \leq D-1$	[-2.048,2.048]	0
Rastrigin	$f_4(x)=10D+\sum [x_i^2-10\cos(2\pi x_i)], 1 \leq i \leq D$	[-5.12,5.12]	0

Among these functions, f_1 and f_2 are unimodal, whereas f_3 and f_4 are multimodal.

The number of nodes $N=40$, dimension of each node $D=30$ [14], and each single test is simulated for 1000 generations and repeated for 100 times. In order to obtain proper accuracy in local range, $w=1/(2*\ln 2)$, $c_1=c_2=0.5+\ln 2$, the upper limit of velocity is $v_{max}=x_{max}$. If x surpasses the domain boundary when updated using Eq. (1), it will be modified to the boundary value. And if velocity is larger than v_{max}, it will be reset to 0.

In WPSO, w decreases from 0.9 to 0.2. In DPSO, chaos factor $c_v=0$, $c_l=0.001$. In IPSO, cooperative coefficient $C=0.8$.

4.2   Results and Comparisons

Table 2. Performance comparison of various PSO algorithms

f	standard	WPSO	DPSO	IPSO	RN-PSO		RD-PSO	
					R	$f(x)$	R	$f(x)$
f_1	1.0997* 1e+02	2.4333* 1e+00	1.1290* 1e+00	1.1081* 1e+00	5	8.6454* 1e-01	5	5.0424* 1e-01
					15	3.4882* 1e-01	15	8.6433* 1e-01
					25	8.6480* 1e-01	25	5.1873* 1e-01
					35	7.3236* 1e-01	35	5.8119* 1e-01
f_2	4.2391* 1e+04	4.6872* 1e+02	4.3112* 1c+02	4.2890 1e+02	5	3.3272* 1e+02	5	2.0848* 1e+02
					15	3.3021* 1e+02	15	3.2960* 1e+02
					25	3.3079* 1c+02	25	1.7107* 1e+02
					35	3.3196* 1e+02	35	3.2957* 1e+02
f_3	9.4379* 1e+03	2.1962* 1e+01	9.8642* 1e+01	9.7579* 1e+01	5	7.0801* 1e+01	5	3.8750* 1e+01
					15	6.1579* 1e+01	15	4.8648* 1e+01
					25	1.1313* 1e+02	25	4.8130* 1e+01
					35	1.0751* 1e+02	35	6.3354* 1e+01
f_4	4.5187* 1e+02	1.0750* 1e+00	1.0475* 1e+00	1.7469* 1e+00	5	1.0169* 1e+00	5	1.0195* 1e+00
					15	1.2231* 1e+00	15	1.0219* 1e+00
					25	9.9054* 1e-01	25	9.9465* 1e-01
					35	1.2258* 1e+00	35	1.2234* 1e+00

Table.2 presents simulation results of our proposed algorithms and comparisons with standard PSO, WPSO, DPSO and IPSO. In particular, R represents the number of neighboring nodes. In order to show the general tendency of R values, we give numerical results of 4 particular R values, i.e. $R=5, 15, 25$ and 35.

Figures in the table show that optimized network structures have much better performance than standard PSO, also better than the comparison algorithms. Another common feature is that RD-PSO (random structure) performs better than RN-PSO (regular network) in general. With same parameters of both network topologies, nodes in RD-PSO have more chance to search for a better solution. This is because in RD-PSO the network structure is relatively flexible, then a node's searching area might cover the whole network. But in RN-PSO, nodes have a fixed network structure that nodes search its vicinity only.

In general, performance worsens with R increasing to the scale close to swarm size. The network gets tightly connected this time as the structure becomes similar to that of the standard algorithm.

5   Conclusion

In this paper, we have proposed two optimization approaches to particle swarm optimization based on complex network analysis. All particles are connected to a number of neighbors that are orderly selected for RN-PSO and random selected for RD-PSO. Experiments and comparisons on four benchmarks prove our methods effective. The performance of regular and random network structures is remarkably superior to that of standard PSO and also better than the compared algorithms.

Acknowledgements. This work is supported in part by the National Natural Science Foundation of China under Grant Nos. 61300239, China Postdoctoral Science Foundation Nos. 2014M551635, and postdoctoral fund of Jiangsu Province under Grant Nos. 1302085B.

References

1. Kennedy, J., Eberhart, R.: Particle Swarm Optimization. In: IEEE International Conference on Neural Networks, pp. 1942–1948. IEEE Press, New York (1995)
2. AlRashidi, M.R., El-Hawary, M.E.: A Survey of Particle Swarm Optimization Applications in Electric Power Systems. IEEE T. Evolut. Comput. 13, 913–918 (2009)
3. Wang, C., Liu, Y., Zhao, Y., Chen, Y.: A Hybrid Topology Scale-free Gaussian-dynamic Particle Swarm Optimization Algorithm Applied to Real Power Loss Minimization. Eng. Appl. Artif. Intel. 32, 63–75 (2014)
4. Jeong, Y.W., Park, J.B., Jang, S.H., Lee, K.Y.: A New Quantum-inspired Binary PSO: Application to Unit Commitment Problems for Power Systems. IEEE T. Power. Syst. 25, 1486–1495 (2010)

5. Eberhart, R., Shi, Y.: Comparing Inertia Weights and Constriction Factors in Particle Swarm Optimization. In: IEEE Congress on Evolutionary Computation, pp. 84–88. IEEE Press, New York (2000)
6. Suganthan, P.N.: Particle Swarm Optimiser with Neighbourhood Operator. In: IEEE Congress on Evolutionary Computation, pp. 195–1962. IEEE Press, New York (1999)
7. Ratnaweera, A., Halgamuge, S., Watson, H.C.: Self-organizing Hierarchical Particle Swarm Optimizer with Time-varying Acceleration Coefficients. IEEE T. Evolut. Comput. 8, 240–255 (2004)
8. Xie, X.F., Zhang, W.J., Yang, Z.L.: Dissipative Particle Swarm Optimization. In: IEEE Congress on Evolutionary Computation, pp. 1456–1461. IEEE Press, New York (2002)
9. Kennedy, J., Mendes, R.: Population Structure and Particle Swarm Performance. In: IEEE Congress on Evolutionary Computation, pp. 1671–1676. IEEE Press, New York (2002)
10. Matsushita, H., Nishio, Y.: Network-Structured Particle Swarm Optimizer with Various Topology and its Behaviors. In: Príncipe, J.C., Miikkulainen, R. (eds.) WSOM 2009. LNCS, vol. 5629, pp. 163–171. Springer, Heidelberg (2009)
11. Matsushita, H., Nishio, Y., Saito, T.: Particle Swarm Optimization with Novel Concept of Complex Network. In: International Symposium on Nonlinear Theory and its Applications, pp. 197–200. IEICE, Tokyo (2010)
12. Gong, Y.J., Zhang, J.: Small-world Particle Swarm Optimization with Topology Adaptation. In: 15th Annual Conference on Genetic and Evolutionary Computation Conference, pp. 25–32. ACM, New York (2013)
13. Wang, X., Li, X., Chen, G.: Complex Network: Theory and applications. Tsinghua University Press (2006) (in Chinese)
14. Zambrano-Bigiarini, M., Clerc, M., Rojas, R.: Standard Particle Swarm Optimisation 2011 at CEC-2013: A baseline for future PSO improvements. In: IEEE Congress on Evolutionary Computation, pp. 2337–2344. IEEE Press, New York (2013)

Fully Learned Multi-swarm Particle Swarm Optimization

Ben Niu[1,2,3,*], Huali Huang[1], Bin Ye[4], Lijing Tan[5,*], and Jane Jing Liang[6]

[1] College of Management, Shenzhen University,
Shenzhen 518060, China
[2] Hefei Institute of Intelligent Machines, Chinese Academy of Sciences,
Hefei 230031, China
[3] Department of Industrial and System Engineering,
The Hong Kong Polytechnic University, Hong Kong
[4] State Grid Anhui Economic Research Institute, Hefei 230022, China
[5] Business Management School, Shenzhen Institute of Information Technology,
Shenzhen 518172, China
[6] School of Electrical Engineering, Zhengzhou University, Zhengzhou 450001, China

Abstract. This paper presents a new variant of PSO, called fully learned multi-swarm particle swarm optimization (FLMPSO) for global optimization. In FLMPSO, the whole population is divided into a number of sub-swarms, in which the learning probability is employed to influence the exemplar of each individual and the center position of the best experience found so far by all the sub-swarms is also used to balance exploration and exploitation. Each particle updates its velocity based on its own historical experience or others relying on the learning probability, and the center position is also applied to adjust its flying. The experimental study on a set of six test functions demonstrates that FLMPSO outperform the others in terms of the convergence efficiency and the accuracy.

Keywords: multi-swarm particle swarm optimization, fully learned, particle swarm optimizer (PSO).

1   Introduction

Particle swarm optimization (PSO), originally proposed by Kennedy and Eberhart [1] [2], has been the family of population-based evolutionary computation techniques. It is motivated from the social simulation of bird flocking. In PSO, each particle adjusts its flying pattern by combining both its own historical experience and its companions' flying experience. The individual has a tendency to adapt the search trajectory combined personal cognition with social interaction among the population over the course of search process.

---

[*] Corresponding Author: Niu Ben, drniuben@gmail.com, Tan Lijing, mstlj@163.com

Y. Tan et al. (Eds.): ICSI 2014, Part I, LNCS 8794, pp. 150–157, 2014.

Through the process of trial and error in the past decades, particle swarm optimization has been developed various kinds of methods and applied to solve different practical problems. According to no free lunch theorems for search, extensive research through a wide variety of improved methods such as parameter selecting [3-4], population topology [5-8], hybridization [9-10] and etc [11-14] were proposed gradually to achieve a better trade-off between the exploration and exploitation. Although the shortcoming of premature has been improved, there is still much work to do. Based on our previous works [15-17], here we introduced a fully learning strategy to facilitate a global search in PSO by incorporating a learning probability to keep the population diversity and multi-swarm mechanism to avoid being trapping into local optimum.

The rest of this paper is organized as following. Section 2 gives a brief description of the basic particle swarm optimization. The fully learned multi-swarm particle swarm optimization is elaborated in Section 3. The experimental studies on FLMPSO including other comparative algorithms and related results are presented in Section 4. Finally, Section 5 concludes this paper.

2   Basic Particle Swarm Optimization

In original PSO, each particle is treated as a potential solution vector in the problem space. And the best position is decided by the best value of the corresponding objective. The best previous experience of all the individuals is recorded and updated. Accordingly each particle integrates its local optimum and global best to update its trajectory. The velocity and position of each dimension for the i th particle are renewed by the following equation, respectively:

$$v_{id} \leftarrow v_{id} + c_1 * r_1 * \left(p_{id} - x_{id} \right) + c_2 * r_2 * \left(p_{gd} - x_{id} \right) \tag{1}$$

$$x_{id} \leftarrow x_{id} + v_{id} \tag{2}$$

where $i = 1,\ldots, ps$, ps is the swarm size. d denotes the corresponding dimension. c_1 and c_2 are two positive constants, r_1 and r_2 are two random numbers in the range [0,1]. p_{id} and p_{gd} are the best previous position of individual and the whole swarm yielding the best fitness value. In addition, v_{id} and x_{id} represent the velocity and the position in the d-dimensional space for the i th particle respectively.

3   Fully Learning Multi-swarm Particle Swarm Optimization

In FLMPSO, the whole population consists of a number of sub-swarms, in which each has the identical individual. And the particle in each sub-swarm interacts by the exchange of individual experience during the course of flight. Meanwhile, the particles belonging to different sub-swarms also cooperate with each other.

Each individual follows the local best or the global optimum as its own exemplar based on the learning probability and the center position of the best experiences found so far by the particle from different sub-swarms is also employed to adjust its flying.

The learning probability Pc_i for i th particle of each sub-swarm is adopted using [14]:

$$Pc_i = 0.05 + 0.45 * \frac{\left(\exp\left(\frac{10(i-1)}{m-1} \right) \right)}{(\exp(10)-1)} \tag{3}$$

where m is the population size of each sub-swarm. And the exemplar is decided by the learning probability compared with a random number. If the learning probability is larger than this random number, the corresponding particle will learn from its own personal experience. Otherwise, it will learn from the best experience of the whole population.

The velocity equation of the i th particle within each sub-swarm is updated as follows:

$$v_{id} \leftarrow v_{id} + c_1 * r_1 * \left(p_{fid} - x_{id} \right) + c_2 * r_2 * \left(p_{cd} - x_{id} \right) \tag{4}$$

where c_1 and c_2 are constant learning factors. r_1 and r_2 are random numbers between 0 and 1. fi defines the exemplar which the particle i should follow. And p_{cd} is the center position of the best experience found so far by the whole group of sub-swarms and d is the dimension.

Table 1. Pseudo code of the proposed algorithm

```
Algorithm FLMPSO
Begin
Initialize the group of sub-swarms and the related parameters.
While (the termination conditions are not met)
  Evaluate the fitness value of each particle.
  Compute the global optimum and the center position.
  For each sub-swarm (do in parallel)
    For each particle i
      Generate a random number and renew the best experience.
      Compare the random number with the learning probability Pc_i.
      Select the exemplar which i th particle should follow.
      Update the velocity and position using Eqs.(4) and (2).
    End for
  End for (Do in parallel)
End while(until a terminate-condition is met)
end
```

Through the description above, the pseudo-code for the FLMPSO algorithm is elaborated as Table 1.

4   Experiments and Results

The set of benchmark functions, which were popularly used in the literatures, will be also introduced to demonstrate the effectiveness of FLMPSO. Details of these benchmark functions are listed in Table 2 and four algorithms from the literatures [3,7,8,14] are also shown as below.

Table 2. Six benchmark functions

Functions	Mathematical Representation	Search Range
Sphere(f_1)	$f_1(x)=\sum_{i=1}^{n} x_i^2$	[-100,100]
Rosenbrock(f_2)	$f_3(x)=\sum_{i=1}^{n-1}((x_i-1)^2+100(x_{i+1}-x_i^2)^2)$	[-2.048,2.048]
Ackley(f_3)	$f_2(x)=20+e-20\exp(-0.2\sqrt{\frac{1}{n}\sum_{i=1}^{n} x_i^2})$ $-\exp(\frac{1}{n}\sum_{i=1}^{n}\cos 2\pi x_i)$	[-32.768,32.768]
Griewank(f_4)	$f_4(x)=1+\frac{1}{4000}\sum_{i=1}^{n} x_i^2 - \prod_{i=1}^{n}\cos(\frac{x_i}{\sqrt{i}})$	[-600,600]
Rastrigin(f_5)	$f_5(x)=\sum_{i=1}^{n}(10-10\cos(2\pi x_i)+x_i^2)$	[-5.12,5.12]
Weierstrass(f_6)	$f_7(x)=\sum_{i=1}^{n}\left(\sum_{k=0}^{20}\left[(0.5)^k\cos(2\pi*3^k(x_i+0.5))\right]\right)$ $-n\sum_{k=0}^{20}\left[(0.5)^k\cos(3^k*\pi)\right]$	[-0.5,0.5]

All the experiments are conducted to compare the algorithms with the same population size of 30, and FLMPSO has 5 sub-swarms which both include 6 particles within each sub-swarm to obtain a fair comparison. Meanwhile each benchmark function runs 20 times and the maximal iteration is set at 1000. In all cases, the inertia weight is linearly decreased from 0.9 to 0.4 [3] except UPSO [8], which adopts the constriction factor $\phi = 0.729$. And c_1 and c_2 are both 2.0. Meanwhile, all the parameters of FLMPSO used in each sub-swarm are as same as those defined above.

Test results on six benchmark functions are presented in the Table 3 and Fig 1. The convergence characteristics in terms of the mean value and standard deviation of the results for each benchmark function are shown. Note that, optimum values obtained are in bold in the following Table.

Table 3. Results on six benchmark functions for 10-D

Algorithm	SPSO	FDR-PSO	UPSO	CLPSO	FLMPSO
f_1	2.207e-028 ±8.328e-028	8.583e-054 ±2.971e-053	4.267e-038 ±7.686e-038	4.518e-027 ±1.519e-026	**1.892e-124** **±8.447e-124**
f_2	4.209e+000 ±1.961e+000	3.312e+000 ±1.887e+000	**3.003e+000** **±1.086e+000**	3.764e+000 ±1.077e+000	7.310e+000 ±1.143e-001
f_3	1.323e+000 ±4.507e-015	1.323e+000 ±1.244e-015	1.323e+000 ±4.556e-016	**1.323e+000** **±1.328e-017**	1.323e+000 ±4.556e-016
f_4	1.011e-001 ±4.103e-002	7.565e-002 ±4.016e-002	3.969e-002 ±1.739e-002	3.594e-003 ±4.510e-003	**0±0**
f_5	4.481e+000 ±1.778e+000	6.119e+000 ±3.390e+000	7.444e+000 ±2.835e+000	4.307e-006 ±4.577e-006	**0±0**
f_6	2.646e+000 ±2.442e+000	6.707e-001 ±6.146e-001	4.631e-001 ±6.604e-001	1.568e-001 ±8.431e-002	**0±0**

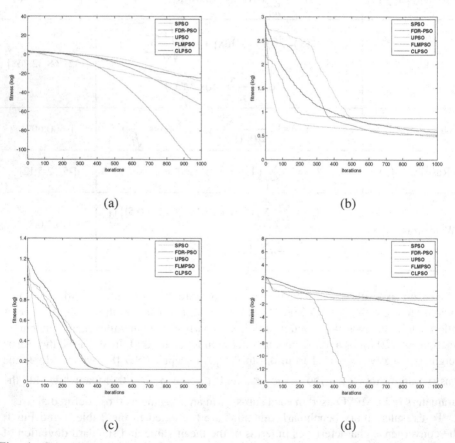

(a)   (b)

(c)   (d)

Fig. 1. Convergence characteristics on 10-dimensional benchmark functions. (a) Sphere function. (b) Rosenbrock function. (c) Ackley function. (d) Griewanks function. (e) Rastrigin function. (f) Weierstrass function.

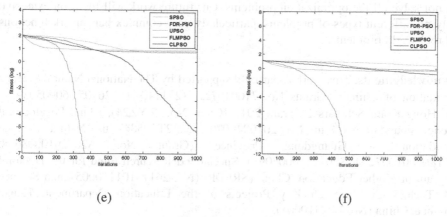

Fig. 1. (*Continued*)

For unimodal problem f_1, as illustrated in Fig 1, FLMPSO is able to find the better minimum within 1000 generations with greatly faster rate. For Rosenbrock function, it still suffers from premature convergence as well as other algorithms. And there is no much difference between FLMPSO and others for 10-D Ackley function. For multimodal functions $f_4 \sim f_6$ as shown above, the proposed algorithm can be able to find the global optimum with the faster speed. However other optimizers seem to get stuck into the premature convergence.

Based on the above results, FLMPSO is easier to find better minimum within 1000 iterations compared with other algorithms. From the figures, we can visually see that FLMPSO has faster convergence speed and better solution accuracy for most problems in our experiment. Overall, the proposed algorithm has a great advantage on global search ability and efficiency for most benchmark functions, except that it seems to have stagnated for Rosenbrock function.

5   Conclusions and Further Work

In this paper, we present an improved multi-swarm particle swarm optimizer with a fully learning scheme called FLMPSO. Different from the original particle swarm optimization, the learning probability is employed to decide whether to follow its own historical experience or to adopt the global best experience from other sub-swarms. Meanwhile, the center position of the best historical experience of the whole population as the comprehensive experience is also employed to avoid getting into local optima. Each particle in the entire population combines the learning experience with its own learning probability to adjust its flight.

To confirm the performance, the proposed algorithm is tested on a set of 10-D test functions as well as other optimizers. As shown in the experimental results, FLMPSO can easily escape from local optima and find the better optimum. In short, the proposed algorithm can outperform others in our simulation. However, FLMPSO also can not solve all the optimization problems. Our future work will focus on extensive study on different types of problems, particularly very complex benchmark functions and real-world problems.

Acknowledgments. This work is partially supported by The National Natural Science Foundation of China （Grants No. 71001072, 71271140, 11226105, 60905039）, The Hong Kong Scholars Program 2012 (Grant No. G-YZ24), China Postdoctoral Science Foundation (Grant No. 20100480705, 2012T50584), the Natural Science Foundation of Guangdong Province (Grant No. S2012010008668, 9451806001002294, S2012040007098), Specialized Research Fund for the Doctoral Program of Higher Education, China (SRFDP) (No. 20114101110005), and Science and Technology Research Key Projects of the Education Department, Henan Province, China (No. 14A410001).

References

1. Kennedy, J., Eberhart, R.: Particle Swarm Optimization. In: Proceedings of the 1995 IEEE International Conference on Neural Networks, pp. 1942–1948 (1995)
2. Eberhart, R., Kennedy, J.: A New Optimizer Using Particle Swarm Theory. In: Proceeding of the Sixth International Symposium on Micro Machine and Human Science, pp. 39–43 (1995)
3. Shi, Y., Eberhart, R.C.: A Modified Particle Swarm Optimizer. In: Proceedings of 1998 IEEE International Conference on Evolutionary Computation, pp. 69–73 (1998)
4. Wu, Z.: An Optimization Algorithm for Particle Swarm with Self-adapted Inertia Weighting Adjustment. International Review on Computers and Software 7(3), 1320–1326 (2012)
5. Kennedy, J., Mendes, R.: Population Structure and Particle Swarm Performance. In: Proceedings of the Congress on Evolutionary Computation, pp. 1671–1676 (2002)
6. Tian, Y.L.: Study on the Topological Structure of the Particle Swarm Algorithm. Journal of Computational Information Systems 9(7), 2737–2745 (2013)
7. Peram, T., Veeramachaneni, K., Mohan, C.K.: Fitness-distance-ratio Based Particle Swarm Optimization. In: Proceedings of the IEEE Swarm Intelligence Symposium, pp. 174–181 (2003)
8. Parsopoulos, K.E., Vrahatis, M.N.: UPSO–A Unified Particle Swarm Optimization Scheme. Lecture Series on Computational Sciences, pp. 868–873 (2004)
9. Shen, H., Zhu, Y.L., Li, J., Zhu, Z.: Hybridization of Particle Swarm Optimization with the K-Means Algorithm for Clustering Analysis. In: Proceedings of IEEE fifth International Conference on Bio-Inspired Computing: Theories and Applications, pp. 531–535 (2010)
10. Wang, Q., Wang, P.H., Su, Z.G.: A Hybrid Search Strategy Based Particle Swarm Optimization Algorithm. In: Proceedings of 2013 IEEE 8th Conference on Industrial Electronics and Applications, pp. 301–306 (2013)
11. Van den Bergh, F., Engelbrecht, A.P.: A Cooperative Approach to Particle Swarm Optimization. IEEE Transactions on Evolutionary Computation 8(3), 225–239 (2004)

12. Chen, H.N., Zhu, Y.L., Hu, K.Y., Ku, T.: PS2O: A Multi-Swarm Optimizer for Discrete Optimization. In: Proceedings of Seventh World Congress on Intelligent Control and Automation, pp. 587–592 (2008)
13. Wang, X.M., Guo, Y.Z., Liu, G.J.: Self-adaptive Particle Swarm Optimization Algorithm with Mutation Operation Based on K-Means. Advanced Materials Research 760–762, 2194–2198 (2013)
14. Liang, J.J., Qin, A.K., Suganthan, P.N., Baskar, S.: Comprehensive Learning Particle Swarm Optimizer for Global Optimization of Multimodal Functions. IEEE Transactions on Evolutionary Computation 10(3), 281–295 (2006)
15. Niu, B., Zhu, Y.L., He, X.X., Wu, H.: MCPSO: A Multi-swarm Cooperative Particle Swarm Optimizer. Applied Mathematics and Computation 185(2), 1050–1062 (2007)
16. Niu, B., Li, L.: An Improved MCPSO with Center Communication. In: Proceedings of 2008 International Conference on Computational Intelligence and Security, pp. 57–61 (2008)
17. Niu, B., Huang, H., Tan, L., Liang, J.J.: Multi-swarm Particle Swarm Optimization with a Center Learning Strategy. In: Tan, Y., Shi, Y., Mo, H. (eds.) ICSI 2013, Part I. LNCS, vol. 7928, pp. 72–78. Springer, Heidelberg (2013)

Using Swarm Intelligence to Search for Circulant Partial Hadamard Matrices

Frederick Kin Hing Phoa, Yuan-Lung Lin, and Tai-Chi Wang

Academia Sinica, Taipei 115, Taiwan
fredphoa@stat.sinica.edu.tw

Abstract. Circulant partial Hadamard matrices are useful in many scientific applications, yet the existence of such matrices is not known in general. Some Hadamard matrices of given orders are found via complete enumeration in the literature, but the searches are too computationally intensive when the orders are large. This paper introduces a search method for circulant partial Hadamard matrices by using natural heuristic algorithm. Slightly deviated from the swarm intelligence based algorithm, this method successfully generates a class of circulant partial Hadamard matrices efficiently. A table of circulant partial Hadamard matrices results are given at the end of this paper.

1 Introduction

With scientific and technological advancements, investigators are becoming more interested in and capable of studying large-scale systems. There is considerable scope for reducing resources used in research by designing more efficient studies. Careful design considerations, even with only minor variation in traditional designs, can lead to a more efficient study in terms of more precise estimate or be able to estimate more effects in the study at the same cost ([12],[11], [14]).

It is well-known that there are some connections between combinatorial and experimental designs. A Hadamard matrix of order n is an $n \times n$ square matrix H with entries (± 1) such that $HH^T = nI_n$, which also suggests that its rows are orthogonal and thus designs based on Hadamard matrices are D-optimal. The studies of Hadamard matrices and their design properties are rich, see [6], [7] and others for details.

[9] pioneered a study of partial Hadamard matrix, which is a $k \times n$ rectangular matrix with entries (± 1) such that $HH^T = kI_n$. [3] introduced a class of $k \times n$ partial Hadamard matrices that are constructed by circulating any row of the matrix k times. Such row is generally called a generating vector. They denote "r-$H(k \times n)$" as a $k \times n$ circulant partial Hadamard matrix such that the numbers of $+1$ and -1 in every row of H differ by r. Table 1 in [3] listed the maximum k for which r-$H(k \times n)$ exists, it followed by several lemmas that describe the relationship among r, k and n.

However, they do not provide the exact generating vectors of ± 1 for each H reported in Table 1. Not to mention some H of large order n, they could only provide possible ranges of maximum k. This motivates us to search for an

Y. Tan et al. (Eds.): ICSI 2014, Part I, LNCS 8794, pp. 158–164, 2014.

efficient algorithm that can be used to find these exact generating vectors, and hopefully this algorithm is efficient enough so that the length n will not be a serious factor in computational burden. It leads to the use of natural heuristic algorithm.

One classical natural heuristic algorithm is particle swarm optimization (PSO), which was first proposed by [8] and then followed up by [13] and [4] among many others. The optimization technique was originally aimed to study social network structure but now is widely used in computational intelligence, industrial optimization, computer science, and engineering research. Textbooks such as [2], [5] and others provide comprehensive introduction of the optimization theories, evolutionary computations, and practical applications of PSO.

It is not trivial to implement a standard PSO algorithm to discrete problems like searching optimal matrices and designs. Not long ago, [10] first proposed a swarm intelligence based (SIB) algorithm for optimizing supersaturated designs, which required minimum column correlation and equal number of $+$ and $-$ (their paper referred it as a balanced design) in each column. The procedure is called "Swarm Intelligence Based Supersaturated Designs" (or in short, SIBSSD) algorithm.

This paper is organized as follows. Section 2 provides a brief review on the background of circulant partial Hadamard matrix (CPHM) and swarm intelligence based (SIB) algorithm. Section 3 contains our proposed algorithm that used to search for the generating vectors such that $r\text{-}H(k \times m)$ can possess the maximum k. Section 4 lists our search results. In the last section, we summarize our work and provide future directions of this work, including additional comments on the use of parallel computing.

2   Some Backgrounds on Circulant Partial Hadamard Matrix and Swarm Intelligence Based Algorithm

Rephrased from [1], we officially define the circulant Hadamard matrix and CPHM here.

Definition 1. *Consider a $k \times n$ matrix $A = \{a_{ij}\}$ with first row $(a_1, a_2, \ldots, a_n)$.*

(a) *A is a circulant if each row after the first is obtained by a right cyclic shift of its predecessor by one position, thus $a_{ij} = a_{j-i \bmod n}$;*
(b) *A is a circulant partial Hadamard matrix if and only if it is circulant and $AA^T = kI_n$;*
(c) *When $n = k$, A is a circulant Hadamard matrix of order n.*

Example 1. We consider a generating vector $\boldsymbol{a} = (-, -, -, +, -, +, +, +)$. Let A be a matrix and $\boldsymbol{a}$ be its first row, which makes $(+, -, -, -, +, -, +, +)$ and $(+, +, -, -, -, +, -, +)$ its second and third rows respectively. A that consists of these three rows is a $0\text{-}H(3 \times 8)$: (i) $r = 0$ because the number of $+$ and $-$ entries in each row are equal; (ii) A has a dimension of 3 rows and 8 columns; (iii) AA^T is a 3×3 diagonal matrix with all diagonal entries 8. In fact, according to Table 1 in [3] and our search, the maximum possible k for A with $r = 0$ and $n = 8$ is 3.

The circulant nature greatly reduces the search space of optimal CPHM from 2^{kn} to 2^n, because the structure of the generating vector of length n governs the matrix properties completely.

The SIB algorithm is a natural heuristic algorithm designed for specific problems in discrete domains. It was first proposed in [10]. To put it simply, for each seed in each iteration after the initialization, the standard procedure of a SIB algorithm includes:

1. Generate candidates via mixing the original seeds with its local best seeds and the global best seed.
2. Choose the best candidates among all by MOVE operation, a decision making step.
3. Update the current seed to the best candidates that selected in previous step, with the possibilities to restart the seed if it is unchanged (trap in the local attractors).

[10] applied the SIBSSD algorithm to search for supersaturated designs under $E(s^2)$ criterion. Most supersaturated designs generated in SIBSSD algorithm reached its theoretical lower bound $E(s^2)$ values; that is to say, they were $E(s^2)$ optimal. [10] further explored the construction to optimal supersaturated designs under D_f criteria for various values of f.

3 SIBCPHM: A SIB Algorithm for Searching Circulant Partial Hadamard Matrix

It is obvious that the optimized target, i.e. the generating vector, is not continuous and thus many natural heuristic algorithms like PSO cannot be directly applicable for finding optimal circulant partial Hadamard matrix. The search space is a large set of entries comprising of ± 1 and the optimization problem is to select a right combination of n entries for the generating vector. To achieve this goal, we apply the SIB algorithm and redefine the update operation in Table 1, and named the procedure SIBCPHM (Swarm Intelligence Based Circulant Partial Hadamard Matrix) algorithm in the rest of the paper. The SIBCPHM algorithm shares some common features with the SIBSSD algorithm, except a major difference in the MOVE operation that will be mentioned below.

Prior to the start of SIBCPHM, users are required to enter some parameters, which include matrix-related parameters $(r; n)$ and SIB-related parameters $(N,$ number of seeds; t, number of maximum iteration; q_{LB}, LB exchange rate; q_{GB}, GB exchange rate). The initial steps (Steps 1 to 4) can be viewed as the zero[th] iteration of SIBCPHM. It starts with N vectors of length n that are created randomly, and the number of + and − in each vector differ by r. Users also need to input t as a stopping criterion. It is not mandatory for the users to input two exchange rates, because $q_{LB} = \lfloor n/3 \rfloor$ and $q_{GB} = \lfloor q_{LB}/2 \rfloor$ are set as default.

Table 1. SIBCPHM Algorithm

1: Randomly generate N vectors of length n as initial particles
2: Evaluate the maximum value of k of each vector
3: Initialize the LB for all vectors
4: Initialize the GB
5: **while** not converge **do**
6:    Update each vector via MOVE operation
7:    Evaluate the maximum value of k of each vector
8:    Update the LB for all vectors
9:    Update the GB
10: **end while**

There is no theoretical reason behind such settings, but it is our experience and observation from simulation results in [10] that these settings work quite well. However, if there are some prior information of optimal settings on exchanging rates from professional opinion, users can manually overwrite these rates.

Following the initialization step, the first iteration starts from the while-loop in Table 1. First, SIBCPHM performs MOVE operation on each vector, which is to update each vector by exchanging some entries with its LB and the GB. More explicitly, there are two intermediate vectors forming in this step. The first intermediate vector is formed via copying q_{LB} entries to the corresponding entries of the vectors from its LB, and the second intermediate vector is formed similarly from the GB. Among the original vector and two intermediate vectors, MOVE selects the vector that leads to the highest k and the selection is viewed as an update.

In order to further enhance the efficiency during the exchange, the selection of entries that used to be exchanged is not random. Prior to the selection, there is a detection on the difference between the original vector and the LB or GB vector. These locations of different entries are then ranked according to the impact on the value of k. The first q_{LB} or q_{GB} entries with the highest value of k are exchanged. Such procedure ensures that the exchange leads to the best possible result. One may wonder how MOVE avoids the vectors being trapped in local attractors. In fact, if the two intermediate vectors are not better than the original vector, MOVE will generate a new vector to restart. In other words, the restart not only allows the vector to jump out from the local attractor, but also to explore more search space.

Following the MOVE operation in each iteration, the maximum values of k of each vector are evaluated again and are compared to those of their individual LB. For some vectors with better maximum k values, their individual LBs are updated. Then the best LB among all is compared to the GB and the GB is updated when the best LB has a higher maximum k value.

The whole SIBCPHM terminates in two conditions. The first condition is when the maximum value of k attains its best possible value. Some of these upper bounds can be found in Table 1 of [3]. The second condition is when the number of iteration reaches t. Once SIBCPHM is stopped, the GB is the generating vector suggested via SIBCPHM.

4   Search Results

The search results of using SIBCPHM are listed in Table 2. The first two columns are the length of the generating vector (or column number of the matrix) and the difference between the numbers of + and −. The third column is the maximum k values obtained from Table 1 of [3], and all the listed results achieved their upper bounds.

Table 2. Results from SIBCPHM

n	r	max k	Generating Vector
8	0	3	$(- - - + - + ++)$
8	2	4	$(- - - - + + - -+)$
12	0	5	$(- - - - + + - + + + -+)$
12	2	6	$(- - - - + - - + + + -+)$
16	0	7	$(- - - - + - - + - + + + - + ++)$
16	2	8	$(- - - - - + - - + + + + - + - ++)$
20	0	7	$(- - - - - + - + - + + - - + + + + - ++)$
20	2	10	$(- - - - + - - - - + - + + + + + - - + + - -+)$
24	0	9	$(- - - - + - - + - - - + + + + - + - + + + - ++)$
24	2	12	$(- - - - - + + + - + + - + - + + + - - - + - -+)$
28	0	9	$(- - - - - + - + - + + - - + + + + + - - + + - + - ++)$
28	2	14	$(- - - - - + - + - - - + + - + + + + - - + - + + + - -+)$
32	0	12	$(- - - - - + - + - - - + + - + + + - + + + + - - - + + - + + -+)$
32	2	14	$(- - - - - + + + - + + + - - + - + + + + - - + - + - - - + - -+)$

The last column is the resulting generating vectors. Notice that these vectors are being cycled so that the longest consecutive − is located in the front. For example, for the first case $n = 8$ and $r = 0$, SIBCPHM may return a generating vector $(+ + + - - - +-)$ and it is trivial that this result is equivalent to the one in the table after three-entry left-shifts. Furthermore, it is also clear that once a generating vector is obtained, any vectors resulted from shifts are still generating vectors for the circulant partial Hadamard matrix of the same size. We list these generating vectors with the longest consecutive − at the front in Table 2, because it looks neat. What is more, it will return the smallest values of decimal counterpart if we treat these vectors as binary codes, which is a common abbreviation method in the field of experimental designs.

This table provides the results only from $n = 8$ to $n = 32$ because the vectors become too long for larger n. In fact, we have successfully obtained the generating vectors up to $n = 52$ for $r = 0$ and $n = 76$ for $r = 2$. For $r = 0$, our result is the largest definitive result reported in [3]; however, this result came up only after a few hours' search. For $r = 2$, our result has gone much further than what [3] has reported in their Table 1. Results for large n are available upon request from the first author.

Moreover, we only report the cases of $r = 0$ and $r = 2$ in this paper, but there are other settings $r > 2$ shown in [3] that we do not report here. First, notice that [3] has derived some theoretical results from very large r but we are not interested in repeating the known results. For the case of $r = 0$, this class is considered as balanced designs in the field of experimental designs and thus it is highly compelling. For the case of $r = 2$, this class possesses maximum k among all different r values. In other words, this class of matrix is able to generate experimental designs with maximum number of two-level factors which are uncorrelated when experiments are analyzed. It is true that there are also some $r = 4$ cases with equivalently maximum k, such as $n = 12$ (generating vector: $(-----+----++))$ and $n = 28$ (generating vector: $(----$ $-+-++-+-++--+++-+++++-+-+++))$. Although we do not report the results in the table, users can still use SIBCPHM to search for their desired cases of r.

5   Discussion and Conclusion

This paper introduces a new SIB algorithm for searching generating vectors, which are the building blocks of circulant partial Hadamard matrices. Due to the discrete nature of the generating vectors, most natural heuristic algorithms fail to be applied. Therefore, we consider a SIB algorithm and introduce SIBCPHM. SIBCPHM consists of a MOVE operation that selects the best outcome among the original vector and two intermediate vectors associated with the LB and GB. This MOVE operation also includes a mechanism that is able to avoid trapping in a local attractor.

We apply SIBCPHM to search for generating vectors of different n and r, and the resulting vectors are reported in Table 2. It is obvious that all our results successfully attain the upper bound k reported in [3], and even better, SIBCPHM is computationally efficient enough to go further for larger n.

In fact, the idea of SIBCPHM is similar to SIBSSD in [10]. They both consist of the MOVE operation for decision making and exchange some discrete components in the MOVE operation. However, one major distinction between them is on the nature of their discrete components. SIBSSD treats the column index as a component. The search space increases exponentially when the dimension increases, but also lowers the search power. On the other hand, SIBCPHM treats the entry of a vector (column) as a component. Since there are only two choices (± 1) for two-level designs or binary matrices, the size of the search space is acceptable even when the dimension is large. Such difference between two SIB algorithms mainly comes from the prior knowledge on the structure of the target. For SIBCPHM, despite the fact that we have some basic knowledge on the structure of circulant partial Hadamard matrices, we can only utilize it and search for its generating vectors. For SIBSSD, the general structure of supersaturated designs (without specific constraints) is unknown, so we allow each particle to go over the whole potential design space.

This work can be extended to the search of many other classes of matrices and designs. For example, one may slightly modify SIBCPHM to search for D-optimal factorial designs with a defined dimension, or an orthogonal array with a defined dimension and a given strength. In addition, our current program is written in R and without parallel computing. One may consider using parallel computing with Graphics Parallel Units (GPUs). The resulting program is certainly much more powerful than ours and more likely to come up with results of very large n within hours.

Acknowledgement. This work was supported by (a) Career Development Award of Academia Sinica (Taiwan) grant number 103-CDA-M04 for Phoa, (b) National Science Council (Taiwan) grant number 102-2628-M-001-002-MY3 for Phoa and Lin, and (c) National Science Council (Taiwan) postdoctoral researcher grant number 102-2811-M-001-148-001 for Wang.

References

1. Chen, R.B., Hsieh, D.N., Hung, Y., Wang, W.: Optimizing Latin hypercube designs by particle swarm. Statistics and Computing 23, 663–676 (2013)
2. Clerc, M.: Particle Swarm Optimization. ISTE, London (2006)
3. Craigen, R., Faucher, G., Low, R., Wares, T.: Circulant partial Hadamard matrices. Linear Algebra and its Applications 439, 3307–3317 (2013)
4. Eberhart, R., Shi, Y.: Particle swarm optimization: developments, application and resources. In: IEEE Proceedings of the Congress on Evolutionary Computation, vol. 1, pp. 81–85 (2001)
5. Engelbrecht, A.P.: Fundamentals of Computational Swarm Intelligence. John Wiley and Sons, Chichester (2005)
6. Geramita, A., Seberry, J.: Quadratic forms, orthogonal designs, and Hadamard matrices. Lecture Notes in Pure and Applied Mathematics Series, vol. 45. Marcel Dekker Inc., New York (1979)
7. Horadam, K.J.: Hadamard matrices and their applications. Priceton University Press, Princeton (2007)
8. Kennedy, J., Eberhart, R.: Particle swarm optimization. In: Proceedings of IEEE International Conferenec on Neural Networks, vol. 4, pp. 1942–1948 (1995)
9. Low, R.M., Stamp, M., Craigen, R., Faucher, G.: Unpredictable binary strings. Congressus Numerantium 177, 65–75 (2005)
10. Phoa, F.K.H., Chen, R.B., Wang, W., Wong, W.K.: Optimizing two-level supersaturated designs by particle swarm techniques. Technometrics (under review)
11. Phoa, F.K.H., Wong, W.K., Xu, H.: The need of considering the interactions in the analysis of screening designs. Journal of Chemometrics 23, 545–553 (2009)
12. Phoa, F.K.H., Xu, H., Wong, W.K.: The use of nonregular fractional factorial designs in combination toxicity studies. Food and Chemical Toxicology 47, 2183–2188 (2009)
13. Shi, Y., Eberhart, R.: A modified particle swarm optimizer. In: Proceedings of IEEE International Conference on Evolutionary Computation, pp. 69–73 (1998)
14. Xu, H., Phoa, F.K.H., Wong, W.K.: Recent developments in nonregular fractional factorial designs. Statistics Survey 3, 18–46 (2009)

High Performance Ant Colony Optimizer (HPACO) for Travelling Salesman Problem (TSP)

Sudip Kumar Sahana[1] and Aruna Jain[2]

[1] Dept of CSE, Birla Institute of Technology, Ranchi, Jharkhand, India
sudipsahana@gmail.com
[2] Dept of CSE, Birla Institute of Technology, Ranchi, Jharkhand, India
jaruna16@rediffmail.com

Abstract. Travelling Salesman Problem (TSP) is a classical combinatorial optimization problem. This problem is NP-hard in nature and is well suited for evaluation of unconventional algorithmic approaches based on natural computation. Ant Colony Optimization (ACO) technique is one of the popular unconventional optimization technique to solve this problem. In this paper, we propose High Performance Ant Colony Optimizer (HPACO) which modifies conventional ACO. The result of implementation shows that our proposed technique has a better performance than the conventional ACO.

Keywords: Travelling Salesman Problem (TSP), Ant Colony Optimization (ACO), Combinatorial Optimization (CO), Pheromone, Meta-heuristics.

1   Introduction

Travelling Salesman Problem, which is NP Hard in nature, is an open question to the researchers to solve in polynomial time. This kind of NP-hard problem can be solved by bio-inspired optimization techniques. Bio-inspired computations are the computational systems that take inspiration from nature. These bio-inspired techniques include artificial neural networks, genetic algorithms, ant colony optimization, artificial life, DNA computing, etc. Jing Zhang [1] made a survey of natural computation for TSP and concluded that ACO is a good performer. Li Wang [2] shows how Ant Colony Algorithm (ACA) can be used in Neural Network based Optimal Method to facilitate distributed computation, and positive feedback. Marco Dorigo [3][4] gave an overview of ACO meta-heuristic inspired by the foraging behavior of real ant. According to many other research works, ACO is widely used for solving complex combinatorial problems [16][17] as it is the most solution specific, goal driven, efficient technique. We have incorporated a new idea in the movement of the ants. Instead of single ant moving randomly, we have considered ants as a group, and applied the effect of group updation on the edges of TSP graph. The local update and global update procedures are modified to influence the whole search procedure.

Y. Tan et al. (Eds.): ICSI 2014, Part I, LNCS 8794, pp. 165–172, 2014.

The remainder of this paper is organized as follows: In section 2 we have discussed the literature review and the problem definition .Section 3 describes the implementation methodology. Section 4 deals with and results and discussions and finally the conclusions are drawn and scope of future study are discussed in Section 6.

2   Literature Review and Problem Definition

In TSP a list of cities and their pair-wise distances are given, and the goal is to find the shortest route that visits each city once and returns to its original city [9]. The most direct solutions try all permutations to check which one is cheapest (using brute force search) having time complexity of O (n!), the factorial of the number of cities. So direct solution becomes impractical even for only 20 cities. One of the earliest applications of dynamic programming is the Held–Karp algorithm [18] that solves the problem in time O $(n^2 2^n)$. Using inclusion–exclusion [19], the problem can be solved in time within a polynomial factor of 2^n and polynomial space. Though a lot of researches have been produced over the years but no perfect algorithm has been devised so far. ACO is one of the popular unconventional optimization techniques to solve TSP. So, we have chosen TSP to validate our proposed High Performance Ant Colony Optimizer. According to research works [20][21]and[22] ACO is widely used for solving complex combinatorial problems as it is the most solution specific, goal driven, efficient technique.

ACS (Ant Colony System) has limitation of the initialization of early stage pheromone and poor convergence speed for the shortest path. The updations (local + global) of ACS [4] are responsible to increase the pheromone level in the suitable path for early convergence which ultimately enhances the performance. So, proper updation is one of the factors for achieving convergence speed towards the goal.

We have designed High Performance Ant Colony Optimizer by modifying the pheromone update procedure and considered three key parameters for our design: (i) the number of computations for the convergence process, (ii) the number of iterations and (iii) the number redundant states.

3   Methodology

Out of the main ACO algorithms presented in the literatures: Ant System (AS) [6], Ant-Q system[8] , ant colony system (ACS)[4] , max–min ant system (MMAS)[5] , rank-based ant system (AS rank)[7], we have chosen Ant Colony System (ACS) for it's satisfactory implementation and rich updation procedures i.e. local updation to diversify the search and global updation to achieve the global best path.

In the conventional ACS, Local Update procedure is governed by the Equation (1) and Global Update procedure is governed by Equation (2) as follows:

$$\tau_{ij} = (1 - \varphi).\tau_{ij} + \varphi . \tau_0 . \tag{1}$$

Where τ_{ij} = updated pheromone value on (i,j) edge by an ant.

$\tau_0 =$ initial pheromone value; φ = pheromone decay.

$$\tau_{ij} = (1-\rho). \tau_{ij} + \sum_{k=1}^{m} \Delta\tau_{ij}^{k} . \qquad (2)$$

Where $\Delta\tau_{ij}^{k}$ is equal to $1/L_k$ if (i , j) belong to tour otherwise zero and ρ = pheromone evaporation.

We have modified the formulae for group local update as in Equation (iii) and global update as Equation (iv)   shown below:

$$\tau_{ij(next)} = (1-\varphi^n)\tau_{ij(current)} + (1- \varphi^n) \tau_0 . \qquad (3)$$

$$\tau_{ij} = \tau_{ij} +b*[(1-\rho). \tau_{ij} + \sum_{k=1}^{m} \Delta\tau_{ij}^{k}] . \qquad (4)$$

Where φ= 0.5 and $\Delta\tau_{ij}^{k}$ is equal to Q/L_k if (i , j) belong to tour otherwise zero.
Q and b are two constants, those influence the pheromone increasing factor.

The working methodology of our proposed HPACO is to first select the start node and searches for the adjacency nodes. The probabilities to visit each edge by the number of ants are calculated depending on their pheromone values and corresponding edges are locally updated. And same process is continued till the destination node is reached.   Paths traced by each ants are then stored in path matrix. If a better path is found then replace with higher path value if path matrix is full. After exploring all promising possible paths, global update is performed on the minimum weighted path and the process is continued till all the ants flow through the minimum weighted path.

4   Results and Discussions

In this section, an example with 5 node TSP for the study of nature of convergence, a comparison between conventional ACO and HPACO for 5 to 9 nodes and test up to 101 nodes along with comparisons with other standard approaches   for the standard instances from the TSP Library are discussed in Section 4.1 ,4.2 and 4.3 respectively.

4.1   Implementation of the High Performance Ant Colony Optimizer (HPACO) for Travelling Salesman Problem Example

The implementation was done on Intel Core 2 Duo Processor T 5750 (2 GHz) machine with 2 GB RAM on Windows7 Professional 32 bit operating system in C language. In Figure 1, a small TSP problem with 5 nodes is presented to show the visual effect of path wise convergence in each iteration. The HPACO is applied to it for finding the shortest optimal tour containing all nodes. Nodes are representing cities and the weights are representing distance between two cities. At starting n (n=100) number of ants are initialized at the source node (node 0).

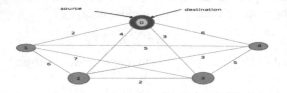

Fig. 1. An example TSP graph (Graph 1) with 5 nodes

The distance between two nodes is considered as the parameter for initial artificial pheromone with a relation of inversely proportional order. Initial pheromone ηij = $1/\delta ij$. δij is the distance between node i and j; also the cost of edge i j.

A summary of all the iterations to show how the number of ants vary path wise in each iteration are shown in Table 1 below.

Table 1. Iteration wise convergence of ants with path length

iterations ──────►		1st	2nd	3rd	4th	5th	6th	7th
Paths	Weights	Ants	Ants	Ants	Ants	Ants	Ants	Ants
0->4->3->2->1->0	21	3	x	x	x	x	x	x
0->4->3->1->2->0	28	1	x	x	x	x	x	x
0->4->2->3->1->0	20	4	x	x	x	x	x	x
0->4->2->1->3->0	25	1	x	x	x	x	x	x
0->4->1->3->2->0	24	3	0	1	x	x	x	x
0->4->1->2->3->0	22	2	2	1	1	x	x	x
0->3->4->2->1->0	19	4	x	x	x	x	x	x
0->3->4->1->2->0	23	2	3	1	1	x	x	x
0->3->2->4->1->0	15	11	73	87	93	97	99	100
0->3->2->1->4->0	22	5	3	1	1	x	x	x
0->3->1->4->2->0	22	3	1	1	x	x	x	x
0->3->1->2->4->0	25	2	1	x	x	x	x	x
0->2->4->3->1->0	21	3	x	x	x	x	x	x
0->2->4->1->3->0	22	3	5	3	1	1	x	x
0->2->3->4->1->0	18	6	1	x	x	x	x	x
0->2->3->1->4->0	24	4	x	x	x	x	x	x
0->2->1->4->3->0	23	2	x	x	x	x	x	x
0->2->1->3->4->0	28	1	x	x	x	x	x	x
0->1->4->3->2->0	18	6	1	1	x	x	x	x
0->1->4->2->3->0	15	10	2	1	1	x	x	x
0->1->3->4->2->0	21	3	x	x	x	x	x	x
0->1->3->2->4->0	20	8	3	2	1	1	x	x
0->1->2->4->3->0	19	5	4	2	1	1	1	x
0->1->2->3->4->0	21	8	x	x	x	x	x	x

4.2   Comparison of ACS and Proposed HPACO

Here under, the performances of our proposed HPACO with respect to conventional ACO are analyzed on TSP examples with 5,6,7,8 and 9 nodes as shown in Figure 1 and Figure 2 respectively.

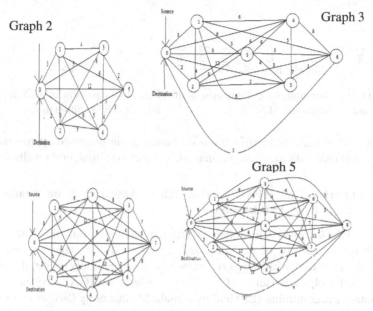

Fig. 2. Example Graphs 2, 3, 4 and 5 with 6, 7, 8 and 9 nodes respectively

Finally, the graph shown in Figure 3 below shows the iteration wise convergence of ants in HPACO and conventional ACS.

After applying both the methods individually on each graph, we have found that HPACO is able to give a certain improvement by means of increasing pheromone to the possible optimal path, so that, it can attract a max number of ants and results in early convergence. The comparison of ACS and HPACO on the basis of time taken for finding the optimal path individually in all the five graphs are shown here under in Table 2.We find that, the time taken by HPACO is less than the conventional ACS approach as shown and graphically represented in Figure 4 as shown below.

Table 2. Comparison of Execution Time of ACS and HPACO on Different Graphs

Graph	No. of links	No. of nodes	Execution time of ACS (millisecond)	Execution time of HPACO (millisecond)
1	10	5	1	1
2	15	6	10	7
3	21	7	72	46
4	28	8	126	84
5	36	9	180	110

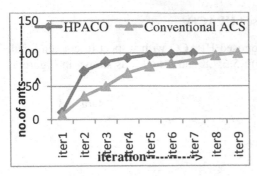

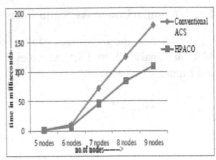

Fig. 3. Overall Iteration wise Comparison of HPACO and Conventional ACS

Fig. 4. Comparison of ACS and Proposed HPACO on execution time

It is evident that, the modifications discussed in our proposed approach exhibits a good performance than the conventional ACS as per the simulation results.

4.3 Comparison of HPACO with other Approaches on Standard TSP Instances from TSP Library

Proposed HPACO have been applied on different standard problems from TSP library to check the optimality of the proposed algorithm. Performance of our proposed HPACO is compared with the performance of other naturally inspired optimization methods: Ant Colony System (ACS), Genetic Algorithm (GA), Hybrid GA (HGA), Evolutionary Programming (EP) and by Simulated Annealing (SA) as shown in Table 3 below and found satisfactory.

Table 3. Comparison of Proposed HPACO for standard instances of TSP

Problem Name	ACS	GA	Hybrid GA	EP	SA	Proposed HPACO	Best known solution from TSP Library
Oliver30	420(423.74)	421(N/A)	(N/A)	420(423.74)	424(N/A)	420(423.75)	420(423.74)
tt48	(N/A)	(N/A)	10,628	(N/A)	(N/A)	10,628(10,631.57)	10,628
Berlin52	(N/A)	(N/A)	7542 (7544.37)	(N/A)	(N/A)	7542(7544.38)	7542 (7544.37)
Eil50	425(427.96)	428(N/A)	(N/A)	426(427.86)	443(N/A)	425(427.97)	425(N/A)
Eil51	426	496	426 (428.87)	(N/A)	(N/A)	426(428.98)	426 (429.98)
Eil75	535(542.31)	545(N/A)	(N/A)	542(549.18)	580(N/A)	535(542.28)	535(N/A)
Eil76	560(562.6)	(N/A)	538 (544.37)	(N/A)	(N/A)	538(545.40)	538 (545.39)
Eil101	672.1	(N/A)	629(640.975)	(N/A)	(N/A)	629(642.32)	629 (642.31)
KroA100	21,282(21,285.44)	21,761(N/A)	21,282	(N/A)	(N/A)	21,282(21,283.85)	21,282(N/A)

Numerical experiments were executed with HPACO, whereas the performance figures for the other algorithms were taken from the literature.Results using EP are from [9] and those using GA are from [10] for KroA100 and [11] for Oliver30, Eil50, and Eil75. Results using SA are from [12]. Oliver30 is from [13], Eil50, Eil75 are from [14], ACS results are taken from (Ant colonies for the travelling salesman problem, TR/IRIDIA/1996-3, *Université Libre de Bruxelles*, Belgium, pp1-10) and the results of Hybrid GA are taken from[15]. A graphical representation for the performance of our HPACO against other techniques on different standard TSP instances Oliver30, Att48, Berlin 52, Eil50, Eil51, Eil75, Eil76, Eil101 and KroA100 are shown in Fig 5.

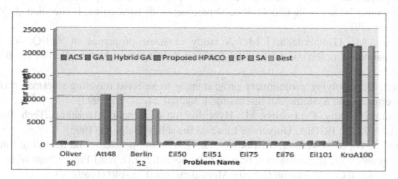

Fig. 5. Graphical representation for standard TSP instances along with proposed HPACO vs. tour length

5   Conclusion and Future Scope

Our proposed High Performance Ant Colony Optimizer technique is capable to converge towards goal earlier than conventional ACS as well as consumes less number of iterations. It is also observed our proposed algorithm is able to achieve the so far known results from the TSP Library TSPLIB: http://www.iwr.uni-heidelberg.de /iwr/comopt/soft/TSPLIB95/TSPLIB.html. By modifying the pheromone updation procedure, we have achieved a better performance than conventional ACS. HPACO can be applied on various applications like: job scheduling problem, time scheduling problem, multicasting, sequential ordering problem, resource constraint project-scheduling problem, open-shop scheduling problem, etc.

References

1. Zhang, J.: Natural Computation for the Traveling Salesman Problem. In: Proceedings of the Second International Conference on Intelligent Computation Technology and Automation (ICICTA), vol. 1, pp. 366–369 (2009)
2. Wang, L., Wang, D., Ding, N.: Research on BP Neural Network Optimal Method Based on Improved Ant Colony Algorithm. In: Second International Conference on Computer Engineering and Applications (ICCEA), vol. 1, pp. 117–121 (2010)
3. Dorigo, M., Socha, K.: An Introduction to Ant Colony Optimization. Technical Report IRIDIA 2006-010 (2006)
4. Dorigo, M., Gambardella, L.: Ant colony system: A cooperative learning approach to the traveling salesman problem. IEEE Transactions on Evolutionary Computation 1(1), 53–66 (1997)

5. Stützle, T., Hoos, H.H.: Max–min ant system. Future Generation Computer Systems 16(8), 889–914 (2000)
6. Dorigo, M., Maniezzo, V., Colorni, A.: Positive feedback as a search strategy. Technical Report 91 (1991)
7. Bullnheimer, B., Hartl, R.F., Strauss, C.: A new rank-based version of the ant system: A computational study. Central European Journal for Operations Research and Economics 7(1), 25–38–016 (1999), Dipartimento di Elettronica, Politecnico di Milano, Italy, (1999)
8. Dorigo, M., Gambardella, L.M.: A study of some properties of Ant-Q. In: Fourth International Conference on Parallel Problem Solving from Nature, Berlin, pp. 656–665 (1996)
9. Fogel, D.: Applying evolutionary programming to selected traveling salesman problems. Cybernetics and Systems: An International Journal 24, 27–36 (1993)
10. Bersini, H., Oury, C., Dorigo, M.: Hybridization of Genetic Algorithms.,Tech. Rep. No. IRIDIA 95-22, IRIDIA, Université Libre de Bruxelles, Belgium (1995)
11. Whitley, D., Starkweather, T., Fuquay, D.: Scheduling problems and travelling salesman: The geneticedge recombination operator. In: Schaffer, J.D. (ed.) Proceedings of the Third International Conference on Genetic Algorithms, pp. 133–140 (1989)
12. Lin, F.–T., Kao, C.–Y., Hsu, C.–C.: Applying the genetic approach to simulated annealing in solving some NP-hard problems. IEEE Transactions on Systems, Man, and Cybernetics 23, 1752–1767 (1993)
13. Oliver, I., Smith, D., Holland, J.R.: A study of permutation crossover operators on the travelling salesman problem. In: Grefensfette, J.J. (ed.) Proceedings of the Second International Conference on Genetic Algorithms, pp. 224–230 (1987)
14. Eilon, S., Watson-Gandy, C.D.T., Christofides, N.: Distribution management: mathematical modeling and practical analysis. Operational Research Quarterly 20, 37–53 (1969)
15. Jayalakshmi, G.A., Sathiamoorthy, S.: A hybrid genetic algorithm – A new approach tosolve traveling salesman problem. International Journal of Computational Engineering Science 2, 339–355 (2001)
16. Sahana, S.K., Jain, A.: An Improved Modular Hybrid Ant Colony Approach for Solving Traveling Salesman Problem. International Journal on Computing (JoC) 1(2), 123–127 (2011), doi: 10.5176-2010-2283_1.249, ISSN: 2010-2283
17. Sahana, S.K., Jain, A., Mustafi, A.: A Comparative Study on Multicast Routing using Dijkstra's, Prims and Ant Colony Systems. International Journal of Computer Engineering & Technology (IJCET) 2(2), 301–310 (2010), ISSN : 0976-6367
18. Johnson, D.S., McGeoch, L.A., Rothberg, E.E.: Asymptotic Experimental Analysis for the Held-Karp Traveling Salesman Bound. In: Proceedings of the Annual ACM-SIAM Symposium on Discrete Algorithms, pp. 341–350 (1996)
19. Karp, R.M.: Dynamic programming meets the principle of inclusion and exclusion. Oper. Res. Lett. 1(2), 49–51 (1982), doi:10.1016/0167-6377(82)90044-X
20. Wu, Z.L., Zhao, N., Ren, G.H., Quan, T.F.: Population declining ACO algorithm and its applications. Expert Systems with Applications 36(3), 6276–6281 (2009)
21. Udomsakdigool, A., Kachitvichyanukul, V.: Multiple colony ant algorithm for job-shop scheduling problem. International Journal of Production Research 46(15), 4155–4175 (2008)
22. Sahana, S.K., Jain, A.: Modified Ant Colony Optimizer (MACO) for the Travelling Salesman Problem. In: Computational Intelligence and Information Technology, CIIT 2012, Chennai, December 3-4. ACEEE Conference Proceedings Series 3 by Elsevier, pp. 267–276 (2012), ISBN:978-93-5107-194-5

A Novel *Physarum*-Based Ant Colony System for Solving the Real-World Traveling Salesman Problem

Yuxiao Lu[1], Yuxin Liu[1], Chao Gao[1], Li Tao[1], and Zili Zhang[1,2,*]

[1] School of Computer and Information Science
Southwest University, Chongqing 400715, China
[2] School of Information Technology, Deakin University, VIC 3217, Australia
zhangzl@swu.edu.cn

Abstract. The solutions to Traveling Salesman Problem can be widely applied in many real-world problems. Ant colony optimization algorithms can provide an approximate solution to a Traveling Salesman Problem. However, most ant colony optimization algorithms suffer premature convergence and low convergence rate. With these observations in mind, a novel ant colony system is proposed, which employs the unique feature of critical tubes reserved in the *Physarum*-inspired mathematical model. A series of experiments are conducted, which are consolidated by two real-world Traveling Salesman Problems. The experimental results show that the proposed new ant colony system outperforms classical ant colony system, genetic algorithm, and particle swarm optimization algorithm in efficiency and robustness.

Keywords: *Physarum*-Inspired Mathematical Model, Real-World Traveling Salesman Problem, Ant Colony System, Meta-Heuristic Algorithm.

1 Introduction

The problem of route design in the real world can be formulated as a Traveling Salesman Problem (TSP). Designing an efficient approach to solve a TSP has great practical significance. Recently, many meta-heuristic algorithms, such as ant colony optimization (ACO) [1], particle swarm optimization (PSO) [2] and genetic algorithm (GA) [3], have been proposed based on the swarm intelligence for solving a TSP. In particular, ACO algorithms are originally designed and have a long tradition in solving a TSP [1,4]. However, ACO algorithms suffer premature convergence and stagnation in generally, and the convergence rate of ACO algorithms are slow [5].

Currently, a unicellular and multi-headed slime mold, *Physarum polycephalum*, shows an ability to form self-adaptive and high efficient networks in biological experiments [6,7]. Moreover, Tero et al. [8] capture the positive feedback mechanism of *Physarum* in foraging and build a mathematical model (PMM). The PMM exhibits a unique feature of critical tubes reserved in the process of network evolution.

---

* Corresponding author.

Y. Tan et al. (Eds.): ICSI 2014, Part I, LNCS 8794, pp. 173–180, 2014.
© Springer International Publishing Switzerland 2014

Taking advantage of this feature, Zhang et al. [9,10] have proposed an optimization strategy for updating the pheromone matrix in ACO algorithms.

For promoting Zhang's work, this paper devotes to answer two questions: 1) Whether the performance of the optimized ACO algorithms based on the PMM, denoted as PMACO algorithms, is better than other meta-heuristic algorithms (e.g., PSO and GA) for solving a TSP? 2) Whether PMACO algorithms can solve the real-world TSP?

The organization of this paper is as follows. Section 2 introduces the formulation and measurements of the real-world TSP. Section 3 presents the basic idea of the optimization strategy for ant colony system (ACS). Section 4 formulates two real-world TSPs and provides comparable results of different algorithms. Section 5 concludes this paper.

2   Problem Statement

In the real world, when people plan to visit several cities, they tend to choose the most economical route. That is, to determine an access-order in which each city should be visited exactly once, and the total distance is minimal. This problem can be formulated as a classical TSP by the following description.

For a complete weighted graph $G = (V, E)$, let nodes $V = \{v_1, v_2, ..., v_n\}$ represent the geographical locations of n cities, and edges $E = \{(v_i, v_j)|v_i, v_j \in V, v_i \neq v_j\}$ represent the set of roads. The distance between two cities v_i and v_j can be denoted as $d(v_i, v_j)$. Then, TSP solution algorithms can be used to find the shortest Hamiltonian circuit x, which will be the optimal route in the real world. The length of x, denoted as S_{min}, is shown in Eq. (1), where $v_{x(i)}$ represents the ith city in the Hamiltonian circuit x, and $v_{x(i)} \in V$.

$$S_{min} = \min(\sum_{i=1}^{n-1} d(v_{x(i)}, v_{x(i+1)}) + d(v_{x(n)}, v_{x(1)})) \tag{1}$$

In order to evaluate the performance of TSP solution algorithms, some measurements are defined as follows:

1. S_{min} stands for the optimal solution to a TSP, which is the length of the shortest Hamiltonian circuit calculated by an algorithm.
2. $S_{average}$ and $S_{variance}$ stand for the average value and the variance of results to a TSP, respectively. They are obtained after C times computation repeatedly, such as $S_{average}$ is calculated as $\sum_{i=1}^{C} S_{min}^{i,steps(k)}/C$, where $S_{min}^{i,steps(k)}$ represents the optimal solution in the kth step for the ith time.

3   The *Physarum*-Based Ant Colony System

The novel ACS algorithm, denoted as PM-ACS (one of the typical PMACO algorithms), proposes an optimization strategy for updating the global pheromone matrix in ACS based on the PMM [9,10]. In the PM-ACS-based TSP, they assume that there is a *Physarum* network with pheromone flows in tubes. Food sources and tubes of the *Physarum* network represent cities and roads connecting two different cities, respectively. After each iteration of ant colony, the amount of pheromone in each tube of the *Physarum* network can be calculated. When updating the global pheromone matrix, PM-ACS considers both of the pheromone released by ants and the flowing pheromone in the *Physarum* network.

Formally, in PM-ACS, the *global pheromone matrix updating rule* in ACS is optimized by appending the amount of flowing pheromone in the *Physarum* network. As shown in Eq. (2), the first two terms come from ACS and the last one is newly added based on the PMM.

$$\tau_{ij} \leftarrow [(1-\rho)\tau_{ij} + \frac{\rho}{S_{global-best}}] + \varepsilon \frac{Q_{ij} \times M}{I_0}, \forall (i,j) \in S_{global-best} \qquad (2)$$

where

$$\varepsilon = 1 - \frac{1}{1 + \lambda^{\frac{Psteps}{2} - (t+1)}} \qquad (3)$$

$$Q_{ij} = \frac{1}{M} \sum_{m=1}^{M} \left| \frac{D_{ij}}{L_{ij}} (p_i^m - p_j^m) \right| \qquad (4)$$

In Eq. (2), ε is defined as an impact factor to measure the effect of flowing pheromone in the *Physarum* network on the final pheromone matrix. As shown in Eq. (3), $Psteps$ stands for the total steps of iteration affected by the PMM, t is the steps of iteration at present, and $\lambda \in (1, 1.2)$. M represents the number of roads in a TSP. I_0 represents the fixed flux flowing in the *Physarum* network.

In one iteration of the PMM, each pair of nodes connected by a tube has an opportunity to be selected as inlet/outlet. When two nodes a and b connected by the mth tube are selected as inlet and outlet nodes, respectively, the pressure on each node p_i^m can be calculated according to the Kirchhoff Law based on Eq. (5).

$$\sum_i \frac{D_{ij}}{L_{ij}} (p_i^m - p_j^m) = \begin{cases} -I_0 & for \ j = a \\ I_0 & for \ j = b \\ 0 & otherwise \end{cases} \qquad (5)$$

where L_{ij} is the length of the tube (i,j), D_{ij} is defined as a measure of the conductivity, which is related to the thickness of the tube.

The above process is repeated until all pairs of nodes in each tube are selected as inlet/outlet nodes once. Then, the flux Q_{ij} through the tube (i,j) is calculated based on Eq. (4). As the iteration goes on, the conductivity of a tube adapts according to the flux based on Eq. (6). Then, the conductivities at the next iteration step will be fed back to Eq. (5), and the flux will be updated based on Eq. (4). Based on the positive feedback mechanism between the conductivity and the flux, the shorter tubes, which we called critical tubes, will become wider and be reserved in the process of network evolution. While, other longer tubes will become narrower and finally disappear.

$$\frac{dD_{ij}}{dt} = \frac{|Q_{ij}|}{1 + |Q_{ij}|} - D_{ij} \qquad (6)$$

Algorithm 1 presents steps of PM-ACS for solving a TSP. The meaning of each parameter can be seen in Table 1.

Algorithm 1. The *Physarum*-Based Ant Colony System (PM-ACS)

1: Initialize parameter $\alpha, \beta, \rho, s, q_0, \lambda, I_0$, *Psteps* and *Tsteps*
2: Initialize pheromone trail τ_0 and conductivity of each tube D_{ij}
3: Set the iteration counter $N := 0$
4: **while** $N < Tsteps$ **do**
5:    **for** $k := 1$ to s **do**
6:        Construct a tour by ant k
7:        Update the local pheromone matrix
8:    **end for**
9:    $best :=$ find the global best ant
10:   $S_{min} :=$ the length of the tour generated by ant $best$
11:   Calculate the flowing pheromone in the *Physarum* network
12:   Update the global pheromone matrix
13:   $N := N + 1$
14: **end while**
15: Output the optimal solution S_{min}

---

4   Experimental Results

4.1   Benchmark Datasets

Two benchmark datasets, eil51 and eil76, are downloaded from the website TSPLIB[1]. All experiments are undertaken in the same experimental environment. The main parameters and their values are listed in Table 1. All results in our experiments are averaged over 50 times.

Table 1. Main parameters and their values used in this paper

Parameter	Explanation	Value
α	The relative importance of the pheromone trail	1
β	The relative importance of the heuristic information	2
ρ	The pheromone evaporation rate	0.8
s	The number of ants	*the number of cities*
q_0	A predefined parameter between 0 and 1	0.1
λ	A parameter determined the value of ε	1.05
I_0	The fixed flux flowing in the *Physarum* network	20
Psteps	The total steps of iteration affected by the PMM	300
Tsteps	The total steps of iteration	300
τ_0	The initial amount of pheromone in each road	1
D_{ij}	The initial value of the conductivity of each tube	1

Figure 1 shows that S_{min} and $S_{average}$ of PM-ACS are the minimum compared with ACS, PSO and GA. It means that PM-ACS has the strongest ability to

---
[1] http://www.iwr.uni-heidelberg.de/groups/comopt/software/TSPLIB95/

exploit the optimal solution. Furthermore, $S_{variance}$ of PM-ACS is the lowest of the four algorithms, which means that PM-ACS has the strongest robustness.

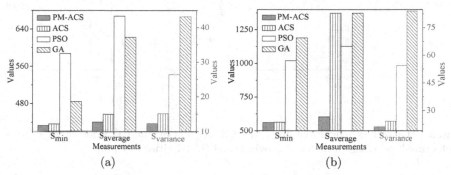

Fig. 1. The efficiency comparison among PM-ACS, ACS, PSO and GA in benchmark datasets: (a) eil51 and (b) eil76

4.2 Empirical Studies

In this section, we consider two real scenarios: How travelers plan their trips to visit 17 prefecture-level cities in Sichuan province and 34 cities in China with the lowest cost.

4.2.1 Formulation of Datasets

First, the latitude and longitude of each city can be extracted from Google Maps. Second, the distance (D_{AB}) between cities A and B is defined as the spherical distance, which can be calculated based on Eq. (7) [11]. Let lat_A and lng_A represent the latitude and longitude of city A, respectively. Then $\varphi_A = lat_A \times \pi/180$ represents the radian of the latitude of city A, and $\lambda_A = lng_A \times \pi/180$ represents the radian of the longitude of city A. R_0 is the radius of earth, π is a symbol in the mathematics, and we set $R_0 = 6378.1370$, $\pi = 3.1416$.

$$D_{AB} = 2R_0 \arcsin\left(\sqrt{sin^2(\frac{\varphi_A - \varphi_B}{2}) + cos(\varphi_A)cos(\varphi_B)sin^2(\frac{\lambda_A - \lambda_B}{2})}\right) \quad (7)$$

4.2.2 Results

Two real-world TSPs are solved by PM-ACS, ACS, PSO and GA, respectively, and values of parameters are the same as Section 4.1. Fig. 2(a) shows that, the four algorithms all can find the optimal solution ($S_{min} = 1445.32$) of 17 prefecture-level cities in Sichuan province, while the average value ($S_{average}$) and variance ($S_{variance}$) of the results calculated by PM-ACS are better than ACS, PSO and GA. The advantage of PM-ACS is more obvious in Fig. 2(b), where S_{min}, $S_{average}$ and $S_{variance}$ of PM-ACS are all better than ACS, PSO and GA. In particular, Fig. 3 illustrates the optimal tour found by PM-ACS for 34 cities' TSP in China.

What's more, Fig. 4(a) and Fig. 4(b) plot the convergent process of $S_{average}$ and $S_{variance}$ with the increment of iterative steps of PM-ACS, ACS, PSO and

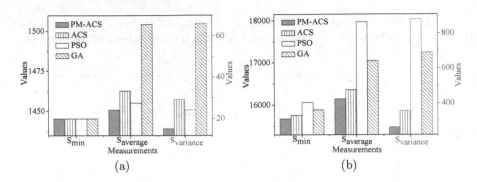

Fig. 2. The results calculated by PM-ACS, ACS, PSO and GA for solving (a) 17 prefecture-level cities in Sichuan province and (b) 34 cities in China

GA for solving the 34 cities' TSP in China, respectively. Fig. 4(a) shows that $S_{average}$ of PM-ACS or ACS decreases more obviously than PSO and GA. Specially, in the earlier iteration, the gap of $S_{average}$ between PM-ACS and ACS is little. With the increment of steps, $S_{average}$ of PM-ACS is better than ACS. The convergent process of $S_{variance}$ in Fig. 4(b) also shows that PM-ACS has the strongest robustness compared with ACS, PSO and GA.

Fig. 3. The illustration of the optimal solution to 34 cities' TSP in China

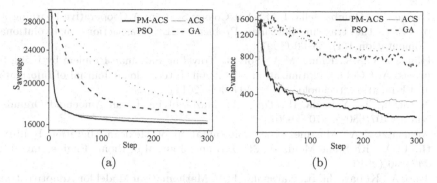

Fig. 4. The results calculated by PM-ACS, ACS, PSO and GA for solving the 34 cities' TSP in China: (a) $S_{average}$ and (b) $S_{variance}$

5 Conclusion

In this paper, we present and estimate a *Physarum*-based ACS (i.e., PM-ACS) that employs the unique feature of critical tubes reserved in the process of network evolution of the PMM. The PM-ACS can enhance the amount of pheromone in critical roads and promote the exploitation of the optimal solution. In order to estimate the efficiency of the optimization strategy for solving a TSP, we compare PM-ACS with traditional meta-heuristic algorithms in benchmark networks. Specially, two real-world TSPs are consolidated for showing the performance of the optimization strategy. Experimental results validate that the optimization strategy is efficient in improving the search ability and robustness of ACO algorithms for solving a TSP.

Acknowledgement. This work was supported by the National Science and Technology Support Program (No. 2012BAD35B08), the Specialized Research Fund for the Doctoral Program of Higher Education (No. 20120182120016), Natural Science Foundation of Chongqing (Nos. cstc2012jjA40013, cstc2013jcyjA400-22), and the Fundamental Research Funds for the Central Universities (Nos. XDJK2012B016, XDJK2012C018, XDJK2013D017).

References

1. Colorni, A., Dorigo, M., Maniezzo, V.: Distributed Optimization by Ant Colonies. In: Proceedings of the First European Conference on Artificial Life, vol. 142, pp. 134–142 (1991)
2. Wang, K.P., Huang, L., Zhou, C.G., Pang, W.: Particle Swarm Optimization for Traveling Salesman Problem. In: Proceedings of the Second International Conference on Machine Learning and Cybernetics, Xi'an, China, vol. 3, pp. 1583–1585 (2003)
3. Chatterjee, S., Carrera, C., Lynch, L.A.: Genetic Algorithms and Traveling Salesman Problems. European Journal of Operational Research 93(3), 490–510 (1996)

180   Y. Lu et al.

4. Dorigo, M., Gambardella, L.M.: Ant Colony System: A Cooperative Learning Approach to the Traveling Salesman Problem. IEEE Transactions on Evolutionary Computation 1(1), 53–66 (1997)
5. Hlaing, Z.C.S.S., Khine, M.A.: Solving Traveling Salesman Problem by Using Improved Ant Colony Optimization Algorithm. International Journal of Information and Education Technology 1(5), 404–409 (2011)
6. Nakagaki, T., Yamada, H., Tóth, A.: Maze-Solving by an Amoeboid Organism. Nature 407(6803), 470 (2000)
7. Adamatzky, A., Martinez, G.J.: Bio-Imitation of Mexican Migration Routes to the USA with Slime Mould on 3D Terrains. Journal of Bionic Engineering 10(2), 242–250 (2013)
8. Tero, A., Kobayashi, R., Nakagaki, T.: A Mathematical Model for Adaptive Transport Network in Path Finding by True Slime Mold. Journal of Theoretical Biology 244(4), 553–564 (2007)
9. Qian, T., Zhang, Z., Gao, C., Wu, Y., Liu, Y.: An Ant Colony System Based on the Physarum Network. In: Tan, Y., Shi, Y., Mo, H. (eds.) ICSI 2013, Part I. LNCS, vol. 7928, pp. 297–305. Springer, Heidelberg (2013)
10. Zhang, Z.L., Gao, C., Liu, Y.X., Qian, T.: A Universal Optimization Strategy for Ant Colony Optimization Algorithms Based on the Physarum-Inspired Mathematical Model. Bioinspiration and Biomimetics 9, 036006 (2014)
11. Wikipedia,
    http://en.wikipedia.org/wiki/Great-circle_distance#cite_note-3

Three New Heuristic Strategies for Solving Travelling Salesman Problem

Yong Xia, Changhe Li*, and Sanyou Zeng

School of Computer Science, China University of Geosciences, Wuhan, 430074, China
455399810@qq.com, changhe.lw@gmail.com, sanyouzeng@gmail.com

Abstract. To solve a travelling salesman problem by evolutionary algorithms, a challenging issue is how to identify promising edges that are in the global optimum. The paper aims to provide solutions to improve existing algorithms and to help researchers to develop new algorithms, by considering such a challenging issue. In this paper, three heuristic strategies are proposed for population based algorithms. The three strategies, which are based on statistical information of population, the knowledge of minimum spanning tree, and the distance between nodes, respectively, are used to guide the search of a population. The three strategies are applied to three existing algorithms and tested on a set of problems. The results show that the algorithms with heuristic search perform better than the originals.

Keywords: Heuristic Strategies, Evolutionary Algorithms, Travelling Salesman Problem.

1 Introduction

Although many population based approaches have been proposed to travelling salesman problem (TSP), most of them focus on specific techniques. They are hardly to be abstracted to a general framework to guide algorithm design. To effectively solve TSP by population based algorithms, a challenging issue is how to identify links/edges that belong to the global optimum during the search process. This paper proposes several general methods to obtain useful information to guide the search of an algorithm. In this paper, three heuristic strategies are proposed. The first strategy is to use the statistical information of a population: for a specific node, the probability of its "next" city is calculated based on the population. The second is to use the knowledge of minimum spanning tree (MST). Experimental results show that links of MST have a very large probability of belonging to the global optimum. The third is to use distance knowledge between nodes where relatively short links have a large probability of being the links in the global optimum. We employ the three strategies to generate three probability matrices to guide the search. The three heuristic strategies are implemented to three existing algorithms. Experimental results show that significant improvement is obtained by using these strategies in comparison with the original algorithms.

---

* Corresponding author.

Y. Tan et al. (Eds.): ICSI 2014, Part I, LNCS 8794, pp. 181–188, 2014.

The paper is organized as follows. Section 2 introduces related algorithms for solving TSP. Section 3 provides the detailed description of the proposed strategies and Section 4 presents the experiment results and discussions. Finally, conclusions are given in Section 5.

2   Related Work for TSP

In the literature of evolutionary computation for TSP, population based algorithms are one of the major research areas. Many operators and strategies have been proposed to address this problem. In the research, some researchers found that the quality of individual solutions in the initial population plays a critical role in determining the quality of the final solution [1]. The method that randomly initializes a population is simple and makes the initial population with good diversity. However, its whole population contains many individuals with bad quality, infeasible solutions sometimes [2]. Methods with a random initial population require a large amount of time to find an optimal solution. Thus, several approaches were proposed to address this issue the claim [1,2].

Nearest neighbor (NN) tour construction heuristic is one of the alternatives for random population seeding in GA, particularly for TSP [3][4]. In the NN technique, individuals in the population seeding are constructed with the nearest city to the current city and such good individuals can refine the subsequent search in the next generation [3]. Though NN works fine, it suffers some critical issues: Several cities are not included in the individuals created initially and have to be inserted at high costs in the end; Neglecting several cities at the population seeding stage leads to severe errors in optimal solution construction and the population diversity is very poor.

Wei et al. [5] proposed a greedy GA (GGA), in which the population seeding is performed using a gene bank (GB). The GB is built by assembling the permutation of n cities based on their distance. In GGA, the population is generated from the GB such that the individuals are of above-average fitness and short length. In GGA, the increase in the number of cities leads to augmented problem complexity and performance degradation. Moreover, a large collection of GB individuals enlarges the cost of computation at each generation. The improved performance of GGA is justified using TSP with a maximum of hundreds cities.

As we mentioned above, the population has very useful information. In fact, heuristic algorithms, e.g., Inverover (GuoTao) algorithm [6] utilizes the information of the population to guide the convergence of population. The inver-over operator has the characteristics of both crossover and mutation. The part of crossover depends on the population. If some edges have a larger frequency appearing among individuals, they will have a higher probability to be generated by inverse operation in the current individual. This indicates that the information of population is important.

3   Heuristic Strategies for TSP

In this section, we will introduce the three general strategies for population based algorithms to identify promising links in detail. Each strategy generates a probability matrix to guide the search of the population. To validate the effectiveness of the proposed strategies, they are implemented into three existing population based algorithms.

Table 1. The percentage of edges that belong to the global optimum in MST and between near cities

Instance	MST	The number of the nearest cities					Instance	MST	The number of the nearest cities				
		1	2	3	4	5			1	2	3	4	5
KROA100	0.768	0.59	0.79	0.87	0.95	0.99	EIL76	0.72	0.579	0.895	0.934	0.987	0.987
EIL101	0.73	0.564	0.861	1	1	1	CH130	0.736	0.554	0.808	0.885	0.923	0.962
CH150	0.765	0.587	0.847	0.933	0.973	0.987	PR76	0.733	0.605	0.829	0.934	0.947	0.974
LIN105	0.779	0.619	0.8	0.857	0.933	0.962	TSP225	0.799	0.587	0.84	0.947	0.964	0.982

3.1 Three Heuristic Strategies

Strategy Based on Topology. An edge between two near cities in geographic space usually has a larger probability of being in the global optimum than those edges between two cities that are far away from each other. The second and fifth columns in Table 1 present percentages of edges between different numbers of nearest cities on eight TSP instances. From the table, it can be seen that more than half of edges in the global optimum are those edges between two nearest cities. The percentage almost increases to 1 when the number of edges between nearest cities increases to five. Therefore, we believe that this information is useful to guide the search of EAs. To implement the idea, we built a probability matrix A_{n*n} (n is the number of cities) according to the city topology of a TSP instance. For a given city i, the nearer city j to i, the larger probability it will be chosen as the next city of i. The Eq. (1) below is used to calculate element a_{ij} in A, where $cost_{ij}$ indicates the distance from node i to node j and n is the number of nodes. Fig. 1 shows a TSP instance with five nodes and distances between nodes, where a_{02} can be obtained by $a_{02} = 1/2^2/(1/2^2 + 1/4^2 + 1/6^2 + 1/10^2) = 0.714$. The probability matrix A is shown in Fig. 2.

Strategy Based on MST. The first and forth columns in Table 1 present the percentages of edges of MST that belong to the global optimum on eight TSP instances. From the results, more than 70 percent of edges of MST are in the global optima for all TSP cases. The results indicate that the knowledge of MST is also useful. Fig. 3 presents one MST of the TSP example in Fig. 1. To build a probability matrix A with the knowledge of MST, links connected to city i are equally treated, i.e., cities connected to city i have the same probability to be selected. For example, cities 0, 1, and 4 have the same probability $(1/3)$ of being chosen as the next city of 2. The probability matrix A is shown in Fig. 4.

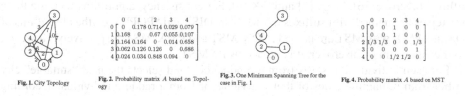

Fig. 1. City Topology

Fig. 2. Probability matrix A based on Topology

Fig. 3. One Minimum Spanning Tree for the case in Fig. 1

Fig. 4. Probability matrix A based on MST

Strategy Based on Population. Generally speaking, for an EA the population will gradually improves as the evolutionary process goes on. The improvement of the population may contain two different aspects. Firstly, more good edges (those brings in improvement) may appear in the population as the search goes on. Secondly, the same good edge may appear in more individuals. Based on this motivation, we count the frequency of an edge that appears in the population. The larger frequency of an edge denotes the more promising it is.

However, we cannot directly use this information to create a probability matrix. To build a matrix A, we need to make some transformations. Intuitively, edges in better individuals should be paid more attention than edges in worse individuals as better individuals have shorter tour lengths than worse individuals. To achieve it, we assign an individual a weight w, which relates to the fitness of that individual. For an individual i, we use Eq. (2) to calculate its relative fitness f'_i, where l_{max} and l_{min} are the maximum and minimum tour length in the population, respectively.

For an individual, to make a relationship between its relative fitness and its weight, we adopted the Eq. (3), where α is a positive value. Eq. (3) indicates that the weight value increases faster for a small fitness than a large fitness. Based on our experimental results, we set α to 4 in this paper. Then, we could calculate element a_{ij} in A. To achieve it, we sum the weights of all individuals, in which edge (i, j) is, then calculate a normalized weight a_{ij}. For example, a population contains three individuals for the instance in Fig.1, i.e., Individual 1: $(0,1,2,3,4)$, tour length: 22; Individual 2: $(0,2,3,4,1)$, tour length: 21; Individual 3: $(0,3,4,2,1)$, tour length: 20. Using Eq.(2) and Eq.(3), we get $w_1=0.58$, $w_2=0.87$, and $w_3=0.96$. Each element in matric A, e.g., a_{23}, is calculated by $a_{23} = (w_1 + w_2)/(2 * (w_1 + w_2 + w_3)) = 0.3$ (edge $(2,3)$ appearing in individuals 1 and 2). The probability matrix is shown in Fig. 5.

$$a_{ij} = \frac{1}{cost_{ij}^2 * \sum_{j=0}^{n-1}(1/cost_{ij}^2)}, i,j = 0,1,\cdots,n-1. \quad (1)$$

$$f'_i = (l_{max} - l_i + 1)/(l_{max} - l_{min} + 1) \quad (2)$$

$$w_i = 2 * (1/(1 + e^{-\alpha * f'_i})) - 1 \quad (3)$$

$$\begin{array}{c} \quad \; 0 \quad 1 \quad \; 2 \quad \; 3 \quad \; 4 \\ \begin{array}{c} 0 \\ 1 \\ 2 \\ 3 \\ 4 \end{array} \left[\begin{array}{ccccc} 0 & 0.5 & 0.18 & 0.2 & 0.12 \\ 0.5 & 0 & 0.32 & 0 & 0.18 \\ 0.18 & 0.32 & 0 & 0.3 & 0.2 \\ 0.2 & 0 & 0.3 & 0 & 0.5 \\ 0.12 & 0.18 & 0.2 & 0.5 & 0 \end{array} \right] \end{array}$$

Fig. 5. Probability matrix based on population

3.2 Strategy Instantiation

In order to test the proposed strategies, we implemented them into three existing algorithms: Inverover [6], PMX [7] and OX [8]. For consistency, algorithms with the three strategies are named with suffixes of "MST", "DIS" and "POP", e.g., the algorithms of OX and PMX with MST are named as OX_MST and PMX_MST, respectively. However, to the algorithm of Inverover, it is named as IO_MST.

For each individual, it starts from a random city and selects the next "suitable" city either from remaining cities of that individual or from a random individual according to a certain probability in the original Inverover algorithm. Here, to implement our strategies into Inverover, we change the behavior of selecting the next city. The choice of the next city is made according to the probability matrix A instead of the original rule. Algorithm 1 shows Inverover with proposed strategies (see steps $8 - 17$ for procedures to select the next city). The matrix A can be obtained by the three strategies depending on the strategy to be used.

Algorithm 1. Inverover with the proposed strategies

---

1. Random initialization of a population P with size of n;
2. **while** (not satisfied termination-condition) **do**
3.   **for** (each individual $S_i \in P$) **do**
4. $S' = S_i$; select (randomly) a city i from S';
5.   **loop**
6. $z \leftarrow 0; p = rand()$;
7.   **while** ($z < n - 1$) **do**
8.   **if** ($p \leq A'_{i0}$) **then** $j = 0$; **break; end if**
9.   **if** ($p > A'_{iz} \&\&p \leq A'_{i(z+1)}$) **then** $j = z + 1$; **break; end if**
10. $z \leftarrow z + 1; \{a'_{ij} = \sum a_{ik}, k = 0, 1, ..., j\}$
11.   **end while**
12.   **if** (the next city or the previous city of city i in S' is j) **then exit** from the loop **end if**;
13.   inverse the section from the next city of city i to the city j in S'; $i = j$;
14.   **end loop**
15.   **If** ($eval(S') \leq eval(S_i)$) **then** $S_i = S'$; **end if**
16.   **end for**
17. **end while**

Table 2. Performance comparison between the proposed algorithms with the three strategies and their original algorithms

Problem	Algorithm	Error	R1	R2	S_ratio	Evals	Algorithm	Error	R1	R2	S_ratio	Evals	Algorithm	Error	R1	R2	S_ratio	Evals
CH150	Inverover	23.8$^{\sim}$±218	1	0.92	0.17	1287507	PMX	3691.8$^{\sim}$±8.2e5	0.43	0.43	0.00	5277534	OX	4572.8$^{\sim}$±2.2e5	1	0.36	0.00	31762522
	IO_DIS	130.9$^{\sim}$±1.1e3	1	0.83	0.00	5429889	PMX_DIS	1172.3^{+}±1.7e4	1	0.68	0.00	22091533	OX_DIS	500.6^{+}±9.9e3	1	0.74	0.00	14887753
	IO_MST	1025.2$^{\sim}$±9.1e3	0.92	0.73	0.00	416258	PMX_MST	1379.0^{+}±2.0e4	1	0.64	0.00	21625808	OX_MST	764.8^{+}±1.7e4	0.99	0.71	0.00	15361670
	IO_POP	11.0^{+}±105	1	0.96	0.43	1000255	PMX_POP	1278.9^{+}±1.8e4	1	0.66	0.00	23839989	OX_POP	686.9^{+}±9.0e3	1	0.71	0.00	10228646
KROC100	Inverover	8.6$^{\sim}$±396	1	0.99	0.80	434672	PMX	6163.5$^{\sim}$±5.6e6	0.57	0.57	0.00	2325874	OX	13135.1$^{\sim}$±3.1e6	1	0.38	0.00	10143379
	IO_DIS	77.3$^{\sim}$±2.4e3	1	0.93	0.03	1606946	PMX_DIS	2332.9^{+}±2.6e5	1	0.76	0.00	7605371	OX_DIS	874.8^{+}±6.8e4	1	0.81	0.00	4788670
	IO_MST	3148.4$^{\sim}$±2.6e5	0.93	0.76	0.00	153784	PMX_MST	2982.7^{+}±2.1e5	1	0.71	0.00	7938812	OX_MST	1532.2^{+}±6.9e4	0.99	0.77	0.00	5653965
	IO_POP	0.0^{+}±0	1	1	1.00	294982	PMX_POP	2667.3^{+}±1.7e5	1	0.72	0.00	8637833	OX_POP	1308.1^{+}±2.6e5	0.99	0.76	0.00	6750488
LIN105	Inverover	4.3$^{\sim}$±133	1	0.99	0.87	443622	PMX	4514.9$^{\sim}$±2.3e6	0.58	0.58	0.00	2670379	OX	9239.6$^{\sim}$±2.7e6	1	0.39	0.00	11653478
	IO_DIS	79.2$^{\sim}$±2.4e3	1	0.93	0.03	1477190	PMX_DIS	1800.9^{+}±1.1e5	1	0.77	0.00	8650150	OX_DIS	863.4^{+}±3.8e4	1	0.85	0.00	5650738
	IO_MST	2545.2$^{\sim}$±9.1e4	0.96	0.74	0.00	210998	PMX_MST	2162.7^{+}±1.2e5	0.99	0.72	0.00	6741161	OX_MST	1452.7^{+}±8.1e4	0.99	0.76	0.00	4739845
	IO_POP	0.0^{+}±0	1	1	1.00	323164	PMX_POP	2014.9^{+}±5.5e4	1	0.73	0.00	9596997	OX_POP	888.2^{+}±7.5e4	0.99	0.79	0.00	7998658
TSP225	Inverover	52.9$^{\sim}$±208	1	0.86	0.00	3278213	PMX	26491.0$^{\sim}$±1.1e6	0.72	0.04	0.00	963889	OX	3285.3$^{\sim}$±9.2e4	1	0.29	0.00	52586126
	IO_DIS	152.8$^{\sim}$±469	1	0.77	0.00	15209708	PMX_DIS	934.8^{+}±8.6e3	1	0.63	0.00	61775716	OX_DIS	445.2^{+}±2.5e3	1	0.71	0.00	40209376
	IO_MST	694.7$^{\sim}$±3.9e3	0.040	0.78	0.00	882910	PMX_MST	1030.0^{+}±1.5e4	1	0.66	0.00	52413648	OX_MST	630.5^{+}±2.3e3	1	0.75	0.00	32203187
	IO_POP	28.4^{+}±103	1	0.89	0.00	2892760	PMX_POP	966.3^{+}±5.6e3	1	0.60	0.00	69385442	OX_POP	651.3^{+}±5.2e3	1	0.59	0.00	60258544

The PMX [7] and OX [8] operators generate two offspring starting with two sub-tours selected from two parents. To embed the proposed strategies into these two algorithms, we select one parent and generate one sub-tour in the way as follows. A city i is randomly selected and chose the next city j according to the selection probability in matrix A. This procedure repeat until the sub-tour makes a circle. Once the sub-tour is obtained, we follow the same steps as PMX or OX to generate one offspring. Due to the space limitation, the pseudo-code of the two algorithms with proposed strategies is not provided.

4   Experimental Study

4.1   Experimental Settings

The parameter settings of the three algorithms are as follows. For algorithms in this paper, the population size was set to 1.2 times of the number of nodes, α in Eq. (3) was set to 4. Parents are directly replaced with their offspring if they are worse than their offspring expect the original PMX algorithm. The original PMX algorithm selects offsprings with the roulette wheel rule. The crossover probability is 0.9 for PMX and OX.

All results are averaged over 30 independent runs. A two-tailed t-test is performed with a 95% confidence interval between the original algorithms and the ones with the proposed strategies. +, -, and $\sim$ denote the results achieved by the proposed algorithms are significantly better than, significantly worse than, and statistically equivalent to the results obtained by the original algorithms, respectively.

4.2 Experimental Results

Table 2 presents the comparison between the proposed algorithms and their original algorithms, where error is the tour length difference between the global optimum and the best solution found by algorithms and the t-test results are associated with the error values, R1 and R2 are the percentage of the number of edges that belong to the global optimum, in the population and in the best solution at the last iteration, respectively. S_ratio is the number of runs where the global optimum is found over the total number of runs. Evals means the average number of fitness evaluations at the end of the run.

By observing Table 2, it can be seen that the error values achieved by IO_POP are significantly better than that of Inverover on most instances, especially in the cases of KROC100 and LIN105 where IO_POP finds the global optimum for each run. The corresponding number of fitness evaluations needed are also much smaller than that of Inverover in all cases. For the other two strategies, the performance are much worse than Inverover on the error values in all cases. This is because that Inverover is already a heuristic algorithm, its performance will get worse if very strong heuristic information is used, such as the strategies based on MST and problem topology. Take the MST heuristic strategy as an example. For each node, only two options on average are available. If those edges are not the edges in the global optimum. This would cause the population to be easily trapped in local optima. The similar issue also exists for the first strategy. Although the error values of IO_DIS get worse than Inverover, all edges in the global optimum appear in the population at the last iteration in almost all cases for IO_DIS (similar observations on the other two algorithms). This indicates that with proper enhancements, the algorithms with the topology strategy may find the global optimum.

Additionally, from Table 2, it can be seen that the algorithms PMX and OX with heuristic guiding perform much better than the corresponding original algorithms, respectively. This is because that PMX and OX are not heuristic based operators. Both algorithms are improved by using these heuristic strategies. However, compared to the results of algorithms based on Inverover, the performance of the PMX and OX based algorithms still needs to be improved.

Fig. 6 presents the evolutionary progress of the number of good edges (links in the global optimum) in the population (left) and in the best solution (middle), and the error value of the best solution. The results on LIN105 are shown only due to space limit (similar behaviors are observed on other instances). By comparing the behavior of each algorithm in Fig. 6, we can have several observations as below:

Firstly, the number of good edges in the best solution achieved by using strategies of MST and problem topology increases much faster than the other two peer algorithms. As discussed above, the strong heuristic information would benefit the search especially during the early search progress. Accordingly, the convergence speed is also faster than

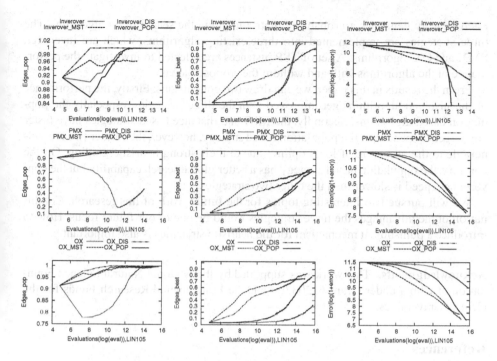

Fig. 6. Comparison between Inverover, PMX and OX and the ones with three strategies on LIN105, where the left and middle graphs are the number of edges that belong to the global optimum obtained in the population and in the best solution, respectively. The error values are shown in the right column.

the later two algorithms. However, this may lead to pre-mature convergence problem as shown in the right graphs in Fig. 6.

Secondly, comparing the results in the left graphs in the second and third rows of Fig. 6, the number of good edges achieved by the proposed algorithms gradually increases nearly to one, which is much better than that achieved by PMX and OX. However, the figure of the origin algorithms initially decreases. The decrease in the number of good edges probably is caused by the lack of heuristic information for PMX and OX, where sub-tours selected from two parents are blindly mapped without any heuristic guiding. Although the number of good edges in the population achieved by OX increases to a similar level of the other three peer algorithms, the number of good edges in the best solution achieved by OX is far less than that of its peer algorithms.

Thirdly, in the first row of Fig. 6 the number of good edges achieved by IO_POP initially decreases as PMX and OX, the figure, however, increases after a certain number of *evals* and eventually reaches almost 1. This results indicate the learning capability of the strategy based on population information.

5   Conclusions and Future Work

This paper proposes three heuristic strategies for EAs to solve TSP. The three strategies are designed based on the knowledge of MST, distance between nodes, and population.

Each strategy generates a probability matrix to guide the search of an algorithm. The three strategies are implemented into three existing algorithms, which are Inverover, PMX and OX algorithms. A set of TSP instances are selected to investigate the performance of the algorithms with and without the proposed ideas.

From the results in this paper, we can draw two conclusions: Firstly, the performance of the involved algorithms get improved with the three strategies, especially the population based algorithms. Secondly, the MST and distance based strategies have faster convergence speed than the population based strategy, however, the former two are vulnerable to the attraction of local optima due to their strong heuristic guiding. On the contrary, the population based strategy has a better global search capability but the convergence speed is slower than the other two strategies.

We will pursue two interesting topics for the future work of this research. Combination of the advantages the three strategies would be one of our future focus. How to introduce a punishment mechanism for the heuristic strategies is also important.

Acknowledgments. This work was supported by the National Natural Science Foundation of China under Grant 61203306 and the Fundamental Research Funds for the Central Universities.

References

1. Togan, V., Daloglu, A.T.: An improved genetic algorithm with initial population strategy and self-adaptive member grouping. Computers & Structures 86, 1204–1218 (2008)
2. Yugay, O., Kim, I., Kim, B., Ko, F.: Hybrid genetic algorithm for solving travelling salesman problem with sorted population. In: Third International Conference on Convergence and Hybrid Information Technology, pp. 1024–1028. IEEE, Busan (2008)
3. Ray, S.S., Bandyopadhyay, S., Pal, S.K.: Genetic operators for combinatorial optimization in tsp and microarray gene ordering. Applied Intelligence 26, 183–195 (2007)
4. Albayrak, M., Allahverdi, N.: Development a new mutation operator to solve the travelling salesman problem by aid of genetic algorithms. Expert Systems with Applications 38, 1313–1320 (2011)
5. Wei, Y., Hu, Y., Gu, K.: Parallel search strategies for tsps using a greedy genetic algorithm. In: Third International Conference on Natural Computation, pp. 786–790. IEEE, Haikou (2007)
6. Tao, G., Michalewicz, Z.: Inver-over operator for the TSP. In: Eiben, A.E., Bäck, T., Schoenauer, M., Schwefel, H.-P. (eds.) PPSN 1998. LNCS, vol. 1498, pp. 803–812. Springer, Heidelberg (1998)
7. Goldberg, D.E., Lingle, R.: Alleles, loci, and the travelling salesman problem. In: Proceedings of the International Conference on Genetic Algorithms and their Applications, pp. 154–159. Lawrence Erlbaum, Pittsburgh (1985)
8. Davis, L.: Applying adaptive algorithms to epistatic domains. In: Proceeding of 9th International Joint Conference on Artificial Intelligence, pp. 162–164. Citeseer, Greece (1985)

A 2-level Approach for the Set Covering Problem: Parameter Tuning of Artificial Bee Colony Algorithm by Using Genetic Algorithm

Broderick Crawford[1,2], Ricardo Soto[1,3], Wenceslao Palma[1], Franklin Johnson[4], Fernando Paredes[5], and Eduardo Olguín[6]

[1] Pontificia Universidad Católica de Valparaíso, Chile
{Broderick.Crawford, Ricardo.Soto, Wenceslao.Palma}@ucv.cl
[2] Universidad Finis Terrae, Chile
[3] Universidad Autónoma de Chile, Chile
[4] Universidad de Playa Ancha, Chile
Franklin.Johnson@upla.cl
[5] Universidad Diego Portales, Chile
Fernando.Paredes@udp.cl
[6] Universidad San Sebastián, Chile
Eduardo.Olguin@uss.cl

Abstract. We present a novel application of the Artificial Bee Colony algorithm to solve the non-unicost Set Covering Problem. The Artificial Bee Colony algorithm is a recent Swarm Metaheuristic technique based on the intelligent foraging behavior of honey bees. We present a 2-level metaheuristic approach where an Artificial Bee Colony Algorithm acts as a low-level metaheuristic and its parameters are set by a higher level Genetic Algorithm.

Keywords: Set Covering Problem, Artificial Bee Colony Algorithm, Swarm Intelligence, Parameter Setting, Genetic Algorithm.

1 Introduction

The Set Covering Problem (SCP) is a classic problem in combinatorial analysis, computer science and theory of computational complexity. It is a problem that has led to the development of fundamental technologies for the field of the approximation algorithms. Also it is one of the problems from the List of 21 Karp's NP-complete problems, its NP-completeness was demonstrated in 1972 [8]. SCP has many applications, including those involving routing, scheduling, stock cutting, electoral redistricting and others important real life situations [5]. Different solving methods have been proposed in the literature for the Set Covering Problem. In this paper we propose a novel application of Artificial Bee Colony (ABC) to solve SCP.

Depending on the algorithm that has been used, the quality of the solution wanted and the complexity of the SCP chosen, it is defined the amount of customization efforts required. Conveniently, this work proposes transferring part of

Y. Tan et al. (Eds.): ICSI 2014, Part I, LNCS 8794, pp. 189–196, 2014.

this customization effort to another metaheuristic (a "high level" metaheuristic) which can handle the task of parameters adjustment for a low level metaheuristic. This approach is considered as a multilevel metaheuristic since there are two metaheuristics covering tasks of parameter setting, for the former, and problem solving, for the latter [11,4]. The main design of the implementation proposed considers a Genetic Algorithm (GA) at parameter setting for a low level metaheuristic (ABC) using an Automatic Parameter Tuning approach. The Automatic Parameter Tuning is carried by an external algorithm which searches for the best parameters in the parameter space in order to tune the solver automatically.

This paper is organized as follows: In Section 2, we explain the problem. In Section 3, we describe our ABC proposal to solve SCP. In section 4 we present the 2-level framework. In section 5 details of implemetation are presented. In Section 6, we present the experimental results obtained. Finally, in Section 7 we conclude the paper and give some perspectives for further research.

2   Set Covering Problem

A general mathematical model of the Set Covering Problem can be formulated as follows:

$$Minimize \ \ Z = \sum_{j=1}^{n} c_j x_j \qquad j = \{1, 2, 3, ..., n\} \tag{1}$$

Subject to:

$$\sum_{j=1}^{n} a_{ij} x_j \geq 1 \qquad i = \{1, 2, 3, ..., m\} \tag{2}$$

$$x_j = \{0, 1\} \tag{3}$$

Equation 1 is the objective function of set covering problem, where c_j is the cost of j-column, and x_j is decision variable. Equation 2 is a constraint to ensure that each row is covered by at least one column where a_{ij} is a constraint coefficient matrix of size m x n whose elements comprise of either "1" or "0". Finally, equation 3 is the integrality constraint in which the value x_j can be "1" if column j is activated (selected) or "0" otherwise.

3   Artificial Bee Colony Algorithm

ABC is one of the most recent algorithms in the domain of the collective intelligence. Created by Dervis Karaboga in 2005, who was motivated by the intelligent behavior observed in the domestic bees to take the process of foraging [7]. ABC is an algorithm of combinatorial optimization based on populations, in which the solutions of the problem of optimization, the sources of food, are modified by the artificial bees, that fungen as operators of variation. The aim of these bees is to discover the food sources with major nectar. In the ABC algorithm, the artificial bees move in a space of multidimensional search choosing sources of nectar depending on its past experience and its companions of beehive or fitting his position. Some bees (exploratory) fly and choose food sources randomly without using experience. When they find a source of major nectar, they memorize his position and forget the previous one. Thus, ABC combines methods of local search and global search, trying to balance the process of the exploration and exploitation of the space of search. The pseudocode of Artificial Bee Colony is showed in Algorithm 1.

Algorithm 1. *ABC()*

---
1. *Initialize Food Sources*;
2. *Evaluate the nectar amount of Food Sources*;
3. **while** (not TerminationCriterion) **do**
4.   *Phase of Workers Bees*;
5.   *Phase of Onlookers Bees*;
6.   *Phase of Scout Bees*;
7.   *Update Optimum()*;
8. **end while**
9. *Return BestSolution*;

The procedure for determining a food source in the neighborhood of a particular food source which depends on the nature of the problem. Karaboga [6] developed the first ABC algorithm for continuous optimization. The method for determining a food source in the neighborhood of a particular food source is based on changing the value of one randomly chosen solution parameter while keeping other parameters unchanged. This is done by adding to the current value of the chosen parameter the product of a uniform variable in [-1, 1] and the difference in values of this parameter for this food source and some other randomly chosen food source. This approach can not be used for discrete optimization problems for which it generates at best a random effect. Singh [10] subsequently proposed a method, which is appropriate for subset selection problems. In his model, to generate a neighboring solution, an object is randomly dropped from the solution and in its place another object, which is not already present in the solution is added. The object to be added is selected from another randomly chosen solution. If there are more than one candidate objects for addition then ties are broken arbitrarily. In this work we use the ABC algorithm described in [1].

4   Multilevel Framework

This section is based on work presented in [11]. Metaheuristics, in their original definition, are solution methods that orchestrate an interaction between local improvement procedures and higher level strategies to create a process capable of escaping from local optima and performing a robust search of a solution space. Over time, these methods have also come to include any procedures that employ strategies for overcoming the trap of local optimality in complex solution spaces, especially those procedures that utilize one or more neighborhood structures as a means of defining admissible moves to transition from one solution to another, or to build or destroy solutions in constructive and destructive processes. A number of the tools and mechanisms that have emerged from the creation of metaheuristic methods have proved to be remarkably effective, so much that metaheuristics have moved into the spotlight in recent years as the preferred line of attack for solving many types of complex problems, particularly those of a combinatorial nature.

Multilevel Metaheuristics can be considered as two or more metaheuristics where a higher level metaheuristic controls de parameters of a lower level one, which is at charge of dealing more directly to the problem. Our ABC algorithm employs four control parameters which are: number of food sources, the value of limit, % Columns to add and % Columns to eliminate. Different combinations of parameters are evaluated by the upper-level GA algorithm relieving the task of manual parameterization. Each individual encodes the parameters of an ABC algorithm generating an ABC instance (see Algorithm 2).

Algorithm 2. *Tuning ABC()*

---

1. *SetUp Genetic Algorithm Population()*;
2. **while** (not TerminationCriterion) **do**
3.   **for** (each GA Individual x_i) **do**
4.   *Run ABC(x_i, cutoff)*;
5.   **end for**
6.   *Apply Genetic Algorithm Operations()*;
7. **end while**
8. *best Individual* ← *select Best Individual()*;
9. *Run ABC(bestIndividual)*;

---

5   Implementation Details

The design proposed for the multilevel implementation is based on documented standards proposed for each metaheuristic. Both metaheuristics are looking to be as close as they can to its origins. In GA the chromosome genes are: *"Food sources"*, it is the number of initial solutions for ABC (which is equal to the number of workers or onlookers bees), it will take values between 50 and 500. The second gene, *"Limit"*, it takes values between 0 and 100. Similarly, the third and fourth genes, *"% Columns to Add"* and *"% Columns Eliminate"*, they take values between 0,01 and 10. The maximum cycle number is set in 1000 iterations. The Genetic Algorithm used to obtain the best values of the ABC parameters was implemented using the Java Genetic Algorithm Package [1](JGAP) version 3.5. The basic behavior of a GA implemented using JGAP contains three main steps: a setup phase, the creation of the initial population and the evolution of the

---

[1] http://jgap.sourceforge.net

population. The GA paremeters were: Number of generations = 20, Population size = 30, Crossover type is Uniform, Crossover rate was 0.4, Mask probability was 0.5, Mutation rate was 0.025, Selector tournament size was 3 and the Tournament selector parameter (p) was 0.75.

6 Experimental Results

The ABC algorithm has been implemented in C in a 2.5 GHz Dual Core with 4 GB RAM computer, running windows 7. ABC has been tested on 65 standard non-unicost SCP instances available from OR Library [2].

Table 1 summarizes the characteristics of each of these sets of instances, each set contains 5 or 10 problems and the column labeled Density shows the percentage of non-zero entries in the matrix of each instance. ABC was executed 30 times on each instance, each trial with a different random seed.

Table 1. Details of the 65 test instances

Instance set	No. of instances	m	n	Cost range	Density (%)	Optimal solution
4	10	200	1000	[1, 100]	2	Known
5	10	200	2000	[1, 100]	2	Known
6	5	200	1000	[1, 100]	5	Known
A	5	300	3000	[1, 100]	2	Known
B	5	300	3000	[1, 100]	5	Known
C	5	400	4000	[1, 100]	2	Known
D	5	400	4000	[1, 100]	5	Known
NRE	5	500	5000	[1, 100]	10	Unknown
NRF	5	500	5000	[1, 100]	20	Unknown
NRG	5	1000	10000	[1, 100]	2	Unknown
NRH	5	1000	10000	[1, 100]	5	Unknown

In comparison with very recent works solving SCP - with Cultural algorithms [3] and Ant Colony + Constraint Programming techniques [2] - our proposal performs best than the SCP instances reported in those works. In order to bring out the efficiency of our proposal the solutions of the complete set of instances have been compared with other swarm metaheuristic. We compared our algorithm solving the complete set of 65 standard non-unicost SCP instances from OR Library with the newest ACO-based algorithm for SCP in the literature: Ant-Cover + Local Search (ANT+LS) [9]. Tables 2 and 3 show the detailed results obtained by the algorithms. Column 2 reports the optimal or the best known solution value of each instance. The third and fourth columns show the best value and the average obtained by our ABC algorithm in 30 runs (trials).

---

[2] http://people.brunel.ac.uk/~mastjjb/jeb/info.html

Table 2. Experimental results - Instances with optimal

Instance	Optimum	Best value found	ABC Avg	ANT-LS Avg	RPD (%)
4.1	429	430	430.5	429	0.35
4.2	512	512	512	512	0
4.3	516	516	516	516	0
4.4	494	494	494	494	0
4.5	512	512	512	512	0
4.6	560	561	561.7	560	0.30
4.7	430	430	430	430	0
4.8	492	493	494	492	0.41
4.9	641	643	645.5	641	0.70
4.10	514	514	514	514	0
5.1	253	254	255	253	0.79
5.2	302	309	310.2	302	2.72
5.3	226	228	228.5	226	1.11
5.4	242	242	242	242	0
5.5	211	211	211	211	0
5.6	213	213	213	213	0
5.7	293	296	296	293	1.02
5.8	288	288	288	288	0
5.9	279	280	280	279	0.36
5.10	265	266	267	265	0.75
6.1	138	140	140.5	138	1.81
6.2	146	146	146	146	0
6.3	145	145	145	145	0
6.4	131	131	131	131	0
6.5	161	161	161	161	0
A.1	253	254	254	253	0.40
A.2	252	254	254	252	0.79
A.3	232	234	234	232.8	0.86
A.4	234	234	234	234	1.10
A.5	236	237	238.6	236	0
B.1	69	69	69	69	0
B.2	76	76	76	76	0
B.3	80	80	80	80	0
B.4	79	79	79	79	0
B.5	72	72	72	72	0
C.1	227	230	231	227	1.76
C.2	219	219	219	219	0
C.3	243	244	244.5	243	0.62
C.4	219	220	224	219	2.28
C.5	215	215	215	215	0
D.1	60	60	60	60	0
D.2	66	67	67	66	1.52
D.3	72	73	73	72	1.39
D.4	62	63	63	62	1.61
D.5	61	62	62	61	1.64

Table 3. Experimental results - Instances with Best Known Solution

Instance	Optimum	Best value found	ABC Avg	ANT-LS Avg	RPD (%)
NRE.1	29	29	29	29	0
NRE.2	30	30	30	30	0
NRE.3	27	27	27	27	0
NRE.4	28	28	28	28	0
NRE.5	28	28	28	28	0
NRF.1	14	14	14	14	0
NRF.2	15	15	15	15	0
NRF.3	14	14	14	14	0
NRF.4	14	14	14	14	0
NRF.5	13	13	13	13.5	0
NRG.1	176	176	176	176	0
NRG.2	154	154	154	155.1	0
NRG.3	166	166	166	167.3	0
NRG.4	168	168	168	168.9	0
NRG.5	168	168	168	168.1	0
NRH.1	63	63	63	64	0
NRH.2	63	63	63	67.9	0
NRH.3	59	59	59	59.4	0
NRH.4	58	58	58	58.7	0
NRH.5	55	55	55	55	0

The next column shows the average values obtained by ANT+LS. The last column shows the Relative Percentage Deviation (RPD) value over the instances tested with ABC. The quality of solutions can be evaluated using the RPD, its value quantifies the deviation of the objective value Z from Z_{opt} which in our case is the best known cost value for each instance. Examining Tables 2 and 3 we observe that ABC is able to find the optimal solution consistently - i.e. in every trial- for 43 of 65 problems. ABC is able to find the best known value in all instances of Table 3. ABC is able to find the best known value in all trials of Table 3. ABC has higher success rate compared to ANT+LS in sets NRE, NRF, NRG and NRH. The *RPD* of BEE is 0,00% and the *RPD* of ANT+LS is 0,86%.

7   Conclusion

A 2-level metaheuristic has been tested on different SCP benchmarks showing to be very effective. We have presented an Artificial Bee Colony Algorithm for the Set Covering Problem where its parameters were tuned using a Genetic Algorithm. We performed experiments throught all ORLIB instances, our approach has demostrated to be very effective, providing an unattended solving method, for quickly producing solutions of a good quality. Experiments shown interesting results in terms of robustness, where using the same parameters for different set of instances giving good results. The promising results of the experiments open up opportunities for further research. The fact that the presented framework is easy to implement, clearly implies that it could also be effectively applied using others metaheuristics. Furthermore, our approach could also be effectively applied to other combinatorial optimization problems.

Acknowledgements. Broderick Crawford is supported by Grant CONICYT/ FONDECYT/ 1140897. Ricardo Soto is supported by Grant CONICYT/ FONDE-CYT/ INICIACION/ 11130459. Fernando Paredes is supported by Grant CONI-CYT/ FONDECYT/ 1130455.

References

1. Crawford, B., Soto, R., Cuesta, R., Paredes, F.: Application of the artificial bee colony algorithm for solving the set covering problem. The Scientific World Journal 2014(189164), 1–8 (2014)
2. Crawford, B., Soto, R., Monfroy, E., Castro, C., Palma, W., Paredes, F.: A hybrid soft computing approach for subset problems. Mathematical Problems in Engineering 2013(716069), 1–12 (2013)
3. Crawford, B., Soto, R., Monfroy, E.: Cultural algorithms for the set covering problem. In: Tan, Y., Shi, Y., Mo, H. (eds.) ICSI 2013, Part II. LNCS, vol. 7929, pp. 27–34. Springer, Heidelberg (2013)
4. Crawford, B., Valenzuela, C., Soto, R., Monfroy, E., Paredes, F.: Parameter tuning of metaheuristics using metaheuristics. Advanced Science Letters 19(12), 3556–3559 (2013), http://www.ingentaconnect.com/content/asp/asl/2013/00000019/00000012/art00026
5. Feo, T.A., Resende, M.G.C.: A probabilistic heuristic for a computationally difficult set covering problem. Operations Research Letters 8(2), 67–71 (1989)
6. Karaboga, D.: An idea based on honey bee swarm for numerical optimization. Technical Report TR06. Computer Engineering Department, Erciyes University, Turkey (2005)
7. Karaboga, D., Basturk, B.: A powerful and efficient algorithm for numerical function optimization: Artificial bee colony (abc) algorithm. J. Global Optimization 39(3), 459–471 (2007)
8. Karp, R.M.: Reducibility among combinatorial problems. In: Miller, R.E., Thatcher, J.W. (eds.) Complexity of Computer Computations. The IBM Research Symposia Series, pp. 85–103. Plenum Press, New York (1972)
9. Ren, Z.G., Feng, Z.R., Ke, L.J., Zhang, Z.J.: New ideas for applying ant colony optimization to the set covering problem. Computers & Industrial Engineering 58(4), 774–784 (2010), http://www.sciencedirect.com/science/article/pii/S0360835210000471
10. Singh, A.: An artificial bee colony algorithm for the leaf-constrained minimum spanning tree problem. Appl. Soft Comput. 9(2), 625–631 (2009), http://dx.doi.org/10.1016/j.asoc.2008.09.001
11. Valenzuela, C., Crawford, B., Soto, R., Monfroy, E., Paredes, F.: A 2-level metaheuristic for the set covering problem. Int. J. Comput. Commun. 7(2), 377–387 (2012)

Hybrid Guided Artificial Bee Colony Algorithm for Numerical Function Optimization

Habib Shah[1], Tutut Herawan[2,3], Rashid Naseem[1], and Rozaida Ghazali[1]

[1] Faculty of Computer Science and Information Technology
Universiti Tun Hussein Onn Malaysia
Parit Raja, 86400 Batu Pahat, Johor, Malaysia
[2] Faculty of Computer Science & Information Technology
University of Malaya
50603 Lembah Pantai, Kuala Lumpur Malaysia
[3] AMCS Research Center, Yogyakarta, Indonesia
{habibshah.uthm,rnsqau}@gmail.com, tutut@um.edu.my,
rozaida@uthm.edu.my

Abstract. Many different earning algorithms used for getting high performance in mathematics and statistical tasks. Recently, an Artificial Bee Colony (ABC) developed by Karaboga is a nature inspired algorithm, which has been shown excellent performance with some standard algorithms. The hybridization and improvement strategy made ABC more attractive to researchers. The two famous improved algorithms are: Guided Artificial Bee Colony (GABC) and Gbest Guided Artificial Bee Colony (GGABC), are used the foraging behaviour of the gbest and guided honey bees for solving optimization tasks. In this paper, GABC and GGABC methods are hybrid and so-called Hybrid Guided Artificial Bee Colony (HGABC) algorithm for strong discovery and utilization processes. The experiment results tested with sets of numerical benchmark functions show that the proposed HGABC algorithm outperforms ABC, PSO, GABC and GGABC algorithms in most of the experiments.

Keywords: Back propagation, Gbest Guided Artificial Bee Colony, Numerical function optimization, Hybrid Guided Artificial Bee Colony algorithm.

1 Introduction

In recent years, the swarm intelligence algorithms have become a research interest to different domain of researchers for solving optimization problems. They are interested in developing the new optimization techniques based on nature collection, intelligence movement, thinking behaviors of Bee Colony [1], ABC [2], Ant Colony Optimization (ACO), Improved ABC algorithm [3], Hybrid Ant Bee Colony (HABC) [4], Gbest Guided Artificial Bee Colony (GGABC) [5], Global Guided Artificial Bee Colony GABC [6-8], multiple Gbest some other hybrid and meta heuristic algorithms. Particle Swarm Optimization (PSO) [9] and its variations have been introduced for solving optimization problems and successfully applied to solve many others real

Y. Tan et al. (Eds.): ICSI 2014, Part I, LNCS 8794, pp. 197–206, 2014.

problems like classification, clustering and prediction [10-13]. Motivated by the foraging behavior of honeybees [14], researchers have initially proposed ABC algorithm for solving various optimization problems. ABC is a relatively new population-
based meta-heuristic approach and is further improved by researchers for many applications [15-16]. To increase the performance of the standard ABC algorithm in exploration and exploitation procedures, researchers improved the standard ABC by using
different strategies like hybridization, modification, gbest, global best, guided so best
so far and other strategies [4, 17-18].

In this paper, to increase the exploration and exploitation process of the standard
ABC algorithm and at the same time for finding optima for numeric function optimi-
zation problems, the two global algorithms, namely Global Artificial Bee Colony
(GABC) and Gbest Guided Artificial Bee Colony (GGABC) are hybrid and so-called
Hybrid Guided Artificial Bee Colony (HGABC) algorithm [6-7]. Further, the perfor-
mance of the proposed HGABC algorithm and four standard algorithms i.e. ABC,
GABC, PSO and GGABC were compared to solve benchmark numerical functions.

2   Gbest Guided Artificial Bee Colony (GGABC) Algorithm

The exploration and exploitation are the famous procedure in the swarm based algorithms, which used for a solution to a given problem. Exploration includes things
captured by terms such as search, variation, risk taking, experimentation, play,
flexibility, discovery, innovation [13, 17-18]. Exploitation includes such things as
refinement, choice, production, efficiency, selection, implementation, execution. To
improve the exploration process, a standard ABC algorithm modified with Gbest
Guided technique called Gbest Guided Artificial Bee Colony (GGABC) by incorporating the information of global best (gbest) solution into the solution search equation
proposed for training multilayer perceptron (MLP) [6]. The new candidate solutions
of both agents are generated by moving the old solution towards (or away from)
another solution selected randomly from the population. Here the employed and
onlookers bee section of standard ABC has modified for improving the exploration
procedure as follow.

$$v_{ij} = x_{ij} + \phi_{ij}(x_{ij} - x_{kj}) + \Psi_{ij}(y_j - x_{i,j}) \; , \tag{1}$$

where y_j is the j_{th} element of the global best solution, v_{ij} is a uniform random number
in $[0, C]$, where C is a nonnegative constant. The value of C can balance the exploitation ability.

3   Guided Artificial Bee Colony (GABC) Algorithm

How to improve the exploration and exploitation processes with balance amount is a
challenge for swarm based algorithm, especially for standard ABC. Guided ABC

algorithm is an improved honey bees inspired technique, which used to increase the exploitation process with balance of ABC [19]. In the standard ABC algorithm, the process of replacing the abandoned food source is simulated by randomly producing a new solution, through the employed and onlooker bee's section with the same strategy as in Equation (2) as follow.

$$v_{ij} = x_{ij} + \phi_{ij}\left(x_{ij} - x_{kj}\right)$$ (2)

The new solutions in scout bee of the standard ABC algorithm are not based on the information of previous best solutions. Researchers used different advance searching strategies like hybridization, improvement, mimetic, global, gbest and so on, with standard ABC. Guided ABC is one of the proposed algorithm [19]. The new modify solution search equation by applying the global best solution to guide the search of scout bees to improve the exploration procedure of standard ABC. In the standard ABC in scout bees new solution generated by using a random approach as in equation (3), thus it is very difficult to generate a new solution that could be placed in the promising region of the search space.

$$x_{ij}^{rand} = x_{ij}^{min} + rand(0,1)(x_j^{max} - x_j^{min}).$$ (3)

Therefore, the GABC algorithm proposed for improving the exploration procedure within scout bee searching strategy [19]. In GABC algorithm, the scout will generate a new solution through global knowledge information ($x_{best,j}$ - the best global food source) to Equation (3). The global best experience will modify Equation (3) with the following best guided strategy as:

$$v_{ij} = x_{ij} + \phi*\left(x_{ij} - x_{kj}\right) + (1-\phi)*(x_{ij} - x_{best\,j}).$$ (4)

The guided ABC will increase the capabilities of the standard ABC algorithm to produce new best solutions located near the feasible area.

4   Hybrid Guided Artificial Bee Colony (HGABC) Algorithm

The Gbest Guided Artificial Bee Colony (GGABC) and Guided Artificial Bee Colony (GABC) are swarm based meta-heuristic algorithms, applied to different problems like engineering, statistical, optimization and to ANN for training and testing purpose and other tasks [4, 6-7]. GABC algorithm has outstanding performance from standard ABC algorithm in terms of exploration process [19]. Furthermore, GGABC has great achievements for increasing exploitation process than standard ABC [5].

Taking the advantage of the GABC and GGABC algorithms to improve and balance the exploitation and exploration procedures, are hybrid called HGABC algorithm. The proposed hybrid approach HGABC has the following modification in employed, onlooker and scout bees phases are as follow.

1) Gbest Guided Employed Bee Phase:
$$v_{ij} = x_{ij} + \phi_{ij}(x_{ij} - x_{kj}) + \Psi_{ij}(y_j - x_{i,j}) \tag{5}$$

2) Gbest Guided Onlookers Bees Phase:
$$v_{ij} = x_{ij} + \phi_{ij}(x_{ij} - x_{kj}) + \Psi_{ij}(y_j - x_{i,j}) \tag{6}$$

3) Guided Scout Bee Phase:
$$v_{ij} = x_{ij} + \phi*(x_{ij} - x_{kj}) + (1-\phi)*(x_{ij} - x_{best\,j}) \tag{7}$$

HGABC algorithm as a hybrid optimization tool provides a population-based search procedure in which individuals called foods positions are modified by global and Gbest artificial bees with time, and the bee's aim is to discover the places of best food sources with high nectar amount and finally the one with the highest nectar. The HGABC algorithm will update the solution step and will convert to best solution based on neighborhood values, though hybridization of guided and best guided strategy. The proposed flow chart of HGABC algorithm is given in Figure 1 as follow.

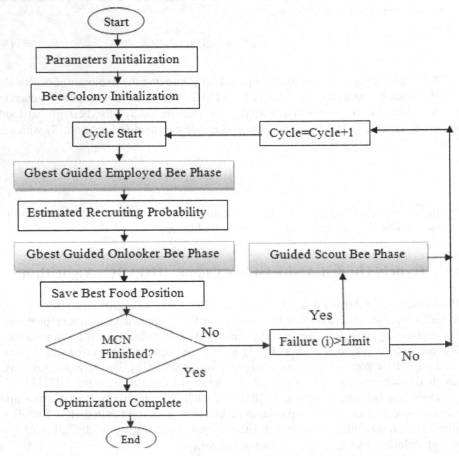

Fig. 1. Flowchart of Hybrid Guided Artificial Bee Colony algorithm

5   Simulation Results and Discussion

In this section, HGABC, PSO, ABC, GGABC and GABC algorithms used for numerical function optimization tasks. The performance of the above mentioned algorithms is calculated by Mean Square Error (MSE) and Standard Deviation by using Matlab 2012b software.

5.1   Six Benchmark Numeric Functions

In order to evaluate the performance of the HGABC for numerical function optimization scheme, simulation experiments performed on Intel Core i7-3612QM @ 2.1Ghz, 6 MB L3 Cache Intel HM76 with 4 GB RAM using Matlab 2012b software. Seven well-known benchmark functions are used to compare the performance of the proposed HGABC algorithm with standard ABC, GABC and GGABC algorithm. These functions contain one unimodal variable-separable function, two unimodal non-separable functions, two multimodal variable separable functions and two multimodal non-separable functions. These are Rosenbrock, Sphere, Rastrigin, Schwefel, Ackley and Griewank functions.

The first benchmark is Rosenbrock function used for simulation, whose global minimum value is 0 at $(1,1,\ldots,1)$. It is a unimodal function with nonseparable variables. Its global optimum is inside a long, narrow, parabolic shaped flat valley. So it is difficult to converge to the global optimum. This problem is repeatedly used to test the performance of the optimization algorithms. The second benchmark function is Sphere used for simulation, whose global minimum value is 0 at $(0,0,\ldots,0)$. It is a unimodal function with separable variables. The third bench function is Rastrigin function used for simulation, whose global minimum value is 0 at $(0,0,\ldots,0)$. This function is based on Sphere function with the addition of cosine modulation to produce many local minima. It is a multimodal function with separable variables.

For finding optimal solutions to this function is that other algorithms easily can be trapped in a local optimum on its way towards the global optimum. The fourth benchmark function is Schwefel used for simulation, whose global minimum value is 0 at $(420.9867,420.9867,\ldots,420.9867)$. The function has a second best minimum far from the global minimum where many search algorithms are trapped. Moreover, the global minimum is near the bounds of the domain. It is also multimodal function with separable variables. The fifth benchmark function is Ackley function whose global minimum value is 0 at $(0,0,\ldots,0)$. It is also multimodal function with non-separable variables. The difficulty of this function is moderated. An algorithm that only uses the gradient steepest descent will be trapped in local optima, but any search strategy that analyses a wider region will be able to cross the valley among the optima and achieve better results.

The six benchmark function is Griewank function whose global minimum value is 0 at $(0,0,\ldots,0)$. Griewank function has a product term that introduces interdependence among the variables. The aim is to overcome the failure of the techniques that optimize each variable independently. It is also multimodal function with non-separable variables.

Initialization range for the, Rosenbrock, Sphere, Rastrigin, Schwefe, Ackley and Griewank benchmark functions are [15, 15], [5.12, 5.12], [15, 15], [500,500], [32.768, 32.768] and [600,600] respectively. The Rosenbrock, Sphere, Rastrigin, Schwefe, Ackley, and Griewank functions are shown f_1, f_2, f_3, f_4, f_5 and f_6 from the equation (8) to (13), respectively.

$$f_1(x) = \sum_{i=1}^{D-1} \left(100\left(x_i^2 - x_{i+1}\right)^2 + \left(1 - x_i\right)^2\right) \tag{8}$$

$$f_2(x) = \sum_{i=1}^{D} \left(x_i^2\right) \tag{9}$$

$$f_3(x) = \sum_{i=1}^{D-1} \left(x_i^2 - 10\cos(2\pi x_i) + 10\right) \tag{10}$$

$$f_4(x) = D * 418.9829 + \sum_{i=1}^{D} - x_i Sin\left(\sqrt{|x_i|}\right) \tag{11}$$

$$f_5(x) = 20 + e - 20e^{\left[-0.2\sqrt{\frac{1}{D}\sum_{i=1}^{D}x_i^2}\right]} - e^{\left[\frac{1}{D}\sum_{i=1}^{D}Cos(2\pi x_i)\right]} \tag{12}$$

$$f_6(x) = \frac{1}{4000}\left(\sum_{i=1}^{D} x_i^2\right) - \left(\prod_{i=1}^{D} \cos\left(\frac{x_i}{\sqrt{i}}\right)\right) + 1 \tag{13}$$

In this experiment, all functions were tested with 30 dimensions and run for 10 times randomly. In the proposed and standard algorithms, the numbers of employed and onlooker bees were half of the population size, and the number of scout bees was selected as one. The abandon limit =100 selected for the above six benchmark function optimization. The standard PSO algorithm was used in this experiment for the optimization task. The comparison of proposed HGABC with GGABC, PSO, GABC and ABC algorithms are discussed based on the simulation results.

Table 1. Six Benchmark Numerical Functions and their details

Function Name	Function's formula	Search Range	$f(x^*)$
Rosenbrock	$f1(x)$	$[-2.048,2.048]^D$	$f(1)=0$
Sphere	$f2(x)$	$[-100, 100]^D$	$f(0)=0$
Rastrigin	$f3(x)$	$[-5.12,5.12]^D$	$f(0)=0$
Schwefel	$f4(x)$	$[-500,500]^D$	$f(420.96)=0$
Ackley	$f5(x)$	$[-32, 32]^D$	$f(0)=0$
Griewank	$f6(x)$	$[-600,600]^D$	$f(0)=0$

Tables 2 shows the numerical function results in terms of MSE and Std of the above mentioned functions through the standard and proposed algorithms. The HGABC algorithm plays a significant character in improving and balancing the exploration and exploitation procedures. The result obtained by HGABC is outstanding compared to other standard algorithms.

Table 2. Results obtained by ABC, GABC, GGABC and HGABC for Numerical Function

Function	MSE/S.D	ABC	GABC	GGABC	PSO	HGABC
f1	MSE	7.49831e-001	2.08e-002	1.68496E-04	1.06450e+002	**1.18E-04**
	Std	5.71665e-001	3.21e-002	1.454cE-04	3.59814e+001	**1.32E-02**
f2	MSE	2.07570e-011	5.96e-017	6.37910E-16	1.11023e-005	**1.40E-09**
	Std	1.63697e-011	1.48e-017	1.203eE-16	9.22993e-006	**1.68E-14**
f3	MSE	6.09826e-001	0	1.34529E-13	8.81223e+001	**1.02E-09**
	Std	7.01110e-001	0	7.966eE-14	1.62872e+001	**3.16E-10**
f4	MSE	3.53639e+002	0	2.227E-15	3.94638e+003	**2.227E-18**
	Std	1.38774e+002	0	1.24E-16	8.44500e+002	**2.24E-16**
f5	MSE	1.33853e-004	3.79e-015	1.34E-12	1.98057e+000	**3.72e-016**
	Std	6.24273e-005	9.17e-016	9.91E-11	3.52026e-001	**2.11E-13**
f6	MSE	6.77262e-006	1.48c-017	1.27305E-15	1.12245e+000	**2.21e-019**
	Std	2.41784e-005	3.91e-017	1.464E-15	3.81335e-002	**3.21e-018**

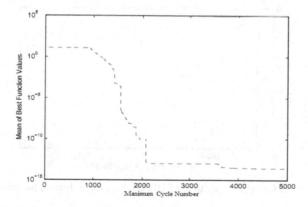

Fig. 2. Convergence curve of HGABC for the Rastrigin function

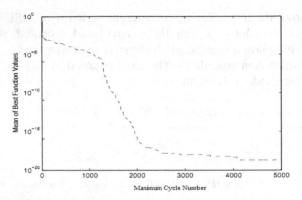

Fig. 3. Convergence curve of HGABC for the Griewank function

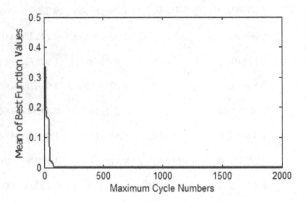

Fig. 4. Convergence curve of HGABC for the Sphere function

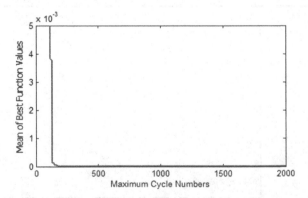

Fig. 5. Convergence curve of HGABC for the Schwefel function

From the Figures 2, 3, 4 and 5 above, the convergence is stable using proposed HGABC learning algorithm for Rastrigin, Sphere, Griewank and Schwefel functions. Overall the performance of HGABC is better than other standard approaches for a numerical function optimization task.

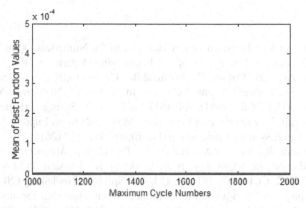

Fig. 6. Convergence curve of HGABC for the Ackley function

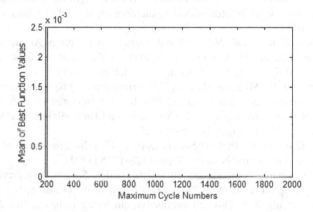

Fig. 7. Convergence curve of HGABC for the Rosenbrock function

Figures 5, 7 and 7 show the convergence performance of the proposed HGABC algorithm. The above simulation results demonstrate, that HGABC algorithm has successfully searched the optima for the above mention numerical function with very less error. The non-linear dynamical behaviour is induced by the bee nature intelligence. Therefore, it leads to the best input, output mapping and an optimum production.

6 Conclusion

The HGABC algorithm has been improved the exploration and exploitation procedure with balance quantity successfully, for numerical function optimization task through guided and gbest strategies. HGABC has the powerful ability of searching global optimal solution than standard ABC, PSO, GGABC and GABC. From the simulation results, the proposed HGABC algorithm can successfully obtain the optimum.

References

1. Karaboga, D.: An Idea based on Honey Bee Swarm for Numerical Optimization. Erciyes University, Engineering Faculty, Computer Engineering Departmen TR06, 1–10 (2005)
2. Karaboga, D., Akay, B., Ozturk, C.: Artificial Bee Colony (ABC) Optimization Algorithm for Training Feed-Forward Neural Networks. In: Torra, V., Narukawa, Y., Yoshida, Y. (eds.) MDAI 2007. LNCS (LNAI), vol. 4617, pp. 318–329. Springer, Heidelberg (2007)
3. Shah, H., Ghazali, R.: Prediction of Earthquake Magnitude by an Improved ABC-MLP. In: Developments in E-systems Engineering (DeSE), pp. 312–317 (2011)
4. Shah, H., Ghazali, R., Nawi, N.M.: Hybrid Ant Bee Colony Algorithm for Volcano Temperature Prediction. In: Chowdhry, B.S., Shaikh, F.K., Hussain, D.M.A., Uqaili, M.A. (eds.) IMTIC 2012. CCIS, vol. 281, pp. 453–465. Springer, Heidelberg (2012)
5. Zhu, G., Kwong, S.: Gbest-guided artificial bee colony algorithm for numerical function optimization. Applied Mathematics and Computation 217, 3166–3173 (2010)
6. Peng, G., et al.: Global artificial bee colony search algorithm for numerical function optimization. In: 2011 Seventh International Conference on Natural Computation (ICNC), pp. 1280–1283 (2011)
7. Shah, H., Ghazali, R., Nawi, N.M.: Global Artificial Bee Colony Algorithm for Boolean Function Classification. In: Selamat, A., Nguyen, N.T., Haron, H. (eds.) ACIIDS 2013, Part I. LNCS, vol. 7802, pp. 12–20. Springer, Heidelberg (2013)
8. Shah, H., Ghazali, R., Mohmad Hassim, Y.M.: Honey Bees Inspired Learning Algorithm: Nature Intelligence Can Predict Natural Disaster. In: Herawan, T., Ghazali, R., Deris, M.M. (eds.) Recent Advances on Soft Computing and Data Mining, SCDM 2014. AISC, vol. 287, pp. 215–225. Springer, Heidelberg (2014)
9. Kennedy, J., Eberhart, R.: Particle Swarm Optimization. In: Proceeding of IEEE International Conference on Neural Network 4, pp. 1942–1948 (1995)
10. Altun, A.: A combination of Genetic Algorithm, Particle Swarm Optimization and Neural Network for palmprint recognition. Neural Computing & Applications, 1–7 (2011)
11. Jun Ying, C., et al.: A PSO-based subtractive clustering technique for designing RBF neural networks. In: IEEE Congress on Evolutionary Computation, CEC 2008. IEEE World Congress on Computational Intelligence, pp. 2047–2052 (2008)
12. Noman, S., et al.: Hybrid Learning Enhancement of RBN Network with Particle Swarm Optimization. In: Hassanien, A.E. (ed.) Foundations of Comput. Intel., vol. 1, pp. 381–397. Springer, Heidelberg (2009)
13. Shah, H., Ghazali, R., Nawi, N.M., Deris, M.M., Herawan, T.: Global Artificial Bee Colony-Levenberq-Marquardt (GABC-LM) Algorithm for Classification. International Journal of Applied Evolutionary Computation (IJAEC) 4, 58–74 (2013)
14. Karaboga, D., Akay, B.: A comparative study of Artificial Bee Colony algorithm. Applied Mathematics and Computation 214, 108–132 (2009)
15. Karaboga, D., Basturk, B.: On the performance of artificial bee colony (ABC) algorithm. Applied Soft Computing 8, 687–697 (2008)
16. Shah, H., Ghazali, R., Nawi, N.M.: Hybrid Global Artificial Bee Colony Algorithm for Classification and Prediction Tasks. Journal of Applied Sciences Research 9, 3328–3337 (2013)
17. Ozturk, C., Karaboga, D.: Hybrid Artificial Bee Colony algorithm for neural network training. In: 2011 IEEE Congress on Evolutionary Computation (CEC), pp. 84–88 (2011)
18. Shah, H., Ghazali, R., Nawi, N.M., Deris, M.M.: G-HABC Algorithm for Training Artificial Neural Networks. International Journal of Applied Metaheuristic Computing 3, 20 (2012)
19. Tuba, M., et al.: Guided artificial bee colony algorithm. Presented at the Proceedings of the 5th European Conference on European Computing Conference, Paris, France (2011)

Classification of DNA Microarrays Using Artificial Bee Colony (ABC) Algorithm

Beatriz Aurora Garro[1], Roberto Antonio Vazquez[2], and Katya Rodríguez[1]

[1] Instituto en Investigaciones en Matemáticas Aplicadas y en Sistemas
Universidad Nacional Autónoma de México
Ciudad Universitaria, México, D.F.
beatriz.garro@iimas.unam.mx, katya.rodriguez@iimas.unam.mx,
[2] Intelligent Systems Group, Facultad de Ingeniería, Universidad La Salle
Benjamin Franklin 47, Condesa, México, D.F., 06140
ravem@lasallistas.org.mx

Abstract. DNA microarrays are a powerful technique in genetic science due to the possibility to analyze the gene expression level of millions of genes at the same time. Using this technique, it is possible to diagnose diseases, identify tumours, select the best treatment to resist illness, detect mutations and prognosis purpose. However, the main problem that arises when DNA microarrays are analyzed with computational intelligent techniques is that the number of genes is too big and the samples are too few. For these reason, it is necessary to apply pre-processing techniques to reduce the dimensionality of DNA microarrays. In this paper, we propose a methodology to select the best set of genes that allow classifying the disease class of a gene expression with a good accuracy using Artificial Bee Colony (ABC) algorithm and distance classifiers. The results are compared against Principal Component Analysis (PCA) technique and others from the literature.

Keywords: DNA microarrays, Artificial Bee Colony (ABC) algorithm, feature selection, pattern classification, PCA technique.

1 Introduction

DNA microarrays are oligonucleotide or DNA molecule lay out in a glass container (the most common). This information allows to analyze the gene expression of millions of genes at the same time from a tissue. Using this technique, it is possible to diagnose diseases, identify tumours, select the best treatment to resist illness, detect mutations and prognosis purpose, etc [1].

For solving some of the previously mentioned problems, the enormous quantity of genes to be analyzed is a disadvantage because many genes could be irrelevant and the time to analyze all data increase. Moreover, the complexity to compute them is high and the number of samples (or patterns) is much lower than the number of genes (characteristics). For these reason, it should be considered to reduce the genes number into a set of the most important that allow to solve a specific problem; this procedure is known as dimensionality reduction or feature selection [2].

Y. Tan et al. (Eds.): ICSI 2014, Part I, LNCS 8794, pp. 207–214, 2014.

There are many works that present different feature selection techniques whose results are applied to classify DNA microarrays data. For example, in [3], the authors use a class prediction model called Logistic Discrimination to solve two cancer problems from DNA microarrays using Partial Least Squares (PLS), Sliced Inverse Regression (SIR) and PCA. An interesting work is [4], where the authors proposed a feature selection technique to classify cancer data using the correlation degree between genes from two kind of leukemia and a weighted vote.

There are many works that apply Evolutionary Algorithms (EA) or swarm-based algorithms in order to select the best genes that represent a specific problem. A good example that apply Least Squares Supported Vector Machines (LS-SVM) to select the best features of DNA microarray dataset and then use PSO to optimize the scaling factors is presented in [5]. In [6], two hybrid algorithms (PSO and Genetic Algorithm (GA)) augmented with Supported Vector Machines (SVM) are presented with the purpose to classify and select the genes of the six kind of cancer problems.

In this paper, we present an approach which focuses on evaluating a gene expression from a DNA microarrays to be classified with a feature selection technique previously applied. Artificial Bee Colony (ABC) algorithm is a good evolutionary technique that will be used to select the best genes. Due to there is not a binary version of ABC, the representation of the individual is binarized to select the subset of genes used to classify the gene expression. Then, the genes selected are classified by an Euclidean and Manhattan distances. In addition, we evaluate the threshold value to know the influence in the values diversity of the solutions. Finally, the accuracy of the proposed methodology is tested over gene expression from a DNA microarrays with leukemia information. Furthermore, the experimental results obtained with ABC algorithm will be compared against those obtained with PCA algorithm. PCA is a classic technique that reduce a complex data set to a lower dimension [7]. This technique has been used for many years to solve unimaginable kinds of problems. Compare new techniques with PCA is a challenge. For this reason, the proposed methodology results are compared with the PCA results.

2   Artificial Bee Colony (ABC) Algorithm

Artificial Bee Colony (ABC) algorithm based on the metaphor of the bees foraging behavior was proposed by Dervis Karaboga [8]. It consists in a population of NB bees (possible solutions) $\mathbf{x}_i \in \mathbb{R}^n, i = 1, \ldots, NB$ represented by the position of the food sources. Three classes of bees are used to achieve the convergence near to the optimal solution: employed bees, onlooker bees and scout bees. These bees have got different tasks in the colony, i. e., in the search space.

Employed bees: Each bee searches for new neighbor food source near of their hive. After that, it compares the food source against the old one using Eq. 1. Then, the best food source is saved in their memory.

$$v_i^j = x_i^j + \phi_i^j \left(x_i^j - x_k^j \right). \tag{1}$$

Where $k \in \{1, 2, ..., NB\}$ and $j \in \{1, 2, ..., n\}$ are randomly chosen indexes and $k \neq i$. ϕ_i^j is a random number between $[-a, a]$.

After that, the bee evaluates the quality of each food source based on the amount of the nectar (the information) i.e. the fitness function is calculated. Finally, it returns to the dancing area in the hive, where the onlookers bees are.

Onlooker bees: This kind of bees watch the dancing of the employed bees so as to know some kind of information such as where the food source can be found, if the nectar is of high quality, as well as the size of the food source. The onlooker bee chooses probabilistically a food source depending on the amount of nectar shown by each employed bee, see Eq. 2.

$$p_i = \frac{fit_i}{\sum\limits_{k=1}^{NB} fit_k} . \tag{2}$$

Where fit_i is the fitness value of the solution i and NB is the number of food sources which are equal to the number of employed bees.

Scout bees: This kind of bees helps the colony to randomly create new solutions when a food source cannot be improved anymore, see Eq. 3. This phenomenon is called "limit" or "abandonment criteria".

$$x_i^j = x_{min}^j + rand\,(0, 1)\left(x_{max}^j - x_{min}^j\right). \tag{3}$$

The pseudo-code of the ABC algorithm is next shown:

```
1.  Initialize the population of solutions x_i∀_i, i = 1, ..., NB.
2.  Evaluate the population x_i∀_i, i = 1, ..., NB.
3.  for cycle = 1 to maximum cycle number MCN do
4.     Produce new solutions v_i from the employed bees by using Eq. 1 and
       evaluate them.
5.     Apply the greedy selection process.
6.     Calculate the probability values p_i for the solutions x_i by Eq. 2.
7.     Produce the new solutions v_i for the onlookers from the solutions x_i
       selected depending on p_i and evaluate them.
8.     Apply the greedy selection process.
9.     Replace the abandoned solutions with a new one randomly produced x_i
       by scout bees using Eq. 3.
10.    Memorize the best solution achieved so far.
11.    cycle = cycle + 1
12. end for
```

---

3   Proposed Methodology

The aim of this work is to select the best subset of genes from a gene expression obtained from a DNA microarray and then apply it to classify and diagnose a specific cancer disease with a good accuracy.

The problem to be solved can be defined as follows: Giving a set of input patterns $\mathbf{X} = \{\mathbf{x}_1, \ldots, \mathbf{x}_p\}, \mathbf{x_i} \in \mathbb{R}^n, i = 1 \ldots, p$ and a set of desired classes $\mathbf{d} = \{d_1, \ldots, d_p\}, d \in \mathbb{N}$, find an subset of genes $G \in \{0,1\}^n$ such that a function defined by $\min(F(\mathbf{X}|_G, \mathbf{d}))$ is minimized.

The food source's position represents the solution to the problem in terms of a subset of genes. This solution is defined with an array $I \in \mathbb{R}^n$. Each individual $I_q, q = 1, \ldots, NB$ is binarized by means of Eq. 4 using a threshold level th in order to select the best set of genes that compose the gene expression defined as $G^k = T_{th}(I^k), k = 1, \ldots, n$; values whose component is set to 1, indicates that this gene will be selected to make up the subset of genes.

$$T_{th}(x) = \left\{ \begin{matrix} 0, x < th \\ 1, x \geq th \end{matrix} \right\}. \tag{4}$$

The aptitude of an individual is computed by means of the classification error (CER) function, defined in Equation 5, that measures how many gene expressions have been incorrectly predicted.

$$F(\mathbf{X}|_G, \mathbf{d}) = \frac{\sum_{i=1}^{p} \left(\left| \underset{k=1}{\overset{K}{\arg\min}} \left(D(\mathbf{x}_i|_G, \mathbf{c}^k) \right) - d_i \right| \right)}{tng}. \tag{5}$$

Where tng is the total number of gene expressions to be classified, D is a distance measure, K is the number of classes and $\mathbf{c}$ is the center of each class. Different distance measures could be applied to classify the gene expression samples, for example classic Euclidean and Manhattan distances given by Equation 6.

$$D(\mathbf{p}, \mathbf{q}) = \sqrt[s]{\sum_{j=1}^{n} |p^j - q^j|^s}. \tag{6}$$

Where $\mathbf{p} \in \mathbb{R}^n$ and $\mathbf{q} \in \mathbb{R}^n$. This equation represents the Manhattan distance when $s = 1$ and the Euclidean distance when $s = 2$.

4   Experimental Results

The proposed methodology was tested applying ABC algorithm with a benchmark high-dimensional biomedical DNA microarray data set [4]. The leukemia ALL-AML database is next described.

Leukemia ALL-AML data set: Consists of 38 bone marrow samples for training (27 ALL and 11 AML), over 7129 probes from 6817 human genes. Also 34 samples testing data are provided, with 20 ALL and 14 AML.

The experiments take into account the threshold for binarizing each individual. To evaluate the threshold value and to know how much affect the values diversity, the threshold value was changed in seven different configurations: 0.1, 0.3, 0.5, 0.7, 0.9, 0.99 and 0.999. These values are labeled with numbers from 1 to 7 in the same sequence.

In order to validate statistically the experimental results, the proposed methodology was executed 30 runs for each distance measure (see Eq. 6) and for each configuration threshold. The database information was previously normalized.

The parameters of ABC algorithm for all the experimentation were: population size ($NB = 40$), maximum number of cycles $MNC = 1000$, limit $l = 100$ and food sources $NB/2$.

Figure 1 shows the learning error evolution during the 1000 iterations for the seven configurations for two distance measures.

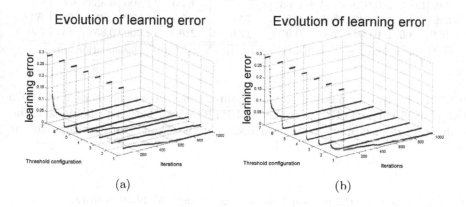

(a)   (b)

Fig. 1. Learning error evolution using different distance classifiers. 1(a) Learning error evolution for Euclidean distance measure. 1(b) Learning error evolution for Manhattan distance measure.

Table 1 shows important information about the experimentation. The results for training classification were above 98.2% where three of this configurations ($th = 0.9, th = 0.99$ and $th = 0.999$) obtained 100% of accuracy. For the case of testing classification, the best results was obtained with configuration five which provided a 79.3% of accuracy. In configuration with the value $th = 0.9$, the minimum genes number was 695. The best threshold configuration using the Euclidean distance and ABC algorithm was $th = 0.3$ with a 100% in training classification and 88.2% in testing classification using three genes whose names are, according to the database: SLC17A2 Solute carrier family 17 (L13258_at), MLC gene (M22919_rna2_at) and FBN2 Fibrillin 2 (U03272_at).

The experimental results using the proposed methodology with the Manhattan distance are shown in Table 2. The average accuracy in training classification for the 30 experiments with six threshold configurations was 100% and above the 73.9% for the case of testing classification. For example, the proposed methodology provides a 100% of training classification for the seven configurations. For testing classification rate, the minimum value was 79.4% and the higher value is above the 91.2%. The best threshold configuration using the Manhattan distance and ABC algorithm was $th = 0.1$ with a 100% in training classification

Table 1. Behavior of the proposed methodology using an Euclidean distance classifier

th	Average accuracy		Average # of genes	Average # of iter.	Best accuracy		# of genes	# of iter.
	Tr. cl.	Te. cl.			Tr. cl.	Te. cl.		
0.1	0.998 ± 0.007	0.751 ± 0.071	434.8	430.0	1.000	0.853	5	318
0.3	0.992 ± 0.012	0.746 ± 0.073	1506.7	724.7	1.000	0.882	3	253
0.5	0.982 ± 0.012	0.769 ± 0.035	2618.3	933.5	1.000	0.824	4	446
0.7	0.996 ± 0.009	0.770 ± 0.011	2054.8	452.4	1.000	0.794	3	335
0.9	1.000 ± 0.000	0.793 ± 0.032	712.0	63.0	1.000	0.853	695	62
0.99	1.000 ± 0.000	0.760 ± 0.054	74.0	57.5	1.000	0.853	79	53
0.999	1.000 ± 0.000	0.706 ± 0.079	12.0	216.1	1.000	0.853	11	759

Tr. cl. = Training classification rate, Te. cl. = Testing classification rate.

and 94.1% for testing classification using three genes whose names are KIAA0075 gene (D38550_at), MYELIN TRANSCRIPTION FACTOR 1 (M96980_at), and Anthracycline resistance associated protein (X95715_at), all according to the database.

Table 2. Behavior of the proposed methodology using a Manhattan distance classifier

th	Average accuracy		Average # of genes	Average # of iter.	Best accuracy		# of genes	# of iter.
	Tr. cl.	Te. cl.			Tr. cl.	Te. cl.		
0.1	0.996 ± 0.009	0.756 ± 0.111	1717.3	503.3	1.000	0.941	3	298
0.3	1.000 ± 0.000	0.778 ± 0.015	4994.7	75.3	1.000	0.794	4938	62
0.5	1.000 ± 0.000	0.777 ± 0.015	3563.8	58.8	1.000	0.794	3557	59
0.7	1.000 ± 0.000	0.778 ± 0.017	2133.2	55.8	1.000	0.824	2154	57
0.9	1.000 ± 0.000	0.783 ± 0.024	710.3	55.1	1.000	0.824	691	54
0.99	1.000 ± 0.000	0.779 ± 0.054	72.4	60.5	1.000	0.912	57	77
0.999	1.000 ± 0.000	0.739 ± 0.093	11.9	225.7	1.000	0.912	9	137

Tr. cl. = Training classification rate, Te. cl. = Testing classification rate.

Comparing the results obtained with ABC algorithm, PCA method was applied to the feature selection and it was tested with Euclidean and Manhattan distances. The parameters of PCA method for the experimentation were seven variance percentage with the values: $0.5, 0.6, 0.7, 0.8, 0.9$ and 1.

The experimental results for PCA using seven variance percentage are shown in Table 3. As can be observed, the results obtained with PCA are not as good as to those obtained with the proposed methodology. Furthermore, if we compare the number of genes selected by the proposed methodology for the best configuration against the number of features used with PCA, the proposed methodology uses less genes than PCA.

Table 3. Behavior of the proposed methodology using PCA technique

% Variance	Euclidean distance		# of features	Manhattan distance		# of features
	Tr. cl.	Te. cl.		Tr. cl.	Te. cl.	
0.5	0.947	0.765	6	0.895	0.794	6
0.6	0.947	0.794	10	0.921	0.765	10
0.7	0.974	0.765	14	0.921	0.735	14
0.8	0.974	0.765	20	0.921	0.765	20
0.9	0.974	0.765	27	0.921	0.765	27
0.95	0.974	0.765	31	0.921	0.794	31
1	0.974	0.765	36	0.921	0.735	36

Tr. cl. = Training classification rate, Te. cl. = Testing classification rate.

Furthermore, in [4] the authors report a training classification of 94.74% and testing classification of 85.29% which are not as good as those obtained with the proposed methodology. In [6] the authors report only a training classification accuracy of 97.38% for PSO algorithm using three genes, and for the case of GA, a percentage of 97.27% was achieved using four genes.

As can be observed from these experiments, the proposed methodology provides the best results using the Manhattan distance combined with ABC algorithm with a threshold $th = 0.1$, a percentage of 100% in training classification and 94.1% for testing classification were achieved, these results were obtained using only three genes.

5   Conclusions

In this paper an approach which focuses on evaluating a gene expression from a DNA microarrays combined with a feature selection previously applied was presented. The features selection task was carried out by means of Artificial Bee Colony (ABC) algorithm using different configurations to binarize each individual or solution and so to know which gene of thousands will be the best to contribute to solve a problem, in this case a classification task. After the feature selection, the Euclidean and Manhattan distances were applied to classify by means of two stages: training and testing.

Concerning to learning error evolution, the experimental results show that Manhattan distance provides a better performance than Euclidean distance. The learning error achieved the best results with few iterations. Talking about the accuracy, it is important to notice that the configuration of ABC algorithm using the Euclidean distance with $th = 0.3$ provides the best classification accuracy with a training percentage of 100% and a testing percentage of 88.2%, using only three genes. The results for Manhattan distance were better than the Euclidean distance results. The best classification accuracy obtained with the proposed methodology using the Manhattan distance was 100% in training classification and 94.1% in testing classification with $th = 0.1$ selecting only three genes.

The results were compared against those obtained with PCA. From these experiments, we observe that the accuracy obtained with PCA are not as good as those obtained with the proposed methodology. Furthermore, if we compare the number of genes, selected by the proposed methodology for the best configuration, against the number of features used by PCA, we observed that our proposed methodology uses less genes than PCA. Furthermore, when the proposed methodology is compared against some results reported in literature, the proposed methodology provides the best results using the Manhattan distance together with ABC algorithm with a threshold of $th = 0.1$ achieving a percentage of 100% in training classification and 94.1%for testing classification, using only three genes.

Nowadays, we are testing different classifier such as artificial neural networks, including spiking neural networks to improve the classification accuracy using ABC algorithm.

Acknowledgments. The authors thank DGAPA, UNAM and Universidad La Salle for the economic support under grants number IN107214 and I-61/12, respectively. Beatriz Garro thanks CONACYT for the posdoctoral scholarship.

References

1. López, M., Mallorquín, P., Vega, M.: Aplicaciones de los microarrays y biochips en salud humana: Informe de vigilancia tecnológica. Genoma España (2005)
2. Fodor, I.: A survey of dimension reduction techniques. Technical report (2002)
3. Dai, J.J., Lieu, L., Rocke, D.: Dimension Reduction for Classification with Gene Expression Microarray Data. Stat. App in Gen. and Mol. Bio. 5(1), 1–21 (2006)
4. Golub, T.R., et al.: Molecular classification of cancer: class discovery and class prediction by gene expression monitoring. Science 286(5439), 531–537 (1999)
5. Tang, E.K., et al.: Feature selection for microarray data using least squares svm and particle swarm optimization. In: Proceedings of IEEE, CIBCB 2005, pp. 1–8 (2005)
6. Alba, E., et al.: Gene selection in cancer classification using pso/svm and ga/svm hybrid algorithms. In: IEEE CEC 2007, pp. 284–290 (2007)
7. Pearson, K.: On lines and planes of closest fit to systems of points in space. Philosophical Magazine 2(6), 559–572 (1901)
8. Karaboga, D.: An idea based on honey bee swarm for numerical optimization. Technical report, Computer Engineering Department, Engineering Faculty, Erciyes University (2005)

Crowding-Distance-Based Multiobjective Artificial Bee Colony Algorithm for PID Parameter Optimization

Xia Zhou[1], Jiong Shen[2], and Yiguo Li[2]

[1] School of Mechanical and Electrical Engineering, Jinling Institute of Technology
Nanjing 211169, P.R.China
zenia77@163.com
[2] School of Energy and Environment, Southeast University
Nanjing 210096, P.R.China
{shenj,lyg}@seu.edu.cn

Abstract. This work presents a crowding-distance(CD)-based multiobjective artificial bee colony algorithm for Proportional-Integral-Derivative (PID) parameter optimization. In the proposed algorithm, a new fitness assignment method is defined based on the nondominated rank and the CD. An archive set is introduced for saving the Pareto optimal solutions, and the CD is also used to wipe off the extra solutions in the archive. The experimental results compared with NSGAII over two test functions show its effectiveness, and the simulation results of PID parameter optimization verify that it is efficient for applications.

Keywords: Crowding-distance, Multiobjective, Artificial bee colony, PID parameter optimization.

1 Introduction

The Proportional-Integral-Derivative (PID) controller is the most accepted controller in the process industry. More than 80% of industrial controllers are implemented based on the PID controller[1]. Objectives of the PID parameter optimization includes getting the best rising time and settling time, minimizing the overshoot and the accumulated error, and etc[2]. Since there are conflicts among the objectives, it is appropriate to solve it by multiobjective optimization method.

In order to solve the PID parameter optimization problem by effective multiobjective algorithm, and taking both the good performance of Crowding Distance(CD)[3] and the efficiency of Artificial Bee Colony(ABC)[4] into consideration, a Crowding-Distance-based Multiobjective Artificial Bee Colony Algorithm(CDMABCA) is proposed in this paper. The rest of this paper is organized as follows: Section 2 describes the proposed algorithm in detail; Section 3 discusses its application in the PID parameter optimization. Finally, the conclusion of the paper is outlined in section 4.

Y. Tan et al. (Eds.): ICSI 2014, Part I, LNCS 8794, pp. 215–222, 2014.

2   Crowding-Distance-Based Multiobjective ABC Algorithm

For convenience of the description, we give some notations in advance.
(1) The size of the employed bees is denoted as n_e. The size of the onlooker bees equals to that of the employed bees.
(2) An external archive set is introduced for saving the Pareto optimal solutions(PF), and the size of the archive is set as n_a.
(3) X_i is the i-th food source, x_{ij} is the j-th value of $X_i(j=1,2...d)$. The upper and lower bounds of the dimension j are denoted as x_{jmax} and x_{jmin}, respectively.

2.1   CD and Its Application in the CDMABCA

The CD is first addressed by Deb[3] for keeping a uniformly spread-out PF. The CD value of the i-th individual is calculated as follows.

$$CD_i = \sum_{j=1}^{m} CD_{ij}, \text{ and } CD_{ij} = \begin{cases} \infty & \text{if } f_{ij} = f_j^{max} \text{ or } f_{ij} = f_j^{min} \\ \dfrac{f_{(i+1).j} - f_{(i-1).j}}{f_j^{max} - f_j^{min}} & \text{otherwise} \end{cases} \tag{1}$$

where f_j^{max} and f_j^{min} are the maximum and minimum value of the j-th objective. $f_{(i+1).j}$ and $f_{(i-1).j}$ are the j-th objective value of the nearest neighbors to the i-th individual in the population.

Besides NSGA-II, the CD has been used as a distribution maintenance method in many other algorithms. In CDMABCA, the CD is not only used in keeping the spread-out of PF, but also used in the fitness assigning.

2.1.1   PF Keeping Based on the CD

As discussed in ref.[5], the CD is rough for it wipes off the extra individuals one time, and it cannot obtain a PF with good uniformity. In CDMABCA, the method proposed in ref.[5] is adopted. The details are described as follows:

Suppose the size of the nondominated set is N, and $N>n_a$. We first calculate the CD values for the members of the nondominated set. Instead of selecting n_a members having the largest CD values, $N-n_a$ members having the smallest CD values are removed one by one, and the CD values for the remaining members of the set are updated after each removal.

2.1.2   Fitness Assignment Based on the CD

With the Pareto-based methods, relations between the solutions are expressed as dominated or not. Select the nondominated individuals in the whole population as the first rank solutions and let its rank equals to one. Select the nondominated individuals in the remaining as the second rank solutions and let its rank equals to two. And so on, until the population is empty. The population is divided into many ranks and each solution belongs to a rank.

It is obvious that the individuals of lower rank are better that of higher rank. But if we simply use the rank value as the fitness value, we cannot distinguish the individuals of the same rank. Here the CD is added. Taking the two aspects into consideration, the fitness value of the i-th individual is evaluated as follows:

$$fit(X_i) = \frac{1}{rank_i + \dfrac{1}{CD_i}} \tag{2}$$

2.2   Description of the Proposed Algorithm

The flow of CDMABCA can be described as follows:

```
Function CDMABCA( )
initialization( );
REPEAT
     nondominated_set_getting( );
     archive_set_proposing( );
     employed_bee_optimization( );
     onlooker_bee_optimization( );
     scout_optimization( );
UNTIL (the stopping criterion is met)
```

The only difference between the CDMABCA and the ABC is that *nondominated_set_getting( )* and *archive_set_proposing( )* are added to the former. In the following, the definitions of each operator are given in detail.

(1) *Initialization*

The initial population is generated by using a random approach. i.e., each food source is generated as follows[4].

$$x_{ij} = x_{jmin} + rand(0,1)* (x_{jmax} - x_{jmin}) \tag{3}$$

where i=1,2...n_e, and j=1,2,..., d.

(2) *Nondominated set getting operator*

In each iteration, the nondominated solutions are acquired by the nondominated set getting operator. Since the efficiency of nondominated set getting has a significant impact on the efficiency of the whole algorithm, the method which has lower computational compexity will be preferred. Please see ref.[6] for the detail.

(3) *Archive set processing operator*

In the first run, nondominated solutions are added to the external archive directly. After that, at the end of each iteration, new generated nondominated solutions are compared with the solutions in the external archive. And the nondominated of the two parts are set as the new archive. If the quantity of the new archive is bigger than n_a, redundant solutions are removed according to the CD which is described in 2.1.1.

(4) *Employed bee optimization*

Just as the ABC algorithm[4], an employed bee takes the local search with the formula (4).

$$v_{ij} = x_{ij} + rand(-1,1)(x_{ij} - x_{kj})$$ (4)

where k and j are produced randomly, $j \in 1,2,...d$, $k \in 1,2,...n_e$, and $k \neq i$.

After local search is performed, the employed bee exploits a new neighbor food source V_i around X_i. Then the new food source will be evaluated and compared with the old one. Here a greedy selection will be performed, and the better food source will be kept in the population.

(5) *Onlooker bee optimization*

An artificial onlooker bee selects a food source depending on the fitness value according to formula(5).

$$P\{T_s(X) = X_i\} = \frac{fit(X_i)}{\sum_{j=1}^{n_e} fit(X_j)}$$ (5)

After the food source is selected, each onlooker bee will perform the same local search approaches and greedy selection procedure as employed bees.

(6) *Scout optimization*

If a food source cannot be further improved through some cycles, then the food source will be abandoned and the corresponded bee would turn out to be a scout[4]. A scout performs random search. It generates a new food source just as the initialization.

(7) *Terminal condition of the algorithm*

Terminal condition of the algorithm is set as arriving at the maximum cycle, or the archive set is unchanged in a continuous iterations.

2.3   Testify of the Proposed Algorithm

This section contains the experimental results obtained by the CDMABCA. In order to validate the performance of the proposed algorithm, it is compared to the NSGAII[3].

The selected test functions are given as follows[3,6].

$$P_1 : \begin{cases} f_1(x) = x \\ f_2(x) = (1 + 10y)[1 - (\dfrac{x}{1 + 10*y})^2 - \dfrac{x}{1 + 10*y} sin(2\pi * 4x)] \\ x, y \in [0,10] \end{cases}$$ (6)

$$P_2 : \begin{cases} f_1 = (1 + g(X_M)) * cos(x_1 * \dfrac{\pi}{2}) * cos(x_2 * \dfrac{\pi}{2}) \\ f_2 = (1 + g(X_M)) .* cos(x_1 * \dfrac{\pi}{2}) .* sin(x_2 * \dfrac{\pi}{2}) \\ f_3 = (1 + g(X_M)) .* sin(x_1 * \dfrac{\pi}{2}) \\ g(X_M) = \sum_{x_i / X_M} (x_i - 0.5)^2, x_i \in [0,1] \end{cases}$$ (7)

In order to make the NSGAII work as the presenter described, the parameters of the algorithm is set originally. The population size and archive size of the NSGAII are both set as 100, and the iteration cycle is set as 100, too. The other parameters of the NSGAII is set according to the values suggested by the developers. In the case of the CDMABCA, there are only four parameters. The population size and the iteration cycle are set according to that of the NSGAII, and they equal to 100 and 50, respectively. The quantity of the archive set equals to that of the population size, and the food source couldn't get optimized in 5 cycles will be abandoned.

Fig.1 shows the comparison of the experimental results of the two algorithms. Each diagram is marked by the algorithm name and the problem number.

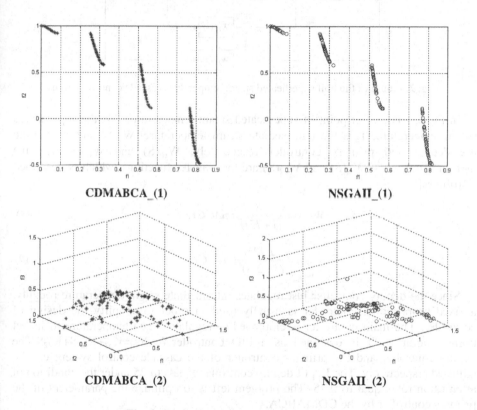

Fig. 1. Comparison of the experimental results

As shown in Fig.1, It is obvious that the CDMABCA performs better at these two benchmark problems. It is probably owing to introduction of improved CD for keeping the spread-out of the PF and the existence of the scouts which performs global random search.

3   Application in PID Parameter Optimization

3.1   PID Controller of Superheated Steam Temperature Control System

A simplified block diagram of superheated steam temperature cascade control system which use spray water injection as deputy controller is shown in Fig.2[7]. Spray water injection control system is extracted to respond quickly, serves as deputy control object; superheated outlet steam temperature which characterized by hysteresis and nonlinear, serves as the primary control object.

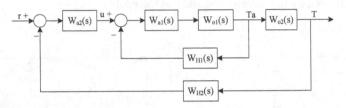

Fig. 2. Figure of the boil superheated steam temperature cascade control system

In Fig.2, r is the set point of superheated steam temperature, T is the superheated steam temperature, T_a is the intermediate steam temperature. $W_{a1}(S)$ and $W_{a2}(S)$ are the deputy and primary controller respectively. $W_{H1}(S)$ and $W_{H2}(S)$ are the measurement units. $W_{o1}(S)$ and $W_{o2}(S)$ are the transfer functions, which is described as follows[8].

$$W_{o1}(s) = \frac{8}{(1+15s)^2}(mA/^{\circ}C) \tag{8}$$

$$W_{o2}(s) = \frac{1.125}{(1+25s)^3}(mA/^{\circ}C) \tag{9}$$

Since the deputy controller just regulates intermediate steam temperature roughly, a fixed proportional controller is usually used in order to simplify the design of controller. The main controller is employed to regulate steam temperature to its set point accurately, it is designed as a PID controller, $Wa2(s)=K_{p2}+K_{i2}/S+K_{d2}S$. The deputy controller and the primary controller of the cascade control system can be adjusted respectively. The K_{p1} of deputy controller equals to 25 under the condition of attenuation rate equal to 0.75. The problem left is to optimize the parameters of the primary controller by the CDMABCA.

3.2   Principle of Multiobjective PID Parameter Optimization

The commonly used performance measures of the PID controller are attenuation rate (ψ), response overshoot (σ), rising time (t_r), settling time (t_s), integral time absolute error(ITAE), and etc. The effectiveness of the PID controller depends on the values of K_p, K_i and K_d. Taking the demands of the primary control object into consideration, we use t_s and ITAE' as the optimization objectives.

$$\min t_s, \text{ITAE}' = \int_{t_s+1}^{+\infty} t|e(t)|dt$$

s.t. $K_{pmin} \le K_{p2} \le K_{pmax}$, $K_{imin} \le K_i 2 \le K_{imax}$, $K_{dmin} \le K_{d2} \le K_{dmax}$, $\psi > 75\%$, $\sigma < 30\%$

where t_s is the settling time with error bound of 5%, and the integral time domain of the ITAE is improved.

3.3   Simulation Results

In this paper, bound value of the parameters are set according to ref.[8]. Which are set as $2 \le K_{p2} \le 4.5$, $0.035 \le K_{i2} \le 0.11$, $30 \le K_{d2} \le 45$. The PID parameters are tuned based on adding a 1mA step disturbance to the set point. With CDMABCA, plenty of Pareto optimal solutions are achieved. Due to limited space, we only list five groups of parameters in Table 3.

Fig.3(a) are the step responses curves of the parameters listed in Table 3. As for comparison, Fig.3(b) shows the step response curve of the best parameters proposed in ref.[7]. It is obvious that the simulation results of the CDMABCA are more satisfying than that of ref.[7] acquired. No matter what the overshoot, or the settling time, the former is better than the latter.

Table 1. parameter and objective values of the selected five groups

k_{p2}	k_{i2}	k_{d2}	$t_{s(s)}$	ITAE'
43.0726	2.4731	0.0476	116	36.755
41.7530	2.2152	0.0350	88	83.610
39.5608	2.1758	0.0351	98	65.244
44.7481	2.5085	0.0427	101	58.919
41.0376	2.3834	0.0453	119	36.665

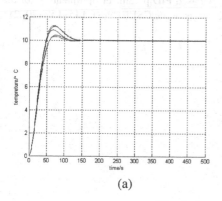

(a)

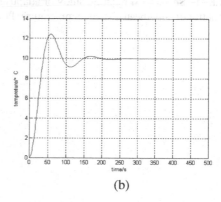

(b)

Fig. 3. Step responses curves

4   Conclusion

In this paper, a novel multiobjective artificial bee colony algorithm for optimizing the PID controller parameters is presented. How to adjust the ABC algorithm suitable for multiobjective optimization by the CD is discussed in detail. The simulation results demonstrate the effectiveness of the proposed algorithm, and it can be applied to other PID parameter tuning or other multiobjective optimization problems.

Acknowledgments. This work is supported by the National Natural Science Foundation of China under No. 51306082.

References

1. Khodabakhshian, A., Hooshmand, R.: A new PID controller design for automatic generation control of hydro power systems. Int. J. Electr. Power Energy Syst. 32, 375–382 (2010)
2. Rani, M.R., Selamat, H., Zamzuri, H., et al.: Multiobjective optimization for PID controller tuning using the global ranking genetic algorithm. International Journal of Innovative Computing, Information and Control 8, 269–284 (2012)
3. Deb, K., Pratap, A., Agarwal, S., et al.: A Fast and Elitist Multiobjective Genetic Algorithm: NSGA-II. IEEE Transactions on Evolutionary Computation 2, 182–197 (2002)
4. Karaboga, D.: An Idea Based on Honey bee Swarm for Numerical Optimization. Technical Report, Computer Engineering Department, Erciyes University, Turkey (2005)
5. Luo, B., Zheng, J., Xie, J., et al.: Dynamic Crowding Distance–A New Diversity Maintenance Strategy for MOEAs. In: The Proceedings of the Fourth International Conference on Natural Computation, pp. 580–585. IEEE (2008)
6. Zhou, A., Qu, B.Y., Li, H., et al.: Multiobjective Evolutionary Algorithms: A Survey of the State-of-the-art. Journal of Swarm and Evolutionary Computation 1, 32–49 (2011)
7. Zhao, L., Ju, G., Lu, J.: An Improved Genetic Algorithm in Multi-objective Optimization and its Application. Proceedings of the CSEE 28(2), 96–102 (2008)
8. Li, M., Shen, J.: Simulating study of adaptive GA-based PID parameter optimization for the control of superheated steam temperature. Proceedings of the CSEE 22(8), 145–149 (2002)

An Adaptive Concentration Selection Model for Spam Detection

Yang Gao, Guyue Mi, and Ying Tan*

Key Laboratory of Machine Perception (MOE), Peking University
Department of Machine Intelligence, School of Electronics Engineering and Computer
Science, Peking University, Beijing, 100871, China
{gaoyang0115,miguyue,ytan}@pku.edu.cn

Abstract. Concentration based feature construction (CFC) approach
has been proposed for spam detection. In the CFC approach, Global concentration (GC) and local concentration (LC) are used independently to
convert emails to 2-dimensional or 2n-dimensional feature vectors. In this
paper, we propose a novel model which selects concentration construction
methods adaptively according to the match between testing samples and
different kinds of concentration features. By determining which concentration construction method is proper for the current sample, the email
is transformed into a corresponding concentration feature vector, which
will be further employed by classification techniques in order to obtain
the corresponding class. The k-nearest neighbor method is introduced
in experiments to evaluate the proposed concentration selection model
on the classic and standard corpora, namely PU1, PU2, PU3 and PUA.
Experimental results demonstrate that the model performs better than
using GC or LC separately, which provides support to the effectiveness
of the proposed model and endows it with application in the real world.

Keywords: Global concentration (GC), local concentration (LC), adaptive concentration selection, spam detection.

1 Introduction

Spam has been a serious problem in the developing of internet. According to the
CYREN internet threats trend report, the average daily spam level for the first
quarter in 2014 was 54 billion emails per day [1]. Large numbers of spam not only
consume many resources online, but also threaten security of the network, especially when they carry viruses and malicious codes. What's more, people usually
take much time to handle spam, which reduces efficiency and productivity.

In the fields of spam detection, intelligent detection methods have been the
most effective way to examine junk mails. On one hand, the intelligent methods
have a higher degree of automation. On the other hand, these methods not
only have high precision and strong robustness, but also can fit email content

---

* Corresponding author.

Y. Tan et al. (Eds.): ICSI 2014, Part I, LNCS 8794, pp. 223–233, 2014.
© Springer International Publishing Switzerland 2014

and users' interests. Now the mainstream of intelligent detection methods can be divided into two categories: machine learning and artificial immune system. Because spam email detection task is a typical classification problem, supervised learning method is general in machine learning fields, such as naive bayes (NB) [2, 3], k-nearest neighbor (KNN) [4,5], support vector machine (SVM) [6], artificial neural networks (ANN) [7, 8] and so on. And in the artificial immune system (AIS), researchers imitate the process of immune cells' recognition to antigen.

In this paper, we propose a model structure, which aims to select concentration methods adaptively. The GC approach transforms each email to a 2-dimensional GC feature vector, which may lose some important information of the email. And the LC approach extracts position-correlated information from each email by mapping it to an 2n-dimensional feature vector, which may get some redundant information. But in our model, we can adjust concentration method adaptively according to distinctive information of different emails.

The remainder of the paper is organized as follows. In Section 2, we introduce the related works. In Section 3, the proposed adaptive concentration selection model is presented in detail. Section 4 gives the detailed experimental setup and results. Finally, we conclude the paper with a detailed discussion.

2   Related Works

This section introduces term selection approaches, concentration-based methods and classifiers that have close relationship with our work.

2.1   Term Selection Approaches

Information Gain. Information gain (IG) [9] is a concept in the information theory, which gives a description of the distance between two probabilities distribution $P(x)$ and $Q(x)$. In the spam detection field, it is utilized to measure the importance of terms. The calculation formula of IG is defined as

$$I(t_i) = \sum_{C \in (C_S, C_L)} \sum_{T \in (t_i, \bar{t}_i)} P(T, C) \log \frac{P(T, C)}{P(T)P(C)} \tag{1}$$

where C indicates an email's class (C_S and C_L are the spam and legitimate email classes) and T denotes the whether term t_i appears in the email or not. And all the probabilities are estimated from the whole data set.

2.2   Concentration-Based Methods

Global Concentration. Global concentration (GC) [10, 11] is an approach inspired from the human immune system, which can transform each email to a 2-dimensional feature vector. The flow chart of GC is described in Fig1. The biological immune system is a complex adaptive system, which has its unique self and non-self cells. Similar to this, the concentration approach proposed has

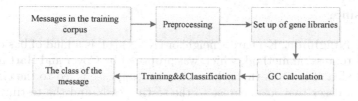

Fig. 1. Construction of GC model

two gene libraries - 'self' and 'non-self' gene libraries. The 'self' gene library is composed of words that present healthy emails. And in contrast, the 'non-self' gene library covers words that can present spam emails. So through the gene libraries, we can calculate global concentration of each email to construct its GC feature vectors.

Local Concentration. Similar to GC, the local concentration (LC) [12, 13] approach also transforms each email to a feature vector. However, the difference between GC and LC is that the LC can provide local information of a document, which can help to 'check' the email microscopically. In the process of LC, it

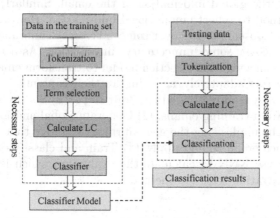

(a) Training phase of the model   (b) Classification phase of the model

Fig. 2. Construction of LC model

mainly covers two parts: the training part and the testing part. And in both parts, tokenization is the first step to pre-process the documents. Then in the term selection step, it chooses the important terms, which can reflect the emails' tendency to spam or non-spam. After calculating the local concentration of each email, every document is represented by a 2n-dimensional feature vectors. Then the feature vectors are transported to the classifier for training or testing.

2.3   Classifier

K-Nearest Neighbor. K-nearest neighbor (KNN) [14] is a kind of basic classification and regression method, which was proposed by Cover and Hart in 1968. The central idea of KNN is that when a new testing case is fed to the classifier, we look for k cases that are nearest to the testing case, and the testing case is classified as the class that those k cases belong to. KNN can be defined as follows

$$y = \arg\max_{C_j} \sum_{X_i \in N_k(x)} I(y_i = c_j), i = 1, 2, \ldots, N; j = 1, 2, \ldots, K \qquad (2)$$

where $I(y_i = c_j)$ is a indicator function, with the value of 1 when $y_i = c_j$, and 0 otherwise, and $(y_i) \in \Upsilon = (c_1, c_2, \ldots, c_k)$. And the special situation that the k is set to 1, KNN degrades to nearest neighbor.

3   Adaptive Concentration Selection Model

3.1   Overview of Our Proposed Model

In global concentration method, we transform an email into a 2-D feature vector, which reflects the global information of the email. Similarly, we use local concentration method to reflect emails' local information. However, global concentration may be too simple to cover some 'necessary' information and local concentration may cover some 'unnecessary' information. As a result, we propose the adaptive concentration selection model to transform emails into global or local feature vectors adaptively, according to their contents.

Our method can be mainly divided into four steps. (1) Set up 'self' and 'non-self' gene library from training emails. (2) Generate global and local concentration vectors of each email, using the gene library. (3) Judge that which concentration method each email should apply. (4) Train and classify on the corpora. In this paper, we use KNN to calculate the evaluation which is the reference standard of concentration selection method.

3.2   Set Up of Gene Libraries

Intuitively, if a word appears mostly in spam emails, it belongs to the 'non-self' gene library largely. Accordingly, a word which can provide more information for spam emails than non-spam emails usually will be put into the 'non-self' gene library, and vice versa. This inspires us to calculate information gain of each word, and sort them in a decent order. Considering the amount of words is too big to build gene library, and most documents contain the same common words, we also discard 95% of the words that appear in all emails, just as the paper does [10].

Algorithm 1. Generation of gene libraries

---

1. Initialize gene libraries, detector DS_s and DS_l to the empty
2. Initialize tendency threshold θ to predefined value
3. Tokenization about the emails
4. **for** each word t_k separated **do**
5.   According to the term selection method, calculate the importance of t_k and the amount of information $I(t_k)$
6. **end for**
7. Sort the terms based on the $I(t)$
8. Expand the gene library with the top $m\%$ terms
9. **for** each term t_i in the gene library **do**
10.   **if** $\|P(t_i|c_l) - P(t_i|c_s)\| > \theta$, $\theta \geq 0$ **then**
11.   **if** $P(t_i|c_l) - P(t_i|c_s) < 0$ **then**
12.   add term t_i to the spam detector set DS_s
13.   **else**
14.   add term t_i to the legitimate detector set DS_l
15.   **end if**
16.   **else**
17.   abandon this term, because it contains little information about those emails
18.   **end if**
19. **end for**

Algorithm 2. Construction of feature vectors based on global concentration

---

1. **for** each term t_j in the email **do**
2.   calculate the matching $M(t_j, DS_s)$ between term t_j with spam detector set;
3.   calculate the matching $M(t_j, DS_l)$ between term t_j with legitimate detector set
4. **end for**
5. According to 3, calculate the concentration of spam detector set SC;
6. According to 4, calculate the concentration of legitimate detector set LC;
7. Combine the above concentration values to construct the global concentration feature vectors $< SC, LC >$

3.3 Construction of Feature Vectors Based on the Immune Concentration

After we have got the gene library, we can construct the feature vectors. According to the generation of detector set, it is obvious that the DS_s can match spam emails and the DS_l can match the legitimate emails with large probability. As a result, the match between two detector sets and emails can reflect the class information of emails, and the two detector sets have complementary advantages with each other, which provides a guarantee for the effectiveness of detection.

$$SC_i = \frac{\sum_{j=1}^{\omega_n} M(t_j, DS_s)}{N_t} \qquad (3)$$

where N_t is the number of distinct terms in the window, and $M(t_j, DS_s)$ is the matching function which is used to measure the matching degree of term t_j and detector DS_s.

$$LC_i = \frac{\sum_{j=1}^{\omega_n} M(t_j, DS_l)}{N_t} \tag{4}$$

where $M(t_j, DS_l)$ is the matching function which is used to measure the matching degree of term t_j and detector DS_l.

Algorithm 3. Construction of feature vectors based on local concentration

1. According to the length of each email and preset number of windows to calculate the value of ω_n
2. Move the ω_n-term sliding window to separate the email, with each moving length being ω_n
3. **for** each moving window **do**
4.   **for** each term in the moving window **do**
5.   calculate the matching $M(t_j, DS_s)$ between term t_j with spam detector set;
6.   calculate the matching $M(t_j, DS_l)$ between term t_j with legitimate detector set;
7.   **end for**
8.   According to 5, calculate the concentration of spam detector set SC_i;
9.   According to 6, calculate the concentration of legitimate detection set LC_i
10. **end for**
11. Combine local concentration values in each sliding window to construct the local concentration feature vector $< (SC_1, LC_1), (SC_2, LC_2), \ldots, (SCn, LCn) >$

$$SC_i = \frac{\sum_{j=1}^{\omega_n} M(t_j, DS_s)}{N_t} = \frac{\sum_{j=1}^{\omega_n} \sum_{d_k \in DS_s} M(t_j, d_k)}{N_t} \tag{5}$$

$$LC_i = \frac{\sum_{j=1}^{\omega_n} M(t_j, DS_l)}{N_t} = \frac{\sum_{j=1}^{\omega_n} \sum_{d_k \in DS_l} M(t_j, d_k)}{N_t} \tag{6}$$

3.4   Implementation of Our Model

Global concentration reflects entire features of emails and the local concentration reflects local characteristics. However, the GC lacks some detailed information and the LC separates the emails quite meticulously. As a result, we propose our model to combine their advantages and make up for their disadvantages. The key point of our model is the evaluation which is used to determine concentration methods. In our paper, we use KNN to calculate the evaluation. As we all know, the main idea of KNN is to count the numbers of neighbors belonging to different kinds of classes. However, if the numbers of different classes are close, it is hard to judge which class the undetermined point belongs to. So in our model, we take use of this characteristic of KNN and adapt it to determine concentration methods.

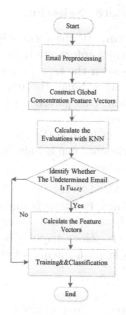

Fig. 3. Implementation of our model

Firstly, after preprocessing, we convert all data to GC feature vectors and use KNN classifier to evaluate them. During the evaluation, if the number of a particular class, which belongs to the neighbors of a undetermined point, is larger than a certain proportion, we can classify the point to this class. But if the number is less than the proportion, we consider this point as a fuzzy one. Secondly, for those fuzzy points, we convert them to LC feature vectors which can reflect their details and evaluate them with KNN classifier again. Thirdly, we manipulate all classification results and assess them with precision, recalls, accuracy and F_1 measure.

4   Experiments

4.1   Experimental Setup

In this paper, experiments were conducted on PU series email corpora, which contains PU1, PU2, PU3 and PUA and were collected and published by Androutsopoulos [15] in 2004. The PU series email corpora were widely utilized in spam detection related experiments. To ensure the objectivity, all the experiments are organized with 10-fold cross validation. At the stage of classification, we choose the KNN method to verify the spam and legitimate emails. Besides, we use recalls, precision, accuracy and F_1 measure to assess the results. Among them, the F_1 measure is taken as the most important evaluating indicator, for its reflection of the recalls and precision. All experiments were conducted on a PC with Intel P7450 CPU and 2G RAM.

4.2   Experiments of Parameter Selection

Proportion of term selection. In the term selection stage, we choose top $m\%$ of the terms according to their information quantity, which decides the size of the gene library. When we screen the terms, on one hand, we need to cut off those noise terms, and on the other hand, the important terms should be held back. In the practical application, this parameter can be adjusted based on the need of time and space complexity.

According to the paper written by Zhu [13], when the parameter m is set to 50%, the performance of experiments can achieve optimal. Therefore, the value of m is set to 50% in our experiments.

Tendency threshold. Tendency function is mainly used to measure the difference between the terms and the two kinds of emails and add corresponding terms to the related detector. In Zhu's paper [13], with the increasing of the tendency threshold θ, the whole performance of the algorithm degrades. As a result, the value of θ is set to 0.

Dimension of feature vectors. In the global concentration method, each email is reconstructed with the self and non-self concentration, which means the dimension is two. And in the local concentration method, this paper adopts variable length sliding window strategy, which means that if we assume N is the number of sliding window, each email is transformed into an $2N$-dimensional feature vector. In this paper, we set the parameter N to 3, according to [10]. As a result, the dimension of local concentration method is 6.

Parameter k in KNN. We have done some experiments to determine the value of parameter k. And the results are shown as follows. As mentioned above, the PU2 is a corpus containing only English emails and the PUA contains not only English emails but also other languages. So the experiments on these two corpora reflect general characteristics. Besides, we find that different experiments based on different values of parameter k perform similarly And as we all know, if the value of k is set too large, the computation complexity will increase. Consequently, we choose a moderate value, which sets the value of k to five.

4.3   Experiments of the Proposed Model

In this paper, we conducted comparison experiments of the model with selection method IG and mainly compares the performance among GC, LC and our model. These experiments are mainly conducted on corpora PU1, PU2, PU3 and PUA using 10-fold cross-validation. The average performance experiments are reported in Table 1 to Table 4.

Compared to GC and LC, the proposed adaptive concentration selection model achieves a better performance on the four corpora. Although in the experiment with PU1, the precision and recall indexes are less than GC or LC, the

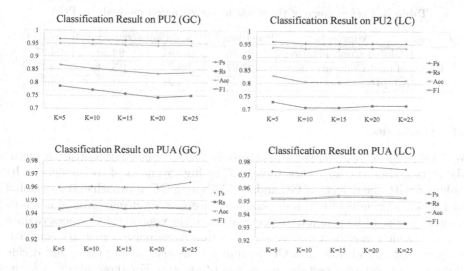

Fig. 4. Classification results on PU2 and PUA

Table 1. Performance of three feature construction methods on PU1

Corpus	Approach	Precision(%)	Recall(%)	Accuracy(%)	$F_1(\%)$
PU1	Global Concentration	95.59	94.37	95.60	94.97
	Local Concentration	96.54	92.92	95.41	94.69
	Adaptive Concentration	96.18	94.17	**95.78**	**95.16**

Table 2. Performance of three feature construction methods on PU2

Corpus	Approach	Precision(%)	Recall(%)	Accuracy(%)	$F_1(\%)$
PU1	Global Concentration	96.74	78.57	**95.07**	**86.71**
	Local Concentration	95.95	72.86	93.80	82.83
	Adaptive Concentration	96.74	78.57	**95.07**	**86.71**

Table 3. Performance of three feature construction methods on PU3

Corpus	Approach	Precision(%)	Recall(%)	Accuracy(%)	$F_1(\%)$
PU1	Global Concentration	96.14	93.57	95.40	94.84
	Local Concentration	96.95	92.86	95.47	94.86
	Adaptive Concentration	96.78	94.07	**95.91**	**95.41**

overall evaluation index, F_1 measure, is better than GC and LC. And we can still conclude that our model performs better on this corpus.

As a result, we can come to a conclusion that the proposed model combines the advantages of GC and LC, and it can enhance the experimental effects so as to classify emails more precisely.

Table 4. Performance of three feature construction methods on PUA

Corpus	Approach	Precision(%)	Recall(%)	Accuracy(%)	$F_1(\%)$
	Global Concentration	95.98	92.81	94.30	94.37
PU1	Local Concentration	97.27	93.33	**95.26**	95.17
	Adaptive Concentration	97.44	94.21	94.65	**95.79**

4.4   Discussion

We have proposed our model for adaptively taking use of concentration methods' feature construction characteristics. The improvement of the model can be explained with the defects of GC and LC. Although GC approach extracts global information of emails into 2-dimensional feature vectors, it may miss some information because of its rough data processing. To the contrary, LC processes data in detail, which may be too excessive to retain some noise terms. By contrast, our proposed model first uses GC feature vectors to evaluate data, and divide all data into two parts: certain classes and fuzzy ones. For those fuzzy ones, the proposed model further takes use of the detailed information based on LC feature vectors and finally we get better performance according to the experimental results. Generally speaking, the model combines both advantages of GC and LC, and avoids large computational complexity of only LC method.

5   Conclusion

In this paper, we present a spam filtering system that combine GC and LC feature construction methods that further makes the system adaptive to different emails. In the stage of feature extraction, we use IG to estimate terms' importance and concentration methods to transform emails into reconstructed feature vectors. And in the classification, according to different characteristics of emails, the system adaptively chooses feature construction methods and the performance is promising.

In the future, we intend to convert emails into variable length future vectors according to the length of emails' messages and study its performance.

Acknowlegements. This work was supported by the National Natural Science Foundation of China under grants number 61375119, 61170057 and 60875080.

References

1. CYREN: Internet threats trend report: April 2014. Tech. rep. (2014)
2. Sahami, M., Dumais, S., Heckerman, D., Horvitz, E.: A bayesian approach to filtering junk e-mail. In: Learning for Text Categorization: Papers from the 1998 Workshop, vol. 62, pp. 98–105. AAAI Technical Report WS-98-05, Madison (1998)
3. Ciltik, A., Gungor, T.: Time-efficient spam e-mail filtering using n-gram models. Pattern Recognition Letters 29(1), 19–33 (2008)

4. Androutsopoulos, I., Paliouras, G., Karkaletsis, V., Sakkis, G., Spyropoulos, C., Stamatopoulos, P.: Learning to filter spam e-mail: A comparison of a naive bayesian and a memory-based approach. Arxiv preprint cs/0009009 (2000)
5. Sakkis, G., Androutsopoulos, I., Paliouras, G., Karkaletsis, V., Spyropoulos, C., Stamatopoulos, P.: A memory-based approach to anti-spam filtering for mailing lists. Information Retrieval 6(1), 49–73 (2003)
6. Drucker, H., Wu, D., Vapnik, V.: Support vector machines for spam categorization. IEEE Transactions on Neural Networks 10(5), 1048–1054 (1999)
7. Clark, J., Koprinska, I., Poon, J.: A neural network based approach to automated e-mail classification. In: Proceedings of the IEEE/WIC International Conference on Web Intelligence, WI 2003, pp. 702–705. IEEE (2003)
8. Wu, C.: Behavior-based spam detection using a hybrid method of rule-based techniques and neural networks. Expert Systems with Applications 36(3), 4321–4330 (2009)
9. Yang, Y.: Noise reduction in a statistical approach to text categorization. In: Proceedings of the 18th Annual International ACM SIGIR Conference on Research and Development in Information Retrieval, pp. 256–263. ACM (1995)
10. Tan, Y., Deng, C., Ruan, G.: Concentration based feature construction approach for spam detection. In: International Joint Conference on Neural Networks, IJCNN 2009, pp. 3088–3093. IEEE (2009)
11. Ruan, G., Tan, Y.: A three-layer back-propagation neural network for spam detection using artificial immune concentration. Soft Computing 14(2), 139–150 (2010)
12. Zhu, Y., Tan, Y.: Extracting discriminative information from e-mail for spam detection inspired by immune system. In: 2010 IEEE Congress on Evolutionary Computation (CEC), pp. 1–7. IEEE (2010)
13. Zhu, Y., Tan, Y.: A local concentration-based feature extraction approach for spam filtering. IEEE Transactions on Information Forensics and Security 6(2), 486–497 (2011)
14. Cover, T., Hart, P.: Nearest neighbor pattern classification. IEEE Transactions on Information Theory 13(1), 21–27 (1967)
15. Androutsopoulos, I., Paliouras, G., Michelakis, E.: Learning to filter unsolicited commercial e-mail. "DEMOKRITOS". National Center for Scientific Research (2004)

Control of Permanent Magnet Synchronous Motor Based on Immune Network Model

Hongwei Mo[1] and Lifang Xu[2]

[1] Automation College, Harbin Engineering University,
150001 Harbin, China
[2] Engineering Training Center, Harbin Engineering University,
150001 Harbin, China
Honwei2004@126.com, mxlfang@163.com

Abstract. Immune control is a kind of intelligent control method which is based on biology immune system. It provides a new way for solving nonlinear, untertain and time variable system. In the paper, for the problem of Permanent Magnet Synchonous Motor (PMSM) speed control, an immune controller based on Varela immune network model is proposed. It uses immune feedback mechanism to construct the immune controller. It is compared with conventional PID controller for PMSM speed control. The Varela immune controller has a smaller starting current to avoid the influence of excessive current to PMSM. Simulation results show the effectiveness and practicability of this method.

Keywords: Varela immune network, permanent magnet synchonous motor, PID control, immune control.

1 Introduction

A closed loop speed control system consisting of PMSM is widely used in aerospace, aviation, machine tools, robots, automation and other fields because of its characteristics such as high precision, wide speed range, excellent dynamic performance and has good application prospects. Researchers carry out a lot of research work created for PMSM AC servo system. Currently, PMSM servo control adapts approximate linear model or linear control. Servo system design uses current speed double-loop structure, in which the inner is current loop, the outer is velocity loop. And control method adapts PI regulator. In order to overcome control deficiencies caused by this approximation linear models, some researchers proposed many new and useful methods or control strategies applying to PMSM servo system, including the synovial variable structure control [1] [2], adaptive control [3] [4], fuzzy control [5], neural networks [6] and so on. But they are still inadequate to PMSM.

In fact, PMSM AC servo system is a typical nonlinear multivariable coupling system. It is easily affected by the unknown disturbance load, friction and magnetic fields at runtime. Approximately linear control is difficult to meet the performance requirements of PMSM. Immune Control is a new intelligent control method. The main principle of the immune control learns from living organisms to address the problems

Y. Tan et al. (Eds.): ICSI 2014, Part I, LNCS 8794, pp. 234–245, 2014.

which cannot be solved well by traditional control methods and provide new ideas for solving the control problems with complexity, nonlinear, time-varying, uncertainty.

In the field of immune control, Takahashi first proposed immune feedback mechanism based on self-tuning immune feedback controller [7]. In[8], the authors used the immune feedback to adaptively adjust the single neuron PID control parameters of a permanent magnet AC servo system. It can enhance the anti-disturbance ability of the system to improve the speed control performance. In literature[9],for PMSM multi-parameter identification, it proposed an immune co-evolutionary particle swarm optimization algorithm for multi-parameter identification and estimation of dq axis inductance motors , the stator winding resistance and rotor flux in order to effectively track parameters change. In [10], in order to solve the uncertainty of the model parameters and motor disturbance problem, an artificial immune controller for PMSM frequency control system was proposed to improve the adaptive capacity and robustness of the system. Varela Varela immune network model is proposed and Stewart in 1990 [11], it was improved for large objects large inertia lag control[12], but also can used for the permanent magnet brushless DC motor (BLDCM) control [13]. Studies show that the network model has good resistance to nonlinear, time delay and parameter adaptive capacity. Immune control is used to design the speed loop of PMSM AC servo system. It needs neither parameter identification, nor the specific form of nonlinear uncertain function. Direct design of controller parameters allows the system to reach the global asymptotic stability.

In this paper, the model is designed to improve the speed of an immune controller for PMSM AC servo system control.

2   Establish Immunity Controller

Varela immune network model [10] can be expressed as a differential equation

$$\frac{\mathrm{d}\,f_i}{\mathrm{d}\,t} = -k_1\,\sigma_i f_i - k_2\,f_i + k_3\,\mathrm{M}(\sigma_i)b_i \tag{1}$$

$$\frac{\mathrm{d}\,b_i}{\mathrm{d}\,t} = -k_4\,b_i + k_5\,\mathrm{P}(\sigma_i)b_i + k_6 \tag{2}$$

where k_1 - k_6 denote the adjustment coefficient of immune network; f_i and b_i denote the concentration of the species cloned antibodies and B cells. The three terms in (1) represent antibody mortality caused by antibody interactions, natural antibody mortality and new antibodies produced by B cells, respectively. The three terms in(2)represent the natural B cells mortality, reproductive rate of B cells and the B cells produced by bone marrow, respectively.

M( ) and P( ) are the functions of reproduction and mature of B cells, respectively. σ_i denotes the sensitivity to network of the ith clone.

$$\sigma_i = \sum_j m_{ij} f_j \tag{3}$$

$$M(\sigma_i) = e^{p_1|\sigma_i|} - e^{p_2|\sigma_i|} \tag{4}$$

$$P(\sigma_i) = e^{p_1|\sigma_i|} - e^{p_2|\sigma_i|} \tag{5}$$

where p_1 and p_2 are the adjustment coefficient exponent; $m_{i,j}$ is Boolean value of the affinity interaction of ith clone and j th clone. The value 1 indicates that there is an affinity effect, 0 means no. h is the species of B_i cell and T_i antibody. Mature and reproductive functions are shown in Fig.1. They are in the form of "bell" shape function.

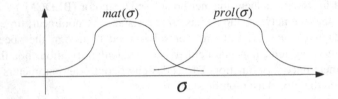

Fig. 1. Ripe function shape

Formulas (1) and (2) to some extent reflect the dynamic process of interaction between the process of artificial immune B cells and antibodies T, but the controller does not reflect the role of antigen.

The basic framework to build immune controller is immune feedback rule. The error in the control system is equivalent to the immune system antigen. In the Varela immune network model above, it is not related to the invasion of antigen. Therefore, to make the network model can be used to control the field, Fu improved it by introducing the role of antigen[11].

When the interaction between antigens is introduced into the body, there would be two different reactions. One is antigen self-replication and reproduction. f_i and a_i represent the ith antibody and antigen. k_7 is the adjustment coefficient of self-replication and reproduction processes. Another one is that antigen will be cleared by antibodies. Assuming antigen clear is mainly accomplished by the meeting of antibodies and antigens. The process is denoted by $f_i a_i$, which is the product of the number of antibodies and antigens. k_7 is the adjustment coefficient antibody removal process, we get

$$\frac{d a_i}{d t} = k_7 a_i - k_8 f_i a_i \tag{6}$$

Combining (1), (2) and (6), we can obtain the kinetic equations of the control model. As the amount of control in the model and the amount of the actual control system have discrepancy, the adjustments based on the actual situation is as follows: (1) The change rate of antigen into the body consists of self-replicating rate of antigen and killing rate of antibody to antigen. But in actual control system, the bias does not have the phenomenon of self-replicating, therefore removing the first term in (1). (2) After the antigen into the body, it stimulates B cell proliferation. $e(t)$ is instead of the antigen a_i and add a new term $k_9 e(t)$ in (2), we have

$$\frac{db_i}{dt} = -k_4 b_i + k_5 P(\sigma_i)b_i + k_6 + k_9 e(t) \tag{7}$$

Let b_i is independent variable of $M(\sigma_i)$. $\dfrac{db_i}{dt}$ is independent variable of $P(\sigma_i)$. The antigen a_i is replaced by the deviation $e(t)$. The concentration b_i of B cells is replaced by control amount $u(t)$.

Because the system is a single input single output control system, through a simple iteration [15], we obtain the Varela immune controller model after adjustment as follows.

$$\begin{cases} \dot{f}(t) = -k_{10}\, f(t) + k_3\, M(u(t))u(t) \\ \ddot{u}(t) = -(k_4 - k_5\, P(u(t))u(t) \dot{}\ \ k_{11}\, \dot{f}(t)e(t) \end{cases} \tag{8}$$

where f_0 indicates small memory antibodies when there is non-invasive antigen. K_m is a constant, $K_m > 0$. p_1 and p_2 are also constant, $p_2 < p_1 < 0$.

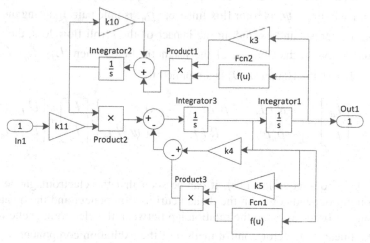

Fig. 2. Block diagram of Varela immune controller

Based on the equation (8) , the improved Varela immune network model, immune controller is built by Matlab Simulink 7.0 .In Figure 2, Fcn1and Fcn2   denote reproduction function $P(\sigma)$ and the mature function M (σ),which are taken as nonlinear function, p1 = -0.1, p2 = -0.10001, reproductive function = 500, mature function M (σ) = 500000.

3   PMSM Speed Control

3.1   PMSM Model

PMSM uses three-phase AC power.   The equivalent of three-phase windings is spatially symmetrical two-phase windings which are mutual difference of 90 °, namely direct axis d and quadrature axis winding q. In the d-q coordinate system, PMSM state equation is:

$$
\begin{bmatrix} i_d \\ i_q \end{bmatrix} = \begin{bmatrix} -R/L_d & p_n \omega_r L_q / L_d \\ -p_n \omega_r L_d / L_q & -R/L_q \end{bmatrix} \cdot \begin{bmatrix} I_d \\ I_q \end{bmatrix} + \begin{bmatrix} 1/L_d \\ & 1/L_q \end{bmatrix} \cdot \begin{bmatrix} U_d \\ U_q - p_n \omega_r \psi_r \end{bmatrix} \tag{9}
$$

Electromagnetic torque equation is:

$$
T_e = 1.5 p_n [\psi_r I_q + (L_d - L_q) I_d I_q] \tag{10}
$$

Among them, I_d and I_q are the currents of d shaft and q shaft, respectively; U_d and U_q are the voltages of d shaft and q shaft , respectively; L_d and L_q are the armature inductances of d shaft and q shaft, respectively; R is the resistance per phase stator winding; ψ_r is rotor flux linkage; p_n is pole pair. Ignoring the magnetic saturation, hysteresis and excluding the impact of the scroll flow loss, the sinusoidal conditions of spatial distribution of the magnetic field, when $L_d = L_q = L$,damping coefficient B = 0, the Equation (9) is simplified to:

$$
\begin{bmatrix} i_d \\ i_q \end{bmatrix} = \begin{bmatrix} -R/L & p_n \omega_r & 0 \\ -p_n \omega_r & -R/L & -p_n \psi_r / L \end{bmatrix} \cdot \begin{bmatrix} I_d \\ I_q \end{bmatrix} + \begin{bmatrix} U_d / L \\ U_q / L \end{bmatrix} \tag{11}
$$

From the torque equation (10), it can be seen that the electromagnetic torque of PMSM output depends only on the d shaft current component and the q shaft current component. In order to make the relationship between the electromagnetic torque and current be linear, the vector control method of the excitation component $I_d = 0$ of the stator current is used here. The advantage of this method is simple to control, and widely applied in practice. Then the electromagnetic torque equation becomes:

$$T_e = 1.5 p_n \psi_r I_q \tag{12}$$

The rotor angular velocity equation:

$$\dot{\omega}_r = 1.5 p_n \psi_r I_q / J - T_L / J \tag{13}$$

where J is the mechanical inertia of the motor, T_L motor load torque.

PMSM starts with no-load, when the load is added, t = 0.1s. The parameters of PMSM model are selected as follows:

$R = 2.875\,\Omega$, $L_d = L_q = L = 0.0085H$, $P_n = 1$, $J = 0.008\,kg.m^2$, $T_L = 4\,N \cdot m$

Transformation equation from a-b-c coordinate system to d-q coordinate system and inverse transformation equation are as follows:

$$\begin{bmatrix} I_d \\ I_q \end{bmatrix} = \frac{2}{3} \begin{bmatrix} \cos\theta & \sin\theta \\ -\sin\theta & \cos\theta \end{bmatrix} \begin{bmatrix} 1 & -1/2 & -1/2 \\ 0 & \sqrt{3}/2 & -\sqrt{3}/2 \end{bmatrix} \begin{bmatrix} i_a \\ i_b \\ i_c \end{bmatrix} \tag{14}$$

$$\begin{bmatrix} i_a \\ i_b \\ i_c \end{bmatrix} = \begin{bmatrix} \cos\theta & -\sin\theta \\ \cos(\theta - 2\pi/3) & -\sin(\theta - 2\pi/3) \\ \cos(\theta + 2\pi/3) & -\sin(\theta + 2\pi/3) \end{bmatrix} \begin{bmatrix} I_d \\ I_q \end{bmatrix} \tag{15}$$

where, i_a, i_b, i_c are three phase stator currents under a-b-c coordinate system, respectively.

4   Simulation and Analysis

4.1   PMSM Speed Control Simulation System

In the experiments, the PMSM speed control system is shown in Figure 3. In the system, PMSM stator is powered by a three-phase SPWM inverter, which uses the double-loop speed control mode, that is, the speed regulator (ASR) as the outer loop, the current regulator (ACR) as the inner loop of. Current regulator is PI controller. Speed controller uses immune controller. Rotor position sensor PG detects the rotor speed and angle. Under the conditions of setting excitation component to be zero, SPWM three-phase voltage modulation signal and current feedback signal are obtained by coordinate transformation 2r/3s and coordinate inverse transform 3s/2r, respectively. Motor starts with no load, t = 0.1s 4 $N \cdot m$ is loaded.

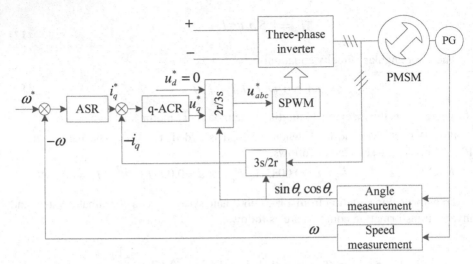

Fig. 3. PMSM speed control system schematics

4.2   Results of Speed Control

The parameters of immune controller and PI controller are selected as the optimal values obtained in several experiments. $k_3 =0.01$, $k_4 =8$, $k_5 =0.000001$, $k_{10} =130$, $k_{11} =20$. Speed loop kp = 1000, ki = 14, current loop kp = 10.7, ki = 10.

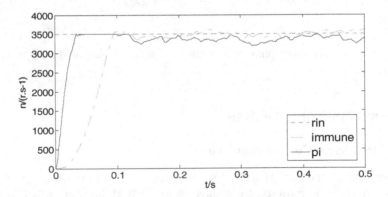

Fig. 4. Speed control response

It can be seen from Figure 4, PMSM starts to load and loads 4 $N \cdot m$ at 0.1 second, the speed appeared fluctuation. Compared with the PID control system, the speed response of PMSM controlled by the immune controller is slightly overshoot and steady-state regulation time is slightly longer, but the steady-state error is relatively small, the speed fluctuation is weak. And it has good capability of speed following.

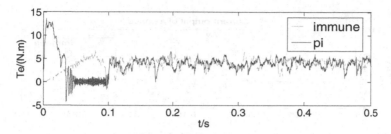

Fig. 5. Electromagnetic torque output

It can be seen from Figure 5, compared with PID controller, the electromagnetic torque output is small at early starting when PMSM controller is under immune controller, which led directly to a longer time for speed reaching to expected value. At 0.1 seconds, the PMSM speed has risen to expectations under the control of the immune controller, then add 4 $N \cdot m$ load has no effect on the output of PMSM electromagnetic torque. The immune controller can effectively suppress the influence of load disturbances with good robustness.

It can be seen from Figure 6, compared with the PID controller, the stator current components I_d of PMSM has a small starting current under the control of immune controller. After the 4 $N \cdot m$ at 0.1 seconds is loaded, the current changes smoothly. The immune controller can overcome overall impact to the control system caused by large starting current of PMSM.

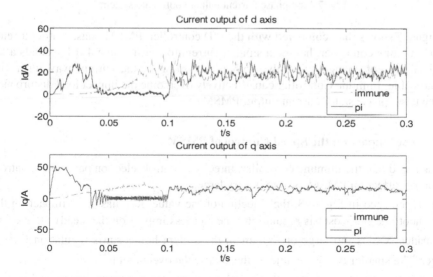

Fig. 6. D-q axis under Stator current output

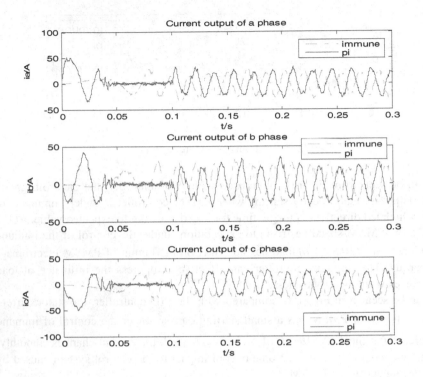

Fig. 7. Abc phase current output coordinate system

Figure 7 shows that, compared with the PID controller, PMSM phase stator currents of the immune controller, having a smaller current to start loading 4 0.1 seconds after the load current change smoothly, without distortion of the current waveform, the output stable. Immune Controller can effectively overcome the impact load disturbance on the three-phase stator current output PMSM

4.3   The Impact on the Speed Control of PMSM

It was found that the immune controller, three key control selection parameter controls the output effect on PMSM system is very large.

It can be seen in Figure 8, the selection of the value k_{10} has great impact on the over shoot and homeostasis regulation time and less impact on the steady-state error. No matter k_{10} is too large or too small, the system homeostasis regulation time is longer. The smaller the k_{10} value is, the greater the overshoot is.

It can be seen from Figure 9, the selection of k_{11} has great impact on system overshoot and homeostasis regulation time, and less impact on the steady-state error. No matter k_{11} is too large or too small, the system homeostasis regulation time is longer. The greater the k_{11} value is, the greater the overshoot is.

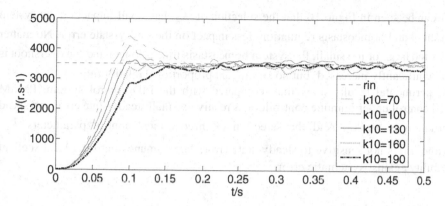

Fig. 8. Impact PMSM speed control response

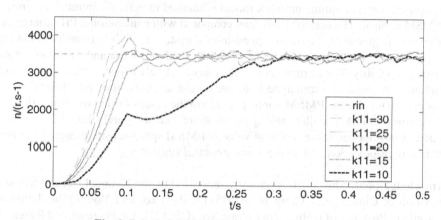

Fig. 9. Pairs influence PMSM speed control response

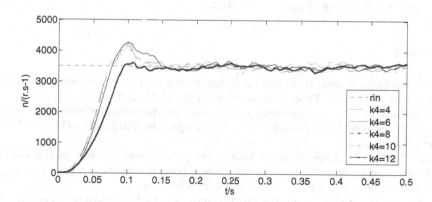

Fig. 10. Pairs influence PMSM speed control response

It can be seen in Figure10 that the selection of k_4 has small impact on the system overshoot and homeostasis regulation, less impact on the steady-state error. No matter k_4 is too large or too small, the system homeostasis time is longer and the overshoot is also significantly increased, but no significant proportional relationship.

Experimental results show that, compared with the PID control system, PMSM speed control under immune controller has relatively small steady-state error and speed fluctuation is weak. And the selection of three critical control parameters k_4, k_{10} and k_{11} is not sensitive to steady-state error. The immune controller has excellent capability against systematic error.

5   Conclusions

In this paper, Varela immune network model is adapted to build the immune controller for PMSM control. The results show that, compared with conventional PID controller, the dynamic response of the motor speed control system is slightly overshoot, but the steady-state error is relatively small, the speed fluctuation is weak and with good speed following capability. The immune controller can effectively suppress the effect of load disturbance on speed, electromagnetic torque output and current output. The selection of key parameters for the PMSM control system to the results further illustrates that the immune controller has excellent ability of anti-system errors. Varela immune controller as a new controller has some practical value in PMSM speed-sensitive control system, as well as high-precision and high accuracy control system.

Acknowledgement. This work is partially supported by the National Natural Science Foundation of China under Grant No.61075113, the Excellent Youth Foundation of Heilongjiang Province of China under Grant No. JC201212, the Fundamental Research Funds for the Central Universities No.HEUCFX041306 and Harbin Excellent Discipline Leader, No.2012RFXXG073.

References

1. Merzoug, M.S., Benalla, H., Louze, L.: Sliding mode control (SMC) of permanent magnet synchronous generators(PSMG). Energy Procedia 18, 43–52 (2002)
2. Li, Z., Hu, G., Cui, J.: Integral sliding mode variable structure control of permanent magnet synchronous motor speed control system. Proceedings of the CSEE 34(3), 431–437 (2014) (in Chinese)
3. Yu, J.P., Ma, Y.M., Chen, B., et al.: Adaptive fuzzy backstepping position tracking control for a permanent magnet synchronous motor. International Journal of Innovative Computing, Information and Control ICIC International 4, 589–1601 (2011)
4. Hou, L., Zhang, H., Liu, X.: Robust passive control of PMSM non speed sensor of adaptive fuzzy sliding mode soft switching. Control and Decision 25(5), 686–690, 694 (2010) (in Chinese)

5. Selvaganesan, N., Saraswathy, R.R.: A simple fuzzy modeling of permanent magnet synchronous generator. Elektrika 11(1), 38–43 (2009)
6. AlHarthi, M.: Control of permanent magnet synchronous motor using artificial neural networks. Advances in Electrical Engineering Systems 1(3), 157–162 (2012)
7. Takahashi, K., Yamada, T.: Application of an immune feedback mechanism to control systems. JSME International Journal, Series C 41(2), 184–191 (1998)
8. Wang, K., Liu, X.: Immune single neuron PID control in permanent magnet AC servo system. Xi'an Jiaotong University 44(4), 76–81 (2010) (in Chinese)
9. Zhen, H., Pu, Z., Kaibo, Z., Li, K., Huaxiu, Z.: Immune co-evolution particle swarm optimization for permanent magnet synchronous motor parameter identification. Source: Zidonghua Xuebao/Acta Automatica Sinica 38(10), 1698–1708 (2012)
10. Liu, G., Yang, L.: Artificial immune controller for PMSM speed regulation system. Year the Document was Publish 2010 Source of the Document Proceedings 2010 IEEE 5th International Conference on Bio-Inspired Computing: Theories and Applications
11. Varela, F.J., Stewart, J.: Dynamics of a class of immune networks Theory Biology, p. 93 (1990)
12. Fu, D.M., Zheng, D.L.: Design and simulation of immune controller based on Varela immune modal. Journal of Beijing Science and Technology University 29(2), 199–203 (2007)
13. Xu, L., Mo, H.: Application of an Immune Controller to Flywheel BLDCM. Journal of South China University of Technology 5(41), 80–86 (2013)

Adaptive Immune-Genetic Algorithm for Fuzzy Job Shop Scheduling Problems

Beibei Chen[1], Shangce Gao[1,2], Shuaiqun Wang[3], and Aorigele Bao[2]

[1] Department of Automation, Donghua University, Shanghai, 201620, China
[2] Faculty of Engineering, University of Toyama, Toyama, 930-8555, Japan
[3] Department of Computer Science, Tongji University, Shanghai, 201804, China

Abstract. In recent years, fuzzy job shop scheduling problems (FJSSP) with fuzzy triangular processing time and fuzzy due date have received an increasing interests because of its flexibility and similarity with practical problems. The objective of FJSSP is to maximize the minimal average customer's degree of satisfaction. In this paper, a novel adaptive immune-genetic algorithm (CAGA) is proposed to solve FJSSP. CAGA manipulates a number of individuals to involve the progresses of clonal proliferation, adaptive genetic mutations and clone selection. The main characteristic of CAGA is the usage of clone proliferation to generate more clones for fitter individuals which undergo the adaptive genetic mutations, thus leading a fast convergence. Moreover, the encoding scheme of CAGA is also properly adapted for FJSSP. Simulation results based on several instances verify the effectiveness of CAGA in terms of search capacity and convergence performance.

Keywords: Job shop scheduling, Fuzzy processing time, Fuzzy due date, Clonal algorithm, Adaptive genetic algorithm.

1 Introduction

Job-shop scheduling problem (JSSP) is one of the well-known hardest combinatorial optimization problems. During the last three decades, this problem has captured the interest of a significant number of researchers [1, 2]. However, some uncertain factors during the production process, such as mechanical failures, delay machining, uncertain job processing time, and fuzzy due date of jobs, strongly affect the arrangement and assessment in the whole scheduling. In this study, the fuzzy job shop scheduling problem (FJSSP) is considered and solved to describe and represent the real-world problems more closely. In the literature, numerous algorithms have been proposed to solve FJSSP. Han et al. [3] studied the single machine scheduling problem with fuzzy due date. Sakawa et al. [4] used genetic algorithm to research the fuzzy shop scheduling problem with fuzzy processing time and fuzzy due date, and proved that the proposed GA gave better results than SA. Ishibuchi et al. [5] studied the fuzzy job shop scheduling problem using GA with fuzzy processing time. Murata et al. [6] reported the validation results for the multi-objective job shop scheduling problem with fuzzy due date. Adamopulos [7] gave another attempt that use the neighborhood search

Y. Tan et al. (Eds.): ICSI 2014, Part I, LNCS 8794, pp. 246–257, 2014.

method to study the single machine scheduling problem with variable processing time and fuzzy due date.

In this paper, an improved adaptive immune-genetic algorithm (CAGA) is proposed to resolve FJSSP. In FJSSP, the fuzzy processing time and fuzzy due date are represented by the triangular fuzzy numbers and the semi-trapezoidal fuzzy numbers respectively. The objective of FJSSP is to maximize the minimal average degree of customers' satisfaction. The main evolutionary procedures of CAGA are the clonal proliferation, adaptive crossover and mutation, and clonal selection. Both search properties of clonal selection algorithms and genetic algorithms are properly emerged, thus resulting in a well balance between the search exploration and exploitation. In addition, the coding scheme is strictly selected for FJSSP. The validation of the performance of CAGA is carried out based on several instances. Experimental results show the superiorities of CAGA over traditional algorithms.

2   Description of the Problem

In FJSSP, the processing tasks set for an available set of machines are allocated in the time to meet the performance indicators [8]. The schedule is for a particular job which can be decomposed under the certain constraints. That is, how to arrange the occupied resources of the operation, the fuzzy processing time, and the order, so that make the customers' minimal average degree of satisfaction maximum for the job's completion time. In general, FJSSP is formulated in the following. Let n different jobs be processed on m different machines, every job has O_j operations. P_{ij} represents the j-th operation of job N_i. O_{ijk} represents the operation P_{ij} be processed on machine M_k. In this paper, the fuzzy processing time of operation O_{ijk} is represented by a triangular fuzzy number $T_{ijk}(t^1_{ijk}〗 t^2_{ijk}〗 t^3_{ijk})$. The semi-trapezoidal fuzzy numbers $D (d^1_i〗 d^2_i)$ represents the fuzzy due date of job N_i. The constraints of the fuzzy schedule model are as follows.

(1) Order constraint expression:

$$S_{ij} \geq S_{i(j-1)} + T_{i(j-1)}, i = 1,2,3,...n, j = 1,2,3,...m; \tag{1}$$

(2) Machine constraint expression:

$$M_{ik} \geq M_{(i-1)k} + T_{(i-1)k}, i = 1,2,3,...n; k = 1,2,...m; \tag{2}$$

(3) Time constraint expression:

$$S_{ij} \geq 0 \quad i = 1,2,3,...n; j = 1,2,...,m; \tag{3}$$

Where S_{ij} represents the start time of the operation j of the job i. M_{ik} denotes the job i is processed on the machine k. T_{ij} denotes the j operation's fuzzy processing time of job i. Eq. (1) shows that a job must be completed before start another process. Eq. (2) represents that each machine can only process a work piece at a time. Eq. (3) indicates that the start time of each operation must be not lower than zero.

2.1   FJSSP with Fuzzy Processing Time and Fuzzy due Date

In the scheduling process, due to the job process is affected by many factors, we use the triangle fuzzy variable to express the processing time. The triangular fuzzy processing time is $T_{ijk}=(t^1_{ijk}, t^2_{ijk}, t^3_{ijk})$, where $t^1_{ijk}, t^2_{ijk}, t^3_{ijk}$ represent the most ideally processing time of the operation, the maximum possible processing time, and the most pessimistic processing time respectively. The triangular fuzzy numbers can truly reflect the actual production conditions and T_{ijk} is shown as follows in Fig.1 (a). The membership function $\mu_{ij}(t)$ is a possibility measure that the operation j of job N_i is completed in the processing time x. The formula (4) of the membership function $\mu_{ij}(t)$ is given as follows.

$$\mu_{ij}(x) = \begin{cases} (x-t^1_{ijk})/(t^2_{ijk}-t^1_{ijk}) & x \in \left[t^1_{ijk}, t^2_{ijk}\right] \\ (t^3_{ijk}-x)/(t^3_{ijk}-t^2_{ijk}) & x \in \left[t^2_{ijk}, t^3_{ijk}\right] \\ 0 & x < t^1_{ijk}, x > t^3_{ijk} \end{cases} \tag{4}$$

The jobs' due date is inseparable with the customers' degree of satisfaction. When only consider the case that tardiness completion is the unwelcome situation, we use the semi-trapezoidal fuzzy numbers $D_i=D\ (d^1_i, d^2_i)$ to represent the fuzzy due date of job N_i. As shown in Fig.1 (b), if N_i is finished below the value d^1_i of the due date, the degree of satisfaction is 1. If the fuzzy completion time is in the window $[d^1_i, d^2_i]$, the degree of satisfaction is expressed with the linear function. c indicates the time it takes to finish the job N_i. $\mu_i(c)$ is the satisfactory degree of the membership function of job N_i. The formula (5) of membership function $\mu_i(c)$ is shown as follows.

$$\mu_i(c) = \begin{cases} 1 & c \le d^1_i \\ (c-d^1_i)/(d^2_i-d^1_i) & c \in \left[d^1_i, d^2_i\right] \\ 0 & c > d^2_i \end{cases} \tag{5}$$

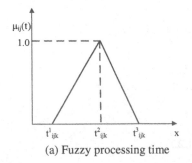
(a) Fuzzy processing time

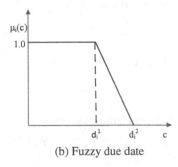
(b) Fuzzy due date

Fig. 1. Fuzzy processing time and Fuzzy due date

2.2   The Objective Function of FJSSP

In this paper, the objective is to find the fittest schedule which minimal average degree of satisfaction is the maximum. As illustrated in Fig. 2, the degree of satisfaction is defined as the ratio between the area S_A (the shade in Fig. 2) which is encircled by the fuzzy completion time's membership function and the fuzzy due date's membership function, and the graphics area S_D which is surrounded by the fuzzy completion time's membership function.

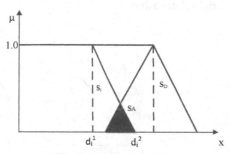

Fig. 2. Satisfaction index

Customer's degree of satisfaction for the job N_i is $AI_i = (S_i \cap S_D)/S_D$, so the average satisfaction of all the jobs is:

$$AI = \sum_{i=1}^{n} AI_i / n = \sum_{i=1}^{n} (S_i \cap S_D)/S_D / n \qquad (6)$$

In Eq. (6), AI_i is the Customer's degree of satisfaction of the job N_i. Fuzzy scheduling must ensure that each job has a certain degree of satisfaction to meet the delivery requirements, thus maximize the minimum average satisfaction index as the goal function. The objective function of FJSSP can be described as follows. In Eq. (7), AI is the all jobs' average satisfaction index.

$$\max\{\min AI \mid I = 1,2,3,\ldots n\} \qquad (7)$$

2.3   Fuzzy Operations

In FJSSP, the fuzzy operations are the key elements. The fuzzy operations involve addition operation, max operation, and the comparison between two fuzzy numbers. Max operation is to determine the job fuzzy starting time of operation. Addition operation is used to determine the fuzzy completion time. Because the processing time is represented by triangular fuzzy numbers, so the process beginning time and completion time are also fuzzy numbers. For two fuzzy numbers $\tilde{A} = (a_1, a_2, a_3)$, $\tilde{B} = (b_1, b_2, b_3)$, the fuzzy operations are described as follows.

(1) Fuzzy addition operation:

$$\tilde{A} + \tilde{B} = ((a_1 + b_1), (a_2 + b_2), (a_3 + b_3)) \qquad (8)$$

(2) Fuzzy max operation:

$$\tilde{A} \vee \tilde{B} = \left(max\left(a_1, b_1\right), max\left(a_2, b_2\right), max\left(a_3, b_3\right) \right) \qquad (9)$$

Here, Fig. 3 shows two triangular fuzzy numbers which take the □ (max) operation in different locations' relationship.

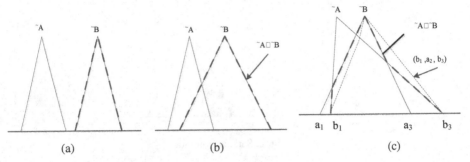

Fig. 3. Max operation of triangular fuzzy numbers

(3) compare operation of triangular fuzzy numbers:
For the comparison operation of triangular fuzzy numbers, the general following three criteria [6] are adopted:

Criterion 1: $C_1\left(\tilde{A}\right) = \left(a_1 + 2a_2 + a_3\right)/4$, we may use the value of C_1 to compare the two triangular fuzzy numbers $\tilde{A}, \tilde{B}$.

Criterion 2: $C_2\left(\tilde{A}\right) = a_2$, when the value C_1 of the two triangular fuzzy numbers are equal, we compare the value C_2.

Criterion 3: $C_3\left(\tilde{A}\right) = a_3 - a_1$, if the value C_1 and C_2 are all equal respectively, we compare the value C_3.

3   Algorithm Design

The genetic algorithm is an essential directly search method that does not rely on the specific issues. It has a strong problem-solving ability and robustness, which can solve the nonlinear optimization problems very well. Therefore, the genetic algorithm in field of the production scheduling has been taken more and more attention [9-13]. On the other hand, immune algorithms, especially the clonal selection algorithm, are inspired by the mechanism of the biological immune system. The basic idea of clonal selection is to select those cells which can recognize the cell antigens, then clone them. In this paper, an attempt of combining clonal selection theory with adaptive genetic algorithm is made to increase the detection and search abilities near the optimum solutions [14], aiming to improve the search efficiency of the algorithm.

3.1   Coding Scheme

Encoding is the conversion process from the solution space to genetic coding. A natural expression of job shop scheduling problem is not easy to be determined because of the working procedure route's restraint. If encoding idea is inappropriate, it will prone to deadlock [15]. In this paper, we use the expression based on the working process. This scheduling encodes based on the process sequence, and all the same jobs are specified as the same decimal notation, and then to be interpreted according to the order they appeared in a given chromosome.

3.2   Clonal Proliferation

The number of clones of each individual:

$$q_i = ceil\left(\alpha * \frac{f(i)}{\sum_{i=1}^{N} f(i)}\right) \tag{10}$$

Here, α represents the cloning coefficient. $f(i)$ denotes the chromosome fitness. $ceil()$ denotes the rounded-off integer. Eq. (10) represents the number of clones of each individual. In order to improve the processing speed, and avoid only to rely on increasing individual scale to achieve optimal solution, the maximum number q_i of clones is set to be $q_i < N/2$, where N is the initial population size.

3.3   Improved Adaptive Crossover and Mutation Operators

The crossover and mutation probability are two important parameters affecting the performance of the algorithm. In this algorithm, the value of the crossover probability P_c and mutation probability P_m can be changed automatically along with the fitness values of individuals in the population. As suggested in [16], we adopt a typical adaptive method to control the above two parameters:

$$P_c = \begin{cases} \dfrac{k_1 \times (f_{max} - f_1)}{f_{max} - f_{avg}} & f_1 \geq f_{avg} \\ k_2 & f_1 < f_{avg} \end{cases} \tag{11}$$

$$P_m = \begin{cases} \dfrac{k_3 \times (f_{max} - f_2)}{f_{max} - f_{avg}} & f_2 \geq f_{avg} \\ k_4 & f_2 < f_{avg} \end{cases} \tag{12}$$

In the above formulas, k_1, k_2, k_3, k_4 are the constants among (0, 1). Eqs. (11) (12) show that when the individual fitness value equals to the maximum fitness value, the crossover and mutation probability is 0, and then the algorithm evolutionary capacity will be limited. Thus we propose the new adaptive crossover and mutation probability functions, as shown in Eqs. (13), (14). When the individual fitness value is equal to the maximum fitness value, it can take crossover and mutation operations according to a certain probability respectively. Therefore the search ability of the algorithm is improved and the search is easier to get out of the local optima.

$$P_c = \begin{cases} P_{c1} + \dfrac{k_1 \times (f_{max} - f_1)}{f_{max} - f_{avg}} & f_1 \geq f_{avg} \\ k_2 & f_1 < f_{avg} \end{cases} \quad (13)$$

$$P_m = \begin{cases} P_{m1} + \dfrac{k_3 \times (f_{max} - f_2)}{f_{max} - f_{avg}} & f_2 \geq f_{avg} \\ k_4 & f_2 < f_{avg} \end{cases} \quad (14)$$

Here, f_1 is the larger fitness value among the two cross individuals. f_2 is the fitness value of an variation individual. f_{avg} is the average fitness value of all the individuals. f_{max} is the largest fitness value of all the individuals. P_{c1} and P_{m1} are initial crossover and mutation probability respectively.

In this study, the crossover operation refers to replace a part of the structure of the two parent individuals to generate two new offspring individuals. The detailed steps of the crossover operation are given as follows.

Step1: According to adaptive crossover probability, we choose two individuals from the population, which are known as the father individuals P1, P2.

Step2: Pick out two non-empty complementary subsets D1, D2 from P1, P2.

Step3: The genes which belong to D2 are picked out from individual P1 to form a new gene fragments H1. The genes belonging to D1 are opted from individual P2 to form a new gene fragments H2. Then, we use H2 to fill into the corresponding location of the gene fragment H1 to obtain an offspring C1.

Step4: Genes which belong to D2 are selected from individual P2 to form a new gene fragments G2. Moreover, choose the genes belonging to D1 from individual P1 to form a new gene fragments G1. Then, we use G1 to fill into the corresponding location of the gene fragment G2 to constitute an offspring C2.

Step5: Finally, compare the fitness value of C1, C2, P1, P2. When the maximum fitness value of C1 and C2 is greater than the maximum fitness value of P1 and P2, thus we consider C1 and C2 are superior to P1 and P2. The offspring C1 and C2 will replace the parents P1 and P2; otherwise, the parents P1 and P2 will be preserved and they will enter the next generation to evolve.

It is clear that the above adaptive crossover will not produce unreasonable chromosome phenotype. The method can improve the algorithm's search capability and the quality of the whole population. Therefore avoid the algorithm to fall into local optimum situation effectively.

After the crossover operation, we use the mutation method with adaptive mutation probability to deal with the population. First, randomly select a parent. Then generate two random integer values, a_1 and a_2, which are both range in $[1, n \times m]$. Finally swap the two genes at location a_1 and a_2 of the parent respectively.

3.4   Clonal Selection Operation

Clonal selection operation chooses the chromosomes which have strong vitality from the current population, to make them have a chance to reproduce, and thereby improve the global convergence and computational efficiency. It is intended to avoid the loss of effective genes. In the vicinity of the optimal solution, clone every parent based on the fitness value. The clonal selection procedure utilizes the elitist reservation strategy. The fittest antibodies Y_i ($i=1,2,...,N$) of all the clones of each parent antibody are firstly selected, i.e., $Y_i = X_{i,j} = min_j\{f(X_{i,1}),...,f(X_{i,j}),...,f(X_{i,m_i})\}$. Then, a hill climbing update rule is used to replace the parent antibodies X_i with selected clones Y_i according to an updating probability P_i.

$$P_i = \begin{cases} 1 & f(Y_i) < f(X_i) \\ exp\left(\dfrac{f(X_i)-f(Y_i)}{k}\right) & f(Y_i) \geq f(X_i) \\ 0 & f(Y_1) \leq f(X_1) \end{cases} \tag{15}$$

Based on Eq. (15), if the fittest antibody of the clones Y_i has worse fitness than its parent antibody X_i, update it with probability "1". As a result, the elites in the offspring have been preserved and enter into the following generation. On the contrary, if Y_i is better than its parent, then update according to an exponential function to maintain the diversity of the population, here $k \in (10,100)$ is a value related to the diversity. Generally, the better the diversity is, the bigger the value of k. Furthermore, in order to save the information of the original population so that the best antibody in the parents could not be replaced, the exponential function is unavailable for parent X_1.

The procedure of CAGA is shown in Fig. 4:

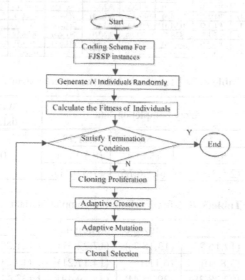

Fig. 4. The implementation procedure of CAGA

4   Experiment and Simulation Results

In this paper, we use 3×3, 6×6, 10×10 job shop scheduling problems as the study instances. The three benchmark instances are taken from the literature [4] and [17].

Table 1. 3×3 fuzzy scheduling and processing data

jobs	Processing machines(fuzzy processing time)			Fuzzy duedate
1	1(2.5,3,3.5)	2(2.5,3,3.5)	3(1.5,2,2.5)	(11,12)
2	1(0.5,1,1.5)	3(4.5,5,5.5)	2(2.5,3,3.5)	(7.5,9)
3	2(2.5,3,3.5)	1(1.5,2,2.5)	3(2.5,3,3.5)	(12,13.5)

Table 2. 6×6 fuzzy scheduling and processing data

jobs	Processing machines(fuzzy processing time)						Fuzzy due date
1	1(5,6,13)	5(3,4,5)	2(1,2,3)	6(3,4,5)	4(2,3,4)	3(2,3,4)	(30,40)
2	1(3,4,5)	2(2,4,5)	3(1,3,5)	6(4,5,6)	4(5,6,7)	5(6,7,8)	(35,40)
3	3(1,2,3)	6(5,6,7)	5(4,5,6)	4(3,4,5)	2(1,2,3)	1(1,2,3)	(20,28)
4	6(2,3,4)	5(1,2,3)	4(2,3,4)	2(2,3,5)	1(3,4,6)	3(3,4,5)	(32,40)
5	6(3,4,5)	5(2,3,4)	4(1,2,3)	3(2,3,4)	2(4,5,6)	1(2,3,4)	(30,35)
6	5(6,7,8)	6(4,5,6)	1(2,3,4)	2(3,4,5)	3(2,3,4)	4(1,2,3)	(40,45)

Table 3. ×10 fuzzy scheduling and processing data

jobs	Processing machines(fuzzy processing time)										Fuzzy due date
1	8(2,3,4)	6(3,5,6)	5(2,4,5)	2(4,5,6)	1(1,2,3)	3(3,5,6)	9(2,3,5)	4(1,2,3)	7(3,4,5)	10(2,3,4)	(45,60)
2	10(2,3,4)	7(2,3,5)	4(2,4,5)	6(1,2,3)	8(4,5,6)	3(2,4,6)	2(2,3,4)	1(1,3,4)	5(2,3,4)	9(3,4,5)	(50,60)
3	6(2,4,5)	9(1,2,3)	10(2,3,5)	8(1,2,4)	1(3,5,6)	7(1,3,4)	4(1,3,5)	2(1,2,4)	5(2,4,5)	3(1,3,5)	(50,65)
4	1(1,2,3)	5(3,4,5)	8(1,3,5)	9(2,4,6)	10(2,4,5)	6(1,2,4)	7(3,4,5)	2(1,3,5)	4(1,3,6)	3(1,3,4)	(50,65)
5	2(2,3,4)	7(1,3,4)	3(1,3,4)	5(1,2,3)	8(1,3,5)	9(2,3,4)	10(3,4,5)	6(1,3,4)	1(3,4,5)	4(1,3,4)	(50,65)
6	4(2,3,4)	2(2,3,4)	3(1,2,3)	5(2,4,5)	6(1,3,4)	8(1,3,4)	7(3,4,5)	9(1,2,3)	10(2,4,5)	1(1,3,4)	(45,60)
7	3(2,3,4)	5(1,4,5)	4(1,3,5)	1(3,4,5)	9(2,3,4)	7(3,4,5)	2(1,2,3)	10(3,5,6)	8(3,5,6)	6(1,2,3)	(45,60)
8	7(3,4,5)	1(1,2,3)	9(3,4,5)	6(2,4,5)	10(1,3,4)	2(2,3,4)	5(1,2,3)	3(2,4,5)	4(3,4,5)	8(2,3,5)	(50,60)
9	9(3,4,5)	4(1,3,4)	10(1,3,5)	2(2,3,4)	3(3,5,6)	6(2,4,5)	8(1,3,4)	1(3,4,5)	5(1,2,3)	7(3,4,5)	(45,60)
10	7(2,4,5)	5(1,2,3)	2(3,4,5)	4(2,3,4)	1(1,2,3)	8(3,4,5)	10(2,4,5)	6(3,4,5)	3(1,2,3)	9(1,2,4)	(50,60)

Table 4. 3×3 fuzzy scheduling completion data

jobs	fuzzy completion data			Average satisfaction
1	(3,4,5)	(5.5,7,8.5)	(9,11,13.5)	
2	(0.5,1,1.5)	(5,6,7)	(8,10,12)	0.579
3	(2.5,3,3.5)	(4.5,6,7.5)	(7.5,9,11)	

Table 5. 6×6 fuzzy scheduling completion data

jobs	fuzzy completion data					
1	(8,10,18)	(12,16,23)	(13,18,26)	(17,22,31)	(21,26,35)	(24,33,45)
2	(3,4,5)	(5,8,10)	(6,11,15)	(14,17,21)	(19,23,28)	(25,30,36)
3	(1,2,3)	(22,28,38)	(29,35,44)	(32,39,49)	(33,41,52)	(34,43,55)
4	(2,3,4)	(7,9,11)	(9,12,15)	(11,15,20)	(15,19,28)	(18,23,33)
5	(5,7,9)	(9,12,15)	(10,14,18)	(12,17,22)	(17,23,32)	(19,26,36)
6	(6,7,8)	(10,12,15)	(12,15,22)	(20,27,37)	(22,30,41)	(23,32,44)

Table 6. 10×10 fuzzy scheduling completion data

job	fuzzy completion data									
1	2,3,4	5,9,11	8,15,19	12,20,25	13,23,32	19,30,41	23,38,51	24,41,57	27,45,62	29,49,66
2	2,3,4	5,7,10	7,14,19	10,16,22	14,21,29	16,25,35	20,32,41	21,36,46	23,39,50	26,43,56
3	2,4,5	4,6,8	7,13,19	8,16,23	11,21,29	14,24,33	15,27,38	23,38,52	25,43,57	26,46,63
4	1,2,3	6,11,14	7,14,19	12,21,29	14,25,34	15,28,38	21,33,43	22,36,48	23,39,54	24,42,58
5	2,3,4	6,10,14	7,13,18	12,23,30	17,32,42	21,35,46	24,39,51	25,42,55	28,46,60	29,49,66
6	2,3,4	4,6,8	5,8,11	10,19,24	11,22,28	15,25,33	18,29,38	19,31,41	21,35,46	23,41,53
7	2,3,4	3,7,9	5,10,14	8,14,19	10,17,23	13,21,28	15,25,32	18,30,40	21,37,48	22,39,51
8	3,4,5	4,6,8	7,10,13	9,14,18	10,17,23	14,23,29	15,25,33	21,34,46	27,45,62	29,48,67
9	3,4,5	4,7,9	5,10,14	7,13,18	10,18,24	13,26,33	16,29,37	19,33,42	20,35,45	24,39,50
10	8,14,19	11,21,27	18,29,37	20,32,42	22,38,49	25,42,54	27,46,59	30,50,64	31,52,67	32,54,71

In CAGA, the initial population size is 100. The initial adaptive cross probability P_{c1} is 0.5. The initial variation probability P_{m1} is 0.1. The cloning coefficient α is 50. The maximum number of iterations MAXGEN=200. The adaptive genetic algorithm (AGA) [17] and genetic algorithms (GA) [4] are also implemented to make a comparison. Each algorithm runs 10 times. Table 4, Table 5, Table 6 recorded the fuzzy scheduling completion data for the above three benchmark FJSSP instances.

From Table 4, we can find that the maximum minimum average degree of satisfaction (0.579) can be obtained every run for the 3×3 problem. This value is better than optimal solution 0.5255 in [17]. Tables 7 and 8 summarize the optimal minimum average degree of satisfaction by using the three algorithms respectively. For the 6×6 problem, the best value by CAGA is 0.773, which is better than the optimal solution 0.69 of the problem in [4]. The convergence curve is depicted in Fig.5 (a). For 10×10 scheduling problems, the objective satisfaction value is 0.961 obtained by CAGA, which is also better than the optimal solution value 0.94 in [4]. The corresponding convergence curve is shown in Fig. 5(b). Besides, Fig. 6(a) and Fig. 6(b) illustrates all convergence curves of CAGA for 10 runs.

Table 7. 6×6 Average customer satisfaction in fuzzy scheduling

Trial	GA	AGA	CAGA	Trial	GA	AGA	CAGA
1	0.617	0.683	0.769	6	0.429	0.687	0.773
2	0.532	0.706	0.748	7	0.652	0.706	0.773
3	0.547	0.703	0.773	8	0.618	0.706	0.769
4	0.547	0.692	0.769	9	0.669	0.694	0.773
5	0.429	0.701	0.773	10	0.627	0.688	0.773

Table 8. 10×10 Average customer satisfaction in fuzzy scheduling

Trial	GA	AGA	CAGA	Trial	GA	AGA	CAGA
1	0.649	0.799	0.921	6	0.713	0.850	0.933
2	0.663	0.802	0.947	7	0.742	0.870	0.947
3	0.692	0.815	0.961	8	0.742	0.855	0.947
4	0.751	0.819	0.961	9	0.751	0.815	0.961
5	0.718	0.821	0.961	10	0.658	0.870	0.961

From the above simulations, we can see that CAGA can achieve better performance than its competitors. The novelty of CAGA is the hybridization of clonal operators with the genetic algorithm. The experimental results verify the effectiveness of such hybridization based on solving several fuzzy job shop scheduling problems.

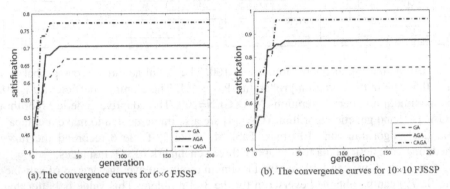

(a).The convergence curves for 6×6 FJSSP   (b). The convergence curves for 10×10 FJSSP

Fig. 5. Maximum Minimum Satisfaction compared during GA, AGA, CAGA

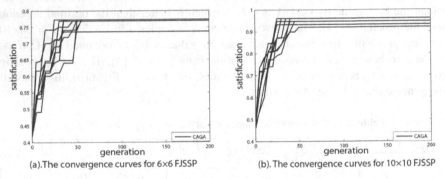

(a).The convergence curves for 6×6 FJSSP   (b). The convergence curves for 10×10 FJSSP

Fig. 6. The Maximum Minimum Satisfaction (CAGA run 10 times)

5   Summary

In this paper, a novel adaptive immune-genetic algorithm (CAGA) is proposed to solve the fuzzy job shop scheduling problems (FJSSP). CAGA manipulates a number of individuals to involve the progresses of clonal proliferation, adaptive genetic mutations and clone selection. The main characteristic of CAGA is the usage of clone proliferation to generate more clones for fitter individuals which undergo the adaptive genetic mutations, thus leading a fast convergence. Moreover, the encoding scheme of CAGA is also properly adapted for FJSSP. Furthermore, the parameters to control crossover and mutation operators are adaptively set. Simulation results based on several instances verify the effectiveness of CAGA in terms of search capacity and convergence performance.

Acknowledgment. This work is partially supported by the National Natural Science Foundation of China (Grants No. 61203325), Shanghai Rising-Star Program (No. 14QA1400100), "Chen Guang" project supported by Shanghai Municipal Education Commission and Shanghai Education Development Foundation (No. 12CG35), Ph.D. Program Foundation of Ministry of Education of China (No. 20120075120004), and the Fundamental Research Funds for the Central Universities (No. 2232013D3-39).

References

1. Davis, L.: Handbook of Genetic Algorithm. Van Nostrand Reinhold (1991)
2. Wang, L.: Shop scheduling with genetic algorithms. Springer Press, Tsinghua University (2003)
3. Han, S., Ishii, H., Fujii, S.: One Machine Scheduling Problem with Fuzzy Due Dates. European Journal of Operational Research 49(1), 1–12 (1994)
4. Sakawa, M., Mori, T.: An Efficient Genetic Algorithm for Job-shop Scheduling Problems with Fuzzy Processing Time and Fuzzy Duedate. Computers & Industrial Engineering 36(2), 325–341 (1999)
5. Ishibuchi, H., Yamamoto, N., Murata, T., Tanaka, H.: Genetic algorithms and neighborhood search algorithms for fuzzy flowshop scheduling problems. Fuzzy Sets and Systems 67(1), 81–100 (1994)
6. Murata, T., Gen, M., Ishibuchi, H.: Multi-objective Scheduling with Fuzzy Due date. Computers& Industrial Engineerig 35(3), 439–442 (1998)
7. Adamopoulos, G.I., Pappis, C.P.: A Neighbourhood-based Hybrid Method for Scheduling with Fuzzy Due dates. International Transactions in Operational Research 5(3), 147–153 (1998)
8. Ju, Q.Y., Zhu, J.Y.: The Study of Dynamic Job Shop Scheduling System Based on Hybrid Genetic Algorithm. Chinese Mechanical Engineering 1(1), 40–43 (2007)
9. Lee, C.Y., Piramuthu, S., Tsai, Y.K.: Job Shop Scheduling with A Genetic Algorithm and Machine Learning. International Journal of Production Research 35(4), 1171–1191 (1997)
10. Chu, P.C., Beasley, J.E.: A Genetic Algorithm for the Multi-dimensional Knapsack Problem. Journal of Heuristic 4(1), 63–68 (1998)
11. Chu, P.C., Beasley, J.E.: A Genetic Algorithm for the Generalized Assignment Problem. Computers & Operations Research 24(1), 17–23 (1997)
12. Hartmann, S.: A Competitive Genetic Algorithm for Resource-constrained Project Scheduling. Naval Research Logistics 45(7), 733–750 (1998)
13. Jiao, L.C., Wang, L.: A Novel Genetic Algorithm Based On Immunity. IEEE Transaction on Systems Man and Cybernetics Part A-Systems and Humans 30(5), 552–556 (2000)
14. Li, Y.Y., Jiao, L.C.: Quantum Immune Clone Algorithm for Solving the SAT Problem. Chinese Journal of Computers 30(2), 176–183 (2007)
15. Li, X., Jiang, C.Y.: Use Genetic Algorithms to Solve Job-shop Production Line Scheduling Problem. System Simulation 13(6), 736–739 (2001)
16. Srinivas, M., Patnaik, L.M.: Adaptive Probabilities of Crossover and Mutation in Genetic Algorithms. IEEE Transactions on Systems, Man and Cybernetics 24(4), 17–26 (1994)
17. Yuan, B., Ying, B.S., Xie, H.: Job shop scheduling under uncertain environment based on Genetic Algorithm. Modern Manufacturing Engineering (10), 52–56 (2012)

A Very Fast Convergent Evolutionary Algorithm for Satisfactory Solutions

Xinchao Zhao[1] and Xingquan Zuo[2]

[1] School of Science,
Beijing University of Posts and Telecommunications, Beijing 100876, China
zhaoxc@bupt.edu.cn
[2] School of Computer Science,
Beijing University of Posts and Telecommunications, Beijing 100876, China
zuoxq@bupt.edu.cn

Abstract. As we know, genetic algorithm converges slowly. It is a natural contradiction when the situation appears with expensive objective function evaluating and satisfactory solutions being adequate. In this paper, a very fast convergent evolutionary algorithm (VFEA) is proposed with inner-outer hypercone crossover, problem dependent and search status involved mutation (PdSiMu). The offsprings produced by hypercone crossover are allowed to be outside the hypercone generated by rotating the parents around their bisectrix. PdSiMu utilizes the problem and evolving information quickly. VFEA is experimentally compared with five competitors based on ten classic 30 dimensional benchmarks. Experimental results indicate that VFEA can reach the accuracy of $10^{-4} - 10^{-1}$ for all the benchmarks within 1500 function evaluations. VFEA arrives significantly better performance than all its competitors with higher solution accuracy and stronger robustness.

Keywords: Very fast evolutionary algorithm, hypercone crossover, PdSiMu, numerical optimization.

1 Introduction

Evolutionary Algorithms (EAs) [1] imitate nature's evolutionary process to solve complex problems. The main idea of Darwinian evolution is *survival of the fittest*. According to this principle, EA employs crossover, mutation and selection operations to generate new solutions. It is simple to execute, and can obtain the global optimal or satisfactory solutions with reasonable computational cost. Consequently, EA has been applied to a very wide range of problems, including function optimization, automatic control, combinatorial optimization, machine learning and other related areas [2]-[10].

As a general problem solver, the semi-blind crossover and mutation operations of EA are random only. This blindness affects the speed of optimization [1]. Low convergence efficiency has become a major obstacle when using EA to real world problems [1]. As the other side of a coin, the semi-blind operations are also obligatory when we are facing the complete black-box problems.

Y. Tan et al. (Eds.): ICSI 2014, Part I, LNCS 8794, pp. 258–266, 2014.

Under the condition of nothing heuristic information available, simultaneously, the problems being also difficult, what we can do is to design the powerful and efficient general operations to accelerate the exploration speed of optimization algorithms.

As we know, population diversity and selection pressure are two interdependent and hostile issues [1] in EA. Their fragile, however, excellent balance, largely determines the performance of EA. EAs, maintaining a diverse population of highly-fit individuals, are capable of adapting quickly to fitness landscape change and well-suited to efficient optimization of multimodal landscapes for promising search areas. Selection pressure is responsible for the finely locating the global/satisfactory solutions. Both issues cooperate and interact each other for the excellent performance of EAs.

This paper mainly focuses on the fast exploration operations for quickly finding the promising areas and solutions. An inner-outer hypercone crossover is presented, whose offsprings are allowed to be outside the hypercone generated by rotating the parents around their bisectrix. A problem and heuristic information involved mutation is proposed. The common features of both genetic operations are fast exploration-oriented in the initial search stage. Tournament selection drives population towards the promising search area(s).

The rest of this paper is organized as follows. Section 2 introduces the related genetic operations and algorithm. Fast and very competitive properties are verified in Section 3. Finally, the paper is concluded in Section 4.

2 Evolutionary Operations and VFGA

Suppose $X_1 = [x_{11}, \cdots, x_{1n}]'$, $Y_1 = [y_{11}, \cdots, y_{1n}]'$ are parental solutions, whose offsprings are X_2, Y_2 after genetic operations.

2.1 Arithmetic Crossover

The arithmetic crossover is usually implemented as follows:

$$X_2 = rX_1 + (1-r)Y_1, \qquad Y_2 = (1-r)X_1 + rY_1 . \tag{1}$$

where r is a uniform random number in $[0, 0.5]$. In terms of geometric perspective, the arithmetic crossover lets offspring vectors being contained in the plane of parental vectors and locating in both sides of the bisectrix of the parents [10]. The arithmetic crossover is illustrated as Fig. 1 in terms of geometrical viewpoint.

2.2 Hypercone Crossover

The fact that the newly generated solution is contained in the plane of parental vectors has a fatal problem, i.e., the inherent, irreconcilable contradiction between the scanned planar region and huge hypervolume of search space.

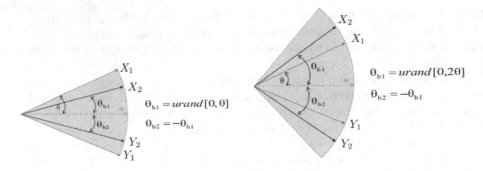

Fig. 1. Geometrical illustration of arithmetic(left)/hypercone(right) crossover

A hypercone crossover operator is proposed for numerical optimization in this paper, which allows the offspring to be outside the plane of parents [10]. The offspring can be limited inside the hypercone resulting from rotating the parents around their bisectrix if the vectors are rotated in a small range of angles. The offspring can also locate outside the hypercone around their bisectrix if the vectors are rotated in an even broad range of angles.

The computing procedure of generating offspring with inner-outer hypercone crossover is presented as follows and illustrated as Fig. 1.

- Calculate the bisectrix B (sectrix curve) of vectors X_1, Y_1;
- Calculate the angle θ between B and any of parents;
- Obtain another vector V orthogonal to B. They construct two bases $[B, V]$ of a plane which contains the bisectrix B;
- Generate an angle θ_b in $[0, 2\theta]$;
- Produce two offspring, $X_2 = [B, V] \cdot e^{\theta_b}$, $Y_2 = [B, V] \cdot e^{-\theta_b}$;

2.3   Gaussian and Cauchy Mutation

In the classic evolutionary programming (EP) [2], one parent (X_i, η_i) creates a single offspring (X_i', η_i') with Eq.(2).

$$x_i'(j) = x_i(j) + \eta_i(j)N_j(0,1), \quad \eta_i'(j) = \eta_i(j)exp(\tau'N(0,1) + \tau N_j(0,1)) . \quad (2)$$

where $x_i(j)$, $x_i'(j)$, $\eta_i(j)$ and $\eta_i'(j)$ denote the j-th component of vectors X_i, X_i', η_i and η_i'. $N(0,1)$ is a normally distributed one-dimensional random number with mean zero and standard deviation one. Factors τ and τ' are commonly set to $(\sqrt{2\sqrt{n}})^{-1}$ and $(\sqrt{2n})^{-1}$.

If the term $N_j(0,1)$ in Eq.(2) is replaced by a Cauchy random variable δ_j as Eq.(3), an EP with Cauchy mutation [2] is obtained.

$$x_i'(j) = x_i(j) + \eta_i(j)\delta_j . \quad (3)$$

2.4   Problem Dependent and Search Status Involved Mutation

Either Gaussian mutation Eq.(2), or Cauchy mutation Eq.(3), has a common insurmountable obstacle, i.e., the search stepsize is nearly independent of problem and the evolving states of algorithm. The above mentioned information can be seen as being included in the stepsize parameter $\eta_i(j)$ to a little extent.

Based on these observations, a problem dependent and search status involved self-adaptive mutation operator (PdSiMu) is proposed.

$$x_i'(j) = x_i(j) + \eta_i(j) * (\text{Max}_j - \text{Min}_j) . \tag{4}$$

where Max_j and Min_j are the maximal and minimal values in the j-th components of all the current solutions, which contains the present searching information of problem and the current evolving information of algorithm.

2.5   Tournament Selection

The above two strategies are global exploration-oriented via large-scale and diverse cooperation. Tournament selection [2,1] is then adopted to drive population to evolve in this paper. For each individual, q opponents are chosen uniformly at random from the newly generated individuals and the best one is put into the next generation.

2.6   Flowchart of VFGA

Elitist strategy is used to ensure algorithm evolve steadily and converge because the proposed exploration-oriented crossover and mutation operations damage the best evolving patterns from time to time.

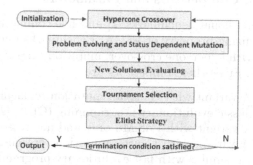

Fig. 2. Flowchart of VFGA

Table 1. Benchmark functions, where $f_1 - f_6$ are unimodal and $f_7 - f_{10}$ are multimodal, whose known optimal values are $f_{min} = 0$

Benchmark functions	Domain	x_{min}
$f_1 = \sum\limits_{i=1}^{n} x_i^2$	$[-100, 100]^n$	$\{0\}^n$
$f_2 = \sum\limits_{i=1}^{n} |x_i| + \prod\limits_{i=1}^{n} |x_i|$	$[-10, 10]^n$	$\{0\}^n$
$f_3 = \sum\limits_{i=1}^{n} \left(\sum_{j=1}^{i} x_j \right)^2$	$[-100, 100]^n$	$\{1\}^n$
$f_4 = \max\limits_{i}\{|x_i|, 1 \leq i \leq n\}$	$[-100, 100]^n$	$\{0\}^n$
$f_5 = \sum\limits_{i=1}^{n} (\lfloor x_i + 0.5 \rfloor)^2$	$[-100, 100]^n$	$\{0\}^n$
$f_6 = \sum\limits_{i=1}^{n} i x_i^4 + random[0, 1)$	$[-1.28, 1.28]^n$	$\{0\}^n$
$f_7 = \sum\limits_{i=1}^{n} \left[x_i^2 - 10\cos(2\pi x_i) + 10 \right]$	$[-5.12, 5.12]^n$	$\{0\}^n$
$f_8 = -20\exp\left[-0.2\sqrt{\frac{1}{n}\sum\limits_{i=1}^{n} x_i^2} \right] -$ $\exp\left(\frac{1}{n}\sum_{i=1}^{n}\cos(2\pi x_i) \right) + 20 + e$	$[-32, 32]^n$	$\{0\}^n$
$f_9 = \frac{1}{4000}\sum\limits_{i=1}^{n} x_i^2 - \prod\limits_{i=1}^{n}\cos\left(\frac{x_i}{\sqrt{i}} \right) + 1$	$[-600, 600]^n$	$\{0\}^n$
$f_{10} = \frac{\pi}{n}\left\{ 10\sin^2(\pi y_1) + \sum\limits_{i=1}^{n-1}(y_i - 1)^2 \cdot \right.$ $\left. \left[1 + 10\sin^2(\pi y_{i+1}) \right] + (y_n - 1)^2 \right\}$ $+ \sum\limits_{i=1}^{n} u(x_i, 10, 100, 4),\ y_i = 1 + \frac{1}{4}(x_i + 1),$ where $u(x_i, a, r, s) = \begin{cases} r(x_i - a)^s, & \text{if } x_i > a \\ 0 & \text{if } -a \leq x_i \leq a \\ r(-x_i - a)^s, & \text{if } x_i < -a. \end{cases}$	$[-50, 50]^n$	$\{1\}^n$

3 Simulation and Experimental Comparisons

3.1 Benchmarks, Competitors and Parameters

Ten classical 30-dimensional benchmarks [2] and five competitors are chosen to verify the fast optimizing property of VFGA. Benchmarks refer to Table 1.

VFEA is an EA with hypercone crossover, mutation PdSiMu and tournament selection operations. Five other competitors are

- EAGau: a VFEA variant with Gaussian mutation replacing PdSiMu, which is similar with classic evolutionary programming (CEP) [2];
- OMEA: a VFEA variant with PdSiMu only and no crossover;
- EACau: a VFEA variant with Cauchy mutation replacing PdSiMu and no crossover, which is similar with fast evolutionary programming (FEP) [2];
- GL25 [6]: solutions are categorized into female parents and male parents and a parent-centric real-parameter crossover operator is proposed;
- CMAES [3]: any new candidate solutions are sampled according to a multivariate normal distribution.

Common parameters for algorithms are: population size is 40, the maximal function evaluation is 1500, independent running time is 50. Crossover and mutation probabilities are 0.6 and 0.4 in EAGau and VFEA, mutation probabilities are 1 in OMEA and EACau. The initial value of step size is 3. The parameters of GL25 and CMAES are the same as the corresponding references.

3.2   Numerical Comparison among Algorithms

Table 2. Results comparison (Mean±Std) among six algorithms

	EAGau	OMEA	EACau
f1	1.48e+1±6.31	6.11e+5±9.15e+3	3.45e+5±4.82e+3
f2	8.92 ±2.05	7.85e+11±3.16e+12	1.18e+4 ±2.29e+4
f3	5.68e+1±2.56e+1	1.17e+5±3.28e+5	4.90e+5±1.00e+5
f4	2.16 ±4.26e-1	8.62e+1±4.49	6.74e+1±3.56
f5	1.68e+1 ±7.89	6.01e+4 ±8.75e+3	3.47e+4 ±3.90e+3
f6	7.83e-1 ±3.17e-1	1.04e+2 ±2.59e+1	2.26e+1 ±5.34
f7	6.53e+1±1.79e+1	4.11e+2±3.34e+1	3.00e+2+1.92e+1
f8	3.57 ±6.55e-1	2.05e+1±2.25e-1	1.92e+1±2.49e-1
f9	5.23e-1±1.33e-1	5.54e+2±6.54e+1	3.90e+2±4.45e+1
f10	1.59 ±3.17e-1	5.02e+8±1.32e+8	8.59e+7±3.15e+7
	VFEA	GL25	CMAES
f1	**2.33e-3±3.43e-3**	6.21e+3±1.54e+3	2.78e+2±1.46e+2
f2	**1.27e-2 ±1.17e-2**	3.52e+1 ±5.68	2.31e+14±1.32e+15
f3	**3.27e-2±5.10e-2**	3.28e+5+6.90e+4	2.19e+5±7.51e+4
f4	**3.78e-2±2.11e-2**	5.05e+1±5.69	2.77e+1±8.64
f5	**0 ±0**	5.67e+3 ±1.74e+3	8.32e+2 ±4.99e+2
f6	**6.35e-3 ±5.13e-3**	2.94 ±1.31	1.17e-1 ±4.19e-2
f7	**3.09e-3±8.80e-3**	2.65e+2±2.02e+2	2.39e+2±4.38e+1
f8	**1.05e-2±8.98e-3**	1.37e+1±1.03	1.94e+1±1.94e-1
f9	**2.36e-4+2.83e-4**	5.52e+1±1.40e+1	9.97 ±4.60
f10	**6.45e-1±1.02e-1**	5.02+6 ±4.49e+6	6.94 ±3.49

Experimental results of Table 2 clearly tell us that algorithm VFEA has significantly better performance than all of its competitors. The closest performance to VFEA is EAGau, however, there is still a clear difference to VFEA, which indicates the excellent property of the proposed PdSiMu mutation. There are great difference of OMEA and EACau with VFEA which illustrates the distinct superiority of the proposed hypercone crossover. The distinctively different performance of the first four algorithms has solidly verified the cooperating and benefiting effects of the hypercone crossover and PdSiMu mutation.

As we know, the state-of-the-art evolutionary algorithms CMAES [3] and GL25 [6] are the milestones of EA and have great performance when optimizing. However, when we are facing is the expensive objective function evaluations and what we are requiring is the roughly satisfactory solutions as the situation of

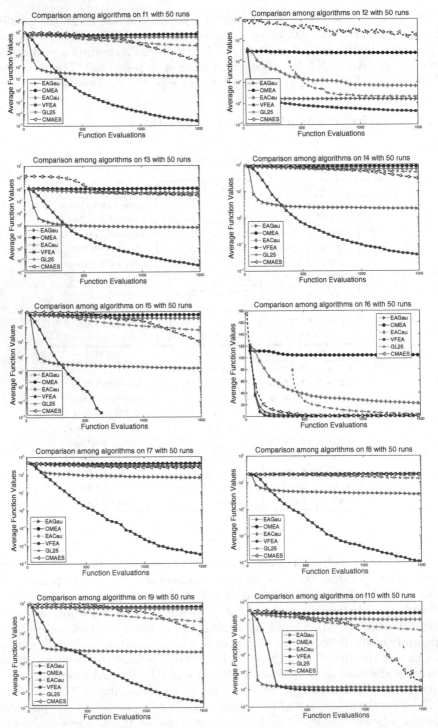

Fig. 3. Evolutionary performance comparisons among algorithms on functions $f_1 - f_{10}$

this paper is considering, it is clear that VFEA shows subversive advantages over both of them.

3.3 Online Evolutionary Performance Comparison

The online evolutionary performance comparisons on unimodal and multimodal functions are plotted in Fig. 3. VFEA performs best and EAGau is following, both of which outperform all their competitors. It needs specifically to say that VFEA has arrived the solution accuracy $10^{-4} - 10^{-1}$ when all its competitors just start or have not started their optimizing process. These phenomena are clearly shown in all the evolving process of functions, whatever the unimodal or the multimodal benchmarks. For example, it can be clearly seen that CMAES just begins to optimize process for functions f_1, f_5, f_9 and f_{10} and VFEA has been reached the precision of $10^{-4} - 10^{-1}$.

4 Discussion and Conclusion

Many state-of-the-art evolutionary algorithms have been proposed recently for various optimization and engineering applications, such as CMAES [3], CoDE [12], JADE [13], Stochastic Ranking [14], NSGA-II [15] and MOEA/D [5]. All of them have very competitive performance with a certain amount of computing costs. However, their performance have not been considered and we do not know their performance with a very limited computing resources. At the same time, there are many questions whose objective function evaluations are very expensive and roughly satisfactory solutions are enough.

Aiming at these analysis, an inner-outer hypercone crossover and a problem dependent and search status involved mutation (PdSiMu) strategy, as well as a fast convergent EA (VFEA), are proposed in this paper. Simulation results show two sides of conclusions. Firstly, both strategies significantly cooperate and benefit each other. Secondly, VFEA greatly outperforms its competitors under the condition of very limited computing costs.

It is easy to see that this is only the very preliminary results on the road of this study. There are some questions to be researched. For example, can it have even more competitive performance when combined with other metaheuristics? What will it perform when it is applied to other questions, such as multi-objective optimization? Its convergent property and other executing mechanism [16] are also interesting questions.

Acknowledgement. This research is supported by Natural Science Foundation of China (61105127, 61375066, 61374204). The first author is supported by Chinese Scholarship Council for his academic visiting to UK.

References

1. De Jong, K.A.: Evolutionary computation: a unified approach. The MIT Press, Cambridge (2007)

2. Yao, X., Liu, Y., Lin, G.M.: Evolutionary Programming Made Faster. IEEE Trans. Evol. Comput. 3(2), 82–102 (1999)
3. Hansen, N., Ostermeier, A.: Completely derandomized self-adaptation in evolution strategies. Evol. Comput. 9(2), 159–195 (2001)
4. Wang, L., Zhang, L., Zheng, D.: An effective hybrid genetic algorithm for flow shop scheduling with limited buffers. Computers & Operations Research 33(10), 2960–2971 (2006)
5. Zhang, Q., Li, H.: MOEA/D: A Multi-objective Evolutionary Algorithm Based on Decomposition. IEEE Trans. Evol. Comput. 11(6), 712–731 (2007)
6. Garcia-Martinez, C., Lozano, M., Herrera, F., Molina, D., Sanchez, A.M.: Global and local real-coded genetic algorithms based on parent-centric crossover operators. Eur. J. Oper. Res. 185(3), 1088–1113 (2008)
7. Gong, M., Fu, B., Jiao, L., Du, H.: Memetic Algorithm for Community Detection in Networks. Phys. Rev. E 84, 056101 (2011)
8. Wang, Y., Cai, Z.: A dynamic hybrid framework for constrained evolutionary optimization. IEEE Transactions on Systems, Man, and Cybernetics, Part B: Cybernetics 42(1), 203–217 (2012)
9. Zhan, Z.-H., Li, J., Cao, J., Zhang, J., Chung, H., Shi, Y.H.: Multiple Populations for Multiple Objectives: A Coevolutionary Technique for Solving Multiobjective Optimization Problems. IEEE Trans. on Cybern. 43(2), 445–463 (2013)
10. Espezua, S., Villanueva, E., Maciel, C.D.: Towards an efficient genetic algorithm optimizer for sequential projection pursuit. Neurocomputing 123, 40–48 (2014)
11. Rahnamayan, S., Tizhoosh, H.R., Salama, M.M.A.: Opposition-based differential evolution. IEEE Trans. Evolut. Comput. 12, 64–79 (2008)
12. Wang, Y., Cai, Z., Zhang, Q.: Differential evolution with composite trial vector generation strategies and control parameters. IEEE Trans. Evolut. Comput. 15(1), 55–66 (2011)
13. Zhang, J., Sanderson, A.C.: JADE: Adaptive differential evolution with optional external archive. IEEE Trans. Evol. Comput. 13(5), 945–958 (2009)
14. Runarsson, T.P., Yao, X.: Stochastic Ranking for Constrained Evolutionary Optimization. IEEE Trans. Evol. Comput. 4(3), 284–294 (2000)
15. Deb, K., Pratap, A., Agarwal, S., Meyarivan, T.: A Fast and Elitist Multiobjective Genetic Algorithm: NSGA-II. IEEE Trans. Evol. Comput. 6(2), 182–197 (2002)
16. Zhao, X.C.: Convergent analysis on evolutionary algorithm with non-uniform mutation. In: Proc. 2008 IEEE Congress on Evolutioanry Computation, Hongkong, pp. 940–945 (2008)

A Novel Quantum Evolutionary Algorithm Based on Dynamic Neighborhood Topology

Feng Qi, Qianqian Feng, Xiyu Liu, and Yinghong Ma

Shandong Normal University, Jinan 250014, China
qfsdnu@126.com

Abstract. A variant of quantum evolutionary algorithm based on dynamic neighborhood topology(DNTQEA) is proposed in this paper. In DNTQEA, the neighborhood of a quantum particle are not fixed but dynamically changed, and the learning mechanism of a quantum particle includes two parts, the global best experience of all quantum particles in population, and the best experiences of its all neighbors, which collectively guide the evolving direction. The experimental results demonstrate the better performance of the DNTQEA in solving combinatorial optimization problems when compared with other quantum evolutionary algorithms.

Keywords: Quantum Evolutionary Algorithm, Particle Swarm Optimization, Dynamic Neighborhood Topology, Algorithm.

1 Introduction

Quantum computing with its powerful computing power has become one of the most focused technology of current science. By considering the quantum information processing, researchers attend to introduce quantum computing mechanism into some traditional optimization algorithms. The combination of quantum computing and evolutionary computation was proposed in [1] named quantum genetic algorithm. In [2][3] quantum-inspired evolutionary algorithms(QEA) are first investigated for a class of combinatorial optimization problems in which quantum rotation gates act as update operators. Many works have tried to improve the performance of QEA. In [4] a new two phase scheme and a new He Gate is proposed for QEA. Reference [5] establishes that QEA is an original algorithm that belongs to the class of estimation of distribution algorithms (EDAs), while the common points and specifics of QEA compared to other EDAs are highlighted. Since proposed, QEA has been applied on several applications in science. Using QEA and Markov Model [6] presents a new method for static VAR which considers existing wing generator voltages and transformer taps as controller to regulate the voltage profile in a distribution system with wind farms. In order to solve the problem of highly non-linear economic load dispatch problem with value point loading [7] proposes an improved real quantum evolutionary algorithm which shows better performance than QEA.

Generally, the structure of the population in evolutionary algorithms is an important parameter. A graph based evolutionary algorithm is proposed in [8]

Y. Tan et al. (Eds.): ICSI 2014, Part I, LNCS 8794, pp. 267–274, 2014.

in which the individuals are located on the nodes of a graph structured population. The effect of variable population structure on Particle Swarm Optimization is investigated in [9]. Random graphs and their performance on several criteria are compared in their work. Similar with the above algorithms, Tayarani [10] proposed a sinusoid size ring structure QEA, experimental results show that the ring structure can be an efficient architecture for an effective Exploration/Exploitation tradeoff and improves the performance of QEA. [11] proposes a dynamic structured interaction among members of population in QEA and the study shows that cellular structure is the best. The structure of the population in above algorithms is fixed and the relationship among individuals never change in evolving process, this characteristics may cause the optimization process trap in local optimums and make the algorithm unstable.

As the similar particle study strategy, many hybrid algorithms by combing QEA and particle swarm optimization(PSO) are proposed . [12] proposed a binary Quantum-behaved PSO algorithm with cooperative approach, the updating method of particle's previous best position and swarm's global best position are performed in each dimension of solution vector to avoid loss some components. The experimental results show that this technique can increase diversity of population and converge more rapidly than other binary algorithms. In [13], a hybrid real-coded QEA is proposed by combing PSO, crossover and mutation. Simulation results show that it performs better in terms of ability to discover the global optimum and convergence speed. [14] uses quantum PSO principles to resolve the satisfiability problem. In [15], a quantum inspired PSO is applied to optimize of simultaneous recurrent neural networks and shows better performance than traditional methods.

From what has been discussed above, this paper proposed a variant of quantum evolutionary algorithm based on dynamic neighborhood topology(DNTQEA), In DNTQEA, the neighborhood of a quantum particle are not fixed but dynamically changed, and the learning mechanism of a quantum particle includes two partsthe global best experience of all quantum particles in populationand the best experiences of its all neighbors, which guide the final evolving direction of the quantum particle.

This paper is organized as follows. Section 2 describes the original QEA. In Section 3 the dynamic structure for QEA and the hybrid algorithm are given. In Section 4 the proposed algorithm is evaluated on some benchmark functions and finally the proposed algorithm is concluded in section 5.

2 Brief Description of Quantum Evolutionary Algorithm

Quantum evolutionary algorithm(QEA) combines quantum mechanism and basic evolutionary algorithm is a kind of probability search algorithm. Its essential characteristics are making full use of the superposition and coherence of quantum state. QEA adopts a new coding party quantum bit code; the concrete form can be described as:

$$\begin{bmatrix} \alpha_1| \ \alpha_2| \ \cdots | \ \alpha_m \\ \beta_1| \ \beta_2| \ \cdots | \ \beta_m \end{bmatrix} \tag{1}$$

In equation (1), $(\alpha_i, \beta i)^T (i = 1, 2, \cdots, m)$, represents a quantum bit and satisfies the following expression $|\alpha^2 + \beta^2| = 1$, m denotes the number of quantum bit; q is called a quantum chromosome used to describe the problem in QEA.

The above representation method has the advantage that it is able to represent any superposition of states, so evolutionary computing with the Q-bit representation has a better characteristic of diversity than classical approaches, since it can represent superposition of states. As α_i^2 or β_i^2 approaches to 1 or 0, the Q-bit chromosome converges to a single state and the property of diversity disappears gradually, and the algorithm converges.

The structure of QEA is described in Table 1 [3]:

Table 1. The details description of QEA

procedure QEA
begin
1. $t \leftarrow 0$;
2.   initialize $Q(t)$;
3.   make $P(t)$ by observing $Q(t)$ states;
4.   evaluate $P(t)$;
5.   store the best solution among $P(t)$;
while(not termination-condition) **do**
begin
6. $t \leftarrow$ t+1;
7.   make $P(t)$ by observing $Q(t-1)$ states;
8.   evaluate $P(t)$;
9.   update $Q(t)$ using quantum gates $U(t)$;
10.   store the best solution among $P(t)$;
end
end

Here, $Q(t)$ represents the population in t generation; $P(t)$ denotes the set of binary solution of t generation. When initializing the population, generally all the quantum bits in quantum chromosomes are initialized to $\sqrt{2}$ which means that all possible superposition states appear in the same probability. In step (3), $Q(t)$ generates $P(t)$ through the observation operation. This process can described as follows: randomly generating a random number between [0, 1], if it is greater than α_i^2, the Q-bit value of corresponding binary solution is 1, otherwise its value is 0. In the step(9), QEA adopts the quantum revolving door $u(t)$ to update $Q(t)$, and mathematical formula can be defined as:

$$\begin{bmatrix} \alpha_i' \\ \beta_i' \end{bmatrix} = \begin{bmatrix} cos(\Delta\theta) & -sin(\Delta\theta) \\ sin(\Delta\theta) & cos(\Delta\theta) \end{bmatrix} \begin{bmatrix} \alpha_i \\ \beta_i \end{bmatrix} \tag{2}$$

Here, $\Delta\theta$ is the rotation angle and controls the speed of convergence and determined by look table given in [3], which shows that these values for $\Delta\theta$ have better performance.

3   Dynamic Neighborhood Topology Based QEA

3.1   Dynamic Neighborhood Topology Structure and Updating Rules

Usually, the structure of the population in improved QEAs [9][10][11] is fixed and the neighbors of each individual are never changed in evolving process, this will cause each individual has fewer learning samples and greatly reduce the diversity of the population. So in order to keep each individual has opportunity to learn from more individuals, avoid trapping in local optimums and existing precocious phenomenon, this paper proposes a strategy based on fitness of each individual to form its neighbors dynamically. This makes the learning samples become diversity and promotes the individual converge to the global optimal position.

In DNTQEA, the neighbor selection rule of each individual is Euclidean distance among individuals. Based on this rule, the neighbors of current individual i can be calculated as follows:

$$\begin{cases} D_i(t) = \{d_{ij}(t) | d_{ij}(t) = ||x_i(t) - x_j(t)||\}, j \neq i, j \in P_s; \\ neighbor_i(t) = arg(min(sort(D_i(t)), n)). \end{cases} \tag{3}$$

Where, $D_i(t)$ denotes the Euclidean distance set between the current individual i and other individuals in population at t generation; P_s denotes the population size; $neighbor_i$ means the neighbors set of the current individual i at t generation; n denotes the number of neighbors belong to the current individual i, its value usually is set to $1/4 \sim 1/3$ of the population size and in DNTQEA, $n = P_s/3$; $sort(A)$ means sorting elements in A from smallest to largest. When executing DNTQEA, if the individual fitness holds the same in continuous T generation, its neighbors need to be reselected. After repeated experiments, when $T = 10$, the algorithm gains the best result, so in DNTQEA, parameter T is set to 10.

The updating method of each individual in [2][3][9][10][11] can be summarized as that each individual only learning from the global best individual or the best one in its neighbors. This method may cause the searching process trapped in local optimums, while in particle swarm optimization(PSO)[15] algorithm, the evolving direction of each particle is decided by its history experience and the global experience. So in DNTQEA, the global best individual and the best individual of neighbors are used to guide the evolving direction of each individual. This updating rule can be described as follows:

$$\begin{cases} q_i^1 \longleftarrow Learn(q_i, g_{best}); \\ q_i^2 \longleftarrow Learn(q_i, n_{best}); \\ q_i' \longleftarrow Random(q_i^1, q_i^2, p); \end{cases} \tag{4}$$

Here, q_i and q_i' denote the ith original individual and the new individual after learning from the g_{best} and n_{best}, respectively; g_{best} means the global best individual in current population; n_{best} denotes the best individual in neighbors of q_i;

$Learn(I_1, I_2)$ means I_1 is updated by predefined quantum gate with reference to I_2; q_i^1 and q_i^2 denote the learning results from g_{best} and n_{best}; $Random(I_1, I_2, p)$ means to choose I_1 with probability p as the final evolving individual.

3.2   Procedure of the Proposed DNTQEA

By combing the dynamic neighborhood topology and updating rule introduced in above subsection, the procedure of the proposed DNTQEA is described in Table 2.

Table 2. The details description of DNTQEA

Procedure DNTQEA
begin
$\quad t = 0;$
1. $\quad$ initialize quantum population $Q(t)$ with the size of P_s;
2. $\quad$ make $X(t)$ by observing the states of $Q(t)$;
3. $\quad$ evaluate $X(t)$;
4. $\quad$ for all binary solutions x_i^t do
$\quad\quad$ begin
5. $\quad\quad$ find neighborhood set N_i in $X(t)$ by definition (3)
6. $\quad\quad$ find binary solution x with best fitness in N_i
7. $\quad\quad$ save x in B_i
$\quad\quad$ end
8. $\quad$ save solution y with best fitness of X_t in g_{best};
9. $\quad$ while(not termination-condition) do
$\quad\quad$ begin
$\quad\quad t = t + 1;$
10. $\quad\quad$ make $X(t)$ by observing the states of $Q(t-1)$;
11. $\quad\quad$ evaluate $X(t)$;
12. $\quad\quad$ update $Q(t)$ based on D_i and g_{best} using Q-gates by rule (4);
13. $\quad\quad$ for all binary solutions x_i^t do
$\quad\quad\quad$ begin
14. $\quad\quad\quad$ find neighborhood set N_i in $X(t)$ by definition (3)
15. $\quad\quad\quad$ select binary solution x with best fitness in N_i
16. $\quad\quad\quad$ if x is fitter than B_i save x in B_i
$\quad\quad\quad$ end
17. $\quad\quad$ select solution y with best fitness of $X(t)$
18. $\quad\quad$ if y is fitter than g_{best} save y in g_{best};
$\quad\quad$ end
end

The pseudo code of DNTQEA is described as below:

1. In initialization step, $[\alpha_{i,k}^0 \ \beta_{i,k}^0]^T$ in q_i^0 are initialized with $1/\sqrt{2}$, where $i = 1, 2, \cdots, P_s$ is the index of the individuals in the population, $k = 1, 2, \cdots, m$, and m is the number of Q-bits in a individual. This initialization means that each Q-bit individual $q_i[0]$ represents the linear superposition of all possible states with equal probability.

2. This step makes a set of binary solutions $X(0) = \{x_i^0 | i = 1, 2, \cdots, P_s\}$ at generation $t = 0$ by observing $Q(0) = \{q_i^0 | i = 1, 2, \cdots, P_s\}$ states, where $X(t)$ at generation t is a random solution of Q-bit population and P_s is population size. Each binary solution, x_i^0 with length m, is formed by selecting each bit using the probability of Q-bit, either $|\alpha_{i,k}^0|^2$ or $|\beta_{i,k}^0|^2$ of q_i^0. The binary bit $x_{i,k}^t$ can be obtained from Q-bit $[\alpha_{i,k}^0 \ \beta_{i,k}^0]^T$ in following way:

$$x_{i,k}^{t} = \begin{cases} 0 \; if \; \; random(0,1) < |\alpha_{i,k}^{t}|^2 \\ 1 \quad\quad\quad otherwise \end{cases} \tag{5}$$

Where $random(0,1)$ is a uniform random number generator.

3. Each binary solution $x_i^0 \in X(0)$ is evaluated to give some measure of its fitness.

4,5,6,7,8. In these steps the neighborhood set N_i of all binary solutions x_i^0 in $X(0)$ is selected by rule 3, meanwhile, the best solution among N_i is stored in B_i and the best solution among $X(0)$ is saved in g_{best}.

9. The **while** loop is terminated when the maximum number of iterations is reached.

10. Observing the binary solutions $X(t)$ from $Q(t-1)$.

11. Evaluating the binary solutions $X(t)$.

12. The quantum individuals are updated using Q-gate based on B_i and g_{best} with updating rule 4.

13. The **for** loop is for all binary solutions $x_i^t(i = 1, 2, \cdots, P_s)$ in population.

14. Finding the neighbors of the binary solution x_i^t.

15. Selecting the best possible solution in N_i and save it to x.

16. If x is fitter than B_i and then replace B_i with x.

17. Finding the best possible solution in $X(t)$ and save it to y.

18. If y is fitter than g_{best} and then replace g_{best} with y.

4   Simulations

The proposed DNTQEA is compared with the original version of QEA and FSQEA [11] which used Cellular structure and a functional population size for QEA. The experimental results are performed for several dimensions (m=50, 100, 250) of Knapsack Problem and 14 numerical benchmark functions. Similar settings with reference [11], the population size of all algorithms is set to 25; termination condition is set for a maximum of 1000 generations. Due to statistical nature of the optimization algorithms, all results are averaged over 30 runs. The parameter of QEA is set to $\Delta\theta = 0.01\pi$ and the parameters of FSQEA are set to the same value with reference [11].

Table 3 summarizes the experimental results on DNTQEA, FSQEA and QEA for Knapsack Problem and 14 benchmark functions (The results for some dimensions are not summarized in Table 3 because of small space of the paper). As it seen in Table 3, DNTQEA has the best results.

5   Conclusions

This paper proposes a variant of quantum evolutionary algorithm based on dynamic neighborhood topology(DNTQEA). In DNTQEA, the neighbors of a quantum particle are dynamically changed by Euclidean distance set between the current individual and other individuals in population. The learning mechanism of a quantum particle contains the global best experience of all quantum

Table 3. Experimental results on Knapsack Problem and 14 numerical benchmark functions for m=100 and m=250

	m=100						m=250					
	DNTQEA		FSQEA		QEA		DNTQEA		FSQEA		QEA	
	Mean	Std	Mean	Std	Mean	Std	Mean	Std	Mean	Std	Mean	Std
KP1	590.28	0.73	562.13	0.92	546.82	11.29	1355.8	15.23	1252.6	17.40	1173	30.19
KP2	438.54	1.14	418.15	1.94	406.91	4.39	1029.55	10.11	994.46	10.31	942.99	20.90
f1	47126	1289.3	45889	1558.8	34437	3984.1	80081	2719	76292	2939	55844	5845.3
f2	-1097.3	66.49	-1287.8	97.97	-2096.3	199.45	-5015	287.18	-5118.9	304.75	-6511	360.74
f3	-13.76	0.08	-16.89	0.14	-17.19	0.09	-16.89	0.08	-17.39	0.11	-17.62	0.11
f4	-27.28	4.19	-30.08	5.24	-39.60	8.48	-123.89	10.19	-134.53	10.16	-154.42	14.04
f5	-1.27e5	12459	-1.43e5	14756	-1.67e5	16807	-5.41e5	20115	-5.60e5	20743	-6.05e5	43232
f6	-21459	2412.8	-23017	2742.1	-36949	4918.7	-0.97e5	4873.2	-1.09e5	5039	-1.44e5	8269.3
f7	38.22	2.27	31.17	2.53	22.04	2.46	55.81	3.28	52.24	4.39	39.43	3.02
f8	54.11	1.21	50.57	1.75	38.19	3.24	97.89	4.89	93.51	5.31	73.50	5.02
f9	-2.198e5	21054	-2.46e5	23972	-4.55e5	68039	-1.15e6	88756	-1.26e6	93119	-1.64e6	1.34e5
f10	-3.561	0.22	-4.2161	0.29	-5.21	0.68	-5.45	0.28	-5.95	0.30	-6.52	0.46
f11	-167.01	5.45	-169.84	7.76	-176.72	3.78	-178.24	1.56	-189.69	2.35	-192.03	1.36
f12	-1.19e7	2.21e6	-1.33e7	2.53e6	-2.56e7	9.20e6	-2.19e8	2.01e7	-2.44e8	2.45e7	-3.09e8	3.59e7
f13	-46873	5104	-49229	5885	-1.08e5	35412	-2.84e5	33298	-3.1e5	35376	-4.81e5	72311
f14	-0.074	0.057	-0.098	0.065	-1.29	1.26	-7.29	1.74	-7.53	1.94	-20.38	6.19

particles in population and the best experiences of its all neighbors, which together guide the evolving direction. The performance of the proposed algorithm is tested on Knapsack Problem and 14 benchmark functions, and simulation results show that DNTQEA is better than other improved QEA and more suitable for solving combinatorial optimization problems problems.

The objective functions which are used here are f1: Schwefel 2.26 [16], f2: Rastrigin [16], f3: Ackley [16], f4: Griewank [16], f5: Penalized 1 [16], f6: Penalized 2 [16], f7: Michalewicz [17], f8: Goldberg [18], f9: Sphere Model [16], f10: Schwefel 2.22 [16], f11: Schwefel 2.21 [16], f12: Dejong [17], f13: Rosenbrock [18], and f14: Kennedy [18].

Acknowledgements. This Research is supported by Natural Science Foundation of China(No.60873058), Natural Science Foundation of Shandong Province (No.ZR2011FM001), Shandong Young Scientists Award Fund(NO.BS2013DX037), Shandong Soft Science Major Project(No.2010RKMA2005) and Natural Science Foundation of China(No.71071090).

References

1. Narayanan, A., Moore, M.: Quantum-inspired Genetic Algorithms. In: Proceedings of IEEE International Conference on Evolutionary Computation, Nagoya, pp. 61–66 (1996)
2. Han, K.H., Kim, J.H.: Genetic Quantum Algorithm and its Application to Combinatorial Optimization Problem. In: Proceedings of the 2000 Congress on Evolutionary Computation, Piscataway, vol. 2, pp. 1354–1360 (2000)
3. Han, K.H., Kim, J.H.: Quantum-inspired Evolutionary Algorithm for a Class of Combinatorial Optimization. IEEE Trans. Evolutionary Computation. 6(6), 580–593 (2000)

4. Han, K., Kim, J.: Quantum-Inspired Evolutionary Algorithms with a New Termination Criterion, He Gate, and Two-Phase Scheme. IEEE Transactions on Evolutionary Computation 8(2), 156–169 (2004)
5. Platel, M.D., Schliebs, S., Kasabov, N.: Quantum-Inspired Evolutionary Algorithm: A Multimodel EDA. IEEE Transactions on Evolutionary Computation 13(6), 1218–1232 (2009)
6. Hong, Y., Pen, K.: Optimal VAR Planning Considering Intermittent Wind Power Using Markov Model and Quantum Evolutionary Algorithm. IEEE Transactions on Power Delivery 25(4), 2987–2996 (2010)
7. Sinha, N., Hazarika, K.M., Paul, S., Shekhar, H., Karmakar, A.A.: Improved Real Quantum Evolutionary Algorithm for Optimum Economic Load Dispatch with Non-convex Loads. In: Panigrahi, B.K., Das, S., Suganthan, P.N., Dash, S.S. (eds.) SEMCCO 2010. LNCS, vol. 6466, pp. 689–700. Springer, Heidelberg (2010)
8. Bryden, K.M., Daniel, A.A., Steven, C., Stephen, J.W.: Graph-Based Evolutionary Algorithms. IEEE Trans. Evol. Comput. 10(5), 550–567 (2006)
9. Kennedy, J., Mendes, R.: Population structure and particle swarm performance. IEEE Proceedings of the Congress on Evolutionary Computation 2, 1671–1676 (2002)
10. Tayarani-N, M.H., Akbarzadeh-T, M.R.: A Sinusoid Size Ring Structure Quantum Evolutionary Algorithm. Proceeding of Cybernetics and Intelligent Systems, 1165–1169 (2008)
11. Najaran, T., Akbarzadeh, T., Mohammad, R.: Improvement of Quantum Evolutionary Algorithm with Functional Sized Population. Applications of Soft Computing 58, 389–398 (2009)
12. Zhao, J., Sun, J., Xu, W.B.: A binary quantum-behaved particle swarm optimization algorithm with cooperative approach. International Journal of Computer Science 10(1), 112–118 (2013)
13. Hossain, A.M., Hossain, K.M., Hashem, M.: A generalized hybrid real-coded quantum evolutionary algorithm based on particle swarm theory with arithmetic crossover. International Journal of Computer Science & Information Technology 2(4), 172–187 (2010)
14. Layeb, A.: A quantum inspried particle swarm algorithm for solving the maximum statisfiability problem. International Journal of Combinatorial Optimization Problems and Informatics 1(1), 13–23 (2010)
15. Kennedy, J., Eberhart, R.: Particle swarm optimization. In: Proceedings of IEEE International Conference on Neural Networks, vol. 4, pp. 1942–1948 (1995)
16. Zhong, W., Liu, J., Xue, M., Jiao, L.: A Multi-agent Genetic Algorithm for Global Numerical Optimization. IEEE Trans. Sys, Man and Cyber. 34, 1128–1141 (2004)
17. Khorsand, A.-R., Akbarzadeh-T, M.-R.: Quantum Gate Optimization in a Meta-Level Genetic Quantum Algorithm. In: IEEE International Conference on Systems, Man and Cybernetics, vol. 4, pp. 3055–3062 (2005)
18. Koumousis, V.K., Katsaras, C.P.: A Saw-Tooth Genetic Algorithm Combining the Effects of Variable Population Size and Reinitialization to Enhance Performance. IEEE Trans. Evol. Comput. 10, 19–28 (2006)

Co-evolutionary Gene Expression Programming and Its Application in Wheat Aphid Population Forecast Modelling

Chaoxue Wang[1], Chunsen Ma[2], Xing Zhang[1], Kai Zhang[1], and Wumei Zhu[1]

[1] School of Information and Control Engineering, Xi'an University of Architecture and Technology, 710055, China
[2] Institute of Plant Protection, Chinese Academy of Agricultural Sciences, Beijing 100193, China
Wb11w@126.com, machunsen@caas.cn

Abstract. A novel approach of function mining algorithm based on co-evolutionary gene expression programming (GEP-DE) which combines gene expression programming (GEP) and differential evolution (DE) was proposed in this paper. GEP-DE divides the function mining process of each generation into 2 phases: in the first phase, GEP focuses on determining the structure of function expression with fixed constant set, and in the second one, DE focuses on optimizing the constant parameters of the function which obtained in the first phase. The control experiments validate the superiority of GEP-DE, and GEP-DE performs excellently in the wheat aphid population forecast problem.

Keywords: Gene expression programming, function mining, differential evolution, co-evolution, wheat aphid population forecast.

1 Introduction

Gene expression programming (GEP) was proposed on the basis of genetic algorithm (GA) and genetic programming (GP) by Portugal scholar Candida in 2001 [1]. It adopts the dual architecture of genotype and phenotype, and retains the advantage of GA and GP, and possesses the characteristics that the algorithm flow is clear, realization is simple and precision is high, and especially it shows the excellent performance in complex function finding and prediction [2-3].

Numerical constants have a great influence on the performance of GEP, where there are three main approaches to handle numerical constants. The first approach does not include any constants in the terminal set, but relies on the spontaneous emergence of necessary constants through the evolutionary process of GEP [1]. The second approach involves the ability to explicitly manipulate random constants by adding a random constant domain DC at the end of the gene [4]. Zuo J et al. proposed the third constants generation approach------MC method [5], where each numerical constant is directly regarded as terminator. This method improves DC method to a certain extent, but it also has some drawbacks, such as the determination of

Y. Tan et al. (Eds.): ICSI 2014, Part I, LNCS 8794, pp. 275–283, 2014.

numerical constants set usually depends on experience, and the interpretability of functions mined is poor.

On the basis of MC method, an improved GEP algorithm, GEP-DE, was proposed in this paper, which embedded different evolution algorithm (DE) [6] into GEP to handle numerical constants. In GEP-DE, the function expression is divided into two parts of structure and constant parameters, and they are respectively optimized by GEP and DE. Control experiments show that the performance of GEP-DE is better than other GEP algorithms proposed in the related literatures. Moreover, GEP-DE performs excellently in the wheat aphid population forecast problem.

2   Introductions of GEP and DE

2.1   Introduction of GEP

Standard GEP algorithm could be defined as a nine-meta group: $GEP = \{C, E, P_0, M, \varphi, \Gamma, \Phi, \Pi, T\}$,where C is the coding means; E is the fitness function; P_0 is the initial population; M is the size of population; φ is the selection operator; Γ is the crossover operator; Φ is the point mutation operator; Π is the string mutation operator; T is the termination condition. In GEP, individual is also called chromosome, which is formed by gene and linked by the link operator. The gene is a linear symbol string which is composed of head and tail. The head involves the variables from the terminator set and the functions from function set, but the tail merely contains the variables from the terminator set. The basic steps of the standard GEP are as follows [1]:

1. Inputting relevant parameters, creating the initial population;
2. Computing the fitness of each individual;
3. If the termination condition is not met, go on the next step, otherwise, terminate the algorithm;
4. Retaining the best individual;
5. Selecting operation;
6. Point mutating operation;
7.String mutating operation (IS transposition, RIS transposition, Gene transposition);
8.Crossover operation (1-point recombination, 2-point recombination, Gene recombination);
9. Go to (2).

2.2   Introduction of DE

DE proposed by Rainer Storn and Kenneth Price in 1995 is a real coding global optimization algorithm based on population evolution [6]. DE performs excellently as soon as it was proposed and has become a very good tool for continuous optimization problem, and is widely used in many fields.

Basic operations of DE are the same as GA, also including mutation, crossover and selection. In mutation operation, DE randomly selects two different individual vectors and subtracts them to generate different vector, and then endows a weight value to the different vector and adds it to the 3rd vector randomly selected to generate mutation vector. Then mutation vector is mixed with target vector to generate experiment vector, this process is named as crossover operation. If the fitness of experiment vector is better than the target vector, the experiment vector will replace the target vector, and this process is called selection operation.

3   Co-evolutionary Gene Expression Programming

The flow chart of Co-evolutionary Gene Expression Programming (GEP-DE) is given as Fig.1.

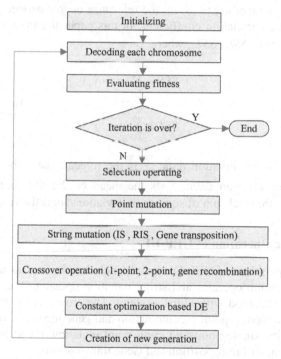

Fig. 1. The flow chart of Co-evolutionary Gene Expression Programming algorithm (GEP-DE)

3.1   Gene Coding and Fitness Evaluation Function

GEP-DE puts a fixed constant set into terminator set, and regards the constant as terminator when creating individual. For instance, function set is $FS = \{+, -, *, /\}$, terminator set is TS= $\{x, y, C\}$, and the constant set C= $\{1.2371, 2.1424, -3.2643, 1.3298, 5.3237\}$. Fig.2 shows a chromosome with single gene, whose head length is 4.

In this example, C is fixed, it forms the terminator set with terminator set together. When creating individual, terminator is randomly selected from terminator set.

a.   -+/*C4xC2yx

b.

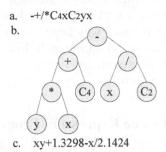

c.   xy+1.3298-x/2.1424

Fig. 2. (a) A chromosome.   (b) Expression tree (ET).   (c) Function expression

In statistics, the method to assess the relevance degree between two groups of data usually uses the correlation coefficient. In this paper the fitness function is devised as: $fitness = R^2 = 1 - SSE / SST$, where

$$SSE = \sum_{j=1}^{m}(y_j - \hat{y}_j)^2 \tag{1}$$

$$SST = \sum_{j=1}^{m}(y_j - \bar{y}_j)^2 \tag{2}$$

Where, y_j is the observation data; $\hat{y}_j$ is the forecast data which is computed with formula and observation data; $\bar{y}$ is the mean of y ; SSE is the residual sum of squares; SST is the total sum of squares of deviations; m is the size of data.

3.2   Genetic Operation of GEP-DE

The genetic operations of GEP-DE mainly include selection operator, crossover operator, point mutation operator and string mutation operator. The selection operator is the tournament method with elitist strategy. The crossover operation includes single point recombination, 2-point recombination and gene recombination. The point mutation includes the single-point and multi-point mutation; the string mutation includes IS transposition, RIS transposition and Gene transposition.

3.3   Constant Optimization Based on DE

The best individuals of each generation in GEP evolution process are passed into DE to be optimized, and the detailed steps are as follows: (a) firstly, elite chromosomes are parsed out to get the number of constant in the expression; (b) initialize the DE population; (c) iteratively run mutation, crossover, selection operation; (d) when program is over, the elite individual with the optimized constant is returned to GEP population.

4   Performance Comparisons and Its Application in Wheat Aphid Population Forecast Modelling

4.1   The Constant Optimization Performance of GEP-DE

Experiment 1: In order to test the performance of numerical constant optimization of GEP-DE, compare GEP-DE with the DC method [4] MC method [5], where target function is the same with the compared literatures, and this function formula is as Eq. (3).

$$N = 5a_n^4 + 4a_n^3 + 3a_n^2 + 2a_n + 1 \tag{3}$$

Where, a_n is a non-negative integer.

Select 10 groups of random test data (a_n=1, 2, 3, 4, 5, 6, 7, 8, 9, 10), and the parameters of this experiment are shown as Tab.1. The experiment results are shown as Tab.2. It can be seen from Tab.2 that the success ratio of GEP-DE reaches to 98%, which is 17% higher than GEP-DC and 8% higher than GEP-MC. All these show that GEP-DE owns competitiveness in the aspect of numerical constant optimization.

Table 1. Parameter settings of experiment 1

parameters	values
Times of run	50
Max evolve generation	100
Size of population	40
Function set	+-*/LEK~SC
Terminator set	A
Link operator	+
The length of head	6
The number of genes	4
Ration of crossover	0.45
Ration of point mutation	0.034
Ration of recombination	0.044
Ration of string mutation	0.3
Max generation of evolution in DE	50
Size of population in DE	20
CR in DE	0.9
F in DE	0.5

(Remarks : L refers to ln function ; E stands for exp(x) ; K refers to log_{10} function ; ~ refers to negative operation   ; S stands for sin function ; C stands for cos function)

Table 2. The results of experiment 1

terms	GEP-DC	GEP-MC	GEP-DE
Times of experiment	100	100	100
Times of success	81	90	98
Ratio of success	81%	90%	98%

Table 3. The descriptions of variables of experiment 2

descriptions	variables
Date	x_1
Daily highest temperature（°C）	x_2
Daily average temperature（°C）	x_3
Daily precipitation（mm）	x_4
Growth stage (GS)	x_5
Number of I-III instar larvae	x_6
Number of IV instar larvae with wing	x_7
Number of IV instar larvae without wing	x_8
Number of adult with wing	x_9
Number of adult without wing	x_{10}
Natural enemies	x_{11}

Table 4. Parameters setting of experiment 2

parameters	Values
Max generation of evolution	2000
Size of population	100
The number of genes	8
Length of gene	23
Link operator	+
Ratio of point mutation	0.31
Ration of 1-point recombination	0.1
Ration of 2-point recombination	0.1
Ration of gene recombination	0.1
Length of IS element	5
Ratio of IS transposition	0.1
Length of RIS element	5
Ratio of RIS transposition	0.1
Max generation of evolution inDE	50
Size of population in DE	20
CR in DE	0.9
F in DE	0.5

4.2   Application of GEP-DE in Wheat Aphid Population Forecast Model

Experiment 2: The data adopted in this paper is from Ref. [7]. We set function set and terminator set respectively as (+, -, *, /, sin, cos, tan, log, $\sqrt{}$, abs, e^x) and (x_1, x_2, x_3, x_4, x_5, x_6, x_7, x_8, x_9, x_{10}, x_{11}, T), where T refers to the total number of wheat aphid, x_1 to x_{11} stand for the variables   shown in Tab.3. Experiment parameters of GEP-DE are shown as Tab.4. The 3 wheat aphid population forecast models about the 3 fields from Ref. [7] mined by GEP-DE are respectively shown as Eq. 4 to Eq.6.

$$T = x_{11}\cos(\cos x_{11} + x_6 + x_{10}) + x_3 + x_4 + x_7 - x_9 + \cos 5 + e^{\sin x_5 \sin e^{x_8}} + \sqrt{|x_1|} + \tan x_2 + \sin(14x_{10}) \qquad (4)$$

$$T = -x_3 + 2x_{10} + 3x_8 + x_7(x_9 + x_2)\sin x_8 + \frac{x_6 x_1}{6 + x_1} + \tan(9 - x_5 + x_3^2) + x_4(0.5 + x_{11}) \qquad (5)$$

$$T = x_{10}^2 + x_{11} + \sin(\sin(\frac{x_3 x_2}{2}) + x_4 + x_1 + \cos(x_6)) + |x_3| - 1 + \frac{x_6}{\log 3 + x_4} \qquad (6)$$

Use above 3 models to forecast the wheat aphid population of the 3 fields in 2004 [7] and compare them with the observation data respectively, and the results are shown as Fig.3 to Fig.5. The results of experiment in Ref. [7] with the same experiment data are shown as Fig.6. By experiment 2, we can find that the forecast results with the model obtained by GEP-DE are closer to the observation data than the results of Ref. [7].

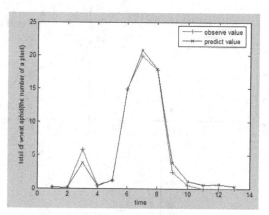

Fig. 3. Compare of forecast values of total of wheat aphid of field 1 with observation values

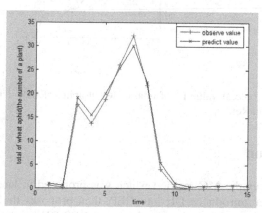

Fig. 4. Compare of forecast values of total of wheat aphid of field 2 with observation values

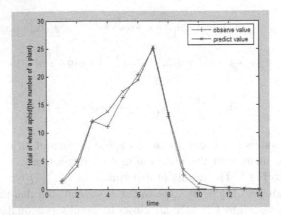

Fig. 5. Compare of forecast values of total of wheat aphid of field 3 with observation values

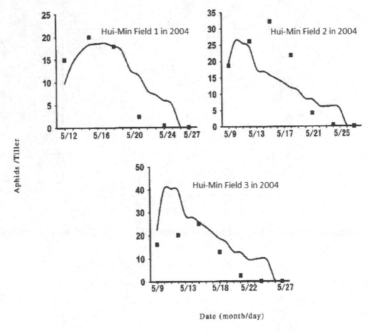

Fig. 6. Compare of forecast value (·) of total of wheat aphid of field 1 to 3 with observation value (solid line) from Ref.[7]

5   Conclusion

This paper improves the MC method which was proposed in latest important literature, and moreover introduces DE into GEP, then proposes the GEP-DE based on co-evolution. The control experiments show that GEP-DE promises excellent performance in constant optimization and overall performance. Lastly GEP-DE is successful applied in the wheat aphid population forecast problem as well.

Acknowledgments. Support from the National Natural Science Foundation of China (NO. 31170393), the Nature Science Foundation of Shaanxi Province (NO. 2012JM8023), and the Nature Science Special Foundation of Department of Education of Shaanxi Province (NO. 12JK0726) are gratefully acknowledged.

References

1. Ferreira, C.: Gene Expression Programming: A New Adaptive Algorithm for Solving Problems. Complex System 13(2), 87–129 (2001)
2. Azamathulla, H.M., Ghani, A.A., Leow, C.S., Chang, C.K., Zakaria, N.A.: Gene-Expression Programming for the Development of a Stage-Discharge Curve of the Pahang River. Water Resources Management 25(11), 2901–2916 (2011)
3. Mousavi, S.M., Aminian, P., Gandomi, A.H., et al.: A new predictive model for compressive strength of HPC using gene expression programming. Advances in Engineering Software 45, 105–114 (2012)
4. Ferreira, C.: Function finding and the creation of numerical constants in gene expression programming. In: The 7th Online World Conference on Soft Computing in Industrial Applications, England, vol. 265 (2002)
5. Zuo, J., Tang, C.J., Li, C., et al.: Time series predication based on gene expression programming. In: The 5th International Conference for Web Information Age (WAIM 2004), Berlin (2004)
6. Storn, R., Price, K.: Differential evolution-a simple and efficient heuristic for global optimization over continuous spaces. Journal of Global Optimization 11(4), 341–359 (1997)
7. Chang, X.Q.: The study of group of sitobion avenae dynamic simulation in field based on AFIDSS(Master thesis). Chinese academy of agriculture, Peking (2006)

Neural Network Intelligent Learning Algorithm for Inter-related Energy Products Applications

Haruna Chiroma[1,2], Sameem Abdul-Kareem[1], Sanah Abdullahi Muaz[1], Abdullah Khan[3], Eka Novita Sari[4], and Tutut Herawan[1]

[1] Faculty of Computer Science & Information Technology
University of Malaya
50603 Lembah Pantai, Kuala Lumpur, Malaysia
[2] Federal College of Education (Technical)
School of Science
Department of Computer Science, Gombe, Nigeria
[3] Faculty of Computer Science and Information Technology
Universiti Tun Hussein Onn Malaysia
Parit Raja, 86400 Batu Pahat, Johor, Malaysia
[4] AMCS Research Center, Yogyakarta, Indonesia
hchiroma@acm.org, {sameem,tutut}@um.edu.my,
abdullahdirvi@gmail.com, samaaz.csc@buk.edu.ng, eka@amcs.co

Abstract. Accurate prediction of energy products future price is required for effective reduction of future price uncertainty as well as risk management. Neural Networks (NNs) are alternative to statistical and mathematical methods of predicting energy product prices. The daily prices of Propane (PPN), Kerosene Type Jet fuel (KTJF), Heating oil (HTO), New York Gasoline (NYGSL), and US Coast Gasoline (USCGSL) interrelated energy products are predicted. The energy products prices are found to be significantly correlated at 0.01 level (2-tailed). In this study, NNs learning algorithms are used to build a model for the accurate prediction of the five (5) energy product price. The aptitudes of the five (5) NNs learning algorithms in the prediction of PPN, KTJF, HTO, NYGSL, and USCGSL are examined and their performances are compared. The five (5) NNs learning algorithms are Gradient Decent with Adaptive learning rate backpropagation (GDANN), Bayesian Regularization (BRNN), Scale Conjugate Gradient backpropagation (SCGNN), Batch training with weight and bias learning rules (BNN), and Levenberg-Marquardt (LMNN). Results suggest that the LMNN and BRNN can be viewed as the best NNs learning algorithms in terms of R^2 and MSE whereas GDANN was found to be the fastest. Results of the research can be use as a guide to reduce the high level of uncertainty about energy products prices, thereby provide a platform for developmental planning that can result in the improvement economic standard.

Keywords: US Coast Gasoline, Heating oil, Propane, Bayesian Regularization, Levenberg-Marquardt.

Y. Tan et al. (Eds.): ICSI 2014, Part I, LNCS 8794, pp. 284–293, 2014.
© Springer International Publishing Switzerland 2014

1   Introduction

The future prices of energy products such as Propane (PPN), Kerosene Type Jet fuel (KTJF), Heating oil (HTO), New York Gasoline (NYGSL), and US Coast Gasoline (USCGSL) are highly uncertain. The uncertainty trailing these energy products prices has succeeded in attracting both domestic and foreign political attention, and this facilitated market ranking [1]. Accurate forecasting of future prices of energy product can effectively be used for risk management as argued by [2].   Sato *et al.* [3] forecast the prices of commodities including Chicken, Coal, Coffee, Copper, Fish, Iron, Maize, Oil (Brent), Peanuts, Pork, Thai Rice, Sugar US, and Uranium using neural network (NN) model. However, only oil is energy product and the effectiveness of the NN model was not evaluated by comparing its performance with another method.

Malliaris and G. Malliaris [2] forecast one month ahead spot prices of crude oil, heating oil, gasoline, natural gas and propane since their spot prices in market are interrelated. Multi linear regression, NNs model, and simple model were applied in each of the energy market to forecast one month future prices of the energy product. Results show the NNs perform better than the statistical models in all markets except for propane market. Wang and Yang [4] examined the probability of predicting crude oil, heating oil, gasoline, and natural gas futures markets within a day using NN, semi parametric function coefficient, nonparametric kernel regression, and generalized autoregressive conditional heteroskedasticity (GARCH). Results indicated only heating oil and natural markets possessed the possibility of being predicted within a day. The NNs was found to outperform the statistical models.

Barunik and Křehlík [5] predict energy product prices using the NN model and its performance was found to be better than the popular heterogeneous autoregressive (HAR) models and autoregressive fractionally integrated (ARFIMA) econometric models. However, these statistical and econometric methods assume normal distribution for input data [6] which makes the statistical methods unsuitable for energy products price prediction because of the non-linear, complex and volatile nature of the energy products, experimental evidence can be found in [7]. Therefore, the comparison of NNs and statistical methods might not provide a fair platform.

Most literature mainly focuses on comparing the architecture of NNs in the domain of energy product price prediction. Recently, Panella *et al.* [8] compared performance of a Mixture of Gaussian NN (MoGNN) with that of RBF, ANFIS, and GARCH which proves the robustness of the MoGNN. Comparing the learning algorithms of NN is limited despite its significance in turning the NNs weights and bias for optimization. In this study, we have chosen a multilayer NN learning algorithms because the recurrent NN structure becomes more complex, thus, further complicates the chosen of the best NN parameters; the computation of the error gradient in a recurrent NN architecture also turns out to be complicated due to   presents of more attractors in the state space of a recurrent NN [7]. In addition, for the recurrent NN to achieve optimal performance, high number of hidden neurons are required in the hidden layer than necessarily needed for learning the problem which is considered as limitation because it reduces the operational efficiency of the recurrent NN [10].

In this paper, we propose neural network intelligent learning algorithm as a useful technique to evaluate and compare the its validity for the prediction of energy products price. The NNs learning algorithms are used to build a model for the prediction of PPN, KTJF, HTO, NYGSL, and USCGSL prices. Subsequently, compare the performances of the learning algorithms in each of the markets.

The rest of this paper is organized as follow. Section 2 describes proposed method. Section 3 describes results and following by discussion. Finally, the conclusion of this work is described in Section 4.

2   Proposed Method

2.1   Neural Network Learning Algorithms

The weights and bias of NNs are iteratively modified during NNs training to minimize error function such as Mean Square Error (MSE) computed using Equation (1) as follow.

$$MSE = \frac{1}{N}\sum_{j=1}^{N}\left(x(j) - y(j)\right)^2 \tag{1}$$

where N, x (j), and y (j) are the total number of predictions made by the model, original observation in the dataset, and the value predicted by the model, respectively. The closer the value of MSE to zero (0), the better is the prediction accuracy. Zero (0) indicates a perfect prediction, which rarely occurs in practice. The most widely use a NN learning algorithm is the BP algorithm which is a gradient-descent technique of minimizing an error function. The synaptic weight (W) in a BP learning algorithm can be updated using Equation (2) as follow.

$$W_{k+1} = W_k + \Delta W_k . \tag{2}$$

Here, k is the iteration in a discrete time and the current weight adaptation is represented by ΔW_k expressed as follow.

$$\Delta W_k = -\eta \frac{\partial ek}{\partial W_k}, \tag{3}$$

where η and $\frac{\partial ek}{\partial W_k}$ are learning rate (typically ranges from 0 to 1) and gradient of the error function to be minimized, respectively. The main drawbacks of the gradient descent BP includes: slow convergence speed and possibility of being trapped in local minima as a result of its iterative nature of solving problem till the error function reaches its minimal level. Appropriate specification of learning rate and momentum determine the success of BP in a large scale problem. Gradient-decent BP is still being applied in many NNs programs. Though, the BP is no more considered as the optimal and efficient learning algorithm. Thus, powerful learning algorithms that are fast in convergence are developed based on heuristic method from the standard steepest descent algorithm referred to as the first category of the fastest learning algorithms. The second category of the fastest learning algorithms were developed based on standard numerical optimization methods such as the Levenberg-Marquardt (LM). Typically, conjugate gradient algorithms converge faster than the variable learning rate BP algorithm, but such results are limited to application domain, implying that the results can differ from a particular problem domain to a different domain. The conjugate gradient algorithms require line search for each iteration, which makes the conjugate gradient to be computationally expensive. The scaled conjugate gradient backpropagation (SCGNN) algorithm was developed in response to the computationally expensive nature of the conjugate gradient so as to speed up convergence. Other alternative learning algorithms includes Gradient Decent with Adaptive learning rate backpropagation (GDANN), Bayesian Regularization (BRNN), Batch training with weight and bias learning rules (BNN), and Levenberg-Marquardt (LMNN). However, LMNN is viewed as the most effective learning algorithm for training a medium sized NNs. Gradient descent is used by the LM to improve on its starting guess for tuning the LMNN parameters [11].

2.2 Energy Product Dataset and Descriptive Statistics

The daily spot prices of HTO, PPN, KTJF, USCGSL, and NYGSL were collected from 9 July, 1992 to 16 October, 2012 source from the Energy Information Administration of the US Department of Energy. The data were freely available, published by the Energy Information Administration of the US Department of Energy. The data were collected on a daily basis since enough data are required for building a robust NNs model. The data comprised of five thousand and ninety (5090) rows and five (5) columns. The data were not normalized to prevent the destruction of the original pattern in the historical data [12]. The descriptive statistics of the data are computed and the results are reported in Table 1. The standard Deviation (Std.D) shown in the last column of Table 1 indicated uniform dispersion among the energy products prices except for PPN.

Table 1. Descriptive statistics of energy products datasets

	N	Min	Max	Mean	Std.D.
PPN	5090	0.2	1.98	0.6969	0.40872
KTJF	5090	0.28	4.81	1.037	0.73589
HTO	5090	0.28	4.08	1.2622	0.88739
NYGSL	5090	0.29	3.67	1.2472	0.82894
USCGSL	5090	0.27	4.87	1.2298	0.82616

Table 2 is a correlation among the energy product prices. The correlation is significant among the HTO, PPN, KTJF, USCGSL, and NYGSL as clearly showed in Table 2. Correlated variables imply that influence of a variable can affect the other variables positively as the case in Table 2. Hair et al. [13] argued that for better prediction, variables in the research data have to be significantly correlated. Thus, HTO, PPN, KTJF, USCGSL can be independent variables whereas NYGSL dependent variable. Also, PPN, KTJF, USCGSL, and NYGSL can be used as independent variables whereas HTO dependent variable. This can also be applied to PPN, KTJF, and USCGSL. Therefore, we compared the NNs learning algorithms in the five different energy markets.

Table 2. Inter correlation matrix of the energy products dataset

	PPN	KTJF	HTO	NYGSL
KTJF	0.702^{**}			
HTO	0.949^{**}	0.745^{**}		
NYGSL	0.937^{**}	0.737^{**}		
USCGSL	0.940^{**}	0.727^{**}	0.984^{**}	0.997^{**}

** Correlation is significant at the 0.01 level (2-tailed).

2.3   Neural Network Model Description

After several trials, our data were partition into training, validation, and test (3562, 764, and 764 samples, respectively). To avoid over-fitting the training data, random sampling was used to partition the dataset. Updating of NNs weights and bias as well as computation of the gradient is performed with the training dataset. To explore the best combination of the activation functions (ACFs), several ACFs are considered: log-sigmoid, linear, soft max, hyperbolic tangent sigmoid, triangular basis, inverse and hard-limit. Several ACFs were tried in the hidden layer neurons while linear ACFs was constantly maintained in the input layer neuron. In the output layer, linear is used to avoid limiting the values in a particular range. Therefore, both input and output layers used linear ACFs throughout the training period. Momentum and learning rate were varied between zero (0) to one (1). Single hidden layer is used since [14] stated that single hidden layer is sufficient to approximate any nonlinear function with arbitrary accuracy. Experimental trials were performed to find the appropriate NNs model with the best MSE, R^2, and convergence speed. The training terminates after six (6) iterations without performance improvement to avoid over-fitting the network. The network architecture with the minimum MSE, highest R^2, and low convergence speed are recorded as the optimal NNs topology. The predictive capabilities of the NNs learning algorithms were evaluated on the test dataset.

3   Results and Discussion

The proposed algorithms were implemented in MATLAB 2013a Neural Network ToolBox on a computer system (HP L1750 model, 4 GB RAM, 232.4 GB HDD, 32-bit OS, Intel (R) Core (TM) 2 Duo CPU @ 3.00 GHz). The number of hidden neurons should not be twice of the independent variables as argued in [15]. Thus, we consider between four (4) to ten (10) ranges of the number of neurons and used to verify for NNs optimal architecture for every learning algorithms. Different ACFs were experimentally tried with corresponding number of hidden neurons. The models with the best results are reported in Tables 3 to 7 and those with poor results are discarded. The best ACFs found for the prediction of LMNN is log-sigmoid, for BNN is a hyperbolic tangent sigmoid, for GDANN is log-sigmoid, for SCGNN is triangular basis and for BRNN is log-sigmoid. Tables 3 to 7 shows performance (Mean Square Error (MSE)) (Regression (R^2)) and convergence speed (Iterations (I) (Time (T) in seconds (Sec.)) for each of the NNs learning algorithms. The momentum and learning rate found to be optimal were 0.3 and 0.6 respectively. The minimum MSE, highest R^2, optimum combinations of I and T are in bold through the Tables.

Table 3. Performance of the prediction of HTO price with different NNs learning algorithm

Learning Method	Performance			Convergence speed		
	Number of hidden neurons			Number of hidden neurons		
	4-8-1	4-7-1	4-6-1	4-8-1	4-7-1	4-6-1
LMNN	0.000178 (0.9959)	**0.000115** (0.9938)	0.00082 (0.9955)	65 (5)	71(5)	**72(4)**
SCGNN	0.00208 (0.8790)	0.00504 (0.9808)	0.00363 (0.9246)	1000 (150)	1000 (117)	1000 (100)
GDANN	2.86 (-0.8471)	3.93 (-0.9142)	0.793 (0.7940)	1(0)	1(0)	1(0)
BNN	8.89 (-0.7898)	5.57 (-0.6456)	8.63 (0.9410)	1000 (27)	1000 (27)	1000 (26)
BRNN	0.00573 **(0.9963)**	0.00635 (0.9961)	0.00669 (0.9958)	217 (25)	230 (22)	293 (22)

From Table 3, it can be deduced that the LMNN algorithm has the lowest MSE and is the fastest in converging to the optimal MSE whereas BRNN achieved better R^2 than the other learning algorithms. These results indicated that the performance of the algorithms in predicting HTO is not consistent because it depends on the performance metrics being considered as the criteria for measuring performance. Though, in this case LMNN can be chosen despite not having the highest R^2 due to its ability to achieve the lowest MSE in the shortest possible time. Seven (7) hidden neurons produce the best MSE result, whereas six (6) hidden neurons is the fastest architecture.

Table 4. Comparison of KTJF predicted by NNs learning algorithm models

Learning Method	Performance			Convergence speed		
	Number of hidden neurons			Number of hidden neurons		
	4-8-1	4-7-1	4-6-1	4-8-1	4-7-1	4-6-1
LMNN	0.102(0.8771)	**0.0873(0.9245)**	0.1(0.88)	47(5)	110(9)	54(2)
SCGNN	1.3000(-0.4111)	34.3000(-0.7662)	3.2(-0.67)	1000 (183)	1000 (168)	1000 (142)
GDANN	0.62(0.7482)	0.744(-0.7108)	0.29(0.72)	**12(2)**	16(2)	31(10)
BNN	15.7(-0.5972)	29(0.7396)	13.6(0.68)	1000 (28)	1000 (27)	1000 (28)
BRNN	0.0896(0.9108)	0.0896(0.91468)	0.098(0.9)	322 (40)	531 (49)	443 (33)

The results of the prediction of KTJF price are reported in Table 4, the LMNN has the minimum MSE and the highest R^2 among the comparison algorithms. The other algorithms such as BRNN, GDANN (4-8-1, 4-6-1) also have competitive values of MSE and R^2 compared to the optimal values. The fastest algorithm is GDANN having the minimum iterations and time of convergence. The performance criteria's indicated that the LMNN is the best in terms of MSE and R^2 whereas convergence speed criteria's shows GDANN outperforms other algorithms. Seven (7) hidden neurons yield the best MSE and R^2 but, the fastest architecture is having eight (8) hidden neurons. This is surprising as less complex structure is expected to be the fastest.

Table 5. The performance of the NNs learning algorithms in the prediction of NYGSL

Learning Method	Performance			Convergence speed		
	Number of hidden neurons			Number of hidden neurons		
	4-8-1	4-7-1	4-6-1	4-8-1	4-7-1	4-6-1
LMNN	0.0044(0.9983)	0.00158(0.9988)	0.00195(0.9987)	39(3)	60(3)	35(1)
SCGNN	1(-0.1631)	18.6(0.4774)	13.5(0.9531)	1000(183)	1000(153)	1000(141)
GDANN	3.22(0.8825)	2.12(0.8311)	0.490(0.9471)	**1(0)**	**1(0)**	**1(0)**
BNN	13(-0.5668)	5.22(0.9338)	0.864(0.8711)	1000(27)	1000(25)	1000(24)
BRNN	0.000025(**0.999**)	**0.0000279**(0.9988)	0.000512(0.9988)	433(48)	954(87)	604(44)

The results of the prediction of NYGSL price are reported in Table 5, showing GDANN as the fastest algorithm to converge to its optimal solution. The BRNN learning algorithm with different architecture is the best predictor with the lowest MSE and the highest R^2 among the comparison algorithms. SCGNN and GDANN are having the poorest values of MSE despite GDANN has a competitive R^2 compared to the promising R^2 value of BRNN, LMNN, BNN (4-7-1, 4-6-1), and SCGNN (4-6-1). In the prediction of NYGSL, the performance exhibited a similar phenomenon to the prediction of HTO and KTJF as consistency is not maintained. In the prediction of the NYGSL we cannot conclude on the best algorithms because the performance exhibited by the algorithms is highly random unlike the case in the prediction of HTO. The BRNN converged to the MSE and R^2 very slow compared to the LMNN, and GDANN speed. The optimal algorithm in this situation depends on the criteria chosen as the priority in selecting the best predictor. If accuracy is the priority, then BRNN can be the best candidate, whereas speed place LMNN above BRNN. Seven (7) hidden neurons have the best MSE value, whereas the architectures with six (6), seven (7), and eight (8) hidden neurons are the fastest. This could probably be caused by memorizing the training data by the algorithms.

Table 6 indicated that GDANN is the fastest to predict PPN price, whereas the MSE of BRNN is the best. The R^2 value of LMNN is the highest compared to the SCGNN, GDANN, BNN, and BRNN R^2 values. The performance exhibited by the algorithms in the prediction of PPN price is not different from that of NYGSL, KTJF, and HTO because consistent performance is not realized. The best algorithm for the prediction of PPN price depends on the performance metrics considered as priority for selecting the optimal algorithm as earlier explained. The algorithms with negative values of R^2 reported in Tables 3 to 6 suggested that the observed price and the predicted once are in opposite directions. Signifying that upward movement of predicted price can influence

Table 6. Results obtained with the NNs learning algorithms in the prediction of PPN

Learning Method	Performance			Convergence speed		
	Number of hidden neurons			Number of hidden neurons		
	4-8-1	4-7-1	4-6-1	4-8-1	4-7-1	4-6-1
LMNN	0.00514(**0.985**)	0.00694(0.9801)	0.00623(0.9820)	74(8)	32(2)	60(4)
SCGNN	0.266(-0.4121)	0.646(-0.8601)	0.686(-0.8592)	1000 (179)	1000 (149)	1000 (136)
GDANN	1.78 (-0.8164)	2.45 (-0.2453)	3.31 (0.7886)	1(0) 1000 (26)	1(0) 1000 (25)	1(0) 1000 (25)
BNN	0.276(-0.28011)	0.211(0.3802)	1.62(-0.4091)			
BRNN	**0.00051**(0.984)	0.00557(0.98231)	0.686(-0.8592)	204(19)	226(21)	385(28)

Table 7. Comparison of USCGSL predicted by NNs learning algorithm models

Learning Method	Performance			Convergence speed		
	Number of hidden neurons			Number of hidden neurons		
	4-8-1	4-7-1	4-6-1	4-8-1	4-7-1	4-6-1
LMNN	0.00264(0.99821)	0.00330(0.99817)	0.00301(0.99624)	103(7)	21(1)	63(3)
SCGNN	0.00598(0.99686)	0.00403(0.99621)	0.00445(0.99818)	37(1)	146(4)	131(1)
GDANN	0.0531(0.95758)	0.0284(0.97383)	0.0484(0.96635)	55(0)	68(0)	94(1)
BNN	0.0246(0.98191)	0.712(0.69757)	0.913(0.96988)	1000(23)	10(0)	12(0)
BRNN	0.00250(**0.99825**)	0.00273(0.99785)	**0.00235**(0.94193)	652(67)	653(57)	563(41)

the observed price to move downward and vice versa. This is not true considering the promising results obtained by other algorithms that show positive R^2 values.

In the prediction of USCGSL price as indicated in Table 7, BRNN have the minimum value of MSE and the highest R^2, though with a different hidden neurons. The fastest algorithm is the GDANN with seven (7) hidden neurons. It seems hidden layer neurons do not always affect the convergence speed of the NNs algorithms based on experimental evidence from Tables 3 to 7. The results do not deviate from similar behavior shown by the prediction of HTO, PPN, NYGSL, and KTJN prices. Small number of iterations do not necessarily imply lower computational time based on evidence from the simulation results. For example, in Table 7, a GDANN converge to a solution in sixty eight (68) iterations, 0 Sec. Whereas, 4 indicated that convergence occurs in thirty five (35) iterations, one (1) Sec. with the LMNN which is considered as the most efficient NN learning algorithm in the literature. The poor performance exhibited by some algorithms can be attributed to the possibility that the algorithms could have been trapped in local minima. The complexity of an NNs affects convergence speed as reported in Tables 3 to 7. The fastest architectures have six (6) hidden neurons with exception in the prediction of KTJN. This is a multitasking experiments performed on the related energy products. We have found from the series of the experiments conducted that the LMNN and BRNN constitute an alternative approaches for prediction in the oil market especially when accuracy is the subject of concern. The objective of the research have been achieved since the idle NN learning algorithms were identified for future prediction of the energy products. Therefore, uncertainty related to the oil market can be reduced to the tolerance level which in turn might stabilize the energy product market. The results obtained do not agree with the results reported by [2]. This could probably be attributed to the fair comparison of our study, unlike the study by [2] that compared NN and statistical methods. The results of this study cannot be generalized to other multi-task problems because the performance of NNs learning algorithm depends on the application domain since the NNs performance differ from domain to domain as argued by [16]. However, the methodology can be modified to be applied on similar datasets or problem. The research presented by [17] differ from the present study in the following ways: Crude oil price was predicted based on the inter-related energy products prices; Genetically optimized NN was applied for the modeling.

4   Conclusion

In this research, the performance of NNs learning algorithms in the energy products price prediction was studied and their performances in terms of MSE, R^2, and convergence speed were compared. BRNN was found to have the best result in the prediction of an HTO price in terms of R^2 whereas LMNN achieved the minimum MSE and converges faster than the SCGNN, GDANN, BRNN, and BNN in predicting HTO price. In the prediction of KTJF price, LMNN performs better than the SCGNN, GDANN, BRNN, and BNN considers MSE and R^2 as performance criteria's. In contrast, GDANN is the algorithm that converges faster than the other NNs learning algorithms. On the other hand, prediction of NYGSL is more effective with BRNN in terms of MSE and R^2, but GDANN is the fastest. BRNN have the minimum MSE whereas LMNN achieved the maximum R^2 in the prediction of PPN price.

The fastest among the learning algorithms in the prediction of PPN price is GDANN despite having the poorest MSE values. BRNN performs better than the SCGNN, GDANN, LMNN, and BNN in the prediction of USCGSL price in terms of MSE and R^2. GDANN recorded the best convergence speed compared to SCGNN, BRNN, LMNN, and BNN. The NNs learning algorithms use for the prediction of energy products prices is not meant to replace the financial experts in the energy sector. Perhaps, is to facilitate accurate decision to be taken by decision makers in order to reach better resolutions that could yield profits for the organization. Investors in the energy sector could rely on our study to suggest future prices of the energy products. This can reduce the high level of uncertainty about energy products prices, thereby provide a platform for developmental planning that can result in the improvement of economic standard.

Acknowledgments. This work is supported by University of Malaya High Impact Research Grant no vote UM.C/625/HIR/MOHE/SC/13/2 from Ministry of Higher Education Malaysia.

References

1. US Department of Energy, Annual Energy Outlook 2004 with Projections to 2025 (2004), http://www.eia.doe.gov/oiaf/archive/aeo04/index.html
2. Malliaris, M.E., Malliaris, S.G.: Forecasting energy product prices. Eur. J. Financ. 14(6), 453–468 (2008)
3. Sato, A., Pichl, L., Kaizoji, T.: Using Neural Networks for Forecasting of Commodity Time Series Trends. In: Madaan, A., Kikuchi, S., Bhalla, S. (eds.) DNIS 2013. LNCS, vol. 7813, pp. 95–102. Springer, Heidelberg (2013)
4. Su, F., Wu, W.: Design and testing of a genetic algorithm neural network in the assessment of gait patterns. Med. Eng. Phys. 22, 67–74 (2000)
5. Barunik, J., Křehlík, T.: Coupling High-Frequency Data with Nonlinear Models in Multiple-Step-Ahead Forecasting of Energy Markets' Volatility (2014), http://dx.doi.org/10.2139/ssrn.2429487
6. Wang, T., Yang, J.: Nonlinearity and intraday efficiency test on energy future markets. Energ. Econ. 32, 496–503 (2010)
7. Shambora, W.E., Rossiter, R.: Are there exploitable inefficiencies in the futures market for oil? Energ. Econ. 29, 18–27 (2007)
8. Panella, M., Barcellona, F., D'Ecclesia, R.L.: Forecasting Energy Commodity Prices Using Neural Networks. Advances in Decision Sciences 2012, Article ID 289810, 26 pages (2012), doi:10.1155/2012/289810
9. Moshiri, S., Foroutan, F.: Forecasting nonlinear crude oil futures prices. Energ. J. 27, 81–95 (2006)
10. Quek, C.M., Kumar, P.N.: Novel recurrent neural network-based prediction system for option trading and hedging. Appl. Intell. 29, 138–151 (2008)
11. Haykin, S.: Neural networks, 2nd edn. Prentice Hall, New Jersey (1999)
12. Jammazi, R., Aloui, C.: Crude oil forecasting: Experimental evidence from wavelet decomposition and neural network modeling. Energ. Econ. 34, 828–841 (2012)
13. Hair, F.J., Black, W.C., Babin, B.J., Anderson, R.E.: Multivariate data analysis. Pearson Prentice Hall, New Jersey (2010)
14. Hornik, K.: Approximation capabilities of multilayer feedforward networks. Neural Netw. 4, 251–257 (1991)
15. Berry, M.J.A., Linoff, G.: Data mining techniques. Willey, New York (1997)
16. Azar, A.T.: Fast neural network learning algorithms for medical applications. Neural Comp. Appl. 23(3-4), 1019–1034 (2013)
17. Chiroma, H., Gital, A.Y.U., Abubakar, A., Usman, M.J., Waziri, U.: Optimization of neural network through genetic algorithm searches for the prediction of international crude oil price based on energy products prices. In: Proceedings of the 11th ACM Conference on Computing Frontiers, p. 27. ACM (2014)

Data-Based State Forecast via Multivariate Grey RBF Neural Network Model

Yejun Guo[1], Qi Kang[1], Lei Wang[1,2], and Qidi Wu[1]

[1] Department of Control Science and Engineering,
Tongji University, Shanghai 201804, China
qkang@tongji.edu.cn
[2] Shanghai Key Laboratory of Financial Information Technology, Shanghai, China

Abstract. This paper presents a multivariable grey neural network (MGM-NN) model for predicting the state of industrial equipments. It combines the merit of MGM model and RBF-NN model on time series forecast. This mode takes the dynamic correlations among multi variables and environment's impact on state of equipment into consideration. The proposed approach is applied to the melt channel state forecast. The results are contrasted to MGM model executed on the same test set. The results show the accuracy and promising application of the proposed model.

Keywords: State Forecast, Multivariate Grey Model, Neural Network, RBF.

1   Introduction

With the rapid development of information technology and automation technology, the modern industrial control system, such as iron-steal smelting industry and auto industry, is becoming more and more integrity and complexity. The current fault diagnosis method cannot meet the requirement these control system. At the same time, we can get any kinds of real-time data from such complexity control system by the continuous development of embedded system and data collection technology. In these contexts, Prognostics and Health Management (PHM), promoted by modern information technology and artificial intelligence (AI), is emerged at the right time, what is a health and fault management solution of high integrity and complexity industrial control system. State forecast is a key technology for PHM and is the comprehensive utilization of all kinds of data, such as monitoring parameter, the status of utilization, current environment and working condition, previous experimental data and experience, relies on some kinds of reasoning technology like mathematics, physics model and artificial intelligence technology to estimate the remaining useful life and the future health of device and system.

Current state forecast algorithm based on data has time-series analysis which applies to stationary stochastic series, the grey model which applies index variation data, the hidden Markov model which applies to random signal, support vector machine (SVM) which applies to small sample and non-linear system, and artificial neural

Y. Tan et al. (Eds.): ICSI 2014, Part I, LNCS 8794, pp. 294–301, 2014.

network which applies to linear and non-linear system[1-6]. Lin et al. [7] apply the Grey forecasting model to forecast accurately the output value of Taiwan's opto-electronics and get the residual error of the forecasting model is lower than 10%. Tseng et al. [8] proved that GM(1,1) grey forecasting model is insufficient for fore-casting time series with seasonality and proposes a hybrid method that combines the GM(1,1) and ratio-to-moving-average deseasonalization method to forecast time series. Mao et al. [9] apply grey model GM(1,1) plus 3-point average technique to forecast motor vehicle fatality risk, and verify this model method is feasible, reliable and highly efficient and also does not need to make any assumptions. Xie et al. [10] proposes a novel discrete grey forecasting model termed DGM model and a series of optimized models of DGM and verify this model can increase the prediction accuracy. Hsu et al. [11] proposed a grey model with factor analysis techniques to deal with the multi-factor forecasting problems and improved multivariable grey forecasting model is more feasible and effective than grey model GM(1,1). He et al. [12] apply the new information multi-variable Grey model NMGM(1,n) to load-strain relation and im-proved that this method is very effective to data processing which have a high per-formance and precision. Zhang et al. [13] combine the radial basis function (RBF) neural network and the adaptive neural fuzzy inference system to forecast short-term load and improved that this model has high precision and can work effectively and overcome the defects of the RBF network. Ture et al. [14] compares time series pre-diction capabilities of three artificial neural networks algorithms-multi layer percep-tron (MLP), RBF, and time delay neural networks (TDNN).

Univariate grey model, lacking of correlation analysis among the multiple signals which cause industrial equipment failure, will result in the loss of some significant information and reduce the accuracy and precision of state forecast; meanwhile cur-rent various state forecasting do not have to consider the actual industrial environment like temperature which has important impact on equipment failure. This paper presents (MGM-RBF) forecast model, which can get a thorough understanding of complexity system by considering various factors from various angel, and can build dynamic relation of various factors, at the same time, also taking the actual environ-ment like temperature into this model, so that it can improve accuracy and precision of state forecast. And verifying this model can achieve more satisfactory result by case simulation.

2   Multivariate Grey RBF Neural Network Model

2.1   Multivariate Grey Forecast Model

Multivariate gray model MGM (1, n) is the nary first-order ordinary differential equa-tions, which model is as follows:

Let $x_i^{(0)}(k)(i = 1,2,...,n)$ been n-ary grey time series which consist by k interac-tive aspects, and generate $x_i^{(1)}(k)(i = 1,2,...,n)$ been a corresponding cumulative sequence.

$$x_i^{(1)}(k) = \sum_{j=1}^k x_i^{(0)}(j) \quad \text{where } k = 1, 2, \ldots, m. \tag{1}$$

MGM (1, n) model is n-ary first-order ordinary differential equations:

$$\frac{dX}{dt} = AX + B. \tag{2}$$

where,

$$A = \begin{bmatrix} a_{11} & a_{12} & \cdots & a_{1n} \\ a_{21} & a_{21} & & a_{2n} \\ & & \cdots & \\ a_{n1} & a_{n1} & \cdots & a_{nn} \end{bmatrix}, B = (b_1, b_2, \ldots, b_n)^T. \tag{3}$$

The continuous time response for equation (3) is:

$$X^{(1)}(t) = e^{At}X^{(1)}(0) + A^{-1}(e^{At} - I) \cdot B. \tag{4}$$

Where: I is unit matrix.
Then, discrete equations (2):

$$x_i^{(0)}(k) = \sum_{j=1}^n \frac{a_{ij}}{2}\left(x_j^{(1)}(k) + x_j^{(1)}(k-1)\right) + b_i. \tag{5}$$

where $a_i = (a_{i1}, a_{i2}, \ldots, a_{in}, b_i)^T$, where $i = 1, 2, \ldots, n$.
Then, get the recognize value $\hat{a}_i$ of a_i by Least Squares:

$$\hat{a}_i = \begin{bmatrix} \hat{a}_{i1} \\ \hat{a}_{i2} \\ \cdots \\ \hat{a}_{in} \\ \hat{b}_i \end{bmatrix} = (L^T L)^{-1} Y_i, i = 1, 2, \ldots, n. \tag{6}$$

where,

$$L = [\frac{1}{2}\left(x_1^{(1)}(m) + x_1^{(1)}(m-1)\right) \ldots \frac{1}{2}\left(x_n^{(1)}(m) + x_n^{(1)}(m-1)\right) \; 1].$$

Then, get the recognize value A and B:

$$\hat{A} = \begin{bmatrix} \hat{a}_{11} & \hat{a}_{12} & \cdots & \hat{a}_{1n} \\ \hat{a}_{21} & \hat{a}_{22} & \cdots & \hat{a}_{2n} \\ & & \cdots & \\ \hat{a}_{n1} & \hat{a}_{n2} & \cdots & \hat{a}_{nn} \end{bmatrix} \quad \hat{B} = \begin{bmatrix} \hat{b}_1 \\ \hat{b}_2 \\ \cdots \\ \hat{b}_n \end{bmatrix}. \tag{7}$$

Finally, get the solution of MGM(1,n) is:

$$\hat{X}^{(1)}(k) = e^{A(k-1)}X^{(1)}(1) + \hat{A}^{-1}(e^{A(k-1)} - I) \cdot \hat{B}, k = 1, 2, \ldots. \tag{8}$$

$$\hat{X}^{(0)}(1) = X^{(0)}(1).$$

$$\hat{X}^{(0)}(k) = \hat{X}^{(1)}(k) - \hat{X}^{(1)}(k-1), k = 2, 3, \ldots. \tag{9}$$

As can be seen from the derivation of the algorithm, when n = 1, the MGM(1,N) degeneration to GM(1,1), and when B = 0, MGM(1,n) is become n GM(1,1) models.

2.2   RBF Neural Network Regression Model

RBF neural network is an efficient feed-forward neural network, which has the best approximation performance and global optimum what other forward neural networks do not have, and simple structure and training in high speed. Radial basis function neural network is a local approximation, which simulates the human brain partial adjustment and mutually covering accepted domains neural network structure. Currently, it has proven to approximate any continuous function with any arbitrary precision. The most basic RBF neural network what is a three-layer feed-forward network with a single hidden layer, where each layer has a completely different role. The input layer, consisting of sensing unit, will connect with the external environments; the role of second layer is to complete the non-linear transformation from the input layer to the hidden layer, in most situations the hidden layer has higher dimension. The third layer is output layer, the role of this layer is provide response to activation model of the input layer [6].

2.3   Multivariate Grey RBF Model

Multivariate grey neural network state forecast model is combine the disadvantage of multivariate grey forecast model in small simple, poor information and multiple inter-related factors and the disadvantage of RBF neural network in non-linear regression. RBF neural network error forecast model, using the state forecast value and the temperature which is the most important environment factors as input to this model, and the error value as output to train RBF neural network, then get the error forecast model. MGM-NN model, MGM-NN uses the error forecast model error forecast to compensate MGM(1,n) result, what can get high accuracy and precision.

3   Case Simulation

3.1   Problem Description

To establish a mathematical model for melt channel is not economic or even impossible, therefore using the historical running data about the performance of melt channel is become the main technical means to forecast the fault of device. Historical running data include the fault information of device, we can learn and mining the internal rule of fault information of device, and we can forecast the running state (including the fault information) by these internal rule to reflect the real fault information of melt channel, and determine whether to use it or not. Therefore we can transform fault forecast to running state forecast (including the fault information), and can also abstract to time serials forecast, following is the mathematical model:

$$y(t + T) = f(y(t), y(t - 1), ..., y(t - n + 1)). \tag{10}$$

where T is the step. In this case T=3.
We can find the approximate f by historical running data.

3.2   Historical Running Data Preprocessing

This paper data from real-time operating of industrial equipment melt ditch, the data preprocessing is shown as follows:

Selection of variables Real-time running state of melting channel, resistor(R) and inductance(X), can be monitored by access to relevant literature, therefore we can turn Voltage(U), Current (I), Active Power (P), Reactive Power (Q) to R and X by electromagnetic principle. Choose temperature (T) as the actual industrial environment factors, because the running state of melting ditch is closely related to temperature.

Processing singular value and normalization in RBF-NN model Find the singular value by Statistical Methods and delete it. In this case, we use following method to do, some processed data shown in Table 1.

Table 1. Some processed data

Date	Temperature(T)	U(V)	I(A)	P(W)	Q(W)	R	X
8.2	1409	210	550	60	90	0.1983	0.3263
8.5	1395	390	1040	230	300	0.2126	0.3089
8.8	1403	210	540	60	90	0.2058	0.3300
8.11	1392	380	1000	210	290	0.2100	0.3167
8.14	1422	360	950	170	260	0.1884	0.3288
8.17	1422	220	600	60	100	0.1667	0.3266
8.20	1407	220	600	60	100	0.1667	0.3266

3.3   Simulation Result

The step of simulation is as follows:
1. Use MGM to forecast the R and X, and get the R_F value and X_F value and $E_R = R_F - R$, $E_X = X_F - X$; the result shown in Table 2.

Table 2. The result of MGM

T	R	X	R_F	X_F	E_R	E_X
1409	0.1983	0.3263	0.2082	0.3175	-0.0099	0.0088
1395	0.2126	0.3089	0.1966	0.3280	0.0160	-0.0192
1403	0.2058	0.3300	0.2136	0.3076	-0.0078	0.0224
1392	0.2100	0.3167	0.2054	0.3321	0.0046	-0.0154
1422	0.1884	0.3288	0.2103	0.3166	-0.0219	0.0123
1422	0.1667	0.3266	0.1852	0.3304	-0.0185	-0.0038
1407	0.1667	0.3266	0.1852	0.3304	-0.0185	-0.0038

2   Take the Temperature (T), R_F (forecast of R), X_F (forecast of X)as input and the E_R (error of R)and E_X (error of X)as output to train RBF-ANN, and get the error matched curve of R and X, shown in Fig.3.

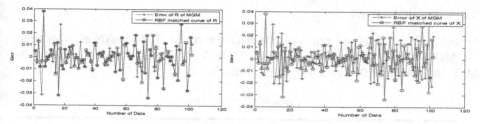

Fig. 3. Error matched curve

3   Use test set to verify MGM-NN model, and compute the Mean Square predict Error (MSE) and Average Relative Error (ARE). The test set data shown in Table II and we have already got the R_F and X_F value by MGM model, following we can get the E_{RF} (error of R forecast) and E_{XF} (error of X forecast) value by RBF model, then get the final value by

$$R_{FF} = R_F + E_{RF}$$
$$X_{FF} = X_F + E_{XF}$$

where R_{FF} is final forecast of R and X_{FF} is final forecast of X.

We can get the E_{RF} and E_{XF} from the RBF model, and the result of them is shown in Table 3.

Table 3. Result of error forecast

R_F	X_F	E_{RF}	E_{XF}	R_{FF}	X_{FF}
0.2082	0.3175	-0.0133	0.0160	0.1950	0.3335
0.1966	0.3280	0.0210	-0.0232	0.2176	0.3048
0.2136	0.3076	-0.0123	0.0261	0.2013	0.3337
0.2054	0.3321	0.0106	-0.0194	0.2160	0.3127
0.2103	0.3166	-0.0266	0.0165	0.1837	0.3331
0.1852	0.3304	-0.0235	-0.0058	0.1617	0.3246
0.1606	0.3277	0.0111	-0.0031	0.1717	0.3246

The result of test set shown in Table 4 and Fig. 4

Table 4. Average relative error rate of test set

$MGM_{RRERate}(\%)$	$MGM_{XRERate}(\%)$	$MGM\text{-}NN_{RRERate}(\%)$	$MGM\text{-}NN_{XRERate}(\%)$
5.2480	3.1602	1.1234	1.094

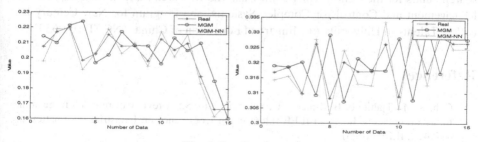

Fig. 4. Results of test set

3.4   Result Analysis

From Fig. 3 and Fig.4, we can conclude that the forecast of MGM-NN model is better the MGM in melt channel, and from the Table 3 and Table 4, we can compute the MSE and ARE by 3.2.1 – 3.2.4 shown in Table 5.

Table 5. MSE and ARE

MGM				MGM-NN			
MSE_R	MSE_X	ARE_R	ARE_X	MSE_R	MSE_X	ARE_R	ARE_X
0.0002	0.0002	5.25%	3.16%	0	0	1.12%	1.09%

From the Table 5, we know that the result of MGM-NN is better than the MGM in both Mean Square Error (MSE) and Average Relative Error (ARE). From the Fig.5, we know that the MGM-NN model can forecast the channel X's and R's trends accurately. Therefore we can draw a conclusion that the MGM-NN forecast model is better than MGM model in melt channel running state forecast and this model can also apply to relative equipment running state forecast.

4   Conclusion

This paper presents a multivariable grey neural network (MGM-NN) model for predicting the state of industrial equipments. It combines the merit of MGM model and RBF-NN model on time series forecast. This mode takes the dynamic correlations among multi variables and environment's impact on state of equipment into consideration. The proposed model is applied to forecast the state of melt channel, and receive better result than MGM model from the case study. On the basis of this work, we can optimize the back-ground value and initial value of MGM-NN and embed the MGM into ANN for further study, and get a more accuracy. SVM is also can apply to small sample data analysis and can receive satisfactory results, so the next research is to mix the SVM and MGM to forecast the equipment running state.

Acknowledgment. This work was supported in part by the National Natural Science Foundation of China (71371142, 61005090, and 61034004), the Fundamental Research Funds for the Central Universities, and the Research Fund of State Key Lab. of Management and Control for Complex Systems, and Program for New Century Excellent Talents in University of Ministry of Education of China （NCET-10-0633）.

References

1. Goh, K.M., Tjahjono, B., Baines, T., Subramaniam, S.: A review of research in manufacturing prognostics. In: 2006 IEEE International Conference on Industrial Informatics, pp. 417–422. IEEE (2006)

2. Wu, D.: Time series prediction for machining errors using support vector regression. In: First International Conference on Intelligent Networks and Intelligent Systems, ICINIS 2008, pp. 27–30. IEEE (2008)
3. Yang, S., Shen, H.: Research and application of machine learning algorithm based on relevance vector machine. Computing Technology and Automation 29(1), 44–47 (2010)
4. Tipping, M.E.: Sparse Bayesian learning and the relevance vector machine. The Journal of Machine Learning Research 1, 211–244 (2001)
5. Li, J., Sun, H.: Prediction control of active power filter based on the grey model. Proceedings of the Chinese Society for Electrical Engineering 22(2), 7–10 (2002)
6. Wei, H., Li, Q.: Gradient learning dynamics of radial basis function networks. Control Theory and Applications 24(3), 112–118 (2007)
7. Lin, C.T., Yang, S.Y.: Forecast of the output value of Taiwan's opto-electronics industry using the grey forecasting model. Technological Forecasting and Social Change 70(2), 177–186 (2003)
8. Tseng, F.M., Yu, H.C., Tzeng, G.H.: Applied hybrid grey model to forecast seasonal time series. Technological Forecasting and Social Change 67(2), 291–302 (2001)
9. Mao, M., Chirwa, E.C.: Application of grey model GM (1, 1) to vehicle fatality risk estimation. Technological Forecasting and Social Change 73(5), 588–605 (2006)
10. Xie, N.M., Liu, S.F.: Discrete grey forecasting model and its optimization. Applied Mathematical Modelling 33(2), 1173–1186 (2009)
11. Hsu, L.C.: Forecasting the output of integrated circuit industry using genetic algorithm based multivariable grey optimization models. Expert Systems with Applications 36(4), 7898–7903 (2009)
12. Luo, Y., Li, J.: Application of Multi-variable Optimizing Grey Model MGM (1, n, q, r) to the Load-strain Relation. In: International Conference on Mechatronics and Automation, ICMA 2009, pp. 4023–4027. IEEE (2009)
13. Yun, Z., Quan, Z., Caixin, S., Shaolan, L., Yuming, L., Yang, S.: RBF neural network and ANFIS-based short-term load forecasting approach in real-time price environment. IEEE Transactions on Power Systems 23(3), 853–858 (2008)
14. Ture, M., Kurt, I.: Comparison of four different time series methods to forecast hepatitis A virus infection. Expert Systems with Applications 31(1), 41–46 (2006)

Evolving Flexible Neural Tree Model
for Portland Cement Hydration Process

Zhifeng Liang[1,2], Bo Yang[1,*], Lin Wang[1,2], Xiaoqian Zhang[1,2], Lei Zhang[1,2],
and Nana He[1,2]

[1] Shandong Provincial Key Laboratory of Network based Intelligent Computing,
Jinan 250022, China
yangbo@ujn.edu.cn
[2] University of Jinan, Jinan 250022, China
{liangzhf,wangplanet,xqZhang1989,
zl.ramiel,little.flying}@Gmail.com

Abstract. The hydration of Portland cement is a complicated process and still
not fully understood. Much effort has been accomplished over the past years to
get the accurate model to simulate the hydration process. However, currently
existing methods using positive derivation from the conditions for physical-
chemical reaction are lack of information in real hydration data. In this paper,
one model based on Flexible Neural Tree (FNT) with acceptable goodness of fit
was applied to the prediction of the cement hydration process from the real microstructure image data of the cement hydration which has been obtained by
Micro Computed Tomography (micro-CT) technology. Been prepared on the
basis of previous research, this paper used probabilistic incremental program
evolution (PIPE) algorithm to optimize the flexible neural tree structure, and
particle swarm optimization (PSO) algorithm to optimize the parameters of the
model. Experimental results show that this method is efficient.

Keywords: Flexible Neural Tree, Probabilistic Incremental Program Evolution,
Particle Swarm Optimization, Cement Hydration.

1 Introduction

Cement has been widely used in many kinds of fields as one of the most important
industrial building material. Meanwhile, to enhance the performance of cement material has strong application value for the national economy and construction. Therefore, it is very significant to study the process of cement hydration in order to understand the hydration principle and forecasting the cement performance [1]. Generally,
hydration data of 28 days are selected as a reference to study the cement microstructure evolution process, because the degree of cement hydration after 28 days maintain
at a low level and the cement performance has been determined basically [2].

---

* Corresponding author.

Y. Tan et al. (Eds.): ICSI 2014, Part I, LNCS 8794, pp. 302–309, 2014.

Manual derivation of hydration kinetic equation is usually adopted in the physical-chemical experimental methods [3] [4]. Single-particle model is proposed firstly in 1967 by Kondo and Kodama [5]. Then, Kondo and Ueda developed this model in 1968 [6]. Subsequently, in 1979, Pommersheim described an integrated reaction-diffusion single-particle model on the basis of the previous work [7]. These models are not quite effective, until 1997, a new model expressed as a single equation which is also founded on the single particle model is proposed by Tomosawa [8].

Although so many achievements have been obtained above, it is still cannot reflect the cement hydration process exactly due to the extreme complexity of the cement hydration. Recent studies show that, intelligent computation are introduced to the reverse modeling and has a good ability to extract rules from experimental data directly. Lin Wang and Bo Yang built a modern mixed kinetic modeling for hydration process based on Gene Expression Programming (GEP) algorithm and Particle Swarm Optimization (PSO) algorithm [9]. Subsequently, they reconstruct the model with Multi-layer Multi-Expression Programming (MMEP).

The advantages of intelligent computation in the function finding and data forecasting prompt us to adopt it to explore the model of cement hydration process. This paper presents a new type of reverse modeling framework using Flexible Neural Tree (FNT) proposed by Chen [10] in 2005 to simulate the process of cement hydration. The structure of FNT is optimized by probabilistic incremental program evolution (PIPE) algorithm and parameters are optimized by PSO. With this model, we can get the 3D microstructure evolution process for cement hydration. The results of simulation show that predicting model achieved by PIPE and PSO based flexible neural tree has better validity than the canonical artificial neural network model and other model.

The rest of the paper is organized as follows: Section 2 raises a description for the FNT model with PIPE and PSO. Section 3 introduces the experimental data capturing from micro-CT and feature selection. The results of the simulation experiment are described in section 4 to show its performance. Section 5 concludes the paper.

2   An Introduction to FNT Model

FNT model is similar to the model of artificial neural network. A tree-structural based encoding method is selected for representing a FNT model. This structure can be evolved expediently using the existing tree-structure-based approaches, like GEP, PIPE, and MEP etc. The parameters used in FNT can be optimized by Genetic Algorithm (GA) and PSO, etc.

Compositions of FNT model include function set F and terminal instruction set T [25]. For example, we use the set of functions $F = \{+_2, +_3 ... +_N\}$ and the set of terminal instruction $T = \{x_1, x_2 ... x_n\}$. With this the FNT model can be expressed as $S = F \cup T = \{+_2, +_3 ... +_N \} \cup \{ x_1, x_2 ... x_n \}$. Instruction $+_N$ is also called a flexible neuron operator with N inputs. An example of flexible neuron using the function $+_N$ and terminal instruction set is given in Fig.1.

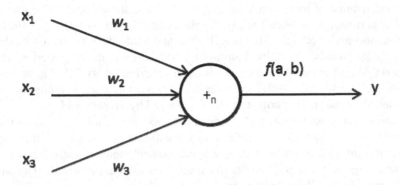

Fig. 1. A flexible neuron operator with the input of x_i. Numbers of flexible neuron operator form a FNT model.

In the process of constructing hybrid neural tree, if a non-leaf node instruction $+_i$ (i = 2, 3 ...n) is selected, $w_1, w_2, ... w_3$ are generated randomly and used for representing the connection weight between node $+_i$ and x_1, x_2... x_i. Meanwhile, two parameters a and b are created randomly as flexible activation function parameters. In this paper, the function is represented by following formulation (1):

$$f(a_i, b_i, x) = e^{-(\frac{x-a_i}{b_i})^2}$$ (1)

The output of $+_i$ is calculated as following formulation (2):

$$out_n = f(a_n, b_n, net_n)$$ (2)

While net_n is the sum of total activation calculated as following formulation (3):

$$net_n = \sum_{j=1}^{n} \omega_j * x_j$$ (3)

Overall output of flexible neural tree can be computed from left to right by depth-first method, recursively. Weight w and parameters a, b are optimized by PSO in our experiment.

FNT model is more flexible when compared with traditional analytical model. Foremost, it allows identifying important input features that are efficient. This means that it has ability of features selection compared with traditional neural network.

2.1   Brief Summary for PIPE

PIPE algorithm was proposed by Salustowicz in 1997 [11]. PIPE "uses a stochastic selection method for successively generating better and better programs according to an adaptive probabilistic prototype tree". No crossover operator is used.

PIPE generates individual according to an underlying probabilistic prototype tree called PPT. Generation-based learning (GBL) and elitist learning (EL) is the main learning algorithm of PIPE. In this paper, GBL is used for learning.

PIPE algorithm can be described as follows:

a) Initialize probabilistic prototype tree.
b) Repeat until termination criteria are met.
    i.   Create population of programs
        a)   Grow PPT if required
    ii.   Evaluate population
        a)   Favor smaller programs if all is equal
    iii.   Update PPT
    iv.   Mutate PPT
    v.   Prune PPT
c) End

2.2 Brief Summary for PSO

Inspired by the flocking and schooling patterns of birds and fish, Particle Swarm Optimization (PSO) which is a population based stochastic optimization technique was invented by Russell Eberhart and James Kennedy in 1995 [12]. PSO shares many similarities with evolutionary computation techniques such as GA.

PSO is initialized with a group of random particles (solutions) and then searches for optima by updating generations. Each particle is updated by following two "best" values at each iteration. The first one is the best solution (fitness) it has achieved so far. This value is called pbest. Another "best" value that is tracked by the particle swarm optimizer is the best value, obtained so far by any particle in the population. This best value is a global best and called gbest. When a particle takes part of the population as its topological neighbors, the best value is a local best and is called lbest. A new velocity for particle i is updated by equation (4):

$$v_i(t+1) = v_i(t) + c_1 * rand() * (p_i(t) - present_i(t)) + c_2 * rand() * (g_i(t) - present_i(t)) \quad (4)$$

v is the particle velocity. Present is the current particle (solution) while pbest and gbest are defined as stated before. Function rand () is a random number between (0,1). Constant c1, c2 are learning factors. Usually c1 = c2 = 2.
The particle updates its positions with following equation (5):

$$present_i(t+1) = present_i(t) + v_i(t+1) \quad (5)$$

Then PSO algorithm description of the procedure can be expressed as follows:
    Step1: Initialize particle for each particle.
    Step2: Calculate fitness value for each particle. Current value is set as the new pBest when the fitness value is better than the best fitness value (pBest) in history.
    Step3: Choose the particle with the best fitness value of all the particles as the gBest.
    Step4: For each particle , calculate particle velocity according equation (4) while update particle position according equation (5)

2.3   Hybrid FNT with PIPE and PSO

In this paper, each individual of PIPE is optimized by PSO to find the tree structure with the best parameters.

At first, set the initial values of parameters used in the PIPE randomly. Assume that, the population of PIPE is 20. At each generation, it created 20 tree structures individual. For each individual, PSO iterate until reach the pre-defined stop point. The fitness function is measured by root mean square error (RMSE). When PSO reach the maximum number of iterations, the individual with the best parameters is stored. After all the 20 individual is optimized by PSO, an individual with the best parameters in this population in first generation has been getting. Then we use PSO to optimize this individual with larger loops with more iteration than before to get a better result. For each iteration of PIPE, repeat the process above until the satisfactory solution is reached.

The optimal design of hybrid FNT model can be described as follows:

Step1: Creat a population with random structures and parameters.

Step2: Optimized the tree structure using PIPE which is described in subsection 2.1.

Step3: Optimized the parameters suing PSO which is described in subsection 2.2.

Step4: repeat Step2 and Step3 above until the satisfactory solution is found

3   Data Acquisition and Feature Selection

In this section, the acquisition and processing of experimental data that used for FNT algorithm will be described.

Microstructure image data of 28 days is chosen as a reference to study the cement hydration. First of all, a large number of micro-CT time-series images in the process of cement hydration according to the standard experiment are obtained. The image of each day is segmented into a small size of 140*140*140. In this paper, the following day's images are used:1D(day), 2D, 3D, 4D, 5D, 6D, 7D, 14D, 21D, and 28D. At the same time, the final training dataset and testing dataset are obtained by a series of image processing like image enhancement and three dimension image registrations. After careful process of cement image, we get a three-dimensional image with that the pixel values are mapped to between 0 and 255.

In this work, a plugin of ImageJ is developed to get the dataset from the observed time-series image. The dataset is consisting of the current pixel values and 26 neighborhood pixel values. These 27 features are used as the input parameter and the pixel value of next time as the output parameter which forming a complete dataset. Then data set split according to the tenfold cross-validation scheme and linear scaling with the output range of <0, 1> is used in order to match nonlinear activation functions domains which were used for FNT.

It is considered very effectively to implement the feature selection using FNT by itself. The feature that selected by FNT is given in following section.

4   Results and Discussion

The flexible neural tree model is applied here in conjunction with cement hydration prediction problems.

The following settings are adopted in the experiment (Table 1).

Table 1. Parameter settings used for the experiment

Parameters	Settings
PIPE	
Population Size	20
Iterations	10000
Mutation rate	0.1
Learning rate	0.01
Epsilon	0.000001
PSO	
Population Size	20
Iterations	500(each individual),1000(best individual)
c_1	2
c_2	2
FNT	
Function set	$\{+_2,+_3,+_4,+_5\}$

The FNT model we get from 900 training data is shown in Fig.2.

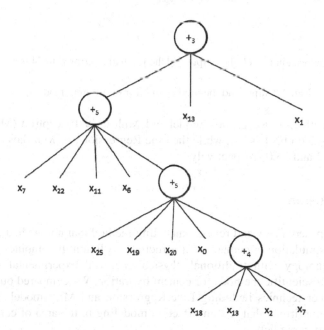

Fig. 2. The FNT model we get for prediction of the cement hydration.

Then the output and the prediction error with 74 testing data are shown in Fig.3.

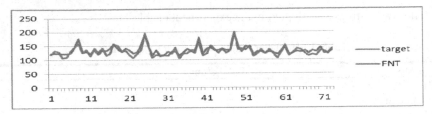

(a)   FNT model: the output and the prediction error with 74 testing data

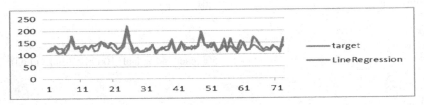

(b) Line Regression model: the output and the prediction error with 74 testing data.

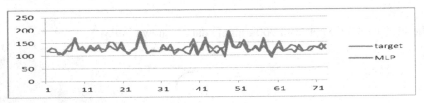

(c) Line Regression model: the output and the prediction error with 74 testing data.

Fig. 3. Output and the prediction error of different models

Compared with Line Regression model and Multilayer Perceptron (MLP) model. The RMSE based on FNT is 9.4, while the Line Regression and Multilayer Perceptron model are 14.32 and 13.08 respectively.

5   Conclusion

This paper proposed a new and reverse modeling method that no one had ever done it before for the simulation of cement microstructure evolution. It combines progressive computer technology with traditional physical-chemical experimental approaches, established new scientific methods for cement hydration. We compared our prediction model with other technics including Line Regression and MLP model. The experimental results confirmed that it's an effective modeling in the area of cement microstructure evolution prediction.

In the future work, we will combine Multi Expression Programming algorithm with Particle Swarm Optimization to get a more accurate model. Moreover, grey value of cement data could be divided into different gray domain to use classification method.

Acknowledgments. This work was supported by National Key Technology Research and Development Program of the Ministry of Science and Technology under Grant 2012BAF12B07-3. National Natural Science Foundation of China under Grant No. 61173078, No. 61203105, No. 61373054, No. 61173079, No. 61070130, No. 81301298. Shandong Provincial Natural Science Foundation, China, under Grant No. ZR2010FM047, No. ZR2012FQ016, No. ZR2012FM010, No. ZR2011FZ001,No. ZR2010FQ028. Jinan Youth Science &Technology Star Project under Grant No. 2013012

References

1. Chen, W., Brouwers, H.J.H.: Mitigating the Effects of System Resolution on Computer Simulation of Portland Cement Hydration. Cem. Concr. Compos. 30, 779–787 (2008)
2. Tennis, P.D., Bhatty, J.I.: Characteristics of Portland and Blended Cements Results of a Survey of Manufacturers. In: Cement Industry Technical Conference, pp. 156–164. IEEE Press, Holly Hill (2006)
3. Tomosawa, F.: Kinetic Hydration Model of Cement. Cem. Concr. 23, 53–57 (1974)
4. Krstulović, R., Dabić, P.: A Conceptual Model of the Cement Hydration Process. Cem. Concr. Res. 30, 693–698 (2000)
5. Kondo, R., Kodama, M.: On the Hydration Kinetics of Cement. Semento Gijutsu Nenpo 21, 77–82 (1967)
6. Kondo, R., Ueda, S.: Kinetics and Mechanisms of the Hydration of Cements. In: Proceedings of the Fifth International Symposium on the Chemistry of Cement, pp. 203–248. Cement Association of Japan, Tokyo (1968)
7. Pommersheim, J.M., Clifton, J.R.: Mathematical Modeling of Tricalcium Silicate Hydration. Cem. Concr. Res. 9, 765–770 (1979)
8. Tomosawa, F.: Development of a Kinetic Model for Hydration of Cement. In: Proceedings of the 10th International Congress on the Chemistry of Cement, pp. 125–137. Amarkai AB and Congrex, Sweden (1997)
9. Wang, L., Yang, B., Zhao, X.Y., Chen, Y.H., Chang, J.: Reverse Extraction of Early-age Hydration Kinetic Equation from Observed Data of Portland Cement. Science China Technological Sciences 53, 1540–1553 (2010)
10. Chen, Y., Yang, B., Dong, J., Abraham, A.: Time-series Forecasting using Flexible Neural Tree Model. Inf. Sci. 174, 219–235 (2005)
11. Salustowicz, R.P., Schmidhuber, J.: Probabilistic Incremental Program Evolution. Evolutionary Computation 5, 123–141 (1995)
12. Kennedy, J., Mendes, R.: Population Structure and Particle Swarm Performance. In: Proceedings of the Congress on Evolutionary Computation, pp. 1671-1676. IEEE Press, Nanjing (2002)

Hybrid Self-configuring Evolutionary Algorithm for Automated Design of Fuzzy Classifier[*]

Maria Semenkina and Eugene Semenkin

Siberian Aerospace University, Krasnoyarsky rabochy avenue 31,
660014, Krasnoyarsk, Russia
semenkina88@mail.ru, eugenesemenkin@yandex.ru

Abstract. For a fuzzy classifier automated design the hybrid self-configuring evolutionary algorithm is proposed. The self-configuring genetic programming algorithm is suggested for the choice of effective fuzzy rule bases. For the tuning of linguistic variables the self-configuring genetic algorithm is used. An additional feature of the proposed approach allows the use of genetic programming for the selection of the most informative combination of problem inputs. The usefulness of the proposed algorithm is demonstrated on benchmark tests and real world problems.

Keywords: Genetic algorithms, genetic-programming, fuzzy classifier, automated design, performance estimation.

1 Introduction

A fuzzy classifier [1] is one of the intelligent information technologies allowing the generation of a fuzzy rule base suitable for interpretation by human experts. The resulting fuzzy rule base that can be written in terms of natural language is the most useful data mining tool for end users that are experts in their area of activity and non-experts in the field of intelligent information technologies. However, the process of fuzzy classifier design and adjustment is rather complex even for experts in fuzzy systems. For the automation of fuzzy classifier implementation we need to consider its design as an optimization problem. This problem consists of two parts: the generation of a fuzzy rule base and the tuning of its linguistic variables. Both are very complicated for standard optimization tools, which makes evolutionary algorithms quite popular in this field [2, 3].

In this paper, we consider a genetic programming algorithm (GP) that automatically designs the fuzzy classifier rule base. It can be a good solution for this problem because of the GP's ability to work with variable length chromosomes. The use of a GP can simplify the implementation of the Pittsburg [3] method and, in this case, it is not required to implement the Michigan [2] method for reducing the dimension of the search space.

---

[*] Research is fulfilled with the support of the Ministry of Education and Science of Russian Federation within State assignment project 140/14.

Y. Tan et al. (Eds.): ICSI 2014, Part I, LNCS 8794, pp. 310–317, 2014.

Applying the GP, one faces the usual problem of determining an effective configuration of the evolutionary algorithm (EA) because the EA's performance depends on the selection of their settings and the tuning of their parameters. The design of the EA consists of choosing variation operators (e.g., selection, recombination, mutation, etc.). Additionally, real valued parameters of the chosen settings (the probability of recombination, the level of mutation, etc.) have to be tuned. The process of the choosing settings and tuning parameters is known to be a time-consuming and complicated task.

Following from the definitions given by Gabriela Ochoa and Marc Schoenauer, organizers of the workshop "Self-tuning, self-configuring and self-generating evolutionary algorithms" (Self* EAs) within PPSN XI [4] we call our algorithms self-configuring because the main idea of the approach relies on automated "selecting and using existing algorithmic components". The self-configuring evolutionary algorithms (Self-CEA) use dynamic adaptation on the population level [5], centralized control techniques [6] for parameter settings with some differences from the usual approaches.

Having conducted numerical experiments, we have found that the proposed approach positively impacts on the algorithms' performance and deserves special attention and further investigation.

The rest of the paper is organized as follows. Section 2 explains the idea of self-configuring evolutionary algorithms. Section 3 describes the proposed method of fuzzy classifier automated design. Section 4 describes the results of the numerical experiments comparing the performance of the proposed approach in solving real-world problems, and in the conclusion we discuss the results.

2   Self-configuring Evolutionary Algorithm

If somebody decides to use evolutionary algorithms for solving real world optimization problems, it will be necessary to choose an effective variant of the algorithm settings such as the kind of selection, recombination and mutation operators. Choosing the right EA settings for each problem is a difficult task even for experts in the field of evolutionary computation. It is the main problem of effective implementation of evolutionary algorithms for end users. We can conclude that it is necessary to find a solution for this problem before suggesting EAs for end users as a tool for automated design in solving real world problems.

We propose using self-configuring evolutionary algorithms (SelfCEA) that do not require any efforts of the end user so that the problem is adjusted automatically. In our algorithm the dynamic adaptation of operators' probabilistic rates on the level of population is used (see Fig.1). Instead of tuning real parameters, variants of settings were used, namely types of selection (fitness proportional, rank-based, and tournament-based with three tournament sizes), crossover (one-point, two-point, as well as equiprobable, fitness proportional, rank-based, and tournament-based uniform crossovers [7]), population control and level of mutation (medium, low, high for all mutation types). Each of these has its own initial probability distribution (see Fig. 2), which are changed as the algorithm executes (see Fig. 3).

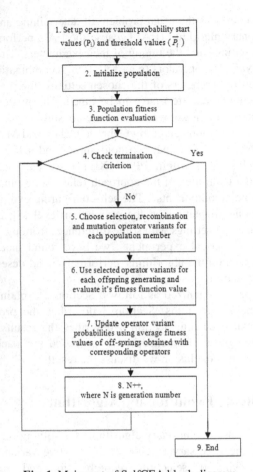

Fig. 1. Main part of SelfCEA block diagram

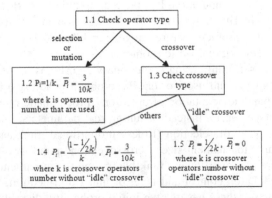

Fig. 2. Flowchart illustrating step 1 in SelfCEA block diagram

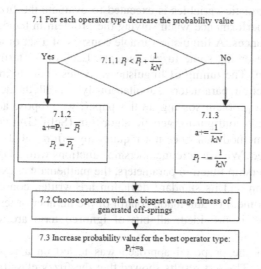

Fig. 3. Flowchart illustrating step 7 in SelfCEA block diagram

This self-configuring technique can be used both for genetic algorithms (SelfCGA) and genetic programming (SelfCGP). In [8] the SelfCGA performance was estimated on 14 test problems from [9]. As a commonly accepted benchmark for GP algorithms is still an "open issue" [10], the symbolic regression problem with 17 test functions borrowed from [9] were used in [7] for testing the self-configuring genetic programming algorithm. Statistical significance was estimated with ANOVA.

Analysing the results of SelfCGA [8] and SelfCGP [7] performance evaluation, we observed that self-configuring evolutionary algorithms demonstrate better reliability than the average reliability of the corresponding single best algorithm. They can be used instead of conventional EA in complex problem solving.

3   Self-configuring Evolutionary Algorithm for Automated Fuzzy Classifier Design

We have to describe our way to model and optimize a rule base for a fuzzy logic system with GP and linguistic variables adjusting with GA.

Usually, a GP algorithm works with a tree representation of solutions, defined by functional and terminal sets, and exploits specific solution transformation operators (selection, crossover, mutation, etc.) until the termination condition is met [11]. The terminal set of our GP includes the terms of the output variable, i.e. class markers. The functional set includes a specific operation for dividing an input variables vector into subvectors or, in other words, for the separation of the examples set into parts through input variable values. It might be that our GP algorithm will ignore some input variables and will not include them in the resulting tree, i.e., a high performance rules base that does not use all problem inputs can be designed. This feature of our approach allows the use of our GP for the selection of the most informative combination of problem inputs.

The tuning of linguistic variables is executed to evaluate the fuzzy system fitness that depends on its performance when solving the problem in hand, e.g., the number of misclassified instances. A linguistic variable consists of a set of terms or linguistic variable values representing some fuzzy concepts. Each of the terms is defined by a membership function. The tuning of linguistic variables consists in the optimization of membership function parameters simultaneously for all the terms of linguistic variables involved in problem solving. In this paper, we propose adjusting linguistic variables with the self-configuring genetic algorithm (SelfCGA) combined with the conjugate gradient method that does not require any efforts of the end user for the problem to be adapted. We use here membership functions with a Gaussian shape. For the coding of membership function parameters, the mathematical expectation value of the Gaussian function and its standard deviation are written consecutively for each term in the chromosome. For automatic control of the number of terms the possibility of the term ignoring is provided: all bits of ignored term are set as 0 with the probability equal to 1/3.

The efficiency of the proposed approach was tested on a representative set of known test problems. The test results showed that the fuzzy classifiers designed with the suggested approach have a small number of rules in comparison with the full rule base. These fuzzy systems have a small enough classification error. This is why we can recommend the developed approach for solving real world problems.

4   Numerical Experiments with Real World Problems

The developed approach was applied to two classification machine learning credit scoring problems from the UCI repository [12] often used to compare the accuracy with various classification models:

- Credit (Australia-1) (14 attributes, 2 classes, 307 examples of the creditworthy customers and 383 examples for the non-creditworthy customers);
- Credit (Germany) (20 attributes, 2 classes, 700 records of the creditworthy customers and 300 records for the non-creditworthy customers).

Both classification problems were solved with fuzzy classifiers designed by hybrid SelfCEA (SelfCEA+FL). This technology was trained on 70% of the instances in the data base and validated on the remaining 30% of the examples. The results of the validations (the portion of correctly classified instances from the test set) averaged for 40 independent runs are given in Table 1 below.

We first compared the fuzzy classifier performance with ANN-based [13, 14] and symbolic regression based [7] classifiers automatically designed by SelfCGP (SelfCGP+ANN and SelfCGP+SRF). As we have observed, the algorithm proposed in this paper demonstrates high performance on both classification tasks.

We then conducted the comparison of our SelfCEA+FL designed classifier with alternative classification techniques. Results for the alternative approaches have been taken from scientific literature. In [15] the performance evaluation results for these two data sets are given for the authors' two-stage genetic programming algorithm (2SGP) specially designed for bank scoring as well as for the following approaches taken from other papers: conventional genetic programming (GP), multilayered perceptron (MLP), classification and regression tree (CART), C4.5 decision trees, k nearest neighbors (k-NN), and linear regression (LR). We have taken additional material for comparison from [16] which includes evaluation data for the authors' automatically designed fuzzy rule based classifier as well as for other approaches found in the literature: the Bayesian approach, boosting, bagging, the random subspace method (RSM), and cooperative coevolution ensemble learning (CCEL). The results obtained are given in Table 1. As can be seen from Table 1, the proposed algorithm demonstrates competitive performance being runner up for both problems.

Table 1. The comparison of classification algorithms

Classifier	Australian credit	German credit
2SGP	**0.9027**	**0.8015**
SelfCEA+FL	0.9022	0.7974
SelfCGP+ANN	0.9022	0.7954
SelfCGP+SRF	0.9022	0.7950
Fuzzy	0.8910	0.7940
C4.5	0.8986	0.7773
CART	0.8986	0.7618
k-NN	0.8744	0.7565
LR	0.8696	0.7837
RSM	0.8660	0.7460
Bagging	0.8470	0.6840
Bayesian	0.8470	0.6790
Boosting	0.7600	0.7000
CCEL	0.7150	0.7151

Our intention was to make it clear whether our approach could give results competitive with alternative techniques without attempting to develop the best tool for bank credit scoring. We used limited computational resources: 500 generations rather than 1000 as, e.g., in [15]. This is necessary to stress that fuzzy classifiers designed by SelfCEA give additionally human interpreted linguistic rules which is not the case for the majority of other algorithms in Table 1. Designed rule bases usually contain 10-15 rules which do not include all given inputs.

Analysis of the data sets shows that input variables can be divided in some groups so that inputs of one group are highly correlated to each other but the correlation between inputs of different groups is weak. There are also inputs weakly correlated with the output. A fuzzy classifier designed with suggested hybrid SelfCEA doesn't usually include inputs of the last kind. Moreover, it usually includes members of every inputs' group but only one input from each, i.e. it doesn't include highly correlated inputs in the rule base. This allows the algorithm to create relatively small rule bases with rather simple rules.

The computational efforts required by the suggested hybrid SelfCEA are high enough but not much higher than ANN-based classifier design requests. This can be considered as an acceptable disadvantage taking into consideration that this approach gives a compact base of easily interpreted rules and not just a computing formulation of a black box as ANN does.

5   Conclusions

A self-configuring genetic programming algorithm and a self-configuring genetic algorithm were hybridized to design fuzzy classifiers. Neither algorithm requires human efforts to be adapted to the problem in hand, which allows the automated design of classifiers. A special way of representing the solution gives the opportunity to create relatively small rule bases with rather simple rules. This makes possible the interpretation of obtained rules by human experts. The quality of classifying is high as well, which was demonstrated through the solutions of two real world classification problems from the area of bank scoring.

The results obtained allow us to conclude that the developed approach is workable and useful and should be further investigated and expanded.

References

1. Ishibuchi, H., Nakashima, T., Murata, T.: Performance Evaluation of Fuzzy Classifier Systems for Multidimensional Pattern Classification Problems. IEEE Trans. on Systems, Man, and Cybernetics 29, 601–618 (1999)
2. Cordón, O., Herrera, F., Hoffmann, F., Magdalena, L.: Genetic Fuzzy Systems: Evolutionary Tuning and Learning of Fuzzy Knowledge Bases. World Scientific, Singapore (2001)
3. Herrera, F.: Genetic Fuzzy Systems: Taxonomy, Current Research Trends and Prospects. Evol. Intel. 1(1), 27–46 (2008)
4. Schaefer, R., Cotta, C., Kołodziej, J., Rudolph, G. (eds.): PPSN XI. LNCS, vol. 6238. Springer, Heidelberg (2010)
5. Meyer-Nieberg, S., Beyer, H.-G.: Self-Adaptation in Evolutionary Algorithms. In: Lobo, F.G., Lima, C.F., Michalewicz, Z. (eds.) Parameter Setting in Evolutionary Algorithm, vol. 54, pp. 47–75. Springer (2007)
6. Gomez, J.: Self Adaptation of Operator Rates in Evolutionary Algorithms. In: Deb, K., Tari, Z. (eds.) GECCO 2004. LNCS, vol. 3102, pp. 1162–1173. Springer, Heidelberg (2004)
7. Semenkin, E., Semenkina, M.: Self-Configuring Genetic Programming Algorithm with Modified Uniform Crossover Operator. In: Proceedings of the IEEE Congress on Evolutionary Computation (IEEE CEC), Brisbane, Australia, pp. 1918–1923 (2012)
8. Semenkin, E., Semenkina, M.: Self-configuring Genetic Algorithm with Modified Uniform Crossover Operator. In: Tan, Y., Shi, Y., Ji, Z. (eds.) ICSI 2012, Part I. LNCS, vol. 7331, pp. 414–421. Springer, Heidelberg (2012)
9. Finck, S., et al.: Real-Parameter Black-Box Optimization Benchmarking 2009. In: Presentation of the noiseless functions. Technical Report Researh Center PPE (2009)
10. O'Neill, M., Vanneschi, L., Gustafson, S., Banzhaf, W.: Open Issues in Genetic Programming. Genetic Programming and Evolvable Machines 11, 339–363 (2010)

11. Poli, R., Langdon, W.B., McPhee, N.F.: A Field Guide to Genetic Programming (2008) (With contributions by J. R. Koza), Published via http://lulu.com and freely available at http://www.gp-field-guide.org.uk
12. Frank, A., Asuncion, A.: UCI Machine Learning Repository. University of California, School of Information and Computer Science, Irvine, CA (2010), http://archive.ics.uci.edu/ml
13. Semenkin, E., Semenkina, M.: Artificial Neural Networks Design with Self-Configuring Genetic Programming Algorithm. In: Filipic, B., Silc, J. (eds.) Bio-inspired Optimization Methods and their Applications: Proceedings of the Fifth International Conference BIOMA 2012, pp. 291–300. Jozef Stefan Institute, Ljubljana (2012)
14. Semenkin, E., Semenkina, M., Panfilov, I.: Neural Network Ensembles Design with Self-Configuring Genetic Programming Algorithm for Solving Computer Security Problems. In: Herrero, Á., et al. (eds.) Int. Joint Conf. CISIS 2012-ICEUTE 2012-SOCO 2012. AISC, vol. 189, pp. 25–32. Springer, Heidelberg (2013)
15. Huang, J.-J., Tzeng, G.-H., Ong, C.-S.: Two-Stage Genetic Programming (2SGP) for the Credit Scoring Model. Applied Mathematics and Computation 174, 1039–1053 (2006)
16. Sergienko, R., Semenkin, E., Bukhtoyarov, V.: Michigan and Pittsburgh Methods Combining for Fuzzy Classifier Generating with Coevolutionary Algorithm for Strategy Adaptation. In: IEEE Congress on Evolutionary Computation. IEEE Press, New Orleans (2011)

The Autonomous Suspending Control Method
for Underwater Unmanned Vehicle
Based on Amendment of Fuzzy Control Rules

Pengfei Peng, Zhigang Chen, and Xiongwei Ren

Navy University of Engineering , WuHan 430033, China

Abstract. For the specific needs of the underwater unmanned vehicle (UUV) in the working environment, the autonomous suspending control method for UUV based on amendment of fuzzy control rules is advanced. According to the traditional fuzzy controller, this method based on particle swarm optimization (PSO) for online search strategies adjusts the amendment of fuzzy control rules in time. This online optimization of fuzzy control method has a faster convergence speed, and can effectively adaptive adjust the motion state of UUV, which carries out the autonomous suspending control within the scope of predetermined depth for the accuracy and robustness . It is illuminated by simulation experiments that this autonomous suspending control method for UUV based on amendment of fuzzy control rules is more effective against uncertain disturbance and serious nonlinear, time change process.

Keywords: Underwater unmanned vehicle, autonomous suspending control, amendment factor, particle swarm optimization.

1   Introduction

When operating the underwater unmanned vehicle (UUV), the good autonomous suspension control is required for their own structural safety and the steady maintenance of underwater depth. Due to the large inertia, the low-speed underwater movement of UUV is coupled nonlinear. Therefore, the autonomous suspension control system of UUV is a kind of large inertia, delay, nonlinear complex control system. For the conventional PID control method[1,2] , the UUV dynamics characteristics is required to know,which should remain unchanged, as small as possible to the external disturbance in particular. However, the actual situation of the particularity and variability in the marine environment makes conventional PID methods to have significant limitations and the uses of ineffective. Fuzzy control is a complex control methods suited for the hard describe with a precise mathematical model and mainly relying on the experience[3,4,5]. So people have to study the fuzzy control applied to a variety Autopilot steering control [6,7]. But the traditional fuzzy controller for selecting a large number of parameters needs to have a wealth of operational experience in the actual sea trials and applications, which the debugging time is often limited. Once the parameter setting is wrong, the results could be disastrous. And the overly complex

Y. Tan et al. (Eds.): ICSI 2014, Part I, LNCS 8794, pp. 318–323, 2014.

parameter adjustment constrains the application of fuzzy control technologies in the underwater motion [8,9,10].

As for the actual requirements, the autonomous suspending control method for UUV based on the amendment of fuzzy control rules is proposed in this paper. Which using particle swarm optimization method to adjust the online amendment factors fuzzy control rules, this method implements the adaptive fuzzy control process. Large number of simulation experiments show that this method based on the amendment of fuzzy control rules is a good kind of autonomous suspending control, especially for the nonlinear state process, uncertainty disturbances and imprecise models in depth deep-sea movements.

2   The Work Principle for Underwater Unmanned Vehicles Autonomous Suspension Control System

The autonomous suspension control system device of underwater unmanned vehicle is mainly composed of tanks, pumps, solenoid valves, oil capsule and related electronic controllers, sensors, fuzzy conversion circuit, power supply. The principle of buoyancy control device is using the pump to fill the bag with oil from the tank, which making the oil bag expand and the underwater unmanned vehicles drainage volume increase, and its buoyancy increases. Conversely, using the pump to draw oil from the bag makes its buoyancy decrease. Thereby changing the buoyancy can make the underwater unmanned vehicle float or sink.

The autonomous suspension control of underwater unmanned vehicle enables the system to stably work underwater at a depth within a predetermined range, especially when it is subjected to external impact or environmental disturbances. With the automatical adjustments of motion state, underwater unmanned vehicle can be ensured to return to a predetermined depth.

3   Autonomous Suspension Control Method Based on Amendment of Fuzzy Control Rules

On the basis of the traditional adaptive fuzzy control method, the fuzzy control rules can be considered for further optimization and adjustment. The search for the amendment factor of fuzzy control rules is a typical nonlinear optimization process. Therefore, in view of PSO effectively applied in parameter optimizations and fuzzy control systems[11,12], it is possible to combine the movement of process simulation with the PSO algorithm. With designing the amendment of fuzzy control rules based on PSO, a new kind of method for autonomous suspension control is proposed.

3.1   The Amendment of Fuzzy Control Rules

The performance of fuzzy controller has great influence for the characteristics of system, which dues to the selected control rules. Among them, the weights between the input variables will directly affect the control rules. Therefore,to introduce a amendment factor to adjust the control rules is needed, which can correct deficiencies in the original set rules to improve the performance of the fuzzy controller.

Fuzzy control rules can be introduced by the following formula:

$$U = -\langle \varepsilon E + (1-\varepsilon)\dot{E} \rangle, \quad \varepsilon \in [0\ 1].$$ (1)

Among them, ε is amendment factor.

When the amendment factor adjusts the size of itself, the weights of the error and error change can easily change. Further, in the actual control system and under different conditions, the weights of the error and error change will have different requirements. Therefore, in different situations or error levels, the introduction of different amendment factors is needed. For example, the discrete domain of the errors is $E = \{-3,-2,-1,0,1,2,3\}$. When the error is small, the amendment factor ε_1 is to correct rules ; When the error is large, the amendment factor ε_2 is to correct rules.

$$U = \begin{cases} -\langle \varepsilon_1 E + (1-\varepsilon_1)\dot{E} \rangle, & E = \pm 1,0 \\ -\langle \varepsilon_2 E + (1-\varepsilon_2)\dot{E} \rangle, & E = \pm 2,\pm 3 \end{cases}.$$ (2)

Among them, $\varepsilon_1, \varepsilon_2 \in [0\ 1]$.

3.2   The Amendment of Fuzzy Control Rules Based on PSO and the Autonomous Suspension Control Method for Underwater Unmanned Vehicle

The preferred amendment factor will help to improve the performance of the actual control system. But only to determine with the experimental testing or experiences must bring a certain blindness, or even difficult to adjust to a set of best parameters. Therefore, the rapid optimization for amendment factor is needed, with the precise adjustment of control rules in time. Here an integral performance index can be used as the system performance evaluation.

$$J(ITAE) = \int_0^\infty t|e(t)|dt = \min.$$ (3)

Wherein, $J(\cdot)$ represents the size of the integral area after the error function weighting. I is integrator, T is time, A means the absolute value, E means that error. The integral performance indicator of $ITAE$ can evaluate the dynamic and static performance of control systems, such as setting evaluations are fast response, regulating in short time, small overshoot and the steady-state small error and so on.

The $ITAE$ is regarded as the objective function, and the optimization process is based on the principle of objective function value decreases gradually. After continuously adjusting the value of the parameters and obtaining the optimized amendment factor, the ideal control rules should come into being.

Because of the complex features for movement of UUV, the search of amendment factor is a typical nonlinear optimization process. With combining the motion control process simulation and particle swarm optimization algorithm, the amendment of fuzzy control rules is designed to carry out the adaptive optimization of fuzzy controller. Therefore, the correction process of fuzzy rules based on particle swarm optimization shown in Figure 1, the specific algorithm steps are as follows:

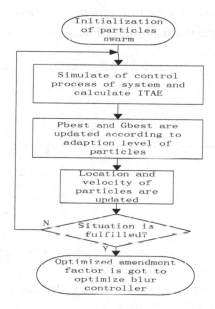

Fig. 1. Flow of rule adjustment with fuzzy control based on PSO

Because of the fuzzy control rules can be adaptively adjusted to optimize, so the autonomous suspending control new method for UUV based on the amendment of fuzzy control rules is put forward in the basis of the traditional fuzzy control algorithm, which shown in Figure 2.

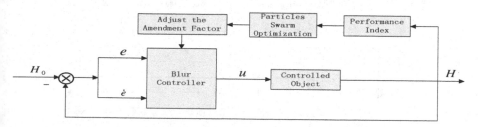

Fig. 2. Principle of adaptation control based on amendment of fuzzy control rules

4   Simulation Analysis

Assumptions: for UUV quality 1520kg, length of 7m, diameter 0.6m, cylindrical shape for the rules, the pumping rate of the charge pump 10 Bull / sec, an initial speed of 13 m/ sec with underwater interference. The suspended motion in the depth of 2300m and control process are simulated by using the designed fuzzy controller. The simulation results show in Figure 3 and Figure 4. In the fuzzy controller, the inertia weight PSO is set to 0.9, and $c_1 = c_2 = 2$. After several iterations, the optimized factor ε is 0.3872, and the control system integrator performance indicator $J(ITAE)$ has reached the minimum at this time.

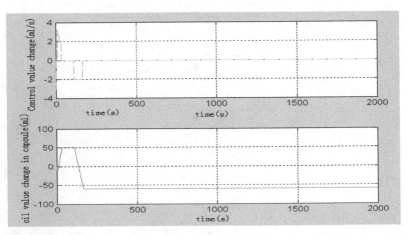

Fig. 3. Control value of suspending movement for UUV based on the amendment of fuzzy control rules(when initial velocity is 13 m/s)

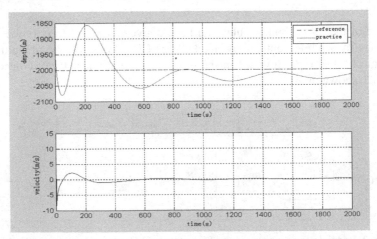

Fig. 4. Depth and velocity value of suspending movement for UUV based on the amendment of fuzzy control rules(when initial velocity is 13 m/s)

By analyzing Figure 3 and Figure 4, it can be see that the autonomous suspending control processes time for UUV based on the amendment of fuzzy control rules is shorter, and the system control variable overshoot and steady-state error could well meet the design requirements.

5   Conclusion

Good autonomous suspension control for UUVs is essential to ensure the normal underwater operation. The amendment algorithm of fuzzy control rules based on modified particle swarm optimization is designed, and then the autonomous suspending control method for UUV based on the amendment of fuzzy control rules is proposed in this paper.The performance of the controller has got a significant improvement. A large number of simulation experiments show that this suspension control method for UUVs is simple, fast convergent, and has broad application prospects in the actual works.

References

1. Qi, J., Zhi-Jie, D.: PID Control Principle and Techniques of Parameter Tuning. Journal of Chongqing Institute of Technology (Natural Science) 22(5), 91–94 (2008)
2. Jin, X., Chen, Z.-B., Ni, X.-X., Lu, Z.-K.: Research of HDRI system based on PID controller. Optical Technique 34(4), 547–551 (2008)
3. Yin, Y.-H., Fan, S.-K., Chen, M.-E.: The Design and Simulation of Adaptive Fuzzy PID Contronller. Fire Control and Command Control 33(7), 96–99 (2008)
4. Hou, Y.-B., Yang, X.-C.: The Methods of Fuzzy Contronller Design. Journal of Xi An University of Science and Technology 23(4), 448–450 (2003)
5. Liu, Y.-T.: The Study of Fuzzy PID Controller Design and Simulation. Journal of Gansu Lianhe University (Natural Sciences) 22(3), 75–78 (2008)
6. Silvestre, C., Pascoal, A.: Depth Control of The INFANTE AUV Using Gain-scheduled Reduced Order Output Feedback. Control Engineering Practice, 1–13 (2006)
7. Richard, D., Dberg, B.: The Development of Autonomous Underwater Vehicles:A Brief Summary,C. In: IEEE International Conference on Robotics and Automation, Korea (2001)
8. Lee, C.C.: Fuzzy Logic in Control System:Fuzzy Logic. IEEE Trans Syst. 20(2), 232–238 (1990)
9. Qin, S.-Y., Chen, F., Zhang, Y.-F.: A Fuzzy Adaptive PID Controller for Longitudinal Attitude Control of A Small UAV. CAAI Transactions on Intelligent Systems 3(2), 121–128 (2008)
10. Zhang, J.-W., Wille, F., Knoll, A.: Fuzzy Logic Rules for Mapping Sensor Data to Robot Control. In: IEEE Proceeding of EUROBOT 1996, pp. 29–38 (1996)
11. Ghoshal, S.P.: Optimizations of PID Gains by Particle Swarm Optimizations in Fuzzy Based Automatic Generation Control. Electric Power Research 72(3), 203–212 (2004)
12. Shi, Y., Eberhart, R.C.: Fuzzy Adaptive Particle Swarm Optimization. In: Proceedings of the IEEE Conference on Evolutionary Computation, pp. 101–106 (2001)

How an Adaptive Learning Rate Benefits Neuro-Fuzzy Reinforcement Learning Systems

Takashi Kuremoto[1,*], Masanao Obayashi[1], Kunikazu Kobayashi[2], and Shingo Mabu1

[1] Graduate School of Science and Engineering, Yamaguchi University
Tokiwadai 2-16-1, Ube, Yamaguchi, 755-8611, Japan
{wu,m.obayas,mabu}@yamaguchi-u.ac.jp
[2] Schoo of Information Science and Technology, Aichi Prefectural University
Ibaragabasama 1522-3, Nagakute-Shi, Aichi, 480-1198, Japan
{wu,m.obayas,mabu}@yamaguchi-u.ac.jp

Abstract. To acquire adaptive behaviors of multiple agents in the unknown environment, several neuro-fuzzy reinforcement learning systems (NFRLSs) have been proposed Kuremoto et al. Meanwhile, to manage the balance between exploration and exploitation in fuzzy reinforcement learning (FRL), an adaptive learning rate (ALR), which adjusting learning rate by considering "fuzzy visit value" of the current state, was proposed by Derhami et al. recently. In this paper, we intend to show how the ALR accelerates some NFRLSs which are reinforcement learning systems with a self-organizing fuzzy neural network (SOFNN) and different learning methods including actor-critic learning (ACL), and Sarsa learning (SL). Simulation results of goal-exploration problems showed the powerful effect of the ALR comparing with the conventional empirical fixed learning rates.

Keywords: Neuro-fuzzy system, swarm behavior, reinforcement learning (RL), multi-agent system (MAS), adaptive learning rate (ALR), goal-exploration problem.

1   Introduction

As an active unsupervised machine learning method, reinforcement learning (RL) has been developed and applied to many fields such as intelligent control and robotics since 1980s [1] – [3]. The learning process of RL is given by the trials of exploration and exploitation of a learner (agent) in unknown or non-deterministic environment. Valuable or adaptive actions which are optimized output of RL systems are obtained according to the modification of action selection policies using the rewards (or punishments) from the environment. Generally, there are four fundamental components in RL: state (observed information from the environment, input); policy (usually using probability selection to keep exploitation to an unknown environment); action (the output of the learner changing the current state) and reward (perceived/obtained by the learner during the transition of the state).

* A part of this work was supported by Grant-in-Aid for Scientific Research (JSPS No. 23500181, No. 25330287, No. 26330254).

Y. Tan et al. (Eds.): ICSI 2014, Part I, LNCS 8794, pp. 324–331, 2014.

In the history of RL research, there are some severe problems need to be solved theoretically: (1) The explosion of state-action space in high dimension problems (the curse of dimensionality); (2) The balance between exploration and exploitation; (3) Learning convergence in partially observable Markov decision process (POMDP).

To tackle the first issue, linear approximation method [2], normalized Gaussian radial basis function classification method [3], fuzzy inference systems [4] – [8], etc. are proposed. Specially, neuro-fuzzy systems with a data-driven type self-organizing fuzzy neural network (SOFNN) proposed in our previous works showed their adaptive state identification ability for different RL algorithms such as actor-critic learning [5] [6], Q-learning, and Sarsa learning [7]. Meanwhile, Derhami et al. proposed to use adaptive learning rate (ALR) and adaptive parameter of state transition function to obtain a suitable balance of exploration and exploitation [8]. Effectiveness of the ALR has been confirmed by its application to a fuzzy controller with Q learning algorithm with simulation results of some benchmark problems such as boat problem and mountain-car problem in [8]. Furthermore, the third problem of RL mentioned above is more serious to a multi-agent system (MAS). Even the exploration environment is stable, the existence of other agents nearby the learner agent is uncertain. In [4] – [8], Kuremoto et al. proposed to calculate the reward of suitable distance between agents to modify the action policy and showed its higher learning convergence comparing with RLs by individuals independently.

In this paper, we adopt Derhami et al.'s ALR to Kuremoto et al.'s neuro-fuzzy reinforcement learning systems to improve the learning performance of agents in MASs. Goal-exploration problems were used in simulation experiments and the comparison between results with conventional empirical fixed learning rate and ALR is reported.

2   Neuro-Fuzzy Reinforcement Learning Systems

Markov decision process (MDP) is used in reinforcement learning (RL) algorithms. Let a state space $\mathbf{X}(x_1, x_2,..., x_n) \in R^n$, an action space $\mathbf{A}(a_1, a_2,..., a_m) \in R^m$, a state transition (from the current stat s_t to the next state s_{t+1}) policy π with probability $P_{s_t s_{t+1}, \pi}^{a_t} : \mathbf{X} \times \mathbf{A} \times \mathbf{X} \in [0, 1]$, and a reward function $R = \sum_{t=0}^{\infty} r_t \in R^1$, where is the reward $r_t \in R^1$ obtained during the state transition. The RL algorithm is to find an optimal policy π^* that can yield the maximum discounted expected reward R^*, i.e. the maximum value function of state $V(\mathbf{X}) \equiv \underset{P(s_{t+1}|s_t, a_t)}{E} \{R\} = \underset{P(s_{t+1}|s_t, a_t)}{E} \{\sum_{t=0}^{\infty} \gamma^t r_t\}$, or the maximum value function of state-action $Q(\mathbf{X}, \mathbf{A}) \equiv E\{V(\mathbf{X}) \mid \pi\}$.

To avoid a mass calculation to whole MDP state transition, and deal with unknown transition probabilities, there is an efficient RL algorithm named temporal difference learning (TD-learning) [2] to yield the maximum value functions. A TD error $\varepsilon \in R^1$ is defined by Eq. (1), and it is used to update state value function in the learning process, for example, Eq. (2).

$$\varepsilon \equiv r_{t+1} + \gamma V^\pi(\mathbf{x}_{t+1}) - V^\pi(\mathbf{x}_t) \tag{1}$$

$$V^\pi(\mathbf{x}_t) \leftarrow V^\pi(\mathbf{x}_t) + \alpha\varepsilon \tag{2}$$

where $0 \leq \gamma \leq 1$ and $0 \leq \alpha \leq 1$ are a damping rate and a learning rate, respectively.

2.1   An Actor-Critic Type Neuro-Fuzzy Reinforcement Learning System

To deal with continuous state space, an actor-critic type neuro-fuzzy reinforcement learning system is proposed by Kuremoto et al. [5] [6]. The system is able to be used as an internal model of an autonomous agent which output a series of adaptive actions in the exploration of the unknown environment. The information processing flow is as follows:

Step 1 Agent observes states x_t from the environment;

Step 2 Agent classifies the inputs into k classes $\phi_k(x_t)$ by Fuzzy net;

Step 3 Agent outputs an action according to the value function $A_m(x_t)$ (i.e. Actor);

Step 4 Agent receives rewards from the environment after the action is executed;

Step 5 TD-error, concerning with the state value function $V(x_t)$ in Critic, is calculated;

Step 6 Modify the action value function using TD-error (Eq. (1)).

Step 7 Return to Step 1if the state is not the terminal state, else end the current trail.

For an n-dimension input state space $\mathbf{x}(x_1(t), x_2(t),..., x_n(t))$, a fuzzy inference net is designed with a hidden layer composed by units of fuzzy membership functions $B_i^k(x_i(t)) = \exp\left\{-\dfrac{\left(x_i(t) - c_i^k\right)^2}{2\sigma_i^{k2}}\right\}$ to classify input states, where c_i^k, σ_i^k denotes the mean and the deviation of ith membership function which corresponding to ith dimension of input $x_i(t)$, respectively.

Let $K(t)$ be the largest number of fuzzy rules, we have Eq. (3) :

$$if\,(\,x_1(t)\,is\,B_1^k(x_1(t))\,,\,....\,,\,x_n(t)\,is\,B_n^k(x_n(t))\,)\,then\,\,\phi_k(\mathbf{x}(t)) = \prod_{i=1}^{n} B_i^k(x_i(t)) \tag{3}$$

where $\phi_k(\mathbf{x}(t))$ means the fitness of the rule R^k for an input set $\mathbf{x}(t)$.

To determine the number of membership functions and rules of fuzzy net, a self-organized fuzzy neural network (SOFNN) which is constructed adaptive membership functions and rules driven by training data and thresholds automatically [5] [6].

The weighted outputs of fuzzy net are used to calculate the value of states (Critic) and actions (Actor) according to Eq. (4) and Eq. (5), respectively.

$$V(\mathbf{x(t)}) = \frac{\sum_k v_k(t)\phi_k(\mathbf{x}(t))}{\sum_k \phi_k(\mathbf{x}(t))} \tag{4}$$

$$A_j(\mathbf{x(t)}) = \frac{\sum_k w_{kj}(t)\phi_k(\mathbf{x}(t))}{\sum_k w_{kj}(\mathbf{x}(t))} \tag{5}$$

Here v_k, w_{kj} are the weighted connections between fuzzy net rules $\phi_k(\mathbf{x}(t))$ and critic function $V(\mathbf{x}(t)) \in R^1$, actor function $A_j(\mathbf{x}(t)) \in R^m$. A_j denotes the jth action selected by agent according to a stochastic policy Eq. (9), where $j = 1, 2, \ldots, m$.

$$P\!\left(a_t = a_j \mid \mathbf{x}(t)\right) = \frac{\exp\!\left(A_j(\mathbf{x}(t))/T\right)}{\sum\limits_m \exp\!\left(A_m(\mathbf{x}(t))/T\right)} \tag{6}$$

Here $T > 0$ is a constant named temperature of Boltzmann distribution.

TD learning rule given by Eqs. (1) and (2) becomes to Eq. (7) and Eq. (8), where β_v, β_w denote learning rates for connections of Critic and Actor.

$$v_k(t+1) = v_k(t) + \beta_v \varepsilon_{TD}(\mathbf{x}(t))\phi_k(\mathbf{x}(t)) \tag{7}$$

$$w_{kj}(t+1) = w_{kj}(t) + \begin{cases} \beta_w \varepsilon_{TD}(\mathbf{x}(t))\phi_k(\mathbf{x}(t)) & a_t = a_j \\ 0 & otherwise \end{cases} \tag{8}$$

When multiple agents explore an unknown environment at the same time, cooperative exploration, i.e., swarm learning may provide more efficient performance comparing with individual learning. Then we give a positive reward r_{swarm} to the agent when Eq. (9) and Eq. (10) is satisfied, or a negative reward r_{swarm} in the opposite.

$$min\_\,dis \le D(\mathbf{x}^p(t), \mathbf{x}^q(t)) \le max\_\,dis \tag{9}$$

$$r_{t+1} = r_t \pm r_{swarm} \tag{10}$$

Here $min\_\,dis, max\_\,dis$, denotes near limit distance, far limit distance between agents in the Euclidean space, respectively.

2.2   Adoption of Adaptive Learning Rate

Learning rates in Eq. (7) and Eq. (8) are decided empirically as fixed values conventionally. However, they need to be reduced to improve the learning convergence during the iterations of exploration process. It is indicated by Derhami et al. that the visited states should use smaller learning rates, meanwhile, less visited states with larger learning rates to modify the exploration policy [8]. So they proposed an adoptive learning rate (ALR) to balance the exploitation/exploration as follows.

Let Φ_t^k be the accumulation firing strength of rule $\phi^k(\mathbf{x}(t))$, the visit value to the state $\mathbf{x}(t)$ is normalized by Eq. (11).

$$FV\!\left(\phi^k(\mathbf{x}(t))\right) = \frac{\sum\limits_{k=1}^{K_t} \phi^k(\mathbf{x}(t))\Phi_t^k}{\sum\limits_{k=1}^{K_t} \Phi_t^k} \tag{11}$$

where $\Phi_t^k = \Phi_{t-1}^k + \phi^k(\mathbf{x}(t))$, $\Phi_{t=0}^k = 0$, $0 \le FV(\mathbf{x}(t)) \le 1$.

The ALR then is given by Eq. (12).

$$\alpha_t^{ALR}(\phi^k(\mathbf{x}(t))) = \min(\frac{\alpha_{\min}K_t}{FV(\mathbf{x}(t))}, \alpha_{\max}) \tag{12}$$

where K_t is the number of fuzzy rules, $\alpha_{\min}, \alpha_{\max}$ are the boundaries of ALR.

2.3   A Sarsa Learning Type Neuro-Fuzzy Reinforcement Learning System

When a RL agent is in a partially observable Markov decision Process (POMDP), there may be different adaptive actions need to be output though the observed states are the same. Actor-critic type neuro-fuzzy reinforcement learning system was hard to find the optimal solution in POMDP [7]. So a sarsa learning (SL) type neuro-fuzzy reinforcement learning system is proposed recently [7].

The mathematical description of Fuzzy net is as same as which in actor-critic learning type system. Two steps states are observed in sarsa learning algorithm [2], and these state-action value functions are given by following:

$$Q(\phi(\mathbf{x}(t)), a_j, \mathbf{w}_j^{Q_t}) = \frac{\sum_k w_{kj}^{Q_t} \phi^k(\mathbf{x}(t))}{\sum_k \phi^k(\mathbf{x}(t))} \tag{13}$$

$$Q(\phi(\mathbf{x}(t+1)), a_j, \mathbf{w}_j^{Q_{t+1}}) = \frac{\sum_k w_{kj}^{Q_{t+1}} \phi^k(\mathbf{x}(t+1))}{\sum_k \phi^k(\mathbf{x}(t+1))} \tag{14}$$

where $k = 1,2,...K^t$ is the number of fuzzy rules, $j = 1,2,...m$ is action number. The learning rule of QL becomes to:

$$w_{kj}^{Q_t} \leftarrow w_{kj}^{Q_t} + \begin{cases} \alpha_t^{ALR} \varepsilon^{sarsa} \phi_k(\mathbf{x}(t)) / \sum_{k=1}^{K} \phi_k(\mathbf{x}(t)) & a_t = a_j \\ 0 & otherwise \end{cases} \tag{15}$$

$$w_{kj}^{Q_{t+1}} \leftarrow w_{kj}^{Q_t+1} + \begin{cases} \alpha_{t+1}^{ALR} \varepsilon^{sarsa} \phi_k(\mathbf{x}(t+1)) / \sum_{k=1}^{K} \phi_k(\mathbf{x}(t+1)) & a_{t+1} = a_j \\ 0 & otherwise \end{cases} \tag{16}$$

where $\alpha_t^{ALR}, \alpha_{t+1}^{ALR}$ are adaptive learning rates described by Eq. (12), TD error is given by Eq. (17).

$$\varepsilon_{TD}^{Q_{t+1}} = r_t + \gamma Q(\phi(\mathbf{x}(t+1)), a_{t+1}, \mathbf{w}_{t+1}^{Q_{t+1}}) - Q(\phi(\mathbf{x}(t)), a_t, \mathbf{w}_t^{Q_t}) \tag{17}$$

3   Simulation Experiments

To confirm how ARL can benefit the learning performance of neuro-fuzzy type reinforcement learning systems, a goal-exploration problem was used to in the simulation experiments (Fig. 1). The problem is assumed that agents explore a goal area in an unknown 2-D square surrounded by walls. All candidate actions may be chosen before the learning processes, and RL algorithms aforementioned find adaptive actions of agent to explore the goal or exploit its experience to move to the goal. Adaptive actions mean that agents choose to avoid obstacles, other agents and wall, and find the shortest path from their start positions to the goal area.

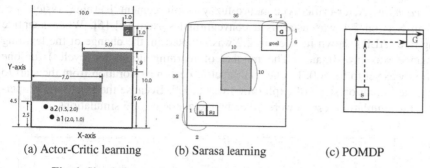

(a) Actor-Critic learning   (b) Sarasa learning   (c) POMDP

Fig. 1. Simulation environments for goal-directed exploration experiment

Agents observed the state of environment by the information of its position (x, y) in Simulation I, and their vicinities in 4 directions: up, down, left, right in Simulation II. In the last case, if the one step ahead is a path that an agent can arrive at it by one time motion, then the value of the direction is 0, oppositely, if there is a wall, or other agent (s), or an obstacle, then the input is 1. So the number of states in the environments of Simulation I is $x \times y$ in a grid scale, and more in the continuous space. In the case of Simulation III, the number of states is 4 bits, 16 states (i.e., 0010, 0011, ...).

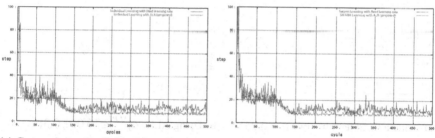

(a) Comparison of individual learning results   (b) Comparison of swarm learning results

Fig. 2. Learning curves of different learning methods by the actor-critic type system

3.1   Simulation I: A Maze-like Environment

An exploration environment is shown in Fig. 1 (a). There were walls on the four sides, obstacles in the exploration area, and a goal area was fixed. Two agents a1 and a2 start from (2.0, 1.0) and (1.5, 2.0) to find the goal area between (9.0, 9.0) to (10.0, 10.0). Each agent observed its position and position of others in the square, and moved 1.0 length per step toward an arbitrary direction. The method to decide the direction of one time movement was given by [5]. Agents did not have any informa-tion of the goal position before they arrived at it. When the distance rewards defined by Eqs. (9) & (10) were not considered, 2 agents learned to search the goal as fast as they can individually (*individual learning*); meanwhile, when agents used the swarm rewards in the same environment to modify their policies, the learning was called *swarm learning*. Actor-critic type neuro-fuzzy reinforcement learning system described in Section 2.1 was used as a conventional system [5] [6]. When adaptive learning rate (ALR) shown in Section 2.2 was adopted in, the change of the learning performance was investigated. The number of learning iterations (cycles) for one simulation was set to be 500. In one trial (cycle), i.e., an exploration from the start to the goal, the limitation steps of exploration was 5,000. Because the stochastic property of RL, the simulation results were given by the averages of 5 simulations.

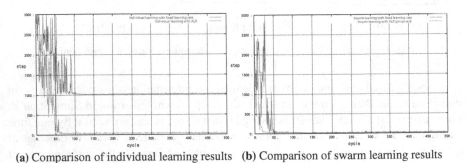

(a) Comparison of individual learning results   (b) Comparison of swarm learning results

Fig. 3. Learning curves of different learning methods by the sarsa learning system

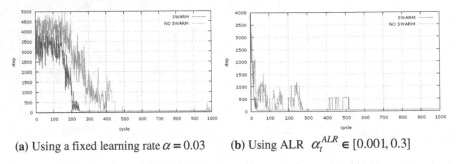

(a) Using a fixed learning rate $\alpha = 0.03$ (b) Using ALR $\alpha_t^{ALR} \in [0.001, 0.3]$

Fig. 4. Learning curves of different learning methods by the sarsa learning system

When we confirmed the learning curves of these different learning methods, not only swarm learning with ALR but also individual learning with ALR showed their prior learning performance comparing with the conventional fixed learning rate. The average results are shown in Fig. 2. In detail, the average lengths of the path explored by agents were 1182.01 using fixed learning rate and individual learning, 41.62 using ALR and individual learning, 26.61 using fixed learning rate and swarm learning, 22.64 using ALR and swarm learning calculated by the last 100 cycles data (from cycle 400 to cycle 500).

3.2   Simulation II: A POMDP Environment

An exploration environment with an obstacle as shown in Fig. 1 (b) was used in Simulation II. The sarsa learning type neuro-fuzzy reinforcement learning system was adopted as the action learning method of autonomous agents. Two agents were input in 4 dimensions, i.e., up, down, left, and right directions with values 0 and 1. The actions were also in 4 directions 1 grid/step ($a_j, j = 1,2,3,4$). The optimal path of the agents should be a transition of a same state (0, 0, 0, 0) even from the start position (2, 2) and (3, 2) to the goal area from (30, 30) to (36, 36) avoiding to crash the 10x10 obstacle in the center of the exploration square. So it was an environment under partially observable Markov decision process (POMDP) and the optimal solution was difficult to be found, alternatively, quasi-optimal solution was available as shown in Fig. 1 (c).

The dramatic improvement for the learning performance by ALR was confirmed with the learning curves of sarsa learning type neuro-fuzzy reinforcement learning system dealing with the POMDP problem as shown in Fig. 4. Swarm learning effects stand out clearly and ALR accelerated learning convergence efficiently.

4   Conclusions

Adaptive learning rate (ALR) was adopted into neuro-fuzzy reinforcement learning systems in this paper. The concept was founded on the balance management of exploration and exploitation of reinforcement learning process. Higher learning rate serves larger modification of value functions dealing with unexplored states. Simulations of goal-navigated exploration problem showed the effectiveness of the proposed method. And adequate distance between agents also accelerated the exploration process. It is expected to apply these effective RL learning systems to autonomous agent in web intelligence or robots in real environment in the future.

References

1. Kaelbling, L.P., Littman, M.L.: Reinforcement Learning: A Survey. J. Artificial Intelligence Research 4, 237–285 (1996)
2. Sutton, R.S., Barto, A.G.: Reinforcement Learning: An Introduction. The MIT Press, Cambridge (1998)
3. Samejima, K., Omori, T.: Adaptive Internal State Space Construction Method for Reinforcement Learning of a Real-World Agent. Neural Networks 12, 1143–1155 (1999)
4. Wang, X.S., Cheng, Y.H., Yi, J.Q.: A fuzzy Actor–Critic Reinforcement Learning Network. Information Sciences 177, 3764–3781 (2007)
5. Kuremoto, T., Obayashi, M., Kobayashi, K.: Adaptive Swarm Behavior Acquisition by a Neuro-Fuzzy System and Reinforcement Learning Algorithm. Intern. J. of Intelligent Computing and Cybernetics 2(4), 724–744 (2009)
6. Kuremoto, T., Yamano, Y., Obayashi, M., Kobayashi, K.: An Improved Internal Model for Swarm Formation and Adaptive Swarm Behavior Acquisition. J. of Circuits, Systems, and Computers 18(8), 1517–1531 (2009)
7. Kuremoto, T., Yamano, Y., Feng, L.-B., Kobayashi, K., Obayashi, M.: A Fuzzy Neural Network with Reinforcement Learning Algorithm for Swarm Learning. In: Zhang, Y. (ed.) Future Computing, Communication, Control and Management. LNEE, vol. 144, pp. 101–108. Springer, Heidelberg (2011)
8. Derhami, V., Majd, V.J., Ahmadabadi, M.N.: Exploration and Exploitation Balance Management in Fuzzy Reinforcement Learning. Fuzzy Sets and Systems 161(4), 578–595 (2010)

Comparison of Applying Centroidal Voronoi Tessellations and Levenberg-Marquardt on Hybrid SP-QPSO Algorithm for High Dimensional Problems

Ghazaleh Taherzadeh[1] and Chu Kiong Loo[2]

[1] School of Information and Communication Technology
Faculty of Science, Environment, Engineering and Technology
Griffith University, Gold Coast campus
Queensland 4111, Australia
[2] Faculty of Computer Science and Information Technology
University of Malaya, Kuala Lumpur, Malaysia
Ghazaleh.taherzadeh@griffithuni.edu.au, lck@fsktm.um.edu.my

Abstract. In this study, different methods entitled Centroidal Voronoi Tessellations and Levenberg-Marquardt applied on SP-QPSO separately to enhance its performance and discovering the optimum point and maximum/ minimum value among the feasible space. Although the results of standard SP-QPSO shows its ability to achieve the best results in each tested problem in local search as well as global search, these two mentioned techniques are applied to compare the performance of managing initialization part versus convergence of agents through the searching procedure respectively. Moreover, because SP-QPSO is tested on low dimensional problems in addition to high dimensional problems SP-QPSO combined with CVT as well as LM, separately, are also tested with the same problems. To confirm the performance of these three algorithms, twelve benchmark functions are engaged to carry out the experiments in 2, 10, 50, 100 and 200 dimensions. Results are explained and compared to indicate the importance of our study.

Keywords: Centroidal Voronoi Tessellations (CVT), Shuffled Complex Evolution with PCA (SP-UCI), Quantum Particle Optimization (QPSO), Levenberg-Marquardt (LM), High Dimensional Benchmark Functions.

1 Introduction

Since many years ago, numerous evolutionary algorithms stimulated the nature and behavior of creatures are introduced by researches and scientists to address the problem of optimization in different areas. Some of the famous and most applicable evolutionary algorithms are Genetic Algorithm (GA) [1], Particle swarm Optimization (PSO) [2], Ant Colony Optimization (ACO) [3] and Shuffled Complex Evolution (SCE) [4].

However, there is no special way to choose the best evolutionary algorithm suited to exploit for each different problem. Therefore, there is no guarantee to reach to the

Y. Tan et al. (Eds.): ICSI 2014, Part I, LNCS 8794, pp. 332–341, 2014.
© Springer International Publishing Switzerland 2014

optimum point using evolutionary algorithms and optimization as they all have their corresponding advantages and disadvantages.

To tackle this problem, hybrid algorithms are proposed to assist and improve the performance of optimization process to discover the best point in the search space. These hybrid evolutionary algorithms take benefit from the advantages of original evolutionary algorithms and fill the gaps in the existing evolutionary algorithms.

Moreover, the other elements which affect the performance of this type of algorithm is the random initialization of points through the search space at the very first stage of process. Accordingly, the most important drawback of random distribution of points is leading to local minima in the beginning of search process. Thus one of the way to effect the result of optimization is to manage the initialization of points all over the search space.

In addition, to ensure the process of optimization and accomplishing the task, convergence of agents and looking for the best point through the search space considered as one of the vital parts in this procedure. To ensure the movement of agents in a right track, there is a need to monitor their position in every iteration and lead them to the optimum results.

In this study, the new proposed hybrid algorithm titled SP-QPSO by the author [5] which shows the promising results to perform in both low dimensional problems as well as high dimensional problem is combined by two previously stated techniques separately. Centroidal Voronoi Tessellations and Levenberg-Marquardt techniques are performed in initialization part to confirm the distribution of points and ensuring the convergence of points in each iteration during optimization process, respectively. To examine the significance of our work, three stated techniques undertake the experience on twelve benchmark functions in 2, 10, 50, 100 and 200 dimensions. The achieved results are compared based on the best values (minimum values) found by the algorithms.

This paper is formed as follow: In section two, a brief review of hybrid SP-QPSO, CVT and LM algorithms can be found. In section three, the proposed approaches have been discussed and the application of CVT and LM on new SP-QPSO algorithm explained in details. Experimental works and achieved results are presented, discussed and analyzed in section four. Lastly, in the final section, conclusion and future works are covered.

2   Research Background

In this section an overview of Hybrid SP-QPSO, Centroidal Voronoi Tessellations (CVT) and Levenberg-Marquardt (LM) algorithms are included and depicted.

2.1   Hybrid SP-QPSO Algorithm

Based on previous researches on proposing a new hybrid algorithm, hybridization can be proposed in variety of ways [6]. Therefore, using high level hybridization and taking the advantages of other algorithms, hybrid SP-QPSO algorithm is offered [5].

Quantum Particle Swarm Optimization (QPSO) [7] and Complex Shuffled Evolution with PCA (SP-UCI) [8] are the two evolutionary algorithms which formed the SP-QPSO. In these two algorithms, agents are called particles and points respectively. Furthermore, the local search is performed by QPSO and the global search is implemented by SP-UCI strategy. According to previous researches [7], QPSO has the promising results in low dimensional problems and SP-UCI performed well in high dimensional problems and avoid population degeneration to accomplish the optimization task. The SP-QPSO algorithm takes these privileges and performs better than QPSO and SP-UCI [5] in both low dimensional problems as well as high dimensional.

In this algorithm, agents are known as point in the beginning. Points are distributed randomly through the search space and the first results are saved and sorted from the smallest value to the largest in an array $D = \{x_i, f_i, i = 1, ..., s\}$. After sorting the values in array D, it is divided into number of complexes of $A_1, ..., A_p$; in each subdivision m points are placed and called simplex. In the next stage, the optimization procedure is applied on each simplex.

$$A^k = \{x_i^k, f_i^k | x_i^k = x_{k+p(j-i)}, f_i^k = f_{k+p(j-i)}, j = 1, ..., m\} \tag{1}$$

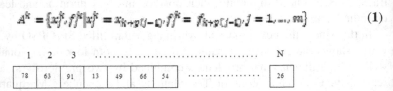

Fig. 1. Dividing D array into sub-arrays

After providing sub-arrays, QPSO strategy applies on each complex to look for the best result over the feasible search space. In QPSO algorithm, Wave function $W(x,t)$ is employed to manage the movements of particles to find out the remaining optimum result. In this strategy, all the particles follow the best point (points) found so far in all the past iterations (*pbest*) individually and also the best point found by all the particles which is known as global best (*gbest*). These movements are specified as follow:

$$p(x) = \begin{cases} x_i(x+1) = p + \beta \times |Mbest_i - x_i(t) \times \ln\left(\frac{1}{u}\right) & \text{when } x \geq 0.5 \\ x_i(x+1) = p - \beta \times |Mbest_i - x_i(t) \times \ln\left(\frac{1}{u}\right) & \text{when } x \leq 0.5 \end{cases} \tag{2}$$

where, β: contraction-expansion coefficient and allocated a value in a range of [0-1], u and k: used in uniform probability distribution. The *Mbest* which is known as mean best is defined as:

$$Mbest = \frac{1}{N} \sum_{d=1}^{N} P_{gd}(t). \tag{3}$$

where, g: index of best particle among all the particles in the population.

One of the advantages in this algorithm is to check and monitor dimensionality of population in each iteration and avoid population degeneration which is accomplished by SP-UCI strategy. Correspondingly to ensure the performance of global search, QPSO is applied to find out the optimum result by employing each particle.

2.2   Centroidal Voronoi Tessellations (CVT)

As distribution of agents over the search space plays vital role and can affect the result of algorithm, variety of distribution techniques are proposed to initialize the agents and start the optimization task [9-10]. One of these strategies which can be applicable in many fields of science and engineering is Centroidal Voronoi Tessellation (CVT). CVT which is a type of Voronoi Tessellation known as one of the best techniques to partition the surface to the sub regions and scatter points to all the sub regions. However, the distribution of points is done in different manner such as: uniform density, non-uniform density, density with peak in middle, density with peak at a corner and etc. [11]. In this work, non- uniform density function is employed.

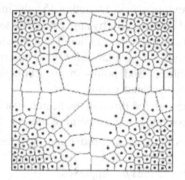

Fig. 2. CVT with non-uniform density function

The representation of non-uniform density is as follow:

Assume: $X = (x_i)_{i=1}^n$ (order set of n sites in where $\Omega \subset \mathbb{R}^N$)   (4)

$$\Omega_i = \{ x \in \Omega \mid \| x - x_i \| \leq \| x - x_j \|, \forall j \neq i \}.$$

where ‖.‖ symbolizes the Euclidean norm in $\mathbb{R}^N$ [12].

Nevertheless, poor distribution of points in the regions is considered as one of the drawbacks of this technique [13].

2.3   Levenberg-Marquardt (LM)

Convergence rate assumed as one the important factors in optimization process to ensure finding of optimum values. Several algorithms are introduced, like gradient descent which has a difficulty to find the convergence in the search space. Levenberg-Marquardt optimizer algorithm (LM) outperforms the other algorithms and shows encouraging performance.

The LM optimizer is classified in the group of efficient local optimizer, expressly in a special condition such that the algorithm found the closest point to global optimum of search space as the best result [14]. LM optimizer also performs similar to the other algorithms such as steepest decent method and Gauss-Newton method [15]. There are several forms of LM optimizer presented to apply on the different type of problems, such as Function Vector, Vector + Jacobian and Funcion + Garadient + Hessian [16].

The basic formula of LM optimizer is defined as follow:

$$f(p + \delta_p) \approx f(p) + J\delta_p,$$

(5)

Where, f known as minimization/ maximization function, p is a parameter vector and $p \in R^m$ as well as J is the Jacobian matrix $\frac{\partial f(p)}{\partial_p}$.

3   Methodology

In this study, based on previous works by author [5], the initialization of points in SP-QPSO algorithm modified individually by applying CVT technique.

Description of our work is explained in detail and the steps required to accomplish the experiments is discussed. The standard SP-QPSO algorithm is as follow:

1) Number of complexes and points in each complex, Number of population size are defined as p, m and $s=pm$.
2) Populations are initialized through the search space.
3) Calculating fitness value of each point in the search space
4) Sort all the values in increasing order and store them in the array $D = \{Xi, f_i, i=1,..., s\}$
5) Dividing the Array D into p complexes
6) QPSO strategy undertake each complex to find the best solution
   a. Initialize QPSO parameters like population size and number of iterations.
   b. Create a swarm consisting of x points from the Y_1^k, ..., Y_q^k in A based on their function value and store in $E^k = \{ Y_i^k, v_i^k, i = 1, ...,q\}$.
   c. Find the optimum value in the search space according to *pbest* and *gbest*.
   d. Compute the *Mbest* for the particle's movement and increase number of iterations in each round.
   e. This sub-routine stops when the required criteria are met.
7) To assist searching process over the rough fitness space Multinormal resampling is applied.
8) Complexes are shuffled and replace the A_1, ..., A_p into D and sort it.
9) This routine is repeated until the stopping criteria is met.

Though applying CVT and LM are illustrated in flowcharts in fig.3 and fig.4 respectively.

Applying CVT on initialization stage helps the algorithm to visit all the regions in the search space and avoiding decoy in local minima at the very beginning of optimization process. Referring to this point, after assigning number of complexes, number of points and population size, the CVT is performed and divided the surface into sub regions.

Besides, LM optimizer assist SP-QPSO algorithm to enhance the convergence rate and ensures all the particles move toward the best results and optimum points. As soon as the global best (gbest) is found by QPSO, LM start to do its function on the local search and improve the convergence of particles.

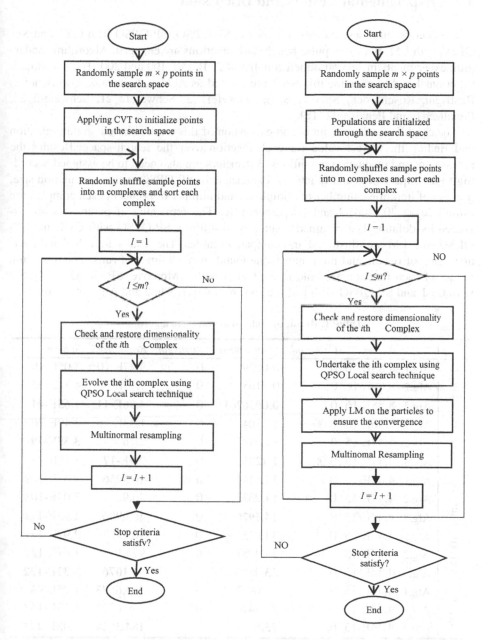

Fig. 3. SP-QPSO with CVT flowchart   **Fig. 4.** SP-QPSO with LM flowchart

By applying CVT and LM optimizer separately on SP-QPSO, its performance is examine and compare with the standard SP-QPSO.

4   Experimental Results and Discussion

In this section, to assess the performance of SP-QPSO, SP-QPSO with CVT and SP-QPSO with LM, twelve popular benchmark functions are engaged. Algorithms undertook experiments in several dimensionality of 2, 10, 50, 100 and 200. The benchmark functions hired to examine this work are entitled as Ackley, Griewank, Quarticnoise, Rastrigin, Rosenbrock, Sphere, Step, Schwefel1_2, Schwefel2_21, Schwefel2_22, Penalized1, and Penalized2 [18].

The conditions in assessment and evaluation of this work is based on minimization and finding the lowest value for each function over the search space. Besides the assessment and evaluation conditions, parameters are also need to be assigned according to the experiences and functions. Parameters of SP-QPSO such as population size, number of iterations, number of complexes and number of points in each complex are initialized as 20, 300, 2 and 20 respectively. The remaining of parameters are assigned by default. The obtained results by SP-QPSO, SP-QPSO with CVT and SP-QPSO with LM are illustrated and compare in tables. The results are calculated average value of results and minimum value found over 20 times of runs. The minimum values of each function are bold in the table.1 and 2. Moreover SP-QPSO, SP-QPSO with CVT and SP-QPSO with LM are shown as Alg.1, Alg.2 and Alg.3 respectively.

Table 1. Result of comparison on three algorithms

		Ackley	*Griewank*	*Quarticnoise*	*Rastrigin*	*Rosenbrock*	*Sphere*
2 Dim	Alg.1	8.88E-16	0	0.0029	0	3.27E-10	3.05E-40
	Alg.2	8.88E-16	0	0.0165	0	1.99E-09	4.81E-33
	Alg.3	8.88E-16	0	**0.0026261**	0	**4.96E-11**	**1.65E-41**
10 Dim	Alg.1	5.36E-10	0.0099	1.1304	0	1.33E-16	2.49E-76
	Alg.2	4.44E-15	**0**	1.2947	0	2.76	**3.35E-79**
	Alg.3	4.44E-15	0.0074	**1.1285**	0	**9.9E-17**	3.79E-78
50 Dim	Alg.1	4.44E-15	0	14.7149	0	46.3956	7.47E-109
	Alg.2	4.44E-15	0	14.2332	0	46.9	7.03E-108
	Alg.3	4.44E-15	0	**14.1926**	0	**46.0906**	**2.93E-111**
100 Dim	Alg.1	4.44E-15	0	34.035	0	96.5787	2.92E-121
	Alg.2	4.44E-15	0	34.3087	0	97	1.41E-121
	Alg.3	4.44E-15	0	**33.9868**	0	**96.1076**	**9.32E-122**
200 Dim	Alg.1	4.44E-15	0	76.0902	0	196.6573	1.95E-93
	Alg.2	4.44E-15	0	76.7474	0	196.6221	6.33E-137
	Alg.3	4.44E-15	0	**75.9438**	0	**184.3924**	**7.08E-143**

Table 2. Result of comparison on three algorithms

	Step	Schwefel	Schwefel1_2	Schwefel2_22	penalized1	penalized2
2 Dim Alg.1	0	2.23E-18	3.52E-23	1.37E-22	2.36E-31	**2.36E-31**
Alg.2	0	5.35E-21	9.16E-19	1.11E-24	2.36E-31	1.35E-32
Alg.3	0	**2.41E-29**	**6.31E-27**	**1.24E-36**	2.36E-31	1.35E-32
10 Dim Alg.1	0	1.52E-15	3.7E-19	1.38E-41	4.71E-32	1.35E-32
Alg.2	0	1.03E-16	1.54E-18	4.15E-43	4.71E-32	1.35E-32
Alg.3	0	**3.17E-19**	**4.25E-26**	**1.98E-48**	4.712E-32	1.35E-32
50 Dim Alg.1	344	7.75E-09	1.65E-06	3.78E-63	9.42E-33	1.35E-32
Alg.2	348	0.0000158	9.36E-07	5E-64	9.42E-33	1.35E-32
Alg.3	**343.0013**	**4.55E-09**	**6.868E-22**	**9.99E-69**	9.42E-33	1.35E-32
100 Dim Alg.1	**1810**	0.0041	1.57E-06	1.26E-78	4.71E-33	1.35E-32
Alg.2	8367.7	0.0033	0.000478	3.29E-76	4.71E-33	1.35E-32
Alg.3	1832.849	**0.00312**	**9.3E-07**	**9.58E-80**	4.71E-33	1.35E-32
200 Dim Alg.1	8367.7	0.042762	93.9405	1.357E-92	2.36E-33	19.4095
Alg.2	8367.7	0.027402	4.5538	1.465E-92	2.356E-33	19.4604
Alg.3	8367.7	**0.022728**	**4.54**	**4.595E-97**	2.356E-33	**19.3976**

Form the overall point of view, all three algorithms performs closely and obtain the results near the optimum points in all functions with different dimensionalities. However, there are some differences between the achieved results which are discussed in this section.

From the statistical point of view and analyze the results in detail, in 2 dimensional experiments on all the functions, all three algorithms perform equal in five functions and SP-QPSO with LM achieved the optimum value except penalized 2 function. In 10 Dimensional experiments on the functions, the completion to find the best value is between SP-QPSO with CVT and SP-QPSO with LM. SP-QPSO with CVT performs better than the other two algorithms in Griewank and Sphere. SP-QPSO with LM performs better than the rest in more functions such as Quarticnoise, Rosenbrock, Schwefel, Schwefel1_21 and Schwefel2_22. In the case of experiments on 50 dimensional functions, SP-QPSO with LM acts better than the other two algorithms and all the three algorithms performs equally on the remaining functions. Like the other cases in 100 dimensional functions, the SP-QPSO with LM reached to the optimum points in six function and all the three algorithms obtained the similar results except one. In 200 dimensional experiments, the SP-QPSO performs well in seven functions and the rest of functions attained the similar value by three algorithms.

According to the results shown in table.1 and table. 2, SP-QPSO with LM algorithm shows promising performance to obtain optimum point and minimum value in most of the cases and experiments. However, referring to the results these three algorithms act close to each other and in some cases the outcomes are equal to each other and there is no differences in some cases. We have also illustrated the convergence graphs to ensure the agents convergence which are not included in this paper.

5   Conclusion

As a conclusion, in this study a new hybrid algorithm entitled SP-QPSO which shows the promising results in both low dimensional problems as well as high dimensional problems is combined with CVT and LM separately to try and enhance its performance. The main idea of this study is to choose the CVT to manage the initialization part or choosing LM to ensure the convergence of agents during the optimization process. In overall, we aim to show the importance of initialization and convergence rate of particles in SP-QPSO algorithm individually. To evaluate these algorithms and show the significance of using CVT in initialization or LM for convergence rate of particles the results are compared. All the algorithms are tested on benchmarks in 2, 10, 50, 100 and 200 dimensions. Results illustrated that the initialization with CVT did not have so much effect on the obtaining the optimum results. Whereas the SP-QPSO with LM shows its ability to find the best fitness value over the search space not only in low dimensions but also in high dimensional functions as well.

6   Future Work

Since this algorithm applied on the benchmark functions and shows its performance in both low dimensional and high dimensional problems, the next work is to apply it on the real world problems and evaluate its performance to ensure its efficiency.

Acknowledgments. This study was funded by University of Malaya Research Grant (UMRG) project RG115-12ICT project title of Creative Learning for Emotional Expression of Robot Partners Using Interactive Particle Swarm Optimization.

References

1. Goldberg, D.E.: Genetic Algorithms in Search, Optimization and Machine Learning. Addison-Wesley Longman Publishing Co. (1989)
2. Kennedy, J., Eberhart, R.: Particle swarm optimization. In: Proceeding of International Conference on Neural Networks, pp. 1942–1948 (1995)
3. Dorigo, M., Caro, G.D.: The Ant Colony Optimization Meta-heuristic. New Ideas in Optimization, pp. 11–32. McGraw-Hill Ltd., UK Maidenhead (1999)
4. Duan, Q.Y., Gupta, V.K., Sorooshian, S.: Shuffled Complex Evolution Approach for Effective and Efficient Global Minimization. Journal of Optimization Theory and Applications 76(3), 501–521 (1993)
5. Taherzadeh, G., Loo, C.K.: A Novel Hybrid SP-QPSO Algorithm Using CVT for High Dimensional Problems. In press in The 2013 World Congress on Global Optimization (WCGO 2013), The Yellow Mountain, China (2013)

6. Talbi, E.G.: A Taxonomy of Hybrid Metaheuristics. Journal of Heuristics 8, 541–546 (2002)
7. Yang, S., Wang, M., Jiao, L.: A Quantum Particle Swarm Optimization. Congress on Evolutionary Computation 1, 320–324 (2004)
8. Chu, W., Gao, X., Sorooshian, S.: A Solution to the Crucial Problem of Population Degeneration in High-dimensional Evolutionary Optimization. IEEE Systems Journal, 5- 3, 362–373 (2011)
9. Du, Q., Gunzburger, M., Ju, L.: Meshfree, Probabilistic Determination of Point Sets and Support Regions for Meshless Computing. Comput. Methods Appl. Mech. Engrg. 191, 1349–1366 (2002)
10. Du, Q., Emelianenko, M., Ju, L.: Convergence of the Lloyd Algorithm for Computing Centroidal Voronoi Tessellations. SIAM Journal on Numerical Analysis 44(1), 102–119 (2006)
11. Du, Q., Gunzburger, M., Ju, L.: Constrained Centroidal Voronoi Tessellations for Surfaces. SIAM Journal on Numerical Analysis 24(5), 1488–1506 (2003)
12. Liu, Y., Wang, W., Levy, B., Sun, F., Yan, D., Lu, L., Yang, C.: On Centroidal Voronoi Tessellation — Energy Smoothness and Fast Computation. ACM Transactions on Graphics 28(4) (2009)
13. Richards, M., Ventura, D.: Choosing a Starting Configuration for Particle Swarm Optimization. In: IEEE International Joint Conference on Neural Networks, pp. 2309–2312 (2004)
14. Katare, S., Kalos, A., West, D.: A Hybrid Swarm Optimizer for Efficient Parameter Estimation. In: Congress on Evolutionary Computation (CEC 2004), vol. 1, pp. 309–315 (2004)
15. Levenberg, K.: A Method for the Solution of Certain Non-linear Problems in Least Squares. Journal of Quarterly of Applied Mathematics 2(2), 164–168 (1944)
16. Gill, P.E., Murray, W.: Algorithms for the Solution of the Nonlinear Least-Squares Problem. SIAM Journal on Numerical Analysis 15(5), 977–992 (1978)
17. Bigiarini, M.: Test Functions for Global Optimization (2012), http://www.rforge.net/doc/packages/hydroPSO/test_functions.html (Cited February 20, 2013)

A Hybrid Extreme Learning Machine Approach for Early Diagnosis of Parkinson's Disease

Yao-Wei Fu, Hui-Ling Chen[*], Su-Jie Chen, Li-Juan Li, Shan-Shan Huang, and Zhen-Nao Cai

College of Physics and Electronic Information
Wenzhou University
Wenzhou, Zhejiang
chenhuiling.jlu@gmail.com

Abstract. In this paper, we explore the potential of kernelized extreme learning machine (KELM) for efficient diagnosis of Parkinson's disease (PD). In the proposed method, the key parameters in KELM are investigated in detail. With the obtained optimal parameters, KELM manages to train the optimal predictive models for PD diagnosis. In order to further improve the performance of KELM models, feature selection techniques are implemented prior to the construction of the classification models. The effectiveness of the proposed method has been rigorously evaluated against the PD data set in terms of classification accuracy, sensitivity, specificity and the area under the ROC (receiver operating characteristic) curve (AUC).

Keywords: Extreme learning machine, Feature Selection, Parkinson's disease diagnosis.

1 Introduction

Parkinson's disease (PD) has become the second most common degenerative disorders of the central nervous system after Alzheimer's disease [1]. Till now, the cause of PD is still unknown, however, it is possible to alleviate symptoms significantly at the onset of the illness in the early stage [2]. Research has shown that approximately 90% of the patients with PD show vocal disorders [3]. The vocal impairment symptoms related with PD are known as dysphonia (inability to produce normal vocal sounds) and dysarthria (difficulty in pronouncing words) [4]. Therefore, it is clearly that the dysphonic indicators play an important role in early diagnosing PD.

Motivated by the pioneer work done in [5], many researchers made use of machine learning techniques to handle the PD diagnosis problem. In [6], AStröm et al. proposed a parallel feed-forward neural network structure for diagnosis of PD, the highest classification accuracy of 91.20% was obtained. In [7], Ozcift et al. combined

---

[*] Correspondence author.

Y. Tan et al. (Eds.): ICSI 2014, Part I, LNCS 8794, pp. 342–349, 2014.

the correlation based feature selection (CFS) algorithm with the RF ensemble classifiers of 30 machine learning algorithms to identify PD, and the best classification accuracy of 87.13% was achieved by the proposed CFS-RF model. In [8], Chen et al. employed the FKNN classifier in combination with the principle component analysis (PCA-FKNN) to diagnose PD, and the best classification accuracy of 96.07% was obtained by the proposed diagnosis system. In this study, an attempt was made to explore the potential of KELM in constructing an automatic diagnostic system for diagnosis of PD. Previous study [5, 8] on PD diagnosis has proven that using dimension reduction before conducting the classification task can improve the diagnosis accuracy. Here, an attempt is made to diagnose PD by using the KELM classifiers in combination with the feature selection methods.

The remainder of this paper is organized as follows. Section 2 offers brief background knowledge on KELM. In section 3 the detailed implementation of the proposed method is presented. Section 4 describes the experimental design. The experimental results and discussions of the proposed approach are presented in Section 5. Finally, Conclusions and recommendations for future work are summarized in Section 6.

2 KELM

This section gives a brief description of extreme learning machine (ELM) and KELM. For more details, one can refer to [9]. The learning steps of the ELM algorithm can be summarized as the following three steps:

Given a training set $\aleph = \{(x_i, t_i) \mid x_i \in R^n, t_i \in R^m, i = 1, 2, \dots, N\}$, an activation function $g(x)$, and the number of hidden neurons $\tilde{N}$,

(1) Randomly assign the input weights w_i and bias b_i, $i = 1, 2, \dots, \tilde{N}$.
(2) Calculate the hidden layer output matrix H.
(3) Calculate the output weight $\beta = H^\dagger T$, $T = [t_1, t_2, \dots, t_n]^T$.

It should be noted that when the feature mapping is unknown to users [10, 11], a kernel matrix for the ELM can be adopted according to the following equation:

$$\Omega_{ELM} = HH^T : \Omega_{ELMi,j} = h(x_i) \cdot h(x_j) = K(x_i, x_j) \tag{1}$$

where $h(x)$ plays the role of mapping the data from the input space to the hidden-layer feature space H. The orthogonal projection method is adopted to calculate the Moore-Penrose generalized inverse of matrix, namely, $H^\dagger = H^T (HH^T)^{-1}$, and a positive constant C is added to the diagonal of HH^T. Now we can write the output function of ELM as

$$f(x) = h\beta = h(x)\mathrm{H}^T (\frac{I}{C} + \mathrm{HH}^T)^{-1}T = \begin{bmatrix} K(x,x_1) \\ \vdots \\ K(x,x_N) \end{bmatrix}^T \left(\frac{I}{C} + \Omega_{ELM}\right)^{-1} T \qquad (2)$$

In this specific kernel implementation of ELM, namely KELM, we can specify the corresponding kernel for ELM model, the hidden layer feature mapping need not to be known to users. In this paper, the Gaussian radial basis function kernel is used, so $K(u,v) = \exp(-\gamma\|u-v\|^2)$. The two main parameters presented in KELM with Gaussian kernel are penalty parameter C and kernel parameter gamma. The parameter C determines the trade-off between the fitting error minimization and the norm of input weights minimization, while the parameter gamma defines the non-linear mapping from the input space to some high-dimensional feature space. Both of them play an important role in model construction.

3   Proposed Hybrid Method for PD Diagnosis

The main objective of the proposed hybrid method is to provide an efficient and accurate diagnosis tool for PD diagnosis. The flowchart of the proposed KELM based diagnosis method is shown in Figure 1. In the proposed methods, feature selection is firstly applied to identify the informative features in PD dataset, after then several feature subsets with top ranked features are fed to KELM model for performance evaluation. In Figure 1, we can see that one main issues of the KELM based method is the choice of the parameter pair. The hybrid method is comprehensively evaluated on the PD dataset in terms AUC, ACC, sensitivity and specificity. The pseudo-code of the proposed method is given bellow.

Pseudo-code for the proposed model
/*Performance estimation by using k-fold CV where $k = 10$*/
Begin
 For $j = 1:k$
   Training set ← k-1 subsets;
   Validation set ← remaining subset;
   Rank features using mRMR, IG, Relief and t-test;
   Train KELM classifier on the reduced training data feature space using different size of feature subset;
   Test the trained KELM models on the validation set;
 EndFor;
 Return the average classification accuracy rates of KELM over jth validation set;
End.

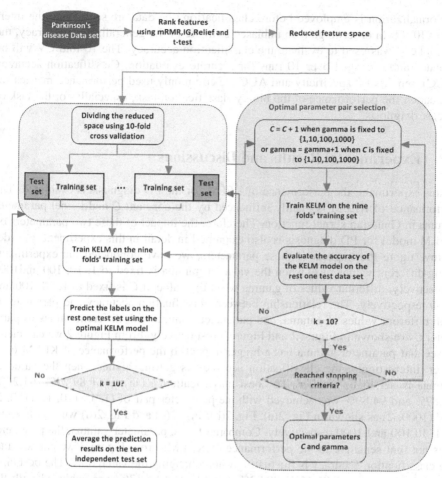

Fig. 1. Overall procedure of the proposed KELM based diagnosis system

4   Experiments Design

The experiment was conducted on the PD data set taken from UCI machine learning repository. The purpose of this data set is to discriminate healthy people from those with PD, given the results of various medical tests carried out on a patient. The whole experiment is conducted in the MATLAB platform, which runs on Windows 7 operating system with AMD Athlon 64 X2 Dual Core Processor 5000+ (2.6 GHz) and 4GB of RAM. mRMR program can be obtained from http://penglab.janelia.org/proj/mRMR/index.htm. The corresponding algorithms of IG and Relief from WEKA tool [12] are called by the main program which is implemented in MATLAB, and we implement the t-test from scratch. For ELM and KELM, the implementation by Huang available from http://www3.ntu.edu.sg/home/egbhuang is used.

Normalization is employed before classification, the data are scaled into the interval of [0, 1]. In order to gain an unbiased estimate of the generalization accuracy, the 10-fold CV was used to evaluate the classification accuracy. The 10-fold CV will be repeated and averaged over 10 runs for accurate evaluation. Classification accuracy (ACC), sensitivity, specificity and AUC are commonly used performance metrics for evaluation the performance of the binary classification task, especially for the task of disease diagnosis.

5   Experimental Results and Discussions

In this experiment, the performance of KELM for the PD diagnosis is examined. The performance of KELM is mainly influenced by the constant C and kernel parameter gamma in Gaussian kernel function. Therefore, the impact of these two parameters on KELM model for PD diagnosis is also examined in detail in this experiment. In order to investigate the impacts of these parameters, we have conducted the experiments using different values of C when the value of gamma is fixed to 1, 10, 100 and 1000 respectively, different values of gamma when the value of C is fixed to 1, 10, 100 and 1000 respectively. The relationship between classification accuracy and parameter C with different values of gamma, and parameter gamma with different values of parameter C are shown in Figure 2 and Figure 3 respectively. From Figure 2 we can clearly see that parameter gamma has a bigger impact to the performance of KELM classifier. Interestingly, the classification accuracy is getting higher when the value of gamma is set to be smaller. The best classification accuracy of 89.79%, 91.26%, 93.87% and 94.89% was achieved with the parameter pair of (1, 1), (10, 1), (100, 2) and (1000, 2) as shown in Fig. 2(a), Fig. 2(b), Fig. 2(c) and Fig. 2(d) when C is equal to 1, 10, 100 and 1000 respectively. Compared to the parameter gamma, the parameter C is not that sensitive to the performance of KELM. From Figure 3 we can see that the classification accuracy is fluctuating when changing the value of C. The best classification accuracy of 96.45%, 89.82%, 87.68% and 86.17% was achieved with the parameter pair of (62, 1), (84, 10), (94, 100) and (48, 1000) as shown in Fig. 3(a), Fig. 3(b), Fig. 3(c) and Fig. 3(d) when gamma is equal to 1, 10, 100 and 1000 respectively. Owing to the best classification accuracy is achieved when C and gamma is set to be 62 and 1 respectively, the optimal parameter pair of (62, 1) is adopted for subsequent analysis.

In order to investigate whether feature selection can further improve the performance of KELM for diagnosis of PD, we further conducted the experiments in the reduced feature space. mRMR, IG, Relief and t-test are implemented to rank the features and the trends of classification accuracy of KELM model over the incremental feature subset. For comparison, the ELM model with the same four feature selection methods are also shown in Figure 4. According to the preliminary experimental results, the ELM model has achieved the best performance with Sine function when the number of the hidden neuron is 67. For convenience, the hidden neuron of 67 is taken for ELM model with Sine function, and the parameter pair of (1, 62) is adopted for KELM. From Figure 4 we can see that feature selection can further improve the

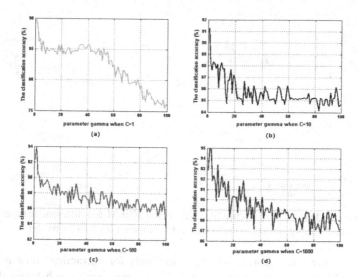

Fig. 2. The relationship between classification accuracy and parameter gamma with different values of parameter C

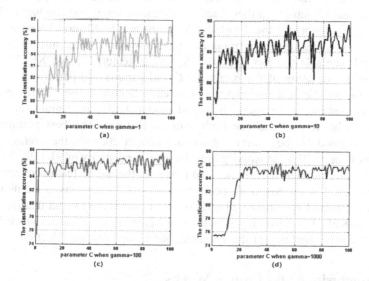

Fig. 3. The relationship between classification accuracy and parameter C with different values of parameter gamma

classification accuracy of the ELM and KELM, except the IG approach. Both ELM and KELM combined with IG achieve the best performance with the feature subset be full with the whole 22 features. It can be also found that the two models coupled with mRMR filter achieve the best classification accuracy with the smallest features among the four feature selectors. Therefore, mRMR has emerged as the promising technique

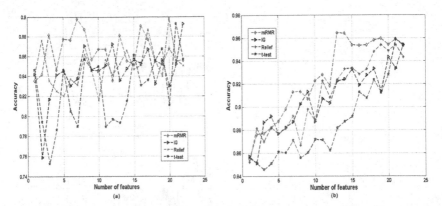

Fig. 4. Trends of classification accuracy of ELM and KELM for different feature subset obtained by different feature selection methods: (a) ELM model (b) KELM model

compared to other three feature selection methods for extracting most informative features. In addition, we can find that KELM still performs much better than ELM with the aid of feature selection.

Since both ELM and KELM are sensitive to the variation of the parameter values on different feature subset, further detailed evaluation should be conducted. For simplicity, here we performed the detailed evaluation for KELM model with the mRMR filter owing to its excellent discriminative ability. We first utilize mRMR to rank the features and then selected top 1, 5, 10, 15, and 20 features. Since KELM model is sensitive to the variation of the parameter C and gamma, we performed the experiment to look for the best parameter pair in each feature subset. The experimental results show that the performance of KELM models built with feature subset size of 15 and 20 is better than the one built with all features. The best performance of KELM is obtained on the feature subset with size of 20, with the average AUC of 94.37%, ACC of 95.97%, sensitivity of 97.61% and specificity of 91.12%.

From the above analysis, we can find that with the aid of feature selection using mRMR, KELM has improved its performance for PD diagnosis in terms of AUC, ACC, sensitivity and specificity. In addition, it is interesting to find that the standard deviation of KELM is becoming smaller than before in most cases, which indicates that KELM has become more robust and reliable through feature selection.

6   Conclusions and Future Works

In this work, we have developed an efficient hybrid method, mRMR-KELM, for addressing PD diagnosis problem. With the aid of the feature selection techniques, especially the mRMR filter, the performance of KELM classifier was improved with much smaller features. The promising performance obtained on the PD dataset has proven that the proposed hybrid method can distinguish well enough between patients with PD and healthy persons. The future investigation will pay much attention to evaluating the proposed method in other medical diagnosis problems.

Acknowledgments. This work is supported by the open project program of Wenzhou University Laboratory under Grant No. 13SK29A.

References

1. de Lau, L.M.L., Breteler, M.M.B.: Epidemiology of Parkinson's disease. The Lancet Neurology 5(6), 52–535 (2006)
2. Singh, N., Pillay, V., Choonara, Y.E.: Advances in the treatment of Parkinson's disease. Progress in Neurobiology 81(1), 29–44 (2007)
3. Ho, A.K., et al.: Speech impairment in a large sample of patients with Parkinson's disease. Behavioural Neurology 11, 131–138 (1998)
4. Baken, R.J., Orlikoff, R.F.: Clinical measurement of speech and voice, 2nd edn. Singular Publishing Group, San Diego (2000)
5. Little, M.A., et al.: Suitability of Dysphonia Measurements for Telemonitoring of Parkinson's Disease. IEEE Transactions on Biomedical Engineering 56(4), 1015–1022 (2009)
6. AStröm, F., Koker, R.: A parallel neural network approach to prediction of Parkinson's Disease. Expert Systems with Applications 38(10), 12470–12474 (2001)
7. Ozcift, A., Gulten, A.: Classifier ensemble construction with rotation forest to improve medical diagnosis performance of machine learning algorithms. Comput. Methods Programs Biomed. 104(3), 443–451 (2011)
8. Chen, H.-L., et al.: An efficient diagnosis system for detection of Parkinson's disease using fuzzy k-nearest neighbor approach. Expert Systems with Applications 40(1), 263–271 (2013)
9. Huang, G.-B., Zhu, Q.-Y., Siew, C.-K.: Extreme learning machine: A new learning scheme of feedforward neural networks. In: IEEE International Joint Conference on Neural Networks, pp. 985–990 (2004)
10. Huang, G.B., et al.: Extreme Learning Machine for Regression and Multiclass Classification. IEEE Transactions on Systems Man and Cybernetics Part B-Cybernetics 42(2), 513–529 (2012)
11. Cheng, C., Tay, W.P., Huang, G.-B.: Extreme learning machines for intrusion detection. In: The International Joint Conference on Neural Networks, pp. 1–8 (2012)
12. Witten, I.H., Frank, E., Hall, M.A.: Data Mining: Practical Machine Learning Tools and Techniques: Practical Machine Learning Tools and Techniques, 3rd edn. Morgan Kaufmann, Burlington (2011)

A Hybrid Approach for Cancer Classification Based on Particle Swarm Optimization and Prior Information

Fei Han, Ya-Qi Wu, and Yu Cui

School of Computer Science and Communication Engineering,
Jiangsu University, Zhenjiang, Jiangsu, China
hanfei@ujs.edu.cn, js.rg.wyq@163.com, allen790336@hotmail.com

Abstract. In this paper, an improved method for cancer classification based on particle swarm optimization (PSO) and prior information is proposed. Firstly, the proposed method uses PSO to implement gene selection. Then, the global search algorithm such as PSO is combined with the local search one such as backpropagation (BP) to model the classifier. Moreover, the prior information extracted from the data is encoded in PSO for better performance. The proposed approach is validated on two publicly available microarray data sets. The experimental results verify that the proposed method selects fewer discriminative genes with comparable performance to the traditional classification approaches.

Keywords: Cancer classification, gene selection, prior information, particle swarm optimization, backpropagation.

1 Introduction

Various methods have been developed for cancer classification based on microarray data obtained by DNA microarray technology [1, 2]. However, the features of microarray data (high-dimensional, small sample, and the existence of inherent noise) make cancer classification a complex issue. Gene selection is a critical step for cancer classification.

ANN (artificial neural network) has been widely used as a classifier, since the neural network with a single nonlinear hidden layer is capable of forming an arbitrarily close approximation of any continuous nonlinear mapping. Since BP (backpropagation) has good local search ability, it is mostly used to train ANN. However, BP is easy to be trapped in some local minima. Some methods used PSO (particle swarm optimization) to train neural networks and these PSO-based ANN methods had nice training performance, fast convergence rate, as well as good predicting ability [3, 4]. Although PSO has good ability of global search, but it is not good at local search. Moreover, PSO suffers from the problem of premature convergence like most of the stochastic search techniques.

In this paper, the proposed approach employs PSO to realize gene selection, and combines PSO and BP to obtain a powerful classifier. Moreover, to obtain better performance, the prior information extracted from BSS/WSS ratio (Between-groups to within-groups sum of squares ratio) [5] of each genes is encoded in PSO. Finally, the proposed approach is validated on two real-world gene expression datasets including

Y. Tan et al. (Eds.): ICSI 2014, Part I, LNCS 8794, pp. 350–356, 2014.

leukemia and colon, and the effectiveness of the proposed approach is verified by the experimental results.

2   Methods

2.1   Particle Swarm Optimization Algorithm

PSO was proposed in 1995 and the adaptive one is then proposed in 1998 [6, 7]. PSO can be stated as initializing a swarm of random particles and finding the optimal solutions by iterations. Each particle will update itself by two optimal positions, *pbest* and *gbest*. This algorithm can be described as follows:

$$V_i(t+1) = w * V_i(t) + c1 * r1 * (P_i(t) - X_i(t)) + c2 * r2 * (P_g(t) - X_i(t))$$

(1)

$$X_i(t+1) = X_i(t) + V_i(t+1)$$

(2)

where $V_i(t)$ is the velocity of the i-th particle in the t-th iteration; $X_i(t)$ is the position of the i-th particle; P_i is the best position achieved by the particle so far; P_g is the best position among all particles in the population; $r1$ and $r2$ are two independently and uniformly distributed random variables with the range of [0,1]; $c1$ and $c2$ are positive constant parameters called accelerated coefficients. w is called the inertia weight that controls the impact of the previous velocity of the particle on its current.

2.2   The PSO-Based Gene Selection and Classifier Optimization Algorithm

Compared with BP, PSO has good ability of global searching. The employment of PSO performing search considering the entire search space, followed by BP performing the search in a small portion of the search space may produce solutions closer to the global minima than the use of each method individually. Many researchers have concentrated on hybrid algorithm basing on PSO and BP [4, 8]. However, prior information was not taken into consideration in these hybrid algorithms. In this study, the BSS/WSS ratio of each gene is considered in the proposed hybrid method.

The detailed steps of the proposed approach are as follows:

Step 1: Data preprocessing is performed with a group-based ensemble gene selection method [9]. As the result of data preprocessing, an informative candidate gene set is obtained. However, this method just focuses on single biomarker correlation degree but neglects the underlying abnormal biomarker association patterns that are responsible for a cancer or cancer type.

Step 2: PSO encoding prior information is employed to select gene subset from the informative candidate gene set. Since the gene selection method in the paper is a wrapper one, each particle of the PSO consists of two parts: candidate gene sets and a set of weights of the ANN. During each iteration, each particle in the swarm is updated according to the performance of the ANN classifier and thus the new gene subset and weights are obtained. If the fitness value reaches the threshold value, or the maximal iterative generations are arrived, the particle with the best fitness value is output, containing the best gene subset and an optimized set of weights of ANN.

Step 3: After the gene subset is determined, BP is adopted to optimize the ANN further whose initial weights are obtained by the above PSO optimization till the training target is achieved.

Step 4: The ultimate ANN trained with BP and the optimal gene subset is applied to the test samples to obtain prediction results.

To reduce searching time of PSO and give the searching a good initial direction, the prior information is obtained from the training samples and encoded into the algorithm [10]. In this paper, the prior information is obtained from BSS/WSS ratio [5] of each gene in the candidate gene subset. The ratio of each gene in the candidate subset is calculated and ranked in descending order, and then the first L genes are taken as most probable selected genes, in which L is the size of the gene subset. The BSS/WSS ratio of j-th gene in the gene subset is calculated as follows:

$$S(j) = \frac{\sum_i \sum_c I(y_i = c)(\overline{x_{cj}} - \overline{x_{.j}})^2}{\sum_i \sum_c I(y_i = c)(x_{ij} - \overline{x_{cj}})^2} \tag{3}$$

$I(.)$ is an indicator function returning one if the condition inside the parentheses is true, otherwise it returns zero. y_i is the class label of the i-th sample. x_{ij} is the expression level of gene j in the i-th sample. $\overline{x_{.j}}$ and $\overline{x_{cj}}$ denote the average expression level of gene j across all tumor samples and across samples belonging to class c only.

Then the prior information is encoded into PSO as follows: First of all, initialize the positions and velocities of a group of particles randomly in the range of [-1, 1]. Every particle is constructed as $[X_{1,......}X_N, X_{N+1},......,X_{N+M}]$ where $[X_{N+1},......,X_{N+M}]$ represent the classifier's weights, and $[X_{1,......}X_N]$ represent the candidate gene subset. If the value of any component in the first N components of a particle is within $[0,+\infty)$, the gene correlated to this component is selected, or else the gene will be cut off. Since the first L genes are selected as the most probable selected genes, each x_i in $[X_{1,......}X_L]$ is initialized within the interval $[0,+\infty)$ and each one in $[X_{L+1,......}X_N]$ is initialized within the interval $(-\infty,0)$.

In each iteration, each particle is valuated according to the fitness function, and the worst particle is replaced by the sorted best particle. The fitness function for i-th particle is given as follows:

$$fitness = 100 * (1 - accuracy(\%)) + E(i) \tag{4}$$

where the $accuracy(\%)$ is the classification accuracy on the training samples basing on the gene subset and the set of classifier's weights offered by the i-th particle which is optimized by PSO alone before using BP. $E(i)$ is the performance of the set of classifier's weights offered by the i-th particle which is optimized by PSO alone before using BP. At the end of the PSO part, the set of classifier's weights offered by the best particle will be taken as the initial weights of the BP network instead of a random set of weights. In details, $E(i)$ is training mean squared error (MSE) for the i-th particle.

Obviously, the smaller the particle's fitness value is, the better performance the selected gene subset and the optimized classifier will obtain.

3   Results and Discussion

Leukemia dataset contains the expression levels of 7,129 genes for 47 patients of acute lymphoblastic leukemia (ALL) and 25 patients of acute myeloid leukemia (AML) [11]. The source of the 7,129 gene expression measurements is publicly available at http://www.broadinstitute.org/cgi-bin/cancer/datasets.cgi/.

Colon cancer dataset consists of expression levels of 62 samples of which 40 samples are colon cancer samples and the remaining are normal samples [11]. Although originally expression levels for 6,000 genes are measured, 4,000 genes out of all the 6,000 genes were removed considering the reliability of measured values in the measured expression levels. The measured expression values of 2,000 genes are publicly available at http://microarray.princeton.edu/oncology/.

In the following experiments, values of parameters in the proposed method are determined by trial and error. The maximal generation of PSO is 20; the weight w is 0.9, the acceleration constants, both $c1$ and $c2$ are 2.0, $r1$ and $r2$ are two random numbers in the range of [0, 1], respectively. Let the maximum velocity be 1.0 and the minimum velocity -1.0. The initial velocities of the initial particles are generated randomly in the range of [0, 1], and the population size is assumed as 50. As for the BP, the maximal generation is assumed as 1000 times. The number of hidden units is assumed to be 5 for the leukemia dataset, while the one is 7 for the colon dataset.

Table 1 lists the LOOCV results of the proposed method and those of previous approaches including Nero-fuzzy ensemble machine (NFEM) [12], Bayesian variable selection (BVS) approach [13], SVM/REF [14], GA/SVM [15] and GA-based approach [14] on the leukemia dataset, and the LOOCV results of the proposed method and those of methods including MFMW [1], SFSW [1], SVM-REF [14] and GA-based method [14] on the colon data. The results show that the proposed method obtains comparatively high prediction performance with the fewer feature genes.

Table 1. Comparison of the LOOCV results of the proposed algorithm with previously reported results on the two data

Leukemia		Colon	
Methods	Accuracy (gene no.)	Methods	Accuracy (gene no.)
NFEM	0.958(20)	MFMW	0.952(6)
BVS	0.972(5)	SFSW	0.903(10)
SVM/REF	0.972(8)	SVM-REF	0.9677(8)
GA/SVM	1(25)	GA/SVM	0.9941(10)
GA-based approach	0.972(12)	GA-based	0.919(8)
The proposed approach	0.972(5)	The proposed approach	0.9677(3)

The 5-fold cross-validation (CV) results on the two data are presented in Table 2. The reported results of PCA/PLS [1], PSO—NB [16], PSO—C4.5 [16], NLN classifier [14] and GA/MRMR [17] on the leukemia data, and the results of QDA/t-store/PLC [18], PSO-SVM [16], PSO—NB [16], PSO—C4.5 [16] and GA/MRMR [17] on the colon data, are listed in Table 2 for comparison. The conclusion as drawn from Table 1 is also suitable for the 5-fold CV case.

Table 2. Comparison of the 5-fold CV results of the proposed algorithm with previously reported results on the two data

Leukemia		Colon	
Methods	Accuracy (gene no.)	Methods	Accuracy (gene no.)
PCA/PLS	0.97(50)	QDA/t-store/PLC	0.919(50)
PSO—NB	0.972(27.8)	PSO-SVM	0.936(25.7)
PSO—C4.5	0.958(26.5)	PSO—NB	0.887(29.4)
NLN classifier	0.853(2)	PSO—C4.5	0.871(28.9)
GA/MRMR	0.945(23)	GA/MRMR	0.867(7)
The proposed approach	0.945(1)	The proposed approach	0.887(3)

To validate the effectiveness of the prior information from the samples encoding into PSO, two experiments are further conducted for comparison using LOOCV on the two datasets respectively. In one experiment, the prior information is encoded into the algorithm, while the other is not. Fig.1 depicts the effectiveness of encoding the prior information on the two datasets respectively. From Fig.1, it is found that the prediction accuracy of the algorithm with the prior information is much higher than the one without the prior information.

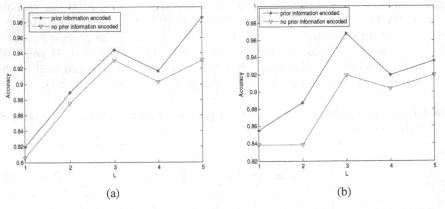

(a)   (b)

Fig. 1. Comparison of algorithms with the prior information and the one without the prior information (a) the prediction accuracy on the leukemia data (b) the prediction accuracy on the colon data

4   Conclusions

In this paper, an improved approach was proposed to perform gene selection and cancer classification. This method focused on three aspects: combining gene selection with classifier optimization, accompanying PSO with BP on optimization of the ANN classifier, and incorporating the prior information obtained from the sample data into PSO. The competitiveness and effectiveness of our approach were validated on two real microarray data sets. Compared with other classical methods, it did not only offer a smaller size of gene subset, but also an effective classifier. However, less attention was paid to the physiological plausibility of the obtained results in this paper. Thus, our feature work will be dedicated to this problem. Besides, our approach will be further validated on more complex multiclass cancer classification problems.

Acknowledgments. This work was supported by the National Natural Science Foundation of China (No.61271385) and the Initial Foundation of Science Research of Jiangsu University (No.07JDG033).

References

1. Wang, H.-Q., Wong, H.-S., Zhu, H., Yip Timothy, T.C.: A neural network-based biomarker association information extraction approach for cancer classification. Journal of Biomedical Informatics 42, 654–666 (2009)
2. Huerta, E.B., Duval, B., Hao, J.-K.: A hybrid LDA and genetic algorithm for gene selection and classification of microarray data. Neurocomputing 73, 2375–2383 (2010)
3. Da, Y., Xiurun, G.: An improved PSO-based ANN with simulated annealing technique. Neurocomputing 63, 527–533 (2005)
4. Geethanjali, M., Mary Raja Slochanal, S., Bhavani, R.: PSO trained ANN-based differential protection scheme for power transformers. Neurocomputing 71, 904–918 (2008)
5. Dudoit, S., Fridlyand, J., Speed, T.P.: Comparison of discrimination methods for classification of tumors using gene expression data. J. Am. Statistical Assoc. 97, 77–87 (2002)
6. Riget, J., Vesterstorm, J.S.: A diversity-guided particle swarm optimizer – the arPSO. EVAlife Technical Report 2:1-13 (2002)
7. Shi, Y.H., Eberhat, R.C.: Parameter selection in particle swarm optimization. In: Porto, V.W., Saravanan, N., Waagen, D., Eiben, A.E. (eds.) EP 1998. LNCS, vol. 1447, pp. 591–600. Springer, Heidelberg (1998)
8. Almeida, L.M., Ludermir, T.B.: A multi-objective memetic and hybrid methodology for optimizing the parameters and performance of artificial neural networks. Neurocomputing 73, 1438–1450 (2010)
9. Huawen, L., Lei, L., Huijie, Z.: Ensemble gene selection by grouping for microarray data classification. Journal of Biomedical Informatics 43, 81–87 (2010)
10. Fei, H., Ling, Q.-H.: A new approach for function approximation incorporating adaptive particle swarm optimization and a priori information. Applied Mathematics and Computation 205, 792–798 (2008)
11. Ruichu, C., Zhifeng, H., Xiaowei, Y.: An efficient gene selection algorithm based on mutual information. Neurocomputing 72, 991–999 (2009)
12. Zhenyu, W., Vasile, P., Yong, X.: Neuro-fuzzy ensemble approach for microarray cancer gene expression data analysis. In: International Symposium on Evolving Fuzzy Systems, vol. 241, pp. 241–246 (2006)
13. Lee, K.E., Sha, N., Dougherty, E.R., Vannucci, M., Mallick, B.K.: Gene selection: A Bayesian variable selection approach. Bioinformatics 19, 90–97 (2003)
14. Ding, S.-C., Juan, L., Qing, Y.: A GA-based optimal gene subset selection method. Machine Learning and Cybernetics 5, 2784–2789 (2004)
15. Cawley, G.C., Talbot, N.L.C.: Gene selection in cancer classification using sparse logistic regression with Bayesian regularization. Bioinformatics 22, 2348–2355 (2006)

16. Xixian, W.: Gene selection and cancer classification based on optimization algorithm and support vector machine. A thesis submitted in partial satisfaction of the Requirements for the degree of Master of Science in Control Science and Engineering in the Graduate School of Hunan University, B.E. 1-65 (2005) (in chinese)

17. Jiexun, L., Hua, S., Hsinchun, C.: Optimal search-based gene aubset selection for gene array cancer classification. IEEE Transactions on Information Technology in Biomedicine 11, 398–405 (2007)

18. Nguyen, D.V., Rocke, D.M.: Tumor classification by partial least squares using microarray gene expression data. Bioinformatics 18, 39–50 (2002)

Grover Algorithm for Multi-objective Searching with Iteration Auto-controlling

Wanning Zhu [1], Hanwu Chen [1,2], Zhihao Liu [1], and Xilin Xue [1]

[1] Department of Computer Science and Engineering, Southeast University,
Nanjing 211189, China
[2] Key Laboratory of Computer Network and Information Integration, Ministry of Education,
Southeast University, Nanjing, 211189, China
hw_chen@seu.edu.cn, {granny025,liuzhtopic}@163.com,
332407606@qq.com

Abstract. This paper presents an improved Grover searching algorithm [1] which can auto-control the iterative processing when the number of target states is unknown. The final amplitude of the target states will be decreased if this number is unknown. So the question is how to perform Grover searching algorithm without the number of target states? As for this question, there are two conventional solutions. One solution tries to find the number of target states before performing the original algorithm. The other solution guesses a random k as the number of target states before performing the original algorithm. Both the two solutions need $O(\sqrt{N})$ additional times Oracle calls than original algorithm and the answer of the first solution is non-deterministic while the second solution needs to check the correctness of the result. Assuming an operator which can judge the sign of the phases of superposition state, based on this technical, this paper shows a novel solution, which can perform Grover searching algorithm even if the number of target states is unknown. This solution only needs adding one gate, which can judge the sign of phase, and one more time Oracle call than the original algorithm.

Keywords: Grover searching algorithm, sign of phase, adaptive control.

1 Introduction

Quantum computation was a novel technical in computer science in recent years. Quantum computation is an interdisciplinary sciences, which can process the eigenstates of a superposition state in parallel. So quantum algorithm becomes quadratic even exponentially better than classical algorithm [2]. In 1996, a quantum searching algorithm for unstructured database was presented by Grover, and the Grover searching algorithm can be quadratic better than any possible classical algorithm. Many results on Grover searching algorithm have been gained so far [3-10].

When we need to do multi-objective searching, we must get the number of the target states [8]. We can explain this as follows. First of all, Grover searching algorithm consists of repeated application of a U operator, which is formed by an Oracle

Y. Tan et al. (Eds.): ICSI 2014, Part I, LNCS 8794, pp. 357–364, 2014.

operator and a Grover operator. As a result, the target states can be measured with high probability. Secondly, the times of iteration must be a suitable number, neither less nor more; otherwise the amplitude of the target states could be decreased. At last, the right number is in connection with the quantity of the target states. So a problem should be solved is how to perform Grover searching algorithm when the quantity is unknown. By the classical way, the time complexity to confirm the number of target states in N elements problem is $O(N)$. By the quantum way, there are two solutions [8-10], but both of them require additional $O(\sqrt{N})$ Oracle calls.

This paper shows a high-performance solution which can operate Grover searching algorithm without the number of target states by judging the sign of phase in X basis $\left\{ |\pm\rangle\,|\,|\pm\rangle = \frac{1}{\sqrt{2}}(|0\rangle\pm|1\rangle) \right\}$. This solution can maximize the probability of target states while it costs only one more Oracle call than the original Grover searching algorithm.

The following text is organized as: a brief introduction of Grover searching algorithm will be showed in the second section. In the third section, an automatically control circuit will be constructed by embedding a phase detecting gate to the Grover searching algorithm circuit. When the new algorithm stops, the probability to get one of the target states will be the same as the original version. The result will be proved by three theorems.

2   Preliminary

Let's introduce the Grover searching algorithm at first. The algorithm is formed by repeated application of a U operator (see Fig. 1).

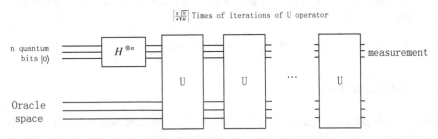

Fig. 1. Grover searching algorithm circuit

It begins with the initial state $|0\rangle^{\otimes n}$, which is transformed to equal superposition state by performing $H^{\otimes n}$. More details shows as the following formula (let $N = 2^n$, and the target is $|t\rangle$):

$$|0\rangle^{\otimes n} \xrightarrow{\ H^{\otimes n}\ } \frac{1}{\left(\sqrt{2}\right)^{n}}\left(|0\rangle+|1\rangle\right)^{\otimes n} = \frac{1}{\sqrt{N}}\sum_{i=0}^{N-1}|i\rangle. \tag{1}$$

We will use $|s\rangle = \dfrac{1}{\sqrt{N}}\displaystyle\sum_{i=0}^{N-1}|i\rangle$ to stand for equal superposition in the following. The U operator in Fig. 1 is formed by an Oracle operator which marks the target states, and a Grover operator that increases the amplitude of the target states.

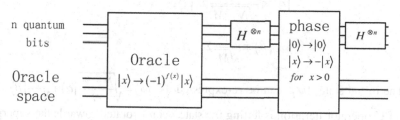

Fig. 2. U operator circuit

The Oracle operator can be expressed by $Oracle = I - 2|t\rangle\langle t|$, which makes the phase of the target inverted. We have a simple form of Grover operator as follow:

$$\begin{aligned} Grover &= H^{\otimes n}\left(2|0\rangle\langle 0|-I\right)^{\otimes n} H^{\otimes n} \\ &= 2H^{\otimes n}\left(2|0\rangle\langle 0|\right)^{\otimes n} II^{\otimes n} - H^{\otimes n}I^{\otimes n}H^{\otimes n} \\ &= 2|s\rangle\langle s|-I. \end{aligned} \tag{2}$$

Let the number of target states is M. By calculation, we know that the times of iterations of U operator is $T = \left\lceil \dfrac{\pi}{4}\sqrt{\dfrac{N}{M}} \right\rceil$. After T times of iterations, the probability of outputting target states is more than $1/2$. Note that the probability could decrease when the times of iterations is either more or less than T. Therefore the amplitude of target states will be sharply decreased if M is unknown.

Quantum counting was used to get the M in Reference [9]. Based on phase estimation, quantum counting can estimate M within accuracy $c\sqrt{M}$ for some constant c. This solution requires additional $O\left(\sqrt{N}\right)$ oracle calls. References [8] and [10] presented another solution which can be called testing algorithm. This solution takes a big enough number k as the number of target states and then performs the Grover searching algorithm. If it is fail to measure the target state, then k decreases. This process will be always repeated until one of the target states is got. There is no doubt that this solution must run several times of Grover searching algorithm. So the solution needs more $O\left(\sqrt{N}\right)$ Oracle calls. Is there a more efficiency solution to operate Grover searching algorithm when M is unknown?

3   Adaptive Iteration Control of Grover Searching Algorithm

Theorem 1: The upper bound of the times of iterations for Grover searching algorithm makes the amplitude of target states maximized for the first time.

Proof: Assume the number of target states is M, in an N items search problem. Let Σ_x indicates a sum over all x which are target states, as well as the Σ'_x indicates a sum over all x which are not target states. Define normalized states as follows:

$$|\alpha\rangle \equiv \frac{1}{\sqrt{N-M}} \Sigma'_x |x\rangle, \tag{3}$$

$$|\beta\rangle \equiv \frac{1}{\sqrt{M}} \Sigma_x |x\rangle. \tag{4}$$

The initial state $|\psi\rangle$ can be re-expressed as $|\psi\rangle = \sqrt{\frac{N-M}{N}}|\alpha\rangle + \sqrt{\frac{M}{N}}|\beta\rangle$. The action of U operator iteration is letting the state vector rotated towards the superposition state $|\beta\rangle$ of target states. According to the analysis of Grover searching algorithm, let the initial angle be $\theta/2$, then It means $\sin(\theta/2) = \sqrt{M/N}$ and $|\psi\rangle = \cos(\theta/2)|\alpha\rangle + \sin(\theta/2)|\beta\rangle$, and the state vector will be rotated by θ at every turn. This fact can be explained by as follows.

Let $|\alpha\rangle$ be horizontal ordinate and $|\beta\rangle$ be the orthogonal one to $|\alpha\rangle$. The Oracle operator reflects the state about $|\alpha\rangle$, which means rotating θ towards $|\alpha\rangle$. The Grover operator reflects the state about $|\psi\rangle$, which means rotating 2θ towards $|\beta\rangle$. So the angle of state vector will be $\theta/2 - \theta + 2\theta = (3/2)\theta$ at last, which means the state vector is rotated by θ towards $|\beta\rangle$ [7].

After k times of iterations, the state is

$$G^k |\psi\rangle = \cos(k\theta + \theta/2)|\alpha\rangle + \sin(k\theta + \theta/2)|\beta\rangle. \tag{5}$$

Notice that the amplitude of target states is $|\sin(k\theta + \theta/2)|$. Obviously the amplitude of target states will not be maximum until $k\theta + \theta/2 \approx \pi/2$. Assume M<<N, so $\theta/2$ can be considered as small enough. As a result, when $k \approx \left\lceil \frac{\pi}{2\theta} \right\rceil$, the amplitude of target states maximizes for the first time. We have $\theta/2 \approx \sin(\theta/2) = \sqrt{M/N}$ by assuming M<<N, so $\left\lceil \frac{\pi}{4}\sqrt{\frac{N}{M}} \right\rceil \approx \left\lceil \frac{\pi}{2\theta} \right\rceil$. Therefore

$\left\lceil \frac{\pi}{4}\sqrt{\frac{N}{M}} \right\rceil$ times of iterations for Grover searching algorithm adequately maximized the amplitude of target states for the first time.

Theorem 2: After performing Oracle operator, when the sum of the phase of all the states turn to negative for the first time, the amplitude of the target states will become maximized for the first time.

Proof: A general superposition state is defined by $|\varphi\rangle = \Sigma_i \alpha_i |i\rangle$, where $|\varphi\rangle$ are normalized. After applying the Oracle operator to $|\varphi\rangle$, $|\varphi\rangle$ is transformed to a new superposition state $|\varphi'\rangle \equiv \Sigma_i (-1)^{f(i)} \alpha_i |i\rangle$, where $f(i) = 1$ for $|i\rangle$ being the target state and $f(i) = 0$ for other conditions. Therefore target states can be expressed as $\Sigma_{i'} - \alpha_{i'} |i'\rangle$, and non-target states can be expressed as $\Sigma_{i''} \alpha_{i''} |i''\rangle$. Applying the Grover operator to the new superposition state, it will become:

$$\left(2|s\rangle\langle s| - I\right)\Sigma_i (-1)^{f(i)} \alpha_i |i\rangle = \frac{2\Sigma_i (-1)^{f(i)} \alpha_i \langle i|i\rangle}{N}\Sigma_i |i\rangle - \left(\Sigma_{i''}\alpha_{i''}|i''\rangle - \Sigma_{i'}\alpha_{i'}|i'\rangle\right). \qquad (6)$$

Notice that $\dfrac{\Sigma_i (-1)^{f(i)} \alpha_i \langle i|i\rangle}{N}$ is the mean of the phase of all the states. Let it be called $\langle \alpha \rangle$ for brief. So Formula (6) can be transformed to the following

$$2\Sigma_i \langle\alpha\rangle|i\rangle - \left(\Sigma_{i''}\alpha_{i''}|i''\rangle - \Sigma_{i'}\alpha_{i'}|i'\rangle\right)$$
$$= \Sigma_{i'} \left(2\langle\alpha\rangle + \alpha_{i'}\right)|i'\rangle + \Sigma_{i''} \left(2\langle\alpha\rangle - \alpha_{i''}\right)|i''\rangle. \qquad (7)$$

The phase of target states is changed to $2\langle\alpha\rangle + \alpha_{i'}$ as formula (4) shows[7][8]. When the initial superposition state is equal superposition state and the number of target states $M \leq N/2$ (notice that Grover searching algorithm has the same restriction), $\langle\alpha\rangle$ is positive obviously. And $\alpha_{i'}$ will become maximum when $\langle\alpha\rangle$ changes to negative for the first time, and then the algorithm stops. According to $\langle\alpha\rangle = \dfrac{\Sigma_i (-1)^{f(i)} \alpha_i \langle i|i\rangle}{N}$, the sign of $\langle\alpha\rangle$ is determined by $\Sigma_i (-1)^{f(i)} \alpha_i$. Therefore after performing Oracle operator, the amplitude of the target states will become maximized for the first time when the sum of the phase of all the states turn to negative for the first time.

According to Theorem 1 and Theorem 2, the times of calling Oracle are $\left\lceil\dfrac{\pi}{4}\sqrt{\dfrac{N}{M}}\right\rceil + 1$, when the sum of the phase of all the states turn to negative for the first time.

Theorem 3: The sign of sum of the phase of all the states in Z basis $\{|0\rangle, |1\rangle\}$ is determined by the sign of phase of $|++...+\rangle$ in X basis.

Proof: This conclusion can be easily got by changing the Z basis to X basis. Grover searching algorithm uses Z basis as the basis, so a product state of n quantum bits can be expressed as:

$$(\alpha_{00}|0\rangle + \alpha_{01}|1\rangle)(\alpha_{10}|0\rangle + \alpha_{11}|1\rangle)\cdots(\alpha_{(n-1)0}|0\rangle + \alpha_{(n-1)1}|1\rangle). \qquad (8)$$

The sum of the phase of all the states above can be calculated by $\Pi_{i=0}^{n-1}\Sigma_{j=0}^{1}\alpha_{ij}$.

The i-th quantum bit can be changed as follow:

$$\alpha_{i0}|0\rangle + \alpha_{i1}|1\rangle = \frac{1}{\sqrt{2}}\big((\alpha_{i0} + \alpha_{i1})|+\rangle + (\alpha_{i0} - \alpha_{i1})|-\rangle\big). \qquad (9)$$

Notice that the phase of $|+\rangle$ is $\frac{1}{\sqrt{2}}\Sigma_{j=0}^{1}\alpha_{ij}$, then the phase of $|++...+\rangle$ is

$\left(\frac{1}{\sqrt{2}}\right)^{n}\Pi_{i=0}^{n-1}\Sigma_{j=0}^{1}\alpha_{ij}$. Obviously $\left(\frac{1}{\sqrt{2}}\right)^{n}$ do not influence the sign of phase; hence

the sign of phase of $|++...+\rangle$ determines the sign of the sum of phase of all states in Z basis.

Assuming that we can construct a phase detecting gate, which has two inputs $|\psi\rangle \otimes |\beta\rangle$. $|\psi\rangle$ is a superposition state, and the other one $|\beta\rangle$ is a constant quantum bit. After applying phase detecting gate, $|\psi\rangle$ always does not change. $|\beta\rangle$ will not change until $\phi_{|++...+\rangle} \leq 0$, which is the phase of $|++...+\rangle$. Obviously this gate is reversible gate. And this gate can be expressed as the following figure.

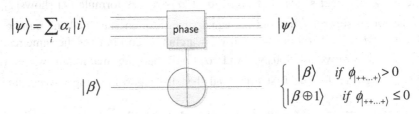

Fig. 3. Phase detecting gate circuit

Insert phase detecting gate to U operator. Let $|\psi\rangle$ be the superposition state after performing Oracle operator and $|\beta\rangle = |0\rangle$, then the U operator is changed to the following form.

As Fig. 4 shows, in every iteration of the U operator, the constant bit will be measured after performing phase detecting gate. If the measuring result is $|0\rangle$, then the algorithm continues. If the measuring result is $|1\rangle$, the algorithm stops.

Analysis of algorithm: when the measuring result is $|1\rangle$, $\phi_{|++...+\rangle} \leq 0$. According to Theorem 3, when $\phi_{|++...+\rangle} \leq 0$, the sum of phase $\Sigma_{i}\alpha_{i} \leq 0$. And according to Theorem 2, when the sign of the sum of phase of all states changes for the first time,

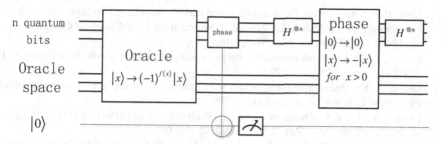

Fig. 4. New U operator circuit of Grover searching auto-control algorithm

the amplitude of target states becomes maximum. There is one more Oracle call, but it does not influence the amplitude because the Oracle operator only changes the sign of phase of the target states. So we can take the conclusion that when the measuring result is $|1\rangle$, the amplitude of target states becomes maximum. Compared with the original Grover searching algorithm, the quantum circuit of the new algorithm only adds one more reversible gate, and the new algorithm only needs one more Oracle call, which means the complexity does not increase. And this new algorithm can operate without knowing the number of target states.

4   Summary

Grover searching algorithm is a quantum searching algorithm for unstructured database with high performance as well as quadratic better than any classical algorithm. Grover searching algorithm is formed by several repeated application of U operator, and the number of iteration during the whole process directly influences the probability of outputting the target states. Whatever the times of iteration is more or less than the optimal number, the probability will decrease. Moreover, the calculation of the optimal number relies on the number of target states M. It cannot output one of the target states with high probability when the number is unknown. In the previous research, the quantum counting and testing algorithm was used to solve this problem. But these solutions need $O\left(\sqrt{N}\right)$ more Oracle calls and cannot get the accurate M as well.

This paper presents a solution that operates Grover searching algorithm without M. The solution can auto-control the iteration by judging the sign of phase of $|++...+\rangle$. This solution can stop the algorithm just when the probability of the target states becomes maximum. Compared with the circuit of the original algorithm, there is only one more reversible gate used to judge phase. And it needs one more Oracle call than the original algorithm.

References

1. Grover, L.K.: A fast quantum mechanical algorithm for database search. In: Proceedings of the 28th Annual ACM Symposium on the Theory of Computing, pp. 212–219. ACM Press, New York (1996)

2. Shor, P.W.: Algorithms for quantum computation: discrete logarithms and factoring. In: Proceedings of the 35th IEEE Annual Symposium on Foundations of Computer Science, pp. 124–134. IEEE Press, New York (1994)
3. Grover, L.K.: Quantum mechanics helps in searching for a needle in a haystack. Phys. Rev. Lett. 79(2), 325–328 (1997)
4. Grover, L.K.: Quantum computers can search rapidly by using almost any transformation. Phys. Rev. Lett. 80(19), 4329–4332 (1998)
5. Long, G.L., Li, Y.S., Zhang, W.L., et al.: Dominant gate imperfection in Grover's quantum search algorithm. Phys. Rev. A 61(4), 042305 (2000)
6. Biron, D., Biham, O., Biham, E., Grassl, M., Lidar, D.A.: Generalized grover search algorithm for arbitrary initial amplitude distribution. In: Williams, C.P. (ed.) QCQC 1998. LNCS, vol. 1509, pp. 140–147. Springer, Heidelberg (1999)
7. Chuang, I.L., Gershenfeld, N., Kubinec, M.: Experimental implementation of fast quantum searching. Phys. Rev. Lett. 80(15), 340–3411 (1998)
8. Boyer, M., Brassard, G., Høyer, P., et al.: Tight bounds on quantum searching. ArXiv: quant-ph/9605034 (1996)
9. Nielsen, M.A., Chuang, I.L.: Quantum Computation and Quantum Information. Cambridge University Press, Cambridge (2000)
10. Daniel, R., Daniel, N., Vladimir, B.: Quantum walks. Acta Phys. Slovaca 61(6), 603–725 (2011)

Pareto Partial Dominance on Two Selected Objectives MOEA on Many-Objective 0/1 Knapsack Problems

Jinlong Li and Mingying Yan

School of Computer Science, The USTC-Birmingham Joint Research Institute in Intelligent Computation and Its Applications (UBRI), University of Science and Technology of China, Hefei, Anhui 230027, China
jlli@ustc.edu.cn, myyan199@mail.ustc.edu.cn

Abstract. In recent years, multi-objective optimization problems (MOPs) have attracted more and more attention, and various approaches have been developed to solve them. This paper proposes a new multi-objective evolutionary algorithm (MOEA), namely Pareto partial dominance on two selected objectives MOEA (PPDSO-MOEA), which calculates dominance between solutions using only two selected objectives when choosing parent population. In the proposed algorithm, two objectives are mainly selected with the first and the second largest distances to the corresponding dimension of the best point. PPDSO-MOEA switches the two-objective combination in every I_g generation to optimize all of the objective functions. The search performance of the proposed method is verified on many-objective 0/1 knapsack problems. State-of-the-art algorithms including PPD-MOEA, MOEA/D, UMOEA/D, and an algorithm selecting objectives with random method (RSO) are considered as rival algorithms. The experimental results show that PPDSO-MOEA outperforms all the four algorithms on most scenarios.

Keywords: Evolutionary Multi-objective Optimization (EMO), Many-objective Optimization, Pareto Partial Dominance.

1 Introduction

Multi-objective optimization problems refer to those problems having more than one objective. The real-world optimization problems that we often face with refer to many objectives, and some objectives conflict with each other. Multi-objective optimization aims at finding a set of well-distributed compromising solutions representing the different trade-offs with respect to the given objective functions. The usefulness of evolutionary algorithms to solve a large number of MOPs has been verified [1,2]. Without loss of generality, a multi-objective optimization problem can be defined as follows [3]:

$$Maximize\ \mathbf{F}(\mathbf{X}) = (f_1(\mathbf{X}), f_2(\mathbf{X}), \ldots, f_m(\mathbf{X}))^T \quad subject\ to\ \mathbf{X} \in \Omega \quad (1)$$

where $\mathbf{X} = (x_1, x_2, \ldots, x_n)^T$ is a solution vector and Ω is the decision space. Furthermore, $f_i(\mathbf{X})$ $(1 \leq i \leq m)$ are the m objective functions involved in the problem. Here, $\mathbf{F} : \Omega \to R^m$ denotes the objective function vectors and R^m is the objective space. One should notice that multi-objective optimization problems with four or more objectives are often referred to as many-objective optimization problems (ManyOPs) [4].

Y. Tan et al. (Eds.): ICSI 2014, Part I, LNCS 8794, pp. 365–373, 2014.

Multi-objective evolutionary algorithms proposed in the past few decades show promising performance when tackling bi-objective or three-objective problems [5], but encounter new difficulties when dealing with ManyOPs. The existing state-of-the-art EMO algorithms scale poorly with the number of objectives. As a result, various approaches for handling ManyOPs by EMO algorithms are proposed.

The indicator-based approaches incorporate the quality indicators in the fitness assignment mechanism to deal with ManyOPs [6]. The aggregation-based approaches optimize the ManyOPs by optimizing a series of scalar functions which aggregate the objective values of original problem [7,8,9]. The reference set based approaches measure the quality of the solutions with the help of a set of reference points [10]. The Pareto-based approaches concentrate on enhancing the convergence and maintaining the diversity of the population at the same time [5], [11]. The objective space reduction approaches are proposed to eliminate the redundant objectives and identify the important objectives by using feature selection techniques or analyzing the relationship among the objectives [12]. Then the search process is employed on the reduced objective set.

Also, [13] presented an algorithm partitioning the objective space based on an analysis of the conflict information obtained from the current Pareto front approximation. It aimed to separate the MOPs into several subproblems in such a way that each of them contains the information to preserve as much as possible the structure of the original problem.

This paper proposes a new method, namely Pareto partial dominance on two selected objectives MOEA (PPDSO-MOEA), to improve the performance of MOEA especially for ManyOPs. We calculate dominance between solutions using only two objective functions selected from m objectives when choosing parent population. Also, we temporally switch the two-objective combination in every I_g generations to optimize all objective functions throughout the entire evolution process. The main process of selecting two objectives can be concluded as the following two steps: Firstly, for every objective, the average distance to the corresponding dimension of the best point is calculated. Secondly, two objective functions with the first and the second largest average distances are chosen. Also, when the above strategy improves less convergence, two objectives are selected with random method. The strategy for choosing objectives here makes the population push on towards the goal of getting better convergence.

The remainder of this paper is organized as follows. Section 2 is devoted to the description of the proposed algorithm. Section 3 presents the test problems, algorithm settings and performance metrics used for performance comparison. The experimental results and discussion are also given in Section 3. Finally, In Section 4, conclusions about the proposed algorithm are drawn, as well as some possible future work.

2   Working of the Proposed Algorithm

We proposes a new method namely Pareto partial dominance on two selected objectives MOEA (PPDSO-MOEA) in this section. Section 2.1 outlines our algorithm, and we detail the sub-algorithms in other sections.

Algorithm 1. PPDSO-MOEA: Main Loop

---

1 **begin**
2      $U_0 \longleftarrow Initialize()$ /*create an initial union population U_0 of size N*/
3      $Evaluate(U_0)$ /* evaluate U_0 */
4   generate the best point R_{point}
5      $A_{set} \longleftarrow U_0$ /*update A_{set} */
6      $ChooseObj(f_{c1}, f_{c2}, A_{set}, R_{point})$ /*choose two objectives with the first and the second largest average distances to the best point R_{point}*/
7   **while** $t \leq G$ **do**
8          $fast - non - dominated - sort(U_t, \{f_{c1}, f_{c2}\})$ /*perform non-dominated sorting on objective set $\{f_{c1}, f_{c2}\}$*/
9          $Crowdingdistance(U_t)$ /*calculate crowding distance/
10         $P_{t+1} \longleftarrow truncation(U_t)$ /* choose N/2 individuals as parent population */
11         $Q_{t+1} \longleftarrow CreateOffspring(P_{t+1})$ /*create offspring population*/
12         $Evaluate(Q_{t+1})$ /*evaluate offspring population*/
13         $UpdateBestPoint(R_{point}, Q_{t+1})$ /*update the best point with the help of Q_{t+1}*/
14         $U_{t+1} \longleftarrow P_{t+1} \cup Q_{t+1}$ /*get the union population*/
15   **if** $t \bmod Ig = 0$ **or** $t = G$ **then**
16             $fast - non - dominated - sort(A_{set} \cup U_{t+1}, F)$ /*perform non-dominated sorting on objective set F */
17             $Crowdingdistance(A_{set} \cup U_{t+1})$ /*calculate crowding distance/
18             $U_{t+1} \longleftarrow truncation(A_{set} \cup U_{t+1})$ /* choose N individuals from $A_{set} \cup U_{t+1}$ */
19             $A_{set} \longleftarrow U_{t+1}$ /*copy U_{t+1} to A_{set}*/
20             $UpdateParameter(cur_{norm}, last_{norm}, A_{set}, duration)$
21   **if** $duration > 5$ **then**
22   choose objectives with random method
23   **else**
24                 $Choose\_Obj(f_{c1}, f_{c2}, A_{set}, R_{point})$ /*choose two objectives with the first and the second largest average distances to the best point R_{point} */
25                 $duration \longleftarrow 0$
26         $t \longleftarrow t + 1$

2.1   Framework of PPDSO-MOEA

As we all know, if the solutions in the population have better values on all objectives, the performance of the population will be better. However, if worse values on some objectives are obtained, worse performance is gained too. So, designing a method improving all objective values is desired. There are many methods measuring the value of a objective, one of which is calculating the average distance to the corresponding dimension of the best point (the best point R_{point} is an array with size m and $R_{point}[i]$ stores the best objective value of the $i-$th objective found so far), which is detailed in Section 2.2. Placing emphasis on improving the values of the objectives who have worse values is needed.

In our proposed method, the main strategy of choosing objectives is choosing two objectives with the first and the second largest average distances to the corresponding

Table 1. The meanings of some notations used in algorithms

name	meaning	name	meaning
N	population size	f_{c1}	the selected objective
G	max generation number	f_{c2}	the selected objective
F	the set of all objectives	I_g	interval generation
m	the number of objectives		

dimension of the best point (we use random method for one time to choose two objectives when the change of metric for convergence is smaller than 10 for five times), and then perform non-dominated sorting by using only those two objectives while choosing parent population, instead of using objective combination in O_{table} which is used in PPD-MOEA [11]. We evolve I_g generations to make those two objectives obtain better values. We temporally switch combination in every I_g generation with the help of the archive set to optimize all objective functions throughout the entire evolution process. Because we make efforts to improve the values of two selected objectives every time, the total convergence of the population will be improved.

An initial union population U_0 of size N (N is the population size) is created and evaluated. Then a best point is generated with the help of U_0. Two objectives: f_{c1} and f_{c2} are chosen with the help of the best point. Line 1 to line 6 of algorithm 1 describe the initial work. The rest of algorithm 1 is described as follows:

1. line 8 $\sim$ 10 implement the following works: $N/2$ non-dominance solutions are selected from U_t as new parent population on the sub-objective set $\{f_{c1}, f_{c2}\}$.
2. line 11 $\sim$ 14 implement the following works: offspring population is generated and evaluated, and the best point is also updated. U_{t+1} is updated as $U_{t+1} = P_{t+1} \cup Q_{t+1}$.
3. line 15 $\sim$ 27 implement the following works: if $t \bmod Ig = 0$ **or** $t = G$ is met, the archive set A_{set} is updated and the two-objective combination is also updated. The function $UpdateParameter(cur_{norm}, last_{norm}, A_{set}, duration)$ implements the following works: firstly, cur_{norm} is set to the norm value of A_{set}. Secondly, if $cur_{norm} - last_{norm} < 10$ is met,then $duration$ is increased by 1, else it is set to 0. Lastly, $last_{norm}$ is updated as cur_{norm}.
4. if $t > G$ is satisfied, the algorithm is terminated, or it will go to 1 and continue.

P_t is the parent population of tth generation, Q_t is the offspring population of tth generation, and U_t is the union of parent and offspring population of tth generation. Set A_{set} is the archive population which is used to maintain Pareto optimal solutions (POS) when the two-objective combination is changed. When new individuals are generated, the best point will be updated. $duration$ is the current times that the norm value keeps a small change. $last_{norm}$ and cur_{norm} are the norm values of the last archive and the current archive. More notations are listed in Table 1.

2.2 The Choosing of Two Objectives

To accelerate the convergence speed, we choose two objectives with the first and the second largest average distances to the corresponding dimension of current best point R_{point}. The work is described in algorithm 2.

The objective function values of each individual are translated using the following function:

$$f_i'(\mathbf{x}) = R_{point}[i] - f_i(\mathbf{x}) . \tag{2}$$

Then the average objective distance of every objective of population A_{set} is calculated as follows:

$$\bar{f}_i' = \frac{\sum_{j=1}^{|A_{set}|} f_i'(\mathbf{X}_j)}{|A_{set}|} . \tag{3}$$

Note that smaller $\bar{f}_i'$ means that the population set has better convergence performance at the i-th objective. Obviously, We aim to choose two objectives with the first and the second largest $\bar{f}'$ to make those two objectives become better.

Algorithm 2. Choose two objective

    Input: A population set, the best points
    Output: Two objectives: f_{c1}, f_{c2}
1 ChooseObj $(f_{c1}, f_{c2}, A_{set}, R_{point})$
2 **begin**
3   **for** $i \leftarrow 1$ to m do
4         $\bar{f}_i' \leftarrow 0$
5   **for** $j \leftarrow 1$ to $|A_{set}|$ do
6            $\bar{f}_i' \leftarrow \bar{f}_i' + (R_{point}[i] - f_i(A_{set_j}))$
7         $\bar{f}_i' \leftarrow \bar{f}_i'/|A_{set}|$ /*the average distance value of every objective is calculated*/
8      $f_{c1} \leftarrow \{f_i | \bar{f}_i' = \max_{j=1...,m}\{\bar{f}_j'\}\}$
9      $f_{c2} \leftarrow \{f_j | \bar{f}_j' = \max_{k=1...m, k\neq i}\{\bar{f}_k'\}\}$

---

3 Experimental Study

3.1 Test Problems and Parameter Settings

State-of-the-art algorithms including PPD-MOEA [11], MOEA/D [7], UMOEA/D [8], and an algorithm selecting objectives with random method (RSO) are considered as rival algorithms.

The search performance of the proposed algorithm is verified on many-objective 0/1 knapsack problems[11] with $m = \{4, 6, 8, 10, 12\}$ and $n = 100$ items, and the feasibility ratio ϕ is set to 0.5. More settings are described as follows:

1. In PPDSO-MOEA, UMOEA/D, PPD-MOEA and RSO, the population size N is set to 200 ($|P_t| = |Q_t| = 100$). The size of the archive population is set to 200. The population size of MOEA/D is the same as the number of weight vectors and cannot be arbitrarily specified. It is controlled by an integer H. More precisely, $\lambda^1, \lambda^2, \ldots, \lambda^N$ are all the weight vectors in which each individual weight takes a value from $\{0, 1/H, 2/H, \ldots, 1\}$. Then, the population size of MOEA/D N is C_{H+m-1}^{m-1}. We use the closest integer to 200 among the possible values as the population size (i.e., 220, 252, 120, 220 and 364 for 4-, 6-, 8-, 10-, 12-objective problems, respectively). Weighted sum approach is used in MOEA/D.

2. In PPDSO-MOEA, MOEA/D and RSO, we adopt uniform crossover with a crossover rate $P_c = 1.0$, and apply bit-flipping mutation with a mutation rate $P_m = 1/n$ (n is the number of items). In PPD-MOEA, two-point crossover with a crossover rate $P_c = 1.0$ and bit-flipping mutation with a mutation rate $P_m = 1/n$ are taken. Moreover, in UMOEA/D, single-point crossover with a crossover rate $P_c = 1.0$ and bitwise mutation with a mutation rate $P_m = 1/n$ are adopted.

3. In our algorithm, I_g varies for different test instance and $I_g = \{30, 10, 5, 10, 5\}$ for $m = \{4, 6, 8, 10, 12\}$, respectively. In PPD-MOEA, $r = 2$, $S_i = 0.35$ and $I_g = \{65, 20, 20, 15, 20\}$ for $m = \{4, 6, 8, 10, 12\}$. And in RSO, $I_g = \{45, 20, 15, 15, 20\}$ for $m = \{4, 6, 8, 10, 12\}$. T (number of the weight vectors in the neighborhood) in MOEA/D and UMOEA/D is set to 10 for all the test instances.

4. Each algorithm performs 30 independent runs for average. An algorithm terminates after 2000 generations.

3.2   Performance Metrics

Finally, to compare the performance of the algorithms on all tested problems, we use some metrics which are detailed later. 0/1 knapsack problems with different number of objective functions are considered as different test problems.

We adopt $Norm$ [14] and MS [15] to evaluate the convergence performance and diversity performance of the obtained POS. Higher value of $Norm$ indicates better convergence to true POS, and higher MS indicates better diversity in POS which can be approximated widely spread Pareto front. Moreover, in order to get statistically sound conclusions, the Wilcoxon signed-rank test [16] at a 0.05 significance level is adopted to test the significance of the differences between assessment results obtained by competing algorithms.

$Norm$ of population $\mathbf{P}$ is calculated by:

$$Norm(\mathbf{P}) = \frac{\sum_{i=1}^{|P|} \rho_i}{|\mathbf{P}|}. \tag{4}$$

Where ρ_i which denotes the norm of the $i - th$ individual $\mathbf{X}_i$ in $\mathbf{P}$ is calculated by

$$\rho_i = \sqrt{\sum_{j=1}^{m} f_j(\mathbf{X}_i)^2}. \tag{5}$$

Table 2. $Norm$ as we increase the number of objective m. Best performance is shown in bold. Considering the high time complexity of UMOEA/D, we did not compare it with other algorithms for $m = \{6, 8, 10, 12\}$. "†" indicates that the result of the considered algorithm is significantly different from that of PPDSO-MOEA at a 0.05 level by the Wilcoxon signed-rank test.

m	UMOEA/D	MOEA/D	RSO	PPD-MOEA	PPDSO-MOEA
4	6445.98(29.83)†	6786.52(21.36)†	6941.78(20.30)†	6953.04(9.23)†	**6956.92(10.20)**
6	-	7443.49(22.54)†	7659.95(46.16)†	7693.40(14.11)†	**7724.43(17.37)**
8	-	8480.86(26.19)†	8681.10(38.68)†	8721.67(23.41)†	**8747.29(25.79)**
10	-	9437.63(51.08)†	9677.94(49.78)†	9697.16(26.34)†	**9718.30(35.49)**
12	-	10146.1(42.70)†	10352.5(74.48)†	10416.8(43.50)†	**10450.9(45.55)**

MS of population **P** is calculated by:

$$MS(\mathbf{P}) = \sqrt{\sum_{i=1}^{m} max\{\|u_i - v_i\| \quad |\mathbf{u}, \mathbf{v} \in \mathbf{P}\}} \, . \tag{6}$$

3.3 Results and Discussion

We use 0/1 Knapsack problems with $m = \{4, 6, 8, 10, 12\}$ objectives. Table 2 and Table 3 show the $Norm$ values and the MS values respectively as we increase the number of objective functions. The values in the tables are the means and standard deviations. "†" in the tables indicates that the result of the considered algorithm is significantly different from that of PPDSO-MOEA at a 0.05 level by the Wilcoxon signed-rank test. Higher value of $Norm$ indicates better convergence to true POS, and higher MS value indicates higher diversity in POS which can be approximated widely spread Pareto front. Considering the high time complexity of UMOEA/D, we did not compare it with other algorithms for $m = \{6, 8, 10, 12\}$. [8] only tests the performance of UMOEA/D for $m = \{2, 3, 4\}$ on 0/1 knapsack problems.

From Table 2, we can see the proposed algorithm, PPDSO-MOEA, has better convergence than PPD-MOEA, MOEA/D, UMOEA/D and RSO for all test instances, which demonstrates the effectiveness of our method. The reason for the better convergence of PPDSO-MOEA is as follows: the main strategy of choosing objectives is choosing two objectives which have the first and the second worst average values to improve them and we use random method for one time to choose two objectives at the appropriate time. Moreover, we temporally switch the two-objective combination in every I_g generation with the help of the archive set and improve those two new worst objectives. In a word, our method for selecting two objectives results in better and better convergence performance.

From Table 3, we can see MOEA/D has better diversity than all other algorithms. PPDSO-MOEA has better diversity than PPD-MOEA for $m = \{8, 10, 12\}$ and slightly worse for $m = \{4, 6\}$. PPDSO-MOEA and RSO share the same diversity for $m = \{6, 8, 10, 12\}$ considering of Wilcoxon signed-rank test. The reason lying in this result is that we put emphasis on improving the convergence. As a result, we get the result with

Table 3. MS as we increase the number of objective m. Best performance is shown in bold. Considering the high time complexity of UMOEA/D, we did not compare it with other algorithms for $m = \{6, 8, 10, 12\}$. "†" indicates that the result of the considered algorithm is significantly different from that of PPDSO-MOEA at a 0.05 level by the Wilcoxon signed-rank test.

m	UMOEA/D	MOEA/D	RSO	PPD-MOEA	PPDSO-MOEA
4	38.209(4.37)†	**62.8786(1.18)**†	43.8704(2.06)†	43.9104(1.31)†	42.1732(1.89)
6	-	**77.8739(1.42)**†	57.5757(4.40)	59.1133(1.94)†	57.7400(2.67)
8	-	**78.4041(2.53)**†	67.3001(4.42)	66.2876(2.71)†	67.0534(1.92)
10	-	**80.3223(5.22)**†	75.5318(4.53)	73.5528(2.52)†	75.0812(3.72)
12	-	**87.6282(3.62)**†	82.9805(4.76)	80.4356(3.73)†	82.6011(2.21)

better convergence and slightly worse diversity. However, PPDSO-MOEA has better diversity than PPD-MOEA for most test problems, which demonstrates that our method for selecting two objectives is more effectiveness than PPD-MOEA.

4 Conclusions

In this paper, we propose a new MOEA algorithm, PPDSO-MOEA (Pareto partial dominance on two selected objectives MOEA), to improve the performance of ManyOPs on many-objective 0/1 knapsack problems. The algorithm calculates dominance between solutions using only two objectives selected from m objectives and reflects the result in parents selection. Those two objectives have the largest distances to the best point. Furthermore, PPDSO-MOEA updates the archive population A_{set} to maintain POS when it changes the two-objective combination. The search performance of PPDSO-MOEA is verified on many-objective 0/1 knapsack problems, with up to 12 objectives. Simulation results for 0/1 Knapsack problems show that significant improvement on $Norm$ was achieved by PPDSO-MOEA. POS obtained by PPDSO-MOEA achieved higher convergence than PPD-MOEA, MOEA/D, UMOEA/D and RSO for all test instances as well as better diversity than PPD-MOEA and UMOEA/D for most test instances.

In the future work, we plan to improve our algorithm and verify its effectiveness on many-objective continuous optimization problems. Also, more Pareto partial dominance need to be explored in the proposed algorithm framework.

References

1. Coello, C.A.C., Lamont, G.B., Veldhuizen, D.A.V.: Evolutionary Algorithms for Solving Multi-Objective Problems. Springer (2007)
2. Deb, K.: Multi-Objective Optimization using Evolutionary Algorithms. John Wiley & Sons, Chichester (2001)
3. K. Miettinen: Nonlinear Multiobjective Optimization. Springer, vol. 12 (1999)
4. Farina, M., Amato, P.: On the Optimal Solution Definition for Many-Criteria Optimization Problems. In: Proceedings of the NAFIPS-FLINT International Conference, pp. 233–238 (2002)
5. Deb, K., Pratap, A., Agarwal, S., Meyarivan, T.: A Fast and Elitist Multiobjective Genetic Algorithm: NSGA-II. IEEE Transactions on Evolutionary Computation 6(2), 182–197 (2002)

6. Zitzler, E., Künzli, S.: Indicator-based selection in multiobjective search. In: Yao, X., et al. (eds.) PPSN VIII 2004. LNCS, vol. 3242, pp. 832–842. Springer, Heidelberg (2004)
7. Zhang, Q., Li, H.: MOEA/D: A multiobjective evolutionary algorithm based on decomposition. IEEE Transactions on Evolutionary Computation 11(6), 712–731 (2007)
8. Tan, Y., Jiao, Y., Li, H., et al.: MOEA/D+ uniform design: A new version of MOEA/D for optimization problems with many objectives. Computers & Operations Research 40(6), 1648–1660 (2013)
9. Liu, H.L., Gu, F., Zhang, Q.: Decomposition of a Multiobjective Optimization Problem into a Number of Simple Multiobjective Subproblems. IEEE Transactions on Evolutionary Computation 18(3), 450–455 (2014)
10. Deb, K., Jain, H.: An Evolutionary Many-Objective Optimization Algorithm Using Reference-point Based Non-dominated Sorting Approach, Part I: Solving Problems with Box Constraints. IEEE Transactions on Evolutionary Computation PP(99) (2013)
11. Sato, H., Aguirre, H.E., Tanaka, K.: Pareto partial dominance moea and hybrid archiving strategy included cdas in many-objective optimization. In: 2010 IEEE Congress on Evolutionary Computation (CEC), pp. 1–8. IEEE (2010)
12. Saxena, D., Duro, J., Tiwari, A., Deb, K., Zhang, Q.: Objective reduction in many-objective optimization: Linear and nonlinear algorithms. IEEE Transactions on Evolutionary Computation 17(1), 77–99 (2013)
13. López Jaimes, A., Coello Coello, C.A., Aguirre, H., et al.: Objective space partitioning using conflict information for solving many-objective problems. Information Sciences 268, 305–327 (2014)
14. Sato, M., Aguirre, H.E., Tanaka, K.: Effects of δ-Similar Elimination and Controlled Elitism in the NSGA-II Multiobjective Evolutionary Algorithm. In: IEEE Congress on Evolutionary Computation, CEC 2006, pp. 1164–1171. IEEE (2006)
15. Zitzler, E.: Evolutionary algorithms for multiobjective optimization: Methods and applications, vol. 63. Shaker, Ithaca (1999)
16. Wikipedia contributors: Wilcoxon signed−rank test. Wikipedia, The Free Encyclopedia. Wikipedia, The Free Encyclopedia (July 1, 2014) (Web July 6, 2014)

Analysis on a Multi-objective Binary Disperse Bacterial Colony Chemotaxis Algorithm and Its Convergence

Tao Feng[1], Zhaozheng Liu[1,2], and Zhigang Lu[1]

[1] Department of Power Electronics for Energy Conservation and Motor Drive,
Yanshan University, Qinhuangdao, Hebei, 066004, China
[2] State Grid Jibei Electric Power Company Limited Qinhuangdao Power Supply Company
Qinhuangdao 066004 China
fengtao@ysu.edu.cn,
liuzhaozheng666@163.com,
zhglu@ysu.edu.cn

Abstract. A simple, convenient and efficient multi-objective binary disperse optimized bacterial colony chemotaxis algorithm (MDOBCC) is proposed, in which the Disp(disperse update mechanism) is defined to handle 0-1 disperse optimization problems. The concept of chemotaxis center is proposed with the item of group and chemotaxis in order to improve the convergence rate of the algorithm. The definition of reference colony is used to retain the elite solution produced during the iteration; the definition of colony spatial radius and density is used to guide the bacteria for determinate variation, thus keeping the algorithm obtain even-distributed Pareto optimum solution set. Furthermore, the derivation analysis is given to prove the convergence of the algorithm and comes to the conclusion of global convergence. The simulate result confirmed the effectiveness of the algorithm.

Keywords: bacterial colony chemotaxis, disperse update mechanism, reference colony, determinate variation.

1 Introduction

In most cases, the objectives of multi-objective disperse optimization problems (MDOP) are mutually conflicted which is different from single-objective optimization, making it difficult to achieve optimization for many objectives simultaneously. There are some efficient intelligent optimized algorithms with solving 0-1 disperse combination optimization problems [1], [2], [3], [4], [5]. Kennedy [1] etc. put forward disperse binary version of PSO algorithm, which uses the Sigmoid function of speed to represent the possibility of location and status change. However, this update mechanism exert restriction on the variation range of speed, lowering the efficiency in dealing with disperse optimization problems. Reference [6] defined the binary arithmetic operation. However, this will increase the calculating cost and consume long time.

This paper proposes a MDOBCC which expect for the fast convergence speed. When dealing with multi-objective disperse optimization problems, Disp is defined,

Y. Tan et al. (Eds.): ICSI 2014, Part I, LNCS 8794, pp. 374–385, 2014.

which can make adjustment according to actual problems, offsetting the shortages of ordinary algorithm's restriction on variable range. In addition, this paper put forward the concept of chemotaxis center which can accelerate the speed for the bacteria in seeking optimal solution. The reference colony, and the colony spatial radius and density are proposed to guide the bacteria for effective variation. So the evenly-distributed Pareto Fronts can be obtained finally.

2   MDOBCC

2.1   Algorithm of Bacteria Colony Chemotaxis

Generally, the Pareto optimal solution [7] of multi-objective optimization problem is a set. As a new functional optimized algorithm inspired by biological behavior, BCC algorithm simultaneously applies the stress reaction of single bacteria under the attractant environment and the interaction of positional information among bacterial colonies for functional optimization. Bremermann [8] and his colleagues [9], [10] studied BCC algorithm firstly. Miiller[11] carried out a further research and proposed the bacterial chemotaxis algorithm on this basis. Reference [11] provides the two-dimension model of bacterial chemotaxis algorithm as well as its optimization process.

2.2   Disperse Update Mechanism

The Disp which was defined in this paper shows as follows:

$$Disp(x) = \begin{cases} \dfrac{x}{k+x} & x \in [0,+\infty) \\ \dfrac{-x}{k-x} & x \in (-\infty,0) \end{cases} \tag{1}$$

$$\rho = \begin{cases} 1 & r \geq Disp(x) \\ 0 & r < Disp(x) \end{cases} \tag{2}$$

where r is the random number within the interval $(0,1)$, and Disp is the possibility for $\rho=1$. The smaller the Disp is, the greater the probability for $\rho=1$. The sign of x represents the direction of bacteria's position. At this moment, x has specific physical significance. This update mechanism is able to self adjust the value of k according to the actual step length without need to exert restriction on the range of independent variables.

2.3   Operators

This paper proposes three operators, such as chemotaxis center, reference colony, colony spatial radius and density, for further improving the performance of the algorithm.

- Chemotaxis center

Fig 1 shows the calculation of bacterial colony chemotaxis center composed by 4 bacteria when there are two target functions. In Fig (a), the open hole represents the previous location of the bacteria while the filled circle means the current location. From the Fig, we can see that 1, 2, and 3 are the bacteria which move to better locations. Then the chemotaxis vector of this bacterial colony is recorded as the average of $d1$, $d2$ and $d3$. The open hole in Fig (b) is the non-dominate center of individual 4. The non-dominate center moves to position A along with the chemotaxis vector d, thus form the chemotaxis center.

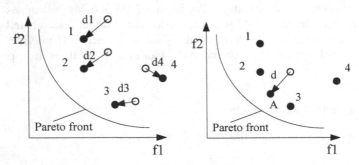

Fig. 1. Calculation of bacterial colony chemotaxis center (a)Move trace (b)Form chemotaxis center

- Reference colony

This paper takes the better location of $\overrightarrow{x}_{new1}$ and $\overrightarrow{x}_{new2}$ as the reference colony. If there's no mutual domination, take $\overrightarrow{x}_{new2}$ as the reference colony.

$$\overrightarrow{x}_{ref} = \overrightarrow{x}_{new1} + a\left(\overrightarrow{x}_{new2} - \overrightarrow{x}_{new1}\right) \tag{3}$$

In this formula, if $\overrightarrow{x}_{new2} \succ \overrightarrow{x}_{new1}$, $a = 1$, otherwise, $a = 0$.

- Colony spatial radius R_{av}

$$R_{av} = \frac{\sum_{i=1}^{n} r_i}{n} \qquad i = 1, 2, 3, \dots n \tag{4}$$

where r_i is the Euclidean distance between individual bacterium, and n is the total amount of Euclidean distance between any two bacteria.

- Colony spatial density η

$$\eta_i = \frac{V_i}{N} \quad i = 1, 2, 3, ...N \tag{5}$$

where V_i is the amount of bacterium i that contains other bacteria within the spatial radius R_{av} and N is the scale of bacterial colony.

The bigger the spatial density is, the more intensive the colony is. Guide the over-crowded bacteria to move to the area with small spatial density for determinate variation. The movement location is the center between target bacterium and its nearest neighbor bacterium.

2.4   MDOBCC of Disperse Property

Extend based on the BCC algorithm and apply it on MDOP. Use the Pareto-dominated concept to evaluate the strengths and weaknesses between two bacteria. Suppose there are k bacterial target functions and n dimensions of discrete variable. The calculation steps are as follows:

1) Initialized bacterial colony generates n bacteria randomly with a steady speed. In this paper, v=1. Elite set is set and good solutions are selected from the bacterial colony generated randomly and put into the elite set.

2) Individual bacterial optimization and compute new location $\overrightarrow{x}_{new1}$;

- Compute the move time τ of bacterial on the new direction, which is determined by probability distribution. Its density function can be expressed as:

$$P(X = \tau) = \frac{1}{T} \ell^{-\tau/T} \tag{6}$$

$$T = \begin{cases} T_0 & if\,(\overrightarrow{x}_{pre} \succ \overrightarrow{x}_{cur})|((\overrightarrow{x}_{pre} \sim \overrightarrow{x}_{cur}) \\ T_0(1+b*\min(\left|\dfrac{f_{pr1}}{l_{pr}}\right|, \left|\dfrac{f_{pr2}}{l_{pr}}\right|)) & if\,\overrightarrow{x}_{pre} \prec \overrightarrow{x}_{cur} \end{cases} \tag{7}$$

- Compute new direction of movement. The included angle between the new direction and original track is subject to Gaussian probability distribution, and the deflection to left and right can be expressed respectively as:

$$P(X = \alpha, v = \mu) = \frac{1}{\sqrt{2\pi}\sigma} \exp\left[-\frac{(\alpha - v)^2}{2\sigma^2}\right]$$

$$P(X = \alpha, v = -\mu) = \frac{1}{\sqrt{2\pi}\sigma} \exp\left[-\frac{(\alpha - v)^2}{2\sigma^2}\right]$$

(8)

where $\alpha \in [0°, 180°]$. The mathematical expectation and variance expression is:

$$(\mu, \sigma) = \begin{cases} \begin{pmatrix} \mu = 62^0 \\ \sigma = 26^0 \end{pmatrix} & if \ (\vec{x}_{pre} \succ \vec{x}_{cur})|((\vec{x}_{pre} \sim \vec{x}_{cur}) \\ \begin{pmatrix} \mu = 62^0(1 - \cos(\theta)) \\ \sigma = 26^0(1 - \cos(\theta)) \end{pmatrix} & if \ x_{pre} \prec x_{cur} \end{cases}$$

(9)

- Compute location $\vec{x}_1$, determined by the formula (10).

$$\vec{x}_1 = \vec{x}_{pre} + v \cdot \tau$$

(10)

- Update bacterial location by using Disp disperse update mechanism and produce the new location $\vec{x}_{new1}$.

3) Bacterial colony optimization and compute new location $\vec{x}_{new2}$

- Compute non-dominant center. Each bacterium has to perceive the surrounding environment and explore whether there's bacteria with better location. If yes, take the center of these bacteria as the non-dominant center $\vec{x}_{cen}$.

$$\vec{x}_{cen} = Aver \ (\vec{x}_j | f(\vec{x}_j) \succ f(\vec{x}_i))$$

(11)

where $Aver(\vec{x}_1, \vec{x}_2, .., \vec{x}_m) = \left(\sum_{j=1}^{m} \vec{x}_j\right) / m$, m is the amount of dominate bacteria,

and $dis(\vec{x}_j, \vec{x}_i)$ is the distance between bacteria i and bacteria j.

- Compute the chemotaxis center $\vec{x}_{cen\_ce}$.

- Compute location $\vec{x}_2$ that is determined by formula (12).

$$\vec{x}_2 = \vec{x}_{pre} - 2U(0,1) \times (\vec{x}_{pre} - \vec{x}_{cen\_che}) \tag{12}$$

Where $U(0,1)$ is the random number within the interval $[0,1]$ and subject to even distribution.

• Update bacterial location by using Disp disperse update mechanism and produce the new location $\vec{x}_{new2}$.

4) The reference colony $\vec{x}_{ref}$ is produced and bacteria move to the new location.

5) Guide the bacteria for determinate variation with the colony spatial radius and density, and select good solutions in the new colony and put into the elite set. At the same moment, remove the poor solutions out of the elite set.

6) Judge whether the terminal condition is satisfied. If yes, end the cycle, otherwise, return to step 2) for further optimization. Finally, the bacteria in the elite set are the Pareto optimal solution.

3   Convergence Analysis

Hypothesis 1. (a) The feasible region Ω of optimal problem is the bounded closed region within R^n; (b) The target function $f_i(x), i = 1, 2, \cdots, k$ is the continuous function in area Ω.

From hypothesis 1, we can see in the formula (13):

$$S = \{(x_1, x_2, \cdots, x_k) \mid \arg \min_{x_i \in \Omega} f_i(x_i)\} \neq \phi \tag{13}$$

Definition 1. Suppose $X_{i_n}(n = 1, 2, \cdots)$ is the random sequence defined in probability space. If there's random variable ε, $\forall \varepsilon > 0$. If $P\{\bigcap_{n=1}^{\infty} \bigcup_{m \geq n} [|X_{i_m} - X_i^*| \geq \varepsilon]\} = 0$, this random sequence is converged to X_i^* at the probability 1.

For $\forall \varepsilon > 0$, if

$$D_0 = \{x \in \Omega \mid |f_i(x) - f_i^*| < \varepsilon\}; D_1 = \Omega \setminus D_0 \tag{14}$$

where $f_i^* = \min\{f_i(x) : x \in \Omega\}$, then n points produced in the algorithm can be divided into two states:

(a) There is at least one point belonging to D_0, recorded as state S_0;

(b) All n points belong to D_1, recorded as state S_1.

The colony sequence $X(1), X(2), \cdots, X(m), \cdots$ is a monotonous one, which means that the Pareto optimal solution set produced by iteration of MODBCC algorithm won't be worse or at least better than the solution of previous iteration.

Lemma 1. Suppose the target function and feasible region of optimal problem satisfy hypothesis 1. If $q_{ij}(i, j = 0,1)$ represents the probability when $X(t)$ belongs to state S_i, $X(t+1)$ belongs to state S_j, then:

(a) For the point set $X(t)$ of any state belongs to S_0, $q_{00} = 1$

(b) For the point set $X(t)$ of any state belongs to S_1, there's a constant $c \in (0,1)$ that enables $q_{11} \leq c$.

Please refer to the reference [12] for prove.

Theorem 1. Suppose $\{X(m)\}$ is the colony sequence generated by MDOBCC algorithm. If the target function and feasible region of optimal problem satisfy hypothesis 1, the colony sequence converges to Pareto Front at the probability 1.

Proof:

For $\forall \varepsilon > 0$, make

$$P_k = P\left\{ \left| f_i\left(X^*(k)\right) - f_i^* \right| > \varepsilon \right\}$$
(15)

then

$$P_k \begin{cases} 0, \exists T \in (0,...,k), X^*(T) \in D_0 \\ \bar{P}_k, X^*(m) \notin D_0, m = 0,1,...,k \end{cases}$$
(16)

From Lemma 1,

$$\sum_{k=1}^{\infty} \bar{P}_k = \sum_{k=1}^{\infty} q_{11} \leq \sum_{k=1}^{\infty} c^k = \frac{c}{1-c} < \infty$$
(17)

and Borel-Cantelli theorem, we can deduce that:

$$P\left\{\bigcap_{n=1}^{\infty}\bigcap_{m\geq n}\left[\left|f_i\left(X^*(k)\right)-f_i^*\right|\geq\varepsilon\right]\right\}=0 \qquad (18)$$

Therefore, the conclusion is correct according to definition 1 and the property of MODBCC algorithm. The proof is finished here.

4   Simulation

Apply MODBCC with and without operators and NSGA-II respectively for simulation, among which NSGA-II applies binary coding to obtain Pareto Fronts. The parameter setting of MODBCC algorithm can refer to [8]. The initial precision is set as 2.0 and end precision is set as $\varepsilon_{end}=10^{-6}$ with a precision update constant of 1.05. The coefficient k of disperse mechanism values 0.1, and the probability of determinate variation $P_m=0.05$. The cross probability of NSGA-II is 0.8, and the variation probability is 0.01. The colony scales of all these three algorithms are 100 and the test function iterates 100 times except for the ZDT series function which iterates 200 times. This paper applies six classical test functions to examine the performance of the algorithm, including the research achievement SCH[13], DEB[14], ZDT1 and ZDT2 [15-16], SRN with constraint condition[17-18]. Binary coding and discretization treatment are applied to test function while the optimization is conducted in the discrete area. The bacteria dimensions of SCH, DEB and SRN, ZDT1 and ZDT2, FON are 10, 20, 180, and 30 respectively. The number of variables are 1, 2, 30, 30, 3and 2. The test functions are all with two objectives.

4.1   Performance Evaluation

In order to evaluate the comprehensive performance of the algorithm, CM[3], SP[3] and RVI performance evaluation indexes are applied in this paper to measure the advantages of the algorithm performance.

- Effective solution proportion index (RVI) refers to the proportion of non-dominant solutions in all solutions, which is used to examine the convergence degree of final solutions. The bigger the RVI index value is, the better performance the algorithm is.

$$RVI=\frac{V_{al-indv}}{N} \qquad (19)$$

where $V_{al-indv}$ means that the non-dominant solutions minus the bacteria amount in same location.

The algorithm carries out for 30 times on each index to obtain their mean value Mean and standard deviation Stdev. Table 1 to 3 shows the Mean and Stdev of CM, SP, and RVI index for NSGA-II, MODBCC with and without operators respectively.

From the table of comparisons, we can see that all indexes of MODBCC with operators are better than that of MODBCC without operators. The performance of MODBCC and NSGA-II is equal to each other in terms of CM index. However, MODBCC algorithm achieves better Mean and Stdev in terms of SP and RVI indexes, indicating its better performance than NSGA-II algorithm.

Table 1. Comparison table of CM index

Standard test function		SCH	ZDT1	ZDT2	FON
MODBCC	Mean	1550	1502	775	1620
(with operators)	Stdev	313	1033	41	342
MODBCC	Mean	3500	1856	26681	19850
(without operators)	Stdev	190	3824	64024	3560
NSGA-II	Mean	1612	1932	569	1120
	Stdev	316	362	165	635

Table 2. Comparison table of SP index

Standard test function		SCH	DEB	ZDT1	ZDT2	FON	SRN
MODBCC	Mean	9300	6300	10500	8600	3700	10600
(with operators)	Stdev	4000	1015	5300	4500	1500	2300
MODBCC	Mean	17700	98000	24500	30000	87800	36700
(without operators)	Stdev	2400	20500	5700	5900	13800	36900
NSGA -II	Mean	16480	173200	73900	27200	240850	24470
	Stdev	1600	10200	25800	4300	57290	5700

Table 3. Comparison table of RVI index

Standard test function		SCH	DEB	ZDT1	ZDT2	FON	SRN
MODBCC (with operators)	Mean	1.000000	0.980000	0.991800	0.992700	0.994000	0.978500
	Stdev	0.000000	0.001667	0.002603	0.002517	0.002020	0.007508
MODBCC (without operators)	Mean	0.920000	0.825714	0.892500	0.874000	0.960000	0.988571
	Stdev	0.018973	0.048819	0.030956	0.034985	0.042426	0.009009
NSGA-II	Mean	0.851250	0.914000	0.896000	0.888000	0.914012	1.000000
	Stdev	0.046904	0.054772	0.011694	0.017321	0.025000	0.000000

4.2   Figures of Simulation Results

In order to further compare the performances among these three algorithms, this paper listed the Figures of simulation results of three algorithms corresponding to SCH, DEB, and FON test functions, shown in Fig 2 to 4. From the simulation Fig, we can see that MODBCC with operators is better than the other two algorithms in view of test problems of both continuous and discontinuous in the front of Pareto, as it can converge to the front of Pareto more effectively and obtain more evenly distribution.

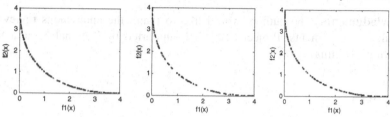

Fig. 2. Simulation result of optimized SCH function (a)MODBCC(with operators) (b)MODBCC(without operators) (c)NSGA-II

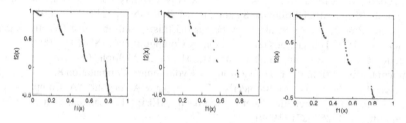

Fig. 3. Simulation result of optimized DEB function (a)MODBCC(with operators) (b)MODBCC(without operators) (c)NSGA-II

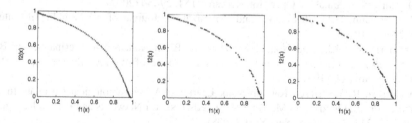

Fig. 4. Simulation result of optimized FON function (a)MODBCC(with operators) (b)MODBCC(without operators) (c)NSGA-II

5   Conclusion

With consideration to the shortages that the ordinary 0-1 disperse mechanism exerts restriction on independent variable range and low efficiency in optimizing disperse problems, this paper defines Disp and put forward a high effective MODBCC. Furthermore, it defines three operators: chemotaxis center, reference colony as well as colony spatial radius and density, which improves the performance of algorithm significantly and accelerates its convergence speed, thus enabling the obtained solution set to distribute evenly on the optimal frontier of Pareto. The simulation results indicate that this algorithm is obvious advantageous and effective in optimizing MODP.

Acknowledgments. The authors would like to thank the anonymous reviewers for their valuable comments. Project 61071201 supported by National Natural Science Foundation of China.

References

1. Kennedy, J., Eberhart, R.C.: A Discrete Binary Version of the Particle Swarm Algorithm. In: The World Multiconference on Systemics, Cybernetics and Informatics, pp. 4104–4108. IEEE Press (1997)
2. Farzaneh, A., Alireza, M., Ashkan, R.K.: A Novel Binary Particle Swarm Optimization Method Using Artificial Immune System. In: EUROCON, pp. 217–220 (2005)
3. Deb, K., Pratap, A., Agarwal, S.: A Fast and Elitist Multi-objective Genetic Algorithm: NSGA-II. IEEE Transactions on Evolutionary Computation 6, 182–197 (2002)
4. Coello, C.A.C., Toscano, G., Salazar, M.: Handling Multiple Objectives with Particle Swarm Optimization. IEEE Transactions on Evolutionary Computation 8, 256–279 (2004)
5. Ziteler, E., Thiele, L.: Multiobjective Evolutionary Algorithms: A Comparative Case Study and the Strength Pareto Approach. IEEE Transactions on Evolutionary Computation 3, 257–271 (1999)
6. Clerc, M.: Discrete Particle Swarm Optimization (2000), Illustrated by the Traveling Salesman Problem, http://www.Mauricecierc.net/2000
7. Chinchuluun, A., Pardalos, P.: A Survey of Recent Developments in Multi-objective Optimization. Annals of Operation Research 154, 29–50 (2007)
8. Bremermann, H.J.: Chemotaxis and Optimization. Journal of the Franklin Institute 297, 397–404 (1974)
9. Bremermann, H.J., Anderson, R.W.: How the Brain Adjusts Synapsesmaybe. In: Boyer, R.S. (ed.) Automated Reasoning: Essays in Honor of Woody Bledsoe, pp. 119–147. Kluwer, Norwell (1991)
10. Anderson, R.W.: Biased Random-walk Learning: A Neurobiological Correlate to Trial-and-error. In: Omidvar, O.M., Dayhoff, J. (eds.) Neural Networks and Pattern Recognition, pp. 221–244. Academic, New York (1998)
11. Miiller, S.D., Marchetto, J., Airaghi, S.: Optimization Based on Bacterial Chemotaxis. IEEE Transactions on Evolutionary Computation, 16–19 (2002)
12. Guo, C.H., Tang, H.W.: Global Convergence Properties of Evolution Strategies. Mathematica Numerica Sinica 23, 106–110 (2001)
13. Schaffer, J.D.: Multiple Objective Optimization with Vector Evaluated Genetic Algorithms. In: Proceedings of the 1st International Conference on Genetic Algorithms, pp. 93–100 (1985)
14. Deb, K.: Multi-Objective Genetic Algorithms: Problem Difficulties and Construction of Test Problems. Evolutionary Computation. 7, 205–230 (1999)

15. Deb, K.: Scalable Test Problems for Evolutionary Multi-objective Optimization, pp. 105–145. Springer, Heidelberg (2005)
16. Yen, G.G., Lu, H.: Dynamic Multiobjective Evolutionary Algorithm: Adaptive Cell Based Rank and Density Estimation. IEEE Transactions on Evolutionary Computation 7, 253–274 (2003)
17. Srinivas, N., Deb, K.: Multiobjective Function Optimization Using Nondominated Sorting Genetic Algorithms. Evolutionary Computation 2, 221–248 (1994)
18. Binh, T.T., Korn, U.: MOBES: A Multiobjective Evolution Strategy for Constrained Optimization Problems. In: Proc. of the IMACS World Congress on Scientific Computation, Berlin, Germany, pp. 357–362 (1997)

Multi-objective PSO Algorithm for Feature Selection Problems with Unreliable Data

Yong Zhang[*], Changhong Xia, Dunwei Gong, and Xiaoyan Sun

School of Information and Electrical Engineering, China University of Mining and Technology,
Xuzhou, Jiangsu, 221116
yongzh401@126.com

Abstract. Feature selection is an important data preprocessing technique in classification problems. This paper focuses on a new feature selection problem, in which sampling data of different features have different reliability degree. First, the problem is modeled as a multi-objective optimization. There two objectives should be optimized simultaneously: reliability and classifying accuracy of feature subset. Then, a multi-objective feature selection method based on particle swarm optimization, called JMOPSO, is proposed by incorporating several effective operators. Finally, experimental results suggest that the proposed JMOPSO is a highly competitive feature selection method for solving the feature selection problem with unreliable data.

Keywords: particle swarm optimization, feature selection, unreliable data, ulti-objective.

1   Introduction

Purpose of a feature selection problem (FSP) is to move the useless features without sacrificing the predictive accuracy. Particle swarm optimization (PSO) is a heuristic search technique that is inspired by the behavior of bird flocks [1]. Due to its advantages such as the simplicity, the fast convergence and the population-based feature, the attention of researchers upon PSO-based feature selection methods is much high in recent years [2-7].

In those studies above, however, they all focused on the case that all sampling data are completely reliable. As we know, equipments that are used to generate feature data often have some degree of error in real world. For the features whose sampling data are from high precision equipments, obviously decision makers should assign high reliability degrees. On the contrary, for those sampling data generated by low precision equipments, their corresponding features should be assigned low reliability degrees. The kind of feature selection problems is general in real life.

In this paper, a novel multi-objective feature selection method based on particle swarm optimization is proposed and implemented for solving the above feature selection problem with unreliable data. First, an encoding strategy based on the

---

[*] Corresponding author.

Y. Tan et al. (Eds.): ICSI 2014, Part I, LNCS 8794, pp. 386–393, 2014.

selection probability of feature is proposed, based on which a feature selection problem with discrete variables is transformed to a continuous problem suitable to PSO. Then, a multi-objective particle swarm approach based on the adaptive jump operator, called JMOPSO, is adopted to optimize the above problem.

The rest of this paper is organized as follows. Section 2 gives multi-objective optimization model of the feature selection problem with unreliable data. Section 3 introduces the proposed multi-objective feature selection method in detail. Section 4 presents experimental results. Finally, Section 5 gives the concluding remarks.

2   Problem Modeling

Like most existing studies [2, 3], we use a binary string to represent a solution. For a set of pending data with D features, if a bit is 1, the corresponding feature is chosen into the feature subset; if 0, it is not. Then, a feature selection problem can be described a combinatorial optimization problem, whose variables only have two values $\{0, 1\}$. Based on it, decision variables of the problem can be represented as follows:

$$X = (x_1, x_2, \cdots, x_D), x_i \in \{0,1\}, i = 1, 2, \cdots, D. \tag{1}$$

Since sampling data of features become unreliable, it is necessary to consider the reliability degree (RD), not just the classifying accuracy (CA), when estimating a feature subset. First, considering the RD, without loss of generality, a value within [0, 1] is used to represent the reliability of a feature. The bigger this value is, the higher the RD value of this feature is; especially, the value 1 means that its corresponding feature is completely credible. So, the RD values of all the D features can be represented by the following vector:

$$E = [e_1, e_2, \cdots, e_D], e_i \in [0,1], i = 1, 2, \cdots, D. \tag{2}$$

For a solution X, furthermore, we use the average of all selected features to estimate the whole reliability degree of X in this paper, as follows:

$$f_1(X) = \frac{\sum\limits_{i=1}^{D} x_i \times e_i}{\sum_{i=1}^{D} x_i} \tag{3}$$

Considering the classifying accuracy (CA), in this paper we adopt the leave-one-out cross-validation (LOOCV) of 1-NN to evaluate the classifying accuracy of a feature subset [3, 5]. In this method, a single datum dat_i from the original data set DAT is selected as a testing sample, and the rest constitute training samples. Then the 1-NN classifier predicts the class of the testing sample by calculating and sorting the distances between the testing sample and the training ones. If the prediction class is right, than the flag of the testing sample $y(dat_i)$ is assigned as 1; otherwise, assigned as 0. This process is repeated so that each datum in the original data set is used once as the testing sample. Finally, the classification accuracy of this feature subset can be calculated by the proportion of correctly predicted samples to all samples:

$$f_2(X) = \frac{1}{|DAT|} \sum_{i=1}^{|DAT|} y(dat_i) \tag{4}$$

Where, $|DAT|$ is the size of original data set DAT.

3   MOPSO-Based Feature Selection Method

This section describes the proposed multi-objective feature selection algorithm. The motivation for this algorithm is to design an effective multi-objective PSO technique, which is suitable to the feature selection problem.

3.1   Encoding of Particles

In this paper, the probability that a feature is chosen into a feature subset is taken as an encoded element, and multiple elements form a particle which represents a candidate solution of the feature selection problem. Taking a data set with D features as an example, the i-th particle in the swarm can been coded by a D-bit real string as follows:

$$X_i = (x_{i,1}, x_{i,2}, \cdots, x_{i,D}), \quad x_{i,j} \in [0,1], \ j = 1, 2, \cdots, D, i = 1, 2, \cdots, N_s, \tag{5}$$

Where N_s is the swarm size, and $x_{i,j} \in [0,1]$ is the probability which the j-th feature is chosen into the feature subset.

For the particle X_i , furthermore, its corresponding decoding solution Z_i is constructed by the following equation:

$$z_{i,j} = \begin{cases} 1, & x_{i,j} \geq rand \\ 0, & \text{otherwise} \end{cases} \tag{6}$$

Where $z_{i,j} = 1$ represents that the j-th feature is chosen into the feature subset Z_i .

3.2   The Adaptive Jump Operator

In order to improve the search capability of JMOPSO, this paper proposes a new method to determine the jump probability of particles on each dimension. Based on it, those stagnation dimensions can jump old positions with a high probability. In previous works [8-10] it has been incorporated in the PSO to improve its performance, by assigning the same jump probability on all the dimension spaces.

In each iteration, the new jump operator first identifies all stagnation dimensions, then each stagnation dimension is assigned a fixed probability to jump old position.

In order to identify those stagnation dimensions, we take the current archive and the swarm as an input, and take a set *SET0* to save the tags of stagnation dimensions. For each dimension space, all the solutions are checked in turn (from 1 to $|Ar|+N_s$). Taking the j-th dimension space as an example, if the values of these solutions on this dimension space all are 1, then this dimension is called to be stagnant, and its tag is saved in the set *SET0*; Similarly, if the values of these solutions on this dimension space all are 0, then it is also called to be stagnant, and its tag is saved in the set *SET0*.

```
Function JUMP:   NEW_X=JUMP (X_i (t)  SET0, D)
//* X_i (t): the i-th particles, SET0: the set to save
tags of all the stagnation dimensions*//
FOR j=1 to D
IF j∈SET0            //*the j-th dimension is stagnant *//
IF X_{i,j} (t)=0 or 1         //*the j-th dimension of the
particle is equal to 1 or 0*//
```

$$X_{i,j}(t) = \begin{cases} \left|1 - X_{i,j}(t)\right|, & rand < 1/N_s \\ X_{i,j}(t), & otherwise. \end{cases}$$

```
ENDIF
ENDIF
ENDFOR
NEW_X←X_{i,j} (t)    //Return the new particle after jump *//
```

By applying the above techniques, all the stagnation dimensions are identified. Then, each particle is checked in turn against all the stagnation dimensions found. Taking the i-th particle $X_i(t)$ as example, if its value on the j-th dimension is equal to 1 or 0, and the j-th dimension also is stagnant, then this dimension of the particle, $X_{i,j}(t)$, will get a chance of $1/N_s$ to jump old value. After jump, the old value 1 becomes 0, and the old value 0 becomes 1. In other words, those died features will be waked up with the $1/N_s$ probability, and those permanent features will also be clear up with the same probability.

3.3   Implement of the Proposed JMOPSO

1) Initialize. First, set relative parameters, including the swarm size N_s, the archive size N_a, and the terminal condition of the swarm. Second, initialize the positions of N_s particles and   the personal best position of each particle.

2) Calculate the objective values of particles based on equations (3), (4) and (6).

3) Update the external archive. In this paper an external archive is introduced to store the non-dominated solutions found during the entire search process. And the crowding distance proposed for multi-objective genetic algorithms [11] is adopted to estimate the diversity of elements among the archive. If the archive has reached its maximal capacity N_a, then the most sparsely spread N_a solutions, i.e., N_a solutions with the largest crowding distance values, are retained in the archive.

4) Update the global best position. In order to exploit sparse areas that include few solutions, the *Gbest* of each particle is selected from the archive on the basis of the solutions' diversity. For any solution in the archive, the higher its crowding distance value is, the higher its probability selected as *Gbest* is.

5) Update the particles' positions and run the adaptive jump operator. In this paper an update method without acceleration coefficients, which is introduced by [12], is used as follows:

$$v_{i,j}(t+1) = w \cdot v_{i,j}(t) + rand \cdot (pb_{i,j}(t) - x_{i,j}(t)) + rand \cdot (gb_{i,j}(t) - x_{i,j}(t))$$
$$x_{i,j}(t+1) = x_{i,j}(t) + v_{i,j}(t+1) \tag{7}$$

Where *rand* is a random number within [0,1]; w is the inertia weight, this paper sets $w=0.4$.; $Pb_i(t) = (pb_{i,1}(t), pb_{i,2}(t), \cdots, pb_{i,D}(t))$ represents the personal best position of the i-th particle; $Gb_i(t) = (gb_{i,1}(t), gb_{i,2}(t), \cdots, gb_{i,D}(t))$ is the global best position of this particle. After that, run the adaptive jump operator to improve the search capability of the swarm.

6) Update the personal best position. For a particle, supposing that its position is $X_i(t+1)$ at the $(t+1)$-th iteration, its *Pbest* value before updated is $Pbest_i(t)$, the novel update strategy is as follows:

$$Pb_i(t+1) = \begin{cases} X_i(t+1), & \text{if } Pb_i(t) \prec X_i(t+1) \\ X_i(t+1), & \text{if } Pb_i(t) \parallel X_i(t+1) \ \& \ | Z(Pb_i(t)) | \geq Z_i(t+1)) | \\ Pb_i(t), & \text{otherwise} \end{cases} \tag{8}$$

where $Z_i(t+1)$ is the decoded solution of $X_i(t+1)$, $|Z|$ represents the number of features in the subset Z; $Pb_i(t) \prec X_i(t+1)$ represents that $X_i(t+1)$ dominates $Pb_i(t)$, and $Pb_i(t) \parallel X_i(t+1)$ represents that both solutions do not dominate each other.

7) Judge whether the algorithm meets termination criterion. If yes, stop the algorithm, and output the archive as finial result; otherwise, go to step 3.

4   Experiment and Analysis

Effectiveness of the proposed algorithm was demonstrated on several real-world benchmark datasets. And for a fair performance comparison, two state-of-the-art multi-objective algorithms are applied in the feature selection problem.

4.1   Preparation Work

To construct an appropriate test datasets, it seems feasible to choose some popular test problems. In this paper, several well known real-world benchmark datasets, including WDBC, Ionosphere and Sonar, are selected. By using the quality vector $E = [e_1, e_2, \cdots, e_D]$ to represent the reliability of each feature, these popular datasets are transformed into bi-objective feature selection problems with unreliable data. Table 1 shows those translated test functions. Note that, the quality vector of each dataset is composed of the elements of E, which begin at the first element 0.7 until the last feature is assigned.

Two state-of-the-art algorithms, the TV-MOPSO algorithm proposed in [13] and the NSGA-II algorithm [11] are selected for performance comparison. In our experiments, the same conditions are used to compare the performance between the proposed algorithm and the other meta-heuristic algorithms, i.e., the size of the swarm=30, the archive size=20, the maximum generation/iteration number=100 and the 1-nearest neighbor method to evaluate the feature subset. Moreover, Table 2 lists values of the rest parameters in detail.

Table 1. Format of test problems

Data sets	No. of samples	No. of classes	No. of features	Quality vector
WDBC	569	2	30	E=[0.7,0.8,0.9,1.0,0.6,1.0,0.6,0.9,1.0 0.9 0.7 0.8 0.4, 0.8 ,1.0 ,0.7, 0.5, 0.9, 1.0, 1.0, 0.6, 1.0, 0.9, 1.0, 0.5, 0.7, 0.8, 0.9, 1.0,
Ionosphere	351	2	34	0.6, 0.7, 1.0, 0.9, 1.0, 0.6, 0.7, 0.8, 0.9, 1.0 ,0.9, 0.7, 0.8,0.9, 1.0 ,0.9, 0.7, 0.8,
Sonar	208	2	60	0.9 ,1.0, 0.9 ,0.7, 0.8,0.9, 1.0, 1.0 , 0.7, 0.8 ,1.0 ,1.0, 1.0].

Table 2. Parameter configurations for selected algorithms

Algorithms	TV-MOPSO	NSGA-II
Parameters	Inertia weight: $w = 0.3*(1-t/T_{max})+0.4$ Cognitive coefficient: $c_1 = (0.5-2.5)\times t/T_{max}+2.5$ Social coefficient: $c_2 = (2.5-0.5)\times t/T_{max}+0.5$ Mutation parameter $b=5$.	Distribution index for SBX is 20; distribution index for polynomial mutation is 20; mutation probability =1/D; crossover probability =0.9; tournament is 2.

In our comparative study, two performance metrics known to multi-objective optimizers are adopted. To evaluate the distribution of solutions throughout the Pareto optimal set found, the spacing metric (SP) [14] is adopted. In order to evaluate the closeness of the obtained Pareto front to the true Pareto front which is unknown in advance, the two-set coverage (SC) [14] is adopted.

4.2 Comparison Results and Analysis

In this subsection, we evaluate the performance of the proposed JMOPSO algorithm by comparing TV-MOPSO [13] and NSGA-II [11]. All the algorithms were set to conduct 30 runs to collect the statistical results for all simulations. The results obtained with respect to each adopted performance metric are shown in Tables 3-5.

For the dataset WDBC, we can see from Table 3 that the proposed algorithm JMOPSO has the best convergence, where 97.40 % and 96.30 % solutions obtained by TV-MOPSO and NSGA-II are dominated by those of JMOPSO respectively. On the other hand, the three algorithms obtain similar average in terms of SP.

For the dataset Ionosphere, we can see from Table 4 that the proposed algorithm JMOPSO still has the best convergence, where 98.77 % and 94.32 % solutions obtained by TV-MOPSO and NSGA-II are dominated by those of JMOPSO respectively; TV-MOPSO has the second best convergence, where 50.49% solutions obtained by NSGA-II are dominated by those of TV-MOPSO. On the other hand, in terms of SP, JMOPSO has also the best average, followed by TV-MOPSO.

For the dataset Sonar, TV-MOPSO finds the solutions with the best distribution; however, it gets the worst convergence. As Table 5 shows, the proposed algorithm JMOPSO still has the best convergence, where 81.67% and 100% solutions obtained by TV-MOPSO and NSGA-II are dominated by those of JMOPSO respectively; TV-MOPSO has the second best convergence, where 72.78% solutions obtained by NSGA-II are dominated by those of TV-MOPSO.

Table 3. Results obtained by the three algorithms on tackling WDBC

Algorithms	SP(Average/Std.)	SC (Average/Std.)		
		(TV-MOPSO,*)	(NSGA-II,*)	(JMOPSO,*)
TV-MOPSO	0.0039/0.002	--	0.2892/0.2898	0.9740/0.0532
NSGA-II	0.0042/0.0038	0.6201/0.3265	--	0.9630/0.1111
JMOPSO	0.0039/0.0016	0.0171/0.0513	0.0101/0.0303	--

Table 4. Results obtained by the three algorithms on tackling Ionosphere

Algorithms	SP(Average/Std.)	SC (Average/Std.)		
		(TV-MOPSO,*)	(NSGA-II,*)	(JMOPSO,*)
TV-MOPSO	0.0055/0.0025	---	0.3471/0.4044	0.9877/ 0.0370
NSGA-II	0.078/0.0071	0.5049/0.3963	---	0.9432/ 0.1338
JMOPSO	0.0052/0.0022	0/0	0/0	---

Table 5. Results obtained by the three algorithms on tackling Sonar

Algorithms	SP(Average/Std.)	SC (Average/Std.)		
		(TV-MOPSO,*)	(NSGA-II,*)	(JMOPSO,*)
TV-MOPSO	0.0050/0.0025	--	0.1202/0.0606	0.8167/0.3018
NSGA-II	0.074/0.0043	0.7278/0.1093	--	1/0
JMOPSO	0.0065/0.003	0.0825/0.1997	0/0	--

5   Conclusion

In this paper, a new feature selection problem with unreliable data is studied, and is modeled as a bi-objective optimization problem. Furthermore, an improved MOPSO-based feature selection algorithm JMOPSO is proposed. By comparing with two well-known multi-objective optimization techniques, the solutions obtained by JMOPSO exhibit certain superior characteristics with respect to the multi-objective performance, as well as the ability of removing redundant features.

Acknowledgment. This work was partially supported by the Natural Science Foundation of Jiangsu province (No.BK2011215), China Postdoctoral Science Foundation funded project (No. 2012M521142), and Jiangsu Planned Projects for Postdoctoral Research Funds (No. 1301009B).

References

1. Kennedy, J., Eberhart, R.C.: Particle swarm optimization. In: Proc. IEEE Int. Conf. Neural Netw., pp. 1942–1948. IEEE (1995)
2. Wang, X., Yang, J., Teng, X., Xia, W., Jensen, R.: Feature selection based onrough sets and particle swarm optimization. Pattern Recognition Letters 28(4), 459–471 (2007)
3. Chuang, L.Y., Yang, C.H., Li, J.C.: Chaotic maps based on binary particle swarm optimization for feature selection. Applied Soft Computing 11(1), 239–248 (2011)
4. Liu, Y.N., Wang, G., Chen, H.L., Dong, H., Zhu, X.D., Wang, S.J.: An improved particle swarm optimization for feature selection. Journal of Bionic Engineering 8, 191–200 (2011)
5. Su, C.T., Chen, K.H., Wang, P.C.: Particle swarm optimization for feature selection with application in obstructive sleep apnea diagnosis. Neural Computing & Applications 21(8), 2087–2096 (2012)
6. Marinakis, Y., Marinaki, M.: A hybridized particle swarm optimization with expanding neighborhood topology for the feature selection problem. In: Blesa, M.J., Blum, C., Festa, P., Roli, A., Sampels, M. (eds.) HM 2013. LNCS, vol. 7919, pp. 37–51. Springer, Heidelberg (2013)
7. Xue, B., Cervante, L., Shang, L., Browne Will, N., Zhang, M.J.: A multi-objective particle swarm optimisation for filter-based feature selection in classification problems. Connection Science 24(2-3), 91–116 (2012)
8. Krohling, R.A., Mendel, E.: Bare bones particle swarm optimization with Gaussian or Cauchy jumps. In: Proc. 11th Conf. CEC, pp. 3285–3291. IEEE (2009)
9. Blackwell, T.: A study of collapse in bare bones particle swarm optimization. IEEE Transactions on Evolutionary Computing 16(3), 354–375 (2012)
10. al-Rifaie, M.M., Blackwell, T.: Bare Bones Particle Swarms with Jumps. In: Dorigo, M., Birattari, M., Blum, C., Christensen, A.L., Engelbrecht, A.P., Groß, R., Stützle, T. (eds.) ANTS 2012. LNCS, vol. 7461, pp. 49–60. Springer, Heidelberg (2012)
11. Deb, K., Pratap, A., Agarwal, S., Meyarivan, T.: A fast and elitist multi-objective genetic algorithm: NSGA-II. IEEE Transactions on Evolutionary Computation 6(2), 182–197 (2012)
12. Coello Coello, C.A., Pulido, G.T., Lechuga, M.S.: Handling multiple objectives with particle swarm optimization. IEEE Transaction Evolutionary Computation 8(3), 256–279 (2004)
13. Tripathi, P.K., Bandyopadhyay, S., Pal, S.K.: Multi-objective particle swarm optimization with time variant inertia and acceleration coefficients. Information Sciences 177(22), 5033–5049 (2007)
14. Zitzler, E., Thiele, L.: Multi-objective evolutionary algorithms: A comparative case study and the strength Pareto approach. IEEE Transactions on Evolutionary Computation 3(4), 257–271 (1999)

Convergence Enhanced Multi-objective Particle Swarm Optimization with Introduction of Quorum-Sensing Inspired Turbulence

Shan Cheng[1], Min-You Chen[2], and Gang Hu[3]

[1] College of Electrical Engineering & New Energy, China Three Gorges University, Yichang 443002, China
[2] State Key Laboratory of Power Transmission Equipment and System Security and New Technology, Chongqing University, Chongqing 400044, China
[3] School of Electrical and Information Engineering, Chongqing University of Science and Technology, Chongqing, 401331
hpucquyzu@ctgu.edu.cn, minyouchen@cqu.edu.cn,
gang.hu.cq@gmail.com

Abstract. Enhancing the convergence property is one of the main goals to achieve when designing a multi-objective particle swarm optimization (MOPSO) algorithm. To promote convergence, a turbulence mechanism derived from the bacteria quorum sensing behavior is introduced and a novel MOPSO (MOPSO-QSIT) is proposed. The inspired turbulence mechanism takes into effect only if the whole current population' velocities are rather small (less than a predefined threshold), which enables to maintain the swarm diversity and avoids declining the swarm evolution. The MOPSO-QSIT algorithm has been tested on a set of benchmark functions and compared with other multi-objective optimization algorithms that are representative of the state-of-the-art. Simulation results illustrate that the proposed algorithm possesses the best convergence performance while keep good diversity performance, and is a competitively effective global optimization tool.

Keywords: Multi-objective optimization, Multi-objective particle swarm optimization, Convergence, Quorum sensing.

1 Introduction

Multi-objective optimization problems (MOPs) with non-commensurable or even conflict-with-each-other objectives are rather common in many scientific and engineering environments. Consequently, multi-objective optimization (MOO) has been extensively studied and widely applied to MOPs to provide more information for supporting decision making tasks [1]. Because of advantages such as fast convergence, fewer parameters to adjust, robust adaptability, and relative simplicity of implement, population based particle swarm optimization (PSO) has been extended for MOO, and such multi-objective PSO (MOPSO) has achieved universal applications to MOPs.

Y. Tan et al. (Eds.): ICSI 2014, Part I, LNCS 8794, pp. 394–403, 2014.

As stated, a high convergence speed is one of the most important features of the PSO. It's a negative effect when premature convergence occurs which result from the rapid loss of the swarm diversity. Especially, such convergence speed may be harmful in the context of MOO, because a PSO-based algorithm may converge to a false Pareto front [2]. Thus, means to maintain the swarm diversity is an important issue to avoid premature convergence.

Recently, incorporation of bio-behaviors into PSO produces a novel approach to strengthen the performance of PSO. Inspired by the phenomenon of chemo taxis in colonies of the bacteria, an improved PSO [3] was presented by analogy to the way that bacteria react to chemo-attractants or chemo-repellents, which not only prevents premature convergence to a high degree, but also keeps a more rapid convergence rate than standard PSO. Similarly, a new PSO algorithm based on the operator of chemo taxis in the bacterial foraging algorithm was presented to avoid fall into the local minimum in the standard PSO algorithm. According to the analysis of biological symbiotic relationship, a two-population PSO model called particle swarm optimization based on parasitic behavior was proposed in [4], the effectiveness of which had been verified. Although some literature is available on PSO with bio-behaviors, few attempts have been made to MOPSO combined with such bio-behaviors. This motivates us to explore the application of bio-behaviors to MOPSO algorithms.

This study introduces a novel MOPSO based on bacteria-quorum-sensing inspired turbulence (MOPSO-QSIT). In the algorithm, for diversity preservation of the swarm and promotion of exploitation, a turbulence mechanism inspired from the bacteria quorum sensing behavior is introduced. The mechanism is first exerting turbulence and then replacement operation in nature and serves as mutation function when the velocity of the swarm is rather small. Besides, each particle's leader is selected dynamically to enrich the exploratory capabilities and an improved NSGA-II crowing distance based sorting (CDS) [5] is incorporated to ensure the good distribution of Pareto solutions on the Pareto Front.

2   MOPSO Algorithm with Introduction of Quorum Sensing Inspired Turbulence

It has long been appreciated that certain groups of bacteria exhibit cooperative behavioral patterns. It seems that intercellular communication likewise can account for such behaviors and the communication was called quorum-sensing (QS) by Fuqua [6]. Early in 1970s, au-to-induction of luminescence was described [7, 8]. Fuqua [6] argued that auto-induction defines an environmental sensing system allowing bacteria to monitor their own population density and the bacteria produces a diffusible compound termed auto-inducer which accumulates in the surrounding environment during growth. In 2007, Higgins et al. [9] discovered that vibrio cholerae, the causative agent of the human disease cholera, also uses quorum sensing mechanism to control pathogenicity and biofilm formation. They also pointed out that cholera activates the expression of virulence factors and forms bio-films when at low cell density, while at high cell density the accumulation of two quorum-sensing auto-inducers re-presses these traits. Therefore, quorum sensing plays an important role in the survival and multiplies of the biology.

2.1   Mechanism of Quorum Sensing Inspired in the MOPSO

PSO is an evolutionary computation technique. During the evolutionary process, the velocity and position update formula of particle i on the dimension $d|_{d=1,2,...D}$ (D stands for the dimension of the decision variables) are described as Eq. (1) and Eq. (2), respectively. As Eq. (1) and Eq. (2) show, when the velocities of the particles in the swarm are almost zero, it becomes unable to create new solutions which might lead the swarm out of this stagnating state. Therefore, creating new particles potentially facilitates the swarm to escape from local optima and to strengthen the exploratory capabilities of PSO [10].

$$v_{id}(t+1) = wv_{id}(t) + c_1 r_1[p_{id} - x_{id}(t)] + c_2 r_2[p_{gd} - x_{id}(t)] \tag{1}$$

$$x_{id}(t+1) = x_{id}(t) + x_{id}(t+1) \tag{2}$$

where t is the current iteration; w is the inertia weight; c_1 and c_2 are acceleration coefficients; r_1 and r_2 are random numbers with uniform distribution between 0 and 1. v_{id} and x_{id} denote the $d^{th}|_{d=1,2,...D}$ dimension of particle i velocity $v_i = [v_{i1}, v_{i2}, ..., v_{iD}]^T$ and position $x_i = [x_{i1}, x_{i2}, ..., x_{iD}]^T$, respectively. p_{id} and p_{gd} indicate the $d^{th}|_{d=1,2,...D}$ dimension of the personal best $p_i = [p_{i1}, p_{i2}, ..., p_{iD}]^T$ and global best $p_g = [p_{g1}, p_{g2}, ..., p_{gD}]^T$, respectively.

Inspired by the quorum sensing phenomenon, QS mechanism is introduced into MOPSO. Once the velocities of the whole swarm are less than a threshold V_{limit}, carry out the quorum sensing inspired turbulence to the swarm. Namely, generate a new swarm P_{tur} with population size N, and then replace 20%N particles in P_{new} with the excellent particles in P_{tur}, where P_{new} denotes the swarm used for next iteration. The quorum sensing inspired turbulence can be described as follows and its scheme is illustrated in Fig. 1.

(1) Read the current population P, non-dominated solution set ND and particle swarm P_{new} for next iteration.

(2) Check whether the velocity of each particle in being less than or not. If yes, go to (3). Else go to (7). Set $i=1$, $n=0$.

(3) Generate a new population P_{tur} according to Eq. (3).

$$x_{tur} = x_{new} + \beta V_{max} sign\left(2\left(rand\left(1, N\right) - 0.5\right)\right) \tag{3}$$

where x_{tur} represents the positions of newly generated swarm P_{tur}, and it is a $D \times N$ matrix; X_{new} is also a matrix which indicates the positions of population P_{new}; β is the turbulence degree and its value is between 0 and 1; V_{max} is a $D \times 1$ matrix represents the max velocity values in each dimension; $sign$ is the sign function; $rand$ is the random operator.

(4) Replace some particles in P_{new} with the excellent particles in P_{tur}. Set $j=1$ and carry out the following operations.

a) Check whether particle j in P_{tur} dominating thenon-dominated solutions in ND, if yes, let $n=n+1$ and replace the $(N-n)^{th}$ particle in P_{new}.

b) Decide the relationship between j and N, and the relationship between n and [0.2N]. If $j<N$ and $n<[0.2N]$, $j=j+1$ and go to a); $j<N$ and $n\geq[0.2N]$, go to (5). If $j\geq N$ and $n<[0.2N]$, $i=i+1$and go to (3); $j\geq N$ and $n>[0.2N]$, go to (5).

(5) Set $P=P_{new}$.

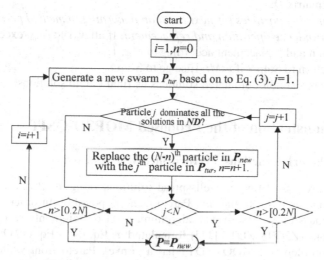

Fig. 1. Scheme of quorum sensing inspired turbulence

2.2   Proposed MOPSO-QSIT

With introduction of quorum sensing inspired turbulence mechanism, the MOPSO-QSIT can be expressed as follows, its flow chart is shown in Fig. 2.

Step 1. *Initialization.* Set the population size N and maximum iteration number M_t. Initialize the population with velocities V and positions X. Set $t=0$.

Step 2. *Selection of leaders.* $t=t+1$. Identify the non-dominated solutions from the current population P and then dynamically select the non-dominated solution with the larges *fitness* value as the leader p_g for each particle according to Eq. (4). Update p_i.

$$fitness = 1 \Big/ \sum_{i=1}^{M} w_i f_i, \quad w_i = \lambda_i \Big/ \sum_{i=1}^{M} \lambda_i, \quad \lambda_i = U(0,1) \tag{4}$$

where M is the number of objectives, and f_i is the ith objective-function. The function $U(0,1)$ generates a uniformly distributed random number within the interval [0,1].

Step 3. *Generation of new particles.* Compute the new velocities v_{new} and positions x_{new} based on the current v and x according to Eq. (5) and (2), where the inertia weight and acceleration term c change with iteration according to Eq. (6) and (7). Combine x_{new} and x together and store them in a temporary list $TempX$.

$$v_{id}(t+1) = wv_{id}(t) + c(r_1(p_{id} - x_{id}(t)) + r_2(p_{gd} - x_{id}(t)) \qquad (5)$$

$$w = 1 + r_3(1 - w_0) \qquad (6)$$

$$c = c_0 + t / Mt \qquad (7)$$

Step 4. *Identification of non-dominated solutions*. Identify the non-dominated solutions from *TempX* and store them in a matrix *ND*, while the dominated solutions are stored in the matrix *D*.

Step 5. *Selecting particles for next iteration according to method presented in* [5] .

Step 6. *Turbulence operation and replacement*. If all $|v_i(t)| < V_{limit}$, execute the turbulence operation and replacement according to Fig. 1.

Step 7. Return to Step 2 if $t<Mt$. Else go to Step 8.

Step 8. Store the non-dominated solutions as final Pareto solutions.

3   Demonstration of the Proposed MOPSO-QSIT

3.1   Benchmark Functions, Performance Metrics Used

MOO is the process of finding well-spread solutions along the Pareto front as diverse as possible and as close to the real Pareto front as possible. In order to demonstrate the effectiveness of MOPSO-QSIT algorithm, a set of commonly recognized benchmark functions (ZDT1~ ZDT4 [11]) formulated as Eq. (8) ~ Eq. (12) have been chosen as test problems of MOO. ZDT1 has a convex Pareto front while ZDT2 has a concave Pareto front. With many MOO algorithms, it's difficult (See Table 2) for ZDT3 to find a diverse set of solutions because of its discreteness. ZDT4 is a multimodal problem and the multiple local Pareto fronts make it difficult for many algorithms to converge to the true Pareto-optimal front (Also see Table 1).

In order to present a quantitative assessment of the performance of MOPSO-QSIT algorithm, two performance metrics [11] are used, namely the generational distance *GD* and the spread *S*. Let *Q* and *P** denote an obtained and a known true Pareto set, respectively. *GD* and *S* are defined as Eq. (13), which measure the closeness of *Q* from *P** and the diversity of the solutions along the Pareto front, respectively. In Eq. (13), *M* is the numbers of objective-functions; D_i is the Euclidean distance between the solution $i \square Q$ and the nearest member of *P**. d_i represents the distance between consecutive solutions in *Q*; d_{ave} is the mean value of all d_i; d^e_m denotes the distance between the extreme solutions of *Q* and *P** along the mth objective. A set of $|P^*|=500$ uniformly distributed Pareto solutions is used to calculate *GD*.

$$X = [x_1, x_2, ..., x_D]^T$$
$$\min \ F(X) = [f_1, f_2]^T \qquad (8)$$
$$s.t.: \ f_1 = x_1$$
$$f_2 = g(x_2, ..., x_D)h(f_1, g(x_2, ..., x_D))$$

$$ZDT1: g(x_2,\cdots,x_D)=1+9\sum_{d=2}^{D}x_d/(D-1),\quad D=30,\quad x_d\in[0,1]$$

$$h(f_1,g(x_2,\cdots,x_D))\sqrt{f_1\bigg/g(x_2,\cdots,x_D)}$$

(9)

$$ZDT2: g(x_2,\cdots,x_D)=1+9\sum_{d=2}^{D}x_d/(D-1),\quad D=30,\quad x_d\in[0,1]$$

$$h(f_1,g(x_2,\cdots,x_D))=1-(f_1\big/g(x_2,\cdots,x_D))^2$$

(10)

$$ZDT3: g(x_2,\cdots,x_D)=1+9\sum_{d=2}^{D}x_d/(D-1),\quad D=30,\quad x_d\in[0,1]$$

$$h(f_1,g(x_2,\cdots,x_D))=1-\sqrt{f_1\big/g(x_2,\cdots,x_D)}-f_1\big/g(x_2,\cdots,x_D)\sin(10\pi f_1)$$

(11)

$$ZDT4: g(x_2,\cdots,x_D)=1+9\sum_{d=2}^{D}x_d/(D-1),\quad D=10\quad x_1\in[0,1]$$

$$h(f_1,g(x_2,\cdots,x_D))=1-\sqrt{f_1\big/g(x_2,\cdots,x_D)}\qquad x_d\in[-5,5],d=2,3,...D$$

(12)

$$\begin{cases} GD=(\sum_{i=1}^{M}D_i^{M})^{1/M}\big/|Q| \\ S=(\sum_{m=1}^{M}d_m^{e}+\sum_{i=1=1}^{|Q|}|d_i-d_{ave}|)\big/\sum_{m=1}^{M}d_m^{e}+|Q|d_{ave} \end{cases}$$

(13)

3.2   Competiveness of MOPSO-QSIT

To demonstrate how competitive it is, the proposed MOPSO-QSIT algorithm is compared with other MOO algorithms that are representative of the state-of-the-art. These algorithms include NSGA-II [11], non-dominated sorting particle swarm optimizer (NSPSO) [12], multi-objective particle swarm optimization (MOPSO) [2], multiobjective particle swarm optimization algorithm based on crowding distance sorting (CDMOPSO)[13], improved niching multi-objective particle swarm optimization (AIPSO) [14] and local search and hybrid diversity strategy based multi-objective particle swarm optimization (LH-MOPSO) [15].

The parameters used are: population/swarm size 100 for NSGA-II, NSPSO, AIPSO and LH-MOPSO, 50 for MOPSO; archive size 100 for MOPSO, number of iterations 250 for NSGA-II, NSPSO and AIPSO, 500 for MOPSO (to keep the number of function evaluations to 25000 for all the algorithms), cross-over probability 0.9 for NSGA-II and CDMOPSO and 0.5 for LH-MOPSO, mutation probability inversely proportional to the chromosome length. The values of c_1 and c_2 have been used as 2 for NSPSO, CDMOPSO and LH-MOPSO. The value of w has been used as 0.3 whereas it has been allowed to decrease from 1.0 to 0.4 for NSPSO and LH-MOPSO. For all test functions handled with MOPSO-QSIT, $w_0=0.3$, $\alpha_0=0.5$, $N=100$, $Mt=250$, $V_{limit}=0.2$. In order to establish repeatability, the proposed algorithm is run 30 times independently. The mean and variance of performance metrics are summarized in Table 1.

Table 1. Mean and variance values of the convergence measure GD and the diversity measure

Algorithm	Index	ZDT1		ZDT2		ZDT3		ZDT4	
		Mean	Var	Mean	Var	Mean	Var	Mean	Var
NSGA-II	GD	0.03348	0.00476	0.07239	0.03169	0.11450	0.00794	0.51305	0.11846
	S	0.68132	0.01335	0.63922	0.00114	0.83195	0.00892	0.96194	0.00114
NSPSO	GD	0.00642	0.00000	0.00951	0.00000	0.00491	0.00000	4.95775	7.43601
	S	1.22979	0.00484	1.16594	0.00768	0.78992	0.00165	0.87046	0.10140
MOPSO	GD	0.00133	0.00000	0.00089	0.00000	0.00418	0.00000	7.37429	5.48286
	S	0.84816	0.00287	0.89292	0.00574	1.22731	0.02925	1.01136	0.00072
CDMOPSO	GD	0.00690	0.00055	0.00692	0.00055	0.00720	0.00099	0.26300	0.02488
	S	0.27400	0.04080	0.27090	0.01820	0.01820	0.01160	0.21780	0.01850
AIPSO	GD	0.00448	0.00000	0.00386	0.00000	0.00612	0.00000	0.14462	0.00709
	S	0.51485	0.00051	0.49909	0.00052	0.50558	0.00028	0.50735	0.00034
LH-MOPSO	GD	0.0021	---	0.0027	---	0.0059	---	0.4811	---
	S	0.4088	---	0.3803	---	0.5607	---	0.4089	---
MOPSO-QSIT	GD	*0.00015*	0.00000	*0.00008*	0.00000	*0.00061*	0.00000	*0.00016*	0.00000
	S	0.58168	0.00066	0.54037	0.00061	*0.48805*	0.00033	0.58360	0.00093

It can be seen that MOPSO-QSIT has resulted in best convergence on all test problems in terms of GD measure, especially on ZDT3 and ZDT4, markedly better than the other algorithms. Meanwhile, MOPSO-QSIT is able to obtain good distribution of the solutions on the Pareto front.

To demonstrate the distribution of the solutions on the final non-dominated front, the results of the four benchmark functions in a single run have been illustrated in Fig. 1. It can be seen that the proposed algorithm performed very well and converged to the Pareto front with a high accuracy while maintaining a good diversity among the Pareto solutions.

3.3   Effect of Introducing Quorum Sensing Inspired Turbulence Mechanism on the Convergence of MOPSO

Sierra etc. [10] argued that the use of a mutation operator or likewise is very important to escape from local optima and to improve the exploratory capabilities of MOPSO. They also pointed that random, unreasonable mutation or mutation-like operation declines the swarm evolution. To verify the effectiveness of introducing quorum sensing inspired turbulence mechanism on the convergence of MOPSO, simulations without the proposed turbulence mechanism are carried out and the results are summarized in Table 2 and Fig. 2. Suppose only the methods for selecting the particles for next iteration are different, and use random (AWPSO [16]), sequential [17] and SE (MRBHPSO-SE [5]), respectively. ZDT3 and ZDT4 have been considered as typical illustrations. It can be seen that the effects on the diversity performance caused by without turbulence mechanism are not as notable as on the convergence. Conclusion can be drawn from Table 2 and Fig. 2 compared with results shown above that, the proposed quorum sensing inspired turbulence mechanism does provides a means of appropriate promotion of diversity and enrichment of the global search capability of MOPSO.

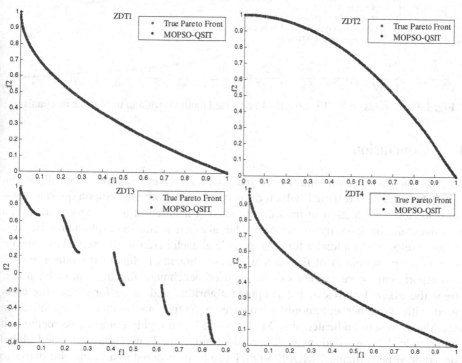

Fig. 2. Final fronts of MOPSO-QSIT on ZDT1~ZDT4 in a single run

Table 2. Mean and variance values of GD and S without turbulence mechanism

Function	Index	Random		Sequential		SE	
		Mean	Var	Mean	Var	Mean	Var
ZDT3	GD	0.65974	0.00019	0.82261	0.00033	0.74400	0.00034
	S	0.99667	0.00023	0.92453	0.00004	0.89491	0.00031
ZDT4	GD	0.17566	0.00002	0.29703	0.00010	0.21267	0.00006
	S	0.94486	0.00038	0.92117	0.00010	0.90563	0.00004

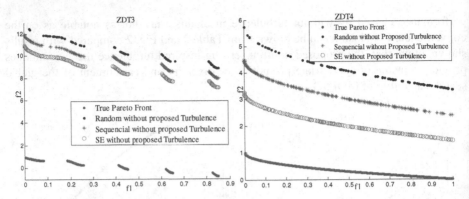

Fig. 3. Final fronts on ZDT3 and ZDT4 performed methods without turbulence mechanism

4   Conclusion

In this paper, a novel multi-objective particle swarm optimization algorithm (MOPSO-QSIT) is presented, which draw inspiration from the bacteria quorum sensing behavior. Introduction of the quorum sensing inspired turbulence mechanism not only maintain the diversity of the swarm but also strengthen the exploitation. Besides, dynamic selection of a leader for each particle at each iteration also served as a means of diversity preservation of the swarm and enrichment of global exploration. Extensive experiments were carried out on a set of benchmark functions in order to examine the major features of the proposed algorithm and its performance was compared with the other commonly recognized effective multi-objective optimization algorithms. Results indicated that MOPSO-QSIT is a highly competitive method; it can provide diverse solutions well spreading along the Prato front with good global convergence performance. MOPSO-QSIT can be considered as a viable alternative to solve multi-objective optimization problems (MOPs). Future work will look into its application to real-world MOPs.

Acknowledgments. This work was supported by the Scientific Research Foundation for Talents of China Three Gorges University (No. KJ2013B077) and by YiChang Administration of Science and Technology via the project "Dynamic Optimal Operation of Grid-connected Micro-grid".

References

1. Li, M., Lin, D., Kou, J.: A Hybrid Niching PSO Enhanced with Recombination-Replacement Crowding Strategy for Multimodal Function Optimization. Appl. Soft. Comput. 12, 975–987 (2012)
2. Coello, C.A.C., Pulido, G.T., Lechuga, M.S.: Handling Multiple Objectives with Particle Swarm Optimization. IEEE Trans. Evol. Comput. 8, 256–279 (2004)

3. Niu, B., Zhu, Y.L., He, X.X., Zeng, X.P.: An Improved Particle Swarm Optimization Based on Bacterial Chemotaxis. In: World Congr. Intell. Control Automa., pp. 3193–3197. Dalian, China (2006)
4. Qin, Q.D., Li, R.J.: Two-population Particle Swarm Optimization Algorithm Based on Bioparasitic Behavior. Control and Decision 26, 548–552 (2011)
5. Chen, M.Y., Cheng, S.: Multi-objective Particle Swarm Optimization Algorithm Based on Random Black Hole Mechanism and Step-by-step Elimination Strategy. Control and Decision 28, 1729–1734, 1740 (2013)
6. Fuqua, W.C., Winans, S.C., Greenberg, E.P.: Quorum Sensing in Bacteria: the LuxR LuxI Family of Cell Density Responsive Transcriptional Regulators. J. Bacteriol. 176, 269–275 (1994)
7. Eberhard, A.: Inhibition and Activation of Bacterial Luciferase Synthesis. J. Bacteriol. 109, 1101–1105 (1972)
8. Nealson, K.H., Platt, T., Hastings, J.: Cellular Control of the Synthesis and Activity of the Bacterial Luminescent System. J. Bacteriol. 104, 313–322 (1970)
9. Higgins, D.A., Pomianek, M.E., Kraml, C.M., Taylor, R.K., Semmelhack, M.F., Bassler, B.L.: TheMmajor Vibrio Cholerae Autoinducer and its Role in Virulence Factor Production. Nature 450, 883–886 (2007)
10. Sierra, M.R., Coello, C.A.C.: Multi-objective Particle Swarm Optimizers: A Survey of the State-of-the-art. Int. J. Comput. Intell. Res. 2, 287–308 (2006)
11. Deb, K., Pratap, A., Agarwal, S., Meyarivan, T.: A Fast and Elitist Multiobjective Genetic Algorithm: NSGA-II. IEEE Trans. Evol. Comput. 6, 182–197 (2002)
12. Li, X.: A Non-dominated Sorting Particle Swarm Optimizer for Multiobjective Optimization. In: Cantú-Paz, E., et al. (eds.) GECCO 2003. LNCS, vol. 2723, pp. 37–48. Springer, Heidelberg (2003)
13. Feng, Y.X., Zheng, B., Li, Z.K.: Exploratory Study of Sorting Particle Swarm Optimizer for Multiobjective Design. Math. Comput. Model 52, 1966–1975 (2010)
14. Huang, P., Yu, J.Y., Yuan, Y.Q.: Improved Niching Multi-objective Particle Swarm Optimization Algorithm. Computer Engineering 37, 1–3 (2011)
15. Jia, S.J., Du, B., Yue, H.: Local Search and Hybrid Diversity Strategy Based Multi-objective Particle Swarm Optimization Algorithm. Control and Decision 27, 813–818 (2012)
16. Mahfouf, M., Chen, M.-Y., Linkens, D.A.: Adaptive Weighted Particle Swarm Optimisation for Multi-objective Optimal Design of Alloy Steels. In: Yao, X., et al. (eds.) PPSN VIII 2004. LNCS, vol. 3242, pp. 762–771. Springer, Heidelberg (2004)
17. Chen, M.Y., Zhang, C.Y., Luo, C.Y.: Adaptive Evolutionary Multi-objective Particle Swarm Optimization Algorithm. Control and Decision 24, 1851–1855, 1864 (2009)

Multiobjective Genetic Method for Community Discovery in Complex Networks

Bingyu Liu, Cuirong Wang, and Cong Wang

College of information science and engineering, Northeastern University,
Shenyang 110819, China
Liuby78@163.com

Abstract. The problem of community structure discovery in complex networks has become one of the hot spots in recent years. This paper proposes a multiobjective genetic algorithm MOGCM to uncover community structure. This method overcomes the limitations of the community detection problems, choosing MinMaxCut and the community fitness as the objective functions. In the experiments, 2 well-known real-life networks are used to validate the performance and the results show that the method successfully detects the communities and it is competitive with state-of-the-art approaches.

Keywords: complex networks, community discovery, genetic method, multiobjective optimization.

1 Introduction

Many real-world systems, such as social networks, biological interaction networks, communications, the world-wide-web and Internet can be described as complex networks. Network nodes represent entities and edges represent the interactions among these entities. Complex network analysis has been one of the most popular research areas in recent years. One of the main problems in the study of complex networks is the detection of community structure. Community structure is an important characteristic of complex networks. The community of a network is also referred to as clusters [1], which is a group of nodes that having dense intra-connections and sparse interconnections [2]. There are several methods currently to discovery community structures [3-8]. The algorithms can be divided into two main categories: heuristic and optimization based methods.

The heuristic methods solve community mining problem based on some intuitive assumptions. One of the most popular methods proposed so far is the Girvan-Newman algorithm [3] which introduces a divisive method that iteratively removes the edge with the greatest betweenness value and use the concept of modularity as criterion to stop the division of a network in sub-networks in their divisive clustering algorithm. Some improved algorithms have also been proposed [9, 10]. These algorithms are based on a foundational measure criterion of community, Modularity. Modularity is a popular quality function of community detection, the larger the Modularity value, the

Y. Tan et al. (Eds.): ICSI 2014, Part I, LNCS 8794, pp. 404–413, 2014.

more accurate the community partition. Thus, community discovery becomes a Modularity value optimization problem. For example, Fast Newman (FN) algorithm [11], Simulated Annealing (SA) algorithm [12], Iterated Tabu Search (ITS) [13], etc. Maximizing the modularity has been proven to be a NP-complete problem. At present, genetic algorithm (GA) has been becoming a type of competitive method in the community mining area due to its effectiveness for solving NP-complete problems. Some other algorithms has been described in [14], it is a survey of community discovery methods in complex networks.

In this paper we propose a new genetic method to discovery communities in a complex network. We apply multiobjective optimization method MOGCM (multiobjective genetic community discovery method) to detect communities, where the first objective is MinMaxCut minimization and the second objective is the community fitness maximization. Because this method returns a set of solutions where each of them correspond to different trade-offs between the two objectives, so do not need a prior knowledge of the number of communities. Experiments on real life networks show that the MOGCM method can find the communities of a complex networks with high accuracy and high modularity, which is the most common metrics.

This paper is organized as follows. Section 2 introduced the community detection problem. Section 3 described the new genetic algorithm, including framework, objective function and operators.

2   Related Works

Community detection can be viewed as an optimization problem. Many single objective optimization techniques have been used to solve this problem.

A single objective optimization problem (Ω, F) can defined as

$$\text{Min F(X), \quad s. t. } X \in \Omega \tag{1}$$

where F(X) is an objective function that needs to be optimized and $\Omega = \{X_1, X_2, ..., X_n\}$ is the set of feasible community structures in a complex network.

One kind of optimization method is to optimize the modularity [15, 16]. Modularity is a quality function introduced by Newman and Girvan in [3] that quantifies the community structure by providing a value for every clustering of a given graph. There are so many optimization methods use the modularity as the objective function, but it may fail to identify modules smaller than a scale which depends on the total number of links in the network and the degree of interconnectedness of the modules, even in case where modules are unambiguously defined.

A general advice for community definition is that the density of nodes inside the same community should be much higher than the density of nodes connecting to the remainder nodes of the graph, this definition pursues two different objectives: maximizing the internal links and minimizing the external links. So we can view the community detection problem as a multiobjective optimization problem. Given a set of quality measures $F_1(S), F_2(S), ..., F_t(S)$, we want to find community S that simultaneously optimizes each quality measure.

The general multiobjective optimization problem $(\Omega, F_1, F_2, \cdots, F_t)$ is posed as follows:

$$\min F_i(S), i = 1, 2, \cdots, t; \text{ s.t } S \in \Omega \tag{2}$$

Since the goal is to optimize a set of competing objectives optimized simultaneously, there is not one unique solution to the problem. Multiobjective optimization aims to generate and select non-dominated solutions through the use of Pareto optimality theory [17]. The solution of the multiobjective optimization problem is a set of Pareto points. These solutions are obtained through the use of Pareto optimality theory and constitute global optimum solutions satisfying all the objectives as best as possible. The Pareto optimal solutions usually include the optimal solutions obtained by single-objective GA when applied to the clustering problem.

Given two solutions S_1 and $S_{2 \in \Omega}$, solution S_1 is said to dominate solution S_2, denoted as $S_1 \prec S_2$, if and only if

$$\forall i : F_i(S_1) \le F_i(S_2) \text{ and } \exists i \text{ s.t. } F_i(S_1) \prec F_i(S_2)$$

Community discovery could be formulated as a multiobjective optimization problem and the framework of Pareto optimality can provide a set of solutions corresponding to the best compromise among the objectives to optimize.

3   Community Discovery Method

In this study, we model a complex network as a graph $G = (V, E)$ where V is a set of vertices and E is a set of edges that connect two elements of V. A network can be described as an adjacency matrix A, where if there is an edge between nodes i and j, then $a_{ij}=1$, otherwise $a_{ij}=0$. Community discovery is to identify a partitioning $\{p_1, p_2, p_n\}$ that maximized the number of connections inside each community and minimizes the number of the links between them. $k_i^{in}(S)$ and $k_i^{out}(S)$ is the internal and external degree of the nodes belonging to a community S.

$$k_i^{in}(S) = \sum_{j \in S} A_{ij} \tag{3}$$

$$k_i^{out}(S) = \sum_{j \in S} A_{ij} \tag{4}$$

A subgraph S is a community in a strong sense if

$$k_i^{in}(S) > k_i^{out}(S), \ \forall i \in S. \tag{5}$$

A subgraph S is a community in a weak sense if

$$\sum_{i \in S} k_i^{in}(S) > \sum_{i \in S} k_i^{out}(S). \tag{6}$$

Thus, each node in a strong community has more connections within the community than with the rest of the network, but in a weak community, the sum of the degrees within the sub graph is larger than the sum of degrees toward the rest of the network. We adopt the concept of weak community.

3.1   Objective Functions

Our aim is to find the community of a complex networks. A community structure S in a network is a set of groups of vertices having a high density of edges among the vertices and a lower density of edges between different groups. We want to find community structure S that simultaneously optimized each quality measure.

Recently, researchers have found so many quality functions, such as the modularity, community score, community fitness, the MinMaxCut value of the community and the Maximum-Out Degree Fraction(ODF). Modularity is the function we described above. Let k be the number of modules found inside a complex network, the modularity is defined as:

$$Q = \sum_{i=1}^{k} \left[\frac{l_i}{m} - \left(\frac{d_i}{2m} \right)^2 \right] \tag{7}$$

where l_i is the total number of edges joining vertices inside the module i, and d_i is the sum of the degree of the nodes of i, m is the total number of links in the network. Though this function Q suffers from resolution limit problems, as shown by Fortunato and Barthelme [18], it still has been widely accepted by the scientific community.

Community score [19] is a global measure of the network division in communities by summing up the local score of each module found. Community fitness [20] is the ratio between the total internal degrees of the nodes belonging to that community and the sum of the total internal and external degrees of the nodes belonging to that community. As mentioned above, $k_i^{in}(S)$ and $k_i^{out}(S)$ is the internal and external degree of the nodes belonging to a community S. The community fitness $P(S)$ of a module S is defined as

$$P(S) = \sum_{i \in S} \frac{k_i^{in}(S)}{\left(k_i^{in}(S) + k_i^{out}(S) \right)^{\alpha}} \tag{8}$$

where α is a positive real-valued parameter controlling the size of the community. MinMaxCut(MMC) tries to maximize similarity of nodes within the same community while minimizing the similarity of nodes belonging to different communities[21]. The MinMaxCut is defined as

$$MMC = \sum_{i=1}^{m} \frac{d^{out}(V_i)}{d^{in}(V_i)} \tag{9}$$

where $d^{in}(V_i) = \sum_{j \neq k, j \in V_i, k \in V_i} A_{jk}$ and $d^{out}(V_i) = \sum_{j \neq k, j \in V_i, k \notin V_i} A_{jk}$, A is the adjacency matrix. The smaller the value of MinMaxCut, the higher the quality of the online community. ODF is the maximum fraction of edges of a node pointing outside the cluster. We choose MMC and the community fitness as the objective function. We use the modularity and Normalized Mutual Information (NMI) to evaluate the quality of the proposed community detection method. Modularity has been described above. NMI is a similarity measure [22]. Given two partitions A and B of a network, let C be the confusion matrix whose element C_{ij} is the number of nodes of community i of the partition A that are also in the community j of the partition B. The normalized mutual information I (A, B) is defined as:

$$I(A,B) = \frac{-2\sum_{i=1}^{C_A} C_{ij} \sum_{j=1}^{C_B} C_{ij} \log\left(\frac{C_{ij}C}{C_i C_j}\right)}{\sum_{i=1}^{C_A} C_i \log\left(\frac{C_i}{C}\right) + \sum_{j=1}^{C_B} C_j \log\left(\frac{C_j}{C}\right)}$$

(10)

where $C_A(C_B)$ is the number of groups in the partition A(B), $C_i(C_j)$ is the sum of the elements of C in row i (column j), and N is the number of nodes. C_{ij} If A=B, I(A,B)=1, else if A and B are completely different, I(A,B)=0.

3.2   Genetic Representation

The algorithm uses the locus-based adjacency representation proposed in [23]. In this representation, each individual chromosome consists of n genes $g_1, g_2, ..., g_n$ and each gene can take values j in the range {1, ..., n}, where n is the number of nodes in the network. If the ith gene assigned a value j, it means there is a link between node i and j. It also means that node i and j will be in the same community. In order to identify all the communities, a further decoding step is necessary.

If the ith gene's value is equal to the kth gene's value, then they are in the same community, else the ith node is in the community same to the ith gene's value. This method can decode rapidly.

3.3   Initialization

Consider the connection between different nodes, assign one of the neighbor's id to the ith gene. Use the neighbor ordered list to ensure this. The ith row of the neighbor ordered list is an ordered list of the ith node. For example, Fig. 1(b) is the neighbor ordered list of Fig. 1(a).

3.4   Crossover

Crossover is the process by which two-selected chromosome with high fitness values exchange part of the genes to generate new pair of chromosomes. Different types of crossover have been used, such as one point crossover, two-point crossover, uniform crossover and multipoint crossover. MOGCM uses two-point crossover. In each chromosome randomly choose two point and exchange genes between them.

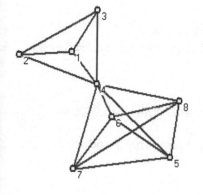

node	Neighbor ordered list						
1	2	3	4				
2	1	3	4				
3	1	2	4				
4	1	2	3	5	6	7	8
5	4	6	7	8			
6	4	5	7	8			
7	4	5	6	8			
8	4	5	6	7			

Fig. 1. (a).a complex network   (b) the neighbor ordered list

Table 1. Example of two-point crossover

Parent1:	2	4	4	5	6	3	4	2
Parent2:	4	5	3	4	6	3	2	4
Offspring1	2	5	3	4	6	3	4	2
Offspring2	4	4	4	5	6	3	2	4

3.5   Mutation

Mutation is the random change of the value of a gene, which is used to prevent premature convergence to local optima. Major ways of mutation are random bit mutation, random gene mutation, creep mutation, and heuristic mutation. MOGCM use random gene mutation, but change the value j of the ith gene randomly may cause a useless exploration of the search space as mentioned above. Thus, use the neighbor ordered list to choose a possible neighbor value to guarantee the generation of the mutated child in which each node is linked only with one of its neighbor.

The new population generated undergoes the further selection, crossover and mutation till the termination criterion is satisfied.

3.6   Algorithm Description

The algorithm calculate the neighbor ordered list D first, then use genetic algorithm to generate the solution and chose the solution with the max modularity. It is descripted as follow:

```
Program1:MOGCM
Model the network as a graph G=(V,E)
Create an adjacent matrix A
Create a neighbor ordered list D
Create a population of random individual
Set the i^th gene's value is one of its neighbor ordered
list value
While j<100
{
  Decode the population and calculate the two fitness
value
  Crossover two chromosomes which has high nondomination
rank
Mutation
}
Return the solutions which have the maximum modularity
value
```

4   Experiments

In this section, we study the effectiveness of our method and compared with other methods using some real world datasets. Our algorithm was implemented into Matlab 2013b. All experiments are conducted on a computer with Intel Core i5 1.7GHz, 4GB RAM.

The Zackary's Karate Club network was generated by Zachary who studied the friendship of 34 members of the karate club over a period of 2 years [24]. During this period, the club divided in two groups almost of the same size because of the disagreements between members. The dolphin network was established by Lusseau, it has 62 nodes each represent one dolphin living in Doubtful Sound, New Zealand [25]. A tie between two dolphins was established by their statistically significant frequent association. The network split naturally into two large groups, the number of ties being 159.

First, compare the MOGCM with GN[3] algorithm. GN algorithm has been explained in the first section.

For each network, the algorithm was executed 15 times. To compare the MOGCM with other method, at each run the solutions with the best value of modularity has been selected. In MOGCM method, the crossover rate was 0.8, the mutation rate was 0.2, the elite reproduction rate was 10% of the population size. GA is a nondeterministic algorithm, due to the random nature of the GA i.e., it may produce a different solution each run. So we calculate the average NMI for each pair of solutions for each network. Because the GA algorithm generates a set of solutions, the MOGCM algorithm chooses the solution with max modularity. Table 2 reported the average NMI value and the modularity values of these 15 runs of MOGCM. A low value indicates that, over many runs, the solutions vary considerably; a high value indicates that the solutions are much more similar to each other. The table shows that the very good performance of MOGCM. The NMI values show that the 15 results of the 15 runs of MOGCM are similar. The communities found by Girvan and Newman split Zackary's Karate Club network in two and misplaces a node, but the MOGCM found the exact communities. MOGCM found two smaller communities with high modularity.

Second, modify the objective function of MOGCM, change the community fitness with community score. Table 3 shows the different between them. Table 2 shows the NMI and modularity obtained by the two different methods, it shows that uses the community and MinMaxCut as the two objective function, can obtain high NMI than use the community score and MinMaxCut. It shows that the community fitness is the better choice here.

Table 2. NMI results and modularity obtained by MOGCM and GN for the real-life data sets

	avg NMI	Avg mod	GN mod
Zackary's Karate Club	0.99	0.418	0.381
Bottlenose Dolphins	0.98	0.496	0.496

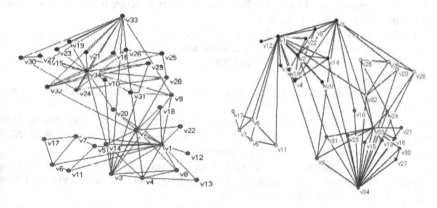

Fig. 2. The community been detected by GN

Fig. 3. The community been detected by MOGCM

Table 3. NMI results obtained by MOGCM and the change

	avg best NMI	Changed avg best NMI
Zackary's Karate Club	0.964	0.803
Bottlenose Dolphins	0.835	0.723

5   Conclusions

In this paper, we proposed an efficient community discovery method and analyzed its performance in terms of MinMaxCut and the community fitness. The optimization of these two objectives allows finding communities such that connections in one community are dense and the connections between communities are sparse. The results on real-world networks have shown our method is more efficient than state-of-art algorithms. Future work aims to extending the method to dynamic networks.

Acknowledgement. This work was supported by National Natural Science Foundation of China (61300195), Natural Science Foundation of Hebei Province (F2014501078) and The General Project of Liaoning Province Department of Education Science Research (L2013099).

References

1. Fortunato, S.: Community detection in graphs. J. Phys. Rep. 486, 75–174 (2010)
2. Girvan, M., Newman, M.E.: Community structure in social and biological networks. Proceedings of the National Academy of Sciences 99(12), 7821–7826 (2002)
3. Newman, M.E.J., Girvan, M.: Finding and evaluating community structure in networks. J. Physics Rev. E 69(2), 026113 (2004)
4. Radicchi, F., Castellano, C., Cecconi, F., Loreto, V., Parisi, D.: Defining and identifying communities in networks. J. Proc. Nat. Acad. 101(9), 2658–2663 (2004)
5. Clauset, A., Newman, M., Moore, C.: Finding community structure in very large networks. J. Phys. Rev. E 70(6), 066111 (2004)
6. Du, J., Korkmaz, E.E., Alhajj, R., Barker, K.: Novel clustering approach that employs genetic algorithm with new representation scheme and multiple objectives. In: Kambayashi, Y., Mohania, M., Wöß, W. (eds.) DaWaK 2004. LNCS, vol. 3181, pp. 219–228. Springer, Heidelberg (2004)
7. Faceli, K., de Carvalho, A.C.P.L.F., de Souto, M.C.P.: Multiobjective clustering ensemble. J. Hybrid Intell. Syst. 4(3), 145–156 (2007)
8. Feng, Z., Xu, X., Yuruk, N., Schweiger, T.A.J.: A novel similarity-based modularity function for graph partitioning. In: Song, I.-Y., Eder, J., Nguyen, T.M. (eds.) DaWaK 2007. LNCS, vol. 4654, pp. 385–396. Springer, Heidelberg (2007)
9. Radicchi, F., Castellano, C., Cecconi, F., Loreto, V., Parisi, D.: Defining and identifying communities in networks. Proceedings of the National Academy of Sciences of the United States of America 101(9), 2658–2663 (2004)
10. Clauset, A., Newman, M., Moore, C.: Finding community structure in very large networks. J. Phys. Rev. E 70(6), 66111 (2004)
11. Newman, M.E.J.: Fast Algorithm for Detecting Community Structure in Networks. Physical Review E 69, 066133 (2004)
12. Guimera, R., Amaral, L.A.N.: Functional Cartography of Complex Metabolic Networks. J. Nature 433, 895–900 (2005)
13. Lü, Z., Huang, W.: Iterated Tabu Search for Identifying Community Structure in Complex Networks. J. Physical Review E 80, 026130 (2009)
14. Coscia, M., Giannotti, F., Pedreschi, D.: A Classification for Community Discovery Methods in Complex Networks. J. Statistical Analysis and Data Mining Journal, Special Issue: Networks 4(5), 512–546 (2011)
15. Saha, S., Bandyopadhyay, S.: A new multiobjective clustering technique based on the concept of stability and symmetry. J. Knowl. Inform. Syst. 23(1), 1–27 (2010)

16. He, D., Wang, Z., Yang, B., Zhou, C.: Genetic algorithm with ensemble learning for detecting community structure in complex networks. In: Computer Sciences and Convergence Information Technology, ICCIT 2009, pp. 702–707. IEEE Press (2009)
17. Firat, A., Chatterjee, S., Yilmaz, M.: Genetic clustering of social networks using random walk. J. Comput. Statist. Data Anal. 51(12), 6285–6294 (2007)
18. Fortunato, S., Barthelemy, M.: Resolution limit in community detection. Proceedings of the National Academy of Sciences 104(1), 36–41 (2007)
19. Pizzuti, C.: GA-net: A genetic algorithm for community detection in social networks. In: Rudolph, G., Jansen, T., Lucas, S., Poloni, C., Beume, N. (eds.) PPSN X 2008. LNCS, vol. 5199, pp. 1081–1090. Springer, Heidelberg (2008)
20. Lancichinetti, A., Fortunato, S., Kertész, J.: Detecting the overlapping and hierarchical community structure of complex networks. J. New J. Phys. 11, 033015 (2009)
21. Demir, G.N., Uyar, A.S., Öğüdücü, S.G.: Graph-based sequence clustering through multiobjective evolutionary algorithms for web recommender systems. In: Proceedings of the 9th Annual Conference on Genetic and Evolutionary Computation, pp. 1943–1950. ACM (2007)
22. Yang, Y.: Information Theory, Inference, and Learning Algorithms. Journal of the American Statistical Association 100(472), 1461–1462 (2005)
23. Park, Y.J., Song, M.S.: A genetic algorithm for clustering problems. In: Proc. 3rd Annu. Conf. Genet. Algorithms, pp. 2–9 (1989)
24. Network DataSets, http://www.personal.umich.edu/mejn/netdata
25. Lusseau, D.: The emergent properties of dolphin social network. Proceedings of the Royal Society of London. Series B. Biological Sciences, S186–S188 (2003)

A Multi-objective Jumping Particle Swarm Optimization Algorithm for the Multicast Routing

Ying Xu[1] and Huanlai Xing[2]

[1] College of Computer Science and Electronic Engineering, Hunan University,
Changsha, 410082, Hunan, P.R. China
[2] School of Information Science and Technology, Southwest Jiaotong University,
Chengdu, 610031, Sichuan, P.R. China
hnxy@hnu.edu.cn, hxx@home.swjtu.edu.cn

Abstract. This paper presents a new multi-objective jumping particle swarm optimization (MOJPSO) algorithm to solve the multi-objective multicast routing problem, which is a well-known NP-hard problem in communication networks. Each particle in the proposed MOJPSO algorithm performs four jumps, i.e. the inertial, cognitive, social and global jumps, in such a way, particles in the swarm follow a guiding particle to move to better positions in the search space. In order to rank the non-dominated solutions obtained to select the best guider of the particle, three different ranking methods, i.e. the random ranking, an entropy-based density ranking, and a fuzzy cardinal priority ranking are investigated in the paper. Experimental results show that MOJPSO is more flexible and effective for exploring the search space to find more non-dominated solutions in the Pareto Front. It has better performance compared with the conventional multi-objective evolutionary algorithm in the literature.

Keywords: Multi-objective Optimization, Jumping Particle Swarm Optimization, Multicast Routing.

1   Introduction

Multicast routing is an important telecommunication technique that simultaneously transfers messages from a source to multiple destinations simultaneously in communication networks. It provides the support for real-time applications, such as the multimedia conference, the distance education, video/audio on demand, etc. In this work, we consider the Multicast Routing Problem (MRP) with multiple quality of service (QoS) requirements. The QoS requirements normally include the link utilization, the end-to-end delay, the delay jitter and the tree cost etc. The underlying model of MRPs is the Steiner tree problem, which is a NP-hard combinatorial optimization problem [1]-[2]. The QoS-based MRPs, i.e., the QoS requirements constrained Steiner tree problem, is thus also NP-hard, which is a challenging optimization problem.

During the past two decades, the QoS-based MRPs have attracted a lot of research attention from both computer communications and operational research ([3]-[5]).

Y. Tan et al. (Eds.): ICSI 2014, Part I, LNCS 8794, pp. 414–423, 2014.
© Springer International Publishing Switzerland 2014

Many efficient meta-heuristic algorithms, including simulated annealing, tabu search, variable neighborhood search, genetic algorithm, ant colony, particle swarm optimization, etc., have been proposed to solve various QoS-based MRPs ([6]-[8]). Recently, the MRPs with multiple QoS constraints have been defined as a multi-objective optimization problem [9]. A variety of multi-objective optimization algorithms based on meta-heuristics have been investigated in the recent literature ([10],[12]).

Particle Swarm Optimization (PSO), a natural-inspired population-based stochastic search method proposed by [13], is an effective meta-heuristic for solving both the combinatorial optimization and the multi-objective optimization problems. PSO simulates social behavior of flocks of birds or schools of fish. In PSO, a population of particles (individuals) moves through a multi-dimensional continuous space with a moving velocity. While PSO was originally designed for solving continuous optimization problems, a variant called Discrete Particle Swarm Optimization (DPSO) has been firstly designed by Kennedy and Eberhart [14]. Since then, many variations of DPSO algorithms have been proposed for different combinatorial problems [15]-[16].

Recently, a DPSO algorithm named Jumping Particle Swarm Optimization (JPSO) algorithm was introduced by [17] to solve combinatorial optimization problems. Without using the concept of velocity, the JPSO algorithm defines different jumps (moves) for particles to move from position to position in a discrete hyper-space[18],[19]. There are four types of moves, the first one is the inertial move, which enables the particle to continue exploring its current position. The other three moves encourage the particle to move towards three different attractors (the best position b_i of the particle i; the best position g_j is of the swarm at iteration j; and the best position $g_{i,j}$ found by the neighbors of particle i at iteration j). Each move has a certain probability given by the weight vector c_x, where $\sum_{x=0}^{3} C_x = 1$. The segment [0, 1] is thus divided into four sub-segments with the variable width subject to a certain c_x. At each generation, a random number r uniformly distributed in [0, 1] is generated to decide which move to be performed. For $r \in [0, c_0)$, the inertial move will be applied. For $r \in [c_0, c_0 + c_1)$, the particle moves towards b_i. For $r \in [c_0 + c_1, c_0 + c_1 + c_2)$, the particle moves towards g_j. For $r \in [c_0 + c_1 + c_2, 1]$, the attractor will be $g_{i,j}$.

In the literature, we only notice one related work in [10], where a hybrid genetic algorithm and particle swarm optimization has been proposed for solving the multi-objective MRP. A quantum-behaved particle swarm optimization (QPSO) algorithm for QoS multicast routing is proposed in [20]. In our recent work [8], a JPSO algorithm has been proposed to tackle the single objective optimization MRP problem. The good performance of JPSO motivated us to further extend our work to solve the multi-objective multicast routing problem.

The rest of the paper is organized as follows. In Section 2, the multi-objective MRP is formally defined. Section 3 describes the proposed JPSO algorithm, which is then evaluated through extensive experimentation in Section 4. Finally, Section 5 concludes the paper and proposes possible future work.

2   The Multi-objective MRP Formulation

A network is modeled as a directed graph $G = (V, E)$ with $|V| = n$ nodes and $|E| = l$ links. As defined in our recent work [11], we use the following notations: $(i \ j) \in E$: the link from node i to node j, $i, j \in V$; $c_{ij} \in R^+$: the cost of link $(i \ j)$; $d_{ij} \in R^+$: the delay of link $(i \ j)$; $z_{ij} \in R^+$: the capacity of link $(i \ j)$; $t_{ij} \in R^+$: the current traffic of link $(i \ j)$; $s \in V$: the source node for a MRP; $R \subseteq V - \{s\}$: the set of destinations, i.e. the multicast group; $r_d \in R$: a destination; $|R|$: the cardinality of R, i.e. the number of destinations, also called the group size; $\phi \in R^+$: the traffic demand; $T(s, R)$: the multicast tree; $p_T(s, r_d) \subseteq T(s, R)$: the path from the source s to a destination $r_d \in R$ in the multicast tree T; $d(p_T(s, r_d))$: the delay of path $p_T(s, r_d)$ is the sum of delays of links along the path, i.e. $d(p_T(s, r_d)) = \sum\limits_{(i,j) \in p_T(s,r_d)} d_{ij}$, $r_d \in R$. By using the above definitions, a multi-objective MRP with five minimizing objectives can then be formulated as follows:

1.   The cost of the multicast tree:

$$C(T) = \phi \cdot \sum_{(i,j) \in T} c_{ij} \tag{1}$$

2.   The maximal end-to-end delay of the multicast tree:

$$DM(T) = Max\{d(p_T(s, r_d))\}, r_d \in R \tag{2}$$

3.   The maximal link utilization:

$$\alpha(T) = Max\left\{\frac{\phi + t_{ij}}{z_{ij}}\right\}, (i, j) \in T \tag{3}$$

4.   The average delay of the multicast tree:

$$DA(T) = \frac{1}{|R|} \sum_{r_d \in R} d\left(p_T(s, r_d)\right) \tag{4}$$

5.   The delay jitter of the multicast tree:

$$DJ(T) = Max\{d(p_T(s, r_d))\} - Min\{d(p_T(s, r_j))\}, r_d, r_j \in R \tag{5}$$

subject to a bandwidth capacity constraint:

$$\phi + t_{ij} \leq z_{ij}, \forall (i, j) \in T(s, R) \tag{6}$$

For a multi-objective optimization problem, a set of trade-off optimal solutions will be obtained, when considering multiple objectives. It means no single solution can be considered superior to the others in the search space. The set of all these Pareto-optimal solutions in X is called the Pareto-optimal Set. Various multi-objective optimization approaches have been proposed in the literature for solving a wide range of optimization problems [21]-[24]. For a recent comprehensive review of numerous multi-objective particle swarm optimization algorithms for a variety of multi-objective optimization problems can be found in [13]. To our knowledge, there is no investigation of JPSO to solve the multi-objective MRP.

3   The Multi-objective Jumping Particle Swarm Optimization (MOJPSO) Algorithm

In this paper, we propose a new jumping particle swarm optimization algorithm to solve the multi-objective multicast routing problem, named MOJPSO. The initial population consists of a fixed number of particles which are random multicast trees. For each particle, a non-dominated solution set NDS_i is maintained to record the best positions of the particle. A non-dominated solution set NDS of the whole swarm is used to track the non-dominated positions during the evolution. After the particle moves to a new position, a local search is used to further explore better neighboring solutions of the particle.

3.1   The Representation of the Multicast Tree

In the proposed MOJPSO algorithm, we use the same encoding method in our previous work in [8] to represent a multicast tree (a solution or position of the particle) for the multi-objective MRPs. In this representation, a multicast tree is represented by an ordered set of $|R|$ paths $\{g_1, g_2, ..., g_{|R|}\}$ from the source node s to each destination $r_d \in R$, $d = 1, ... |R|$.

3.2   The Four Jumps in MOJPSO

In the proposed MOJPSO, depending on where the random number r falls within the interval $[0, 1]$, a specific jump is defined to influence the how the particle moves from its current position to a new one. At each generation, we define four jumps as follows:

1) If $r \in [0, c_0)$, the particle moves around its current position. In MOJPSO, after a destination is randomly selected, the path from the source to the chosen destination will be replaced by a random path from the k-shortest [25] path set. We set $k = 25$ in our MOJPSO algorithms.

2) For r falls in the three different intervals, i.e. $[c_0, c_0+ c_1)$, $[c_0+ c_1, c_0+ c_1+ c_2)$ and $[c_0+ c_1+ c_2, 1]$, the attractor is b_i, g_j and $g_{i,j}$, respectively. Once the attractor is selected, a jumping operation will be performed. Here, a multicast tree is represented by a binary vector with $n = |V|$ bits (see [8]). Each bit in the vector takes a value of 1 if the corresponding node is in the multicast tree, 0 otherwise. At each step, each bit of the current multicast tree's binary vector will be changed to the same value as that in the binary vector of the attractor. Then the Prim's algorithm [26] is used to generate a minimum spanning tree for the given nodes.

This operator repeats until a new feasible tree is generated or the two binary vectors are the same.

3.3   The Ranking Methods in MOJPSO

The key issue in MOJPSO is the selection of the guiding particles (attractors). How to select the attractor, i.e. b_i, g_j and $g_{i,j}$, will affect the performance of the proposed algorithm. Each particle will obtain a non-dominated solution set NDS_i, how to rank these solutions in NDS_i to select the best solution b_i, as well as the selection of global best solution g_j from NDS and $g_{i,j}$ from $NeiNDS_i$ (the non-dominated neighboring particles set of the current particle at generation j). There are some ranking approaches in the area of multi-objective optimization, including the fitness sharing [27], the crowding distance [22], the nitched sharing [28], the fuzzy cardinal priority [21], and the entropy-based strategy [10], etc. In our proposed JPSO, we implement three simple ranking methods as follows:

1)   The entropy-based density ranking: A multicast tree (solution) can be thought of as an uncertain system based on entropy optimization principles of Shannon [29], and the entropy of the mth locus (the path from the source to the mth destination) of the multicast tree is defined as:

$$H_m(N) = -\sum_{t=1}^{S} p_{tm} ln(p_{tm}) \tag{9}$$

where p_{tm} denotes the probability that the tth symbol appears the mth locus, and it can be calculated:

$$p_{tm} = c/N \tag{10}$$

where N is the number of solutions, c denotes the total number of the tth path appears at the mth locus among N solutions, S is the number of paths. The entropy-based density ranking is calculated as follows:

-- For $N = 2$ with two multicast trees T_i and T_j, the entropy of T_i for the mth locus is $H_m(2) = -p_m ln(p_m)$, where if Ti and Tj have the same path to the mth destination, $p_m = 2/2 = 1$ according to Equation (10); otherwise $p_m = 1/2$.

-- The mean entropy of T_i and T_j :

$$H_{i,j}(2) = \frac{1}{|R|} \sum_{m=1}^{|R|} H_m(2) \tag{11}$$

-- Similarity $S_{i,j} \in [0,1]$ between T_i and T_j is decided by:

$$S_{i,j} = \frac{1}{1 + H_{i,j}(2)} \tag{12}$$

-- Density of T_i is:

$$C(i) = \frac{count_{j \in pop, s_{i,j} > \lambda}}{|P|} \tag{13}$$

where λ is the similarity constant, P is the current population, $|P|$ is the population size, and $count_{j\in P, s_{i,j}>\lambda}$ is the number of solutions whose similarity to T_i exceeds λ.

2) *The fuzzy cardinal priority ranking:* In order to rank the solutions in the non-dominated solution set, the normalized membership function β_j is calculated to provide the fuzzy cardinal priority ranking of each non-dominated solution j. The solution with the maximal value of β_j is considered as the best compromise solution. β_j is defined as [21]:

$$\beta_j = \frac{\sum_{i=1}^{N_{obj}} u_i^j}{\sum_{i=1}^{N_{obj}} \sum_{j=1}^{M} u_i^j} \tag{14}$$

where Nob_j is the number of objectives for the multi-objective optimization problem, M is the number of non-dominated solutions in the *NDS*. u_i^j is given by:

$$u_i^j = \frac{f_i^{max} - f_i^j}{f_i^{max} - f_i^{min}} \tag{15}$$

where f_i^{max} and f_i^{min} are the maximum and minimum value of the ith objective, respectively, f_i^j is the ith objective value for the jth solution in the *NDS*.

3) *The random ranking:* A solution is randomly selected in the *NDS*.

3.4   The Local Search in MOJPSO

In each generation, after each particle jumping to a new position, a local search is performed to further enhance the solution. At each step, it randomly flips a bit of a node in the binary vector of the multicast tree which represents the solution using the Prim's algorithm to generate a minimum spanning tree on the given nodes. This operation repeats until a new feasible tree is generated.

4   Performance Evaluation

4.1   Simulation Environment

We have carried out a large amount simulations to test our algorithms on some benchmark and random networks. Six variants of our MOJPSO have been evaluated and then compared with a traditional multi-objective evolutionary algorithm in [9].

1) MOJPSO-DR/ MOJPSO-DR-LS, the proposed algorithm with the entropy-based density ranking without or with the local search

2) MOJPSO-FR/MOJPSO-FR-LS, the proposed algorithm with the fuzzy cardinal priority ranking without or with the local search

3) MOJPSO-RR/MOJPSO-RR-LS, the proposed algorithm with the random ranking without or with the local search

4) MOEA/MOEA-LS, a multi-objective evolutionary algorithm without or with the local search.

We use the same multicast routing simulator to generate the network topologies as in [10]. For a fair comparison, the population size $|P|$ is set as 50 in the variants of MOJPSO and MOEA.

4.2   The Impact of the Local Search

In this group of experiments, we test the effectiveness of the local search procedure designed in this paper upon a set of random networks. Four MRP instances are generated on two random networks of $|V| =100$ with different group sizes (the number of destination nodes $|R| = 20\%*|V|$ and $|R| = 30\%*|V|$). In order to visually compare the Pareto optimal front obtained by variant algorithms of MOJPSO in 20 runs, the running time for each run is 160 seconds. we consider only two objectives, namely the cost and the delay as defined in Section 2, in Figure 1. It shows the local search can further enhance the search to find better non-dominated solutions. It is interesting to see that MOJPSO with three different ranking methods have similar performance.

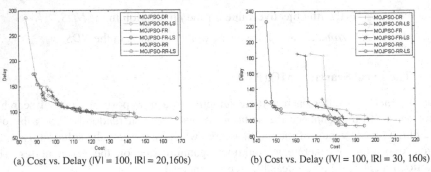

(a) Cost vs. Delay ($|V| = 100$, $|R| = 20,160s$)   (b) Cost vs. Delay ($|V| = 100$, $|R| = 30, 160s$)

Fig. 1. The Pareto-front of the solutions found by each algorithm for the random networks

4.3   The Effectiveness of Different Ranking Methods

Finally, we test MOJPSO with different ranking methods described in Section 3.3 and compare these variants of MOJPSO with MOEA upon these random networks with respect to all five objectives defined in Section 2. For each network, we run 20 times for each network. The same running time, i.e. 160 seconds is set for all algorithms in each run. We plot all the non-dominated solutions found by these algorithms in 20 runs in Figure 2. It can be seen that the performance of MOJPSO with three different ranking methods are competitive during the search. MOJPSO with different ranking methods can obtained more non-dominated solutions than the conventional MOEA. For the instance in Figure 2, three MOJPSO algorithms can find better Pareto non-dominated solution set than that of MOEA, which demonstrate the effectiveness and flexibility of our proposed MOJPSO.

5 Conclusions

In this paper, we propose a new Multi-Objective Jumping Particle Swarm Optimization algorithm MOJPSO for solving the multi-objective multicast routing problems in communication networks. Four jumps which guide the particles to move during the evolutionary procedure have been designed. Three ranking methods, including the entropy-based density ranking, the fuzzy cardinal priority ranking and the random ranking, have been investigated to rank non-dominated solutions for selecting the best guiders. Experimental results show our proposed MOJPSO algorithms perform better than the conventional MOEA by finding more and better non-dominated solutions. We thus conclude that MOJPSO is a competitively approach for solving the multi-objective multicast routing problem. In our future work, we intend to extend our MOJPSO algorithm to solve other multi-objective optimization problems.

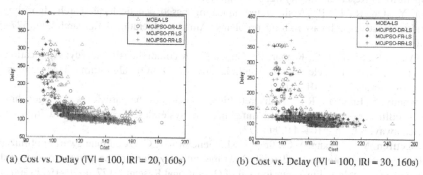

(a) Cost vs. Delay (|V| = 100, |R| = 20, 160s) (b) Cost vs. Delay (|V| = 100, |R| = 30, 160s)

Fig. 2. The non-dominated solutions found by each algorithm for the random networks

Acknowledgments. This research is supported by Natural Science Foundation of China (NSFC project No. 61202289 and 61272396) and the project of the support plan for young teachers in Hunan University, China (Ref. 531107021137).

References

1. Hwang, F.K., Richards, D.S.: Steiner tree problems. Networks 22, 55–89 (1992)
2. Drake, D.E., Hougardy, S.: On approximation algorithms for the terminal Steiner tree problem. Information Processing Letters 89(1), 15–18 (2004)
3. Diot, C., Dabbous, W., Crowcroft, J.: Multipoint communication: A survey of protocols, functions, and mechanisms. IEEE J SEL AREA COMM 15, 277–290 (1997)
4. Yeo, C.K., Lee, B.S., Er, M.H.: A survey of application level multicast techniques. Computer Communications 27(15), 1547–1568 (2004)
5. Oliveira, C., Pardalos, P.M.: A survey of combinatorial optimization problems in multicast routing. Computers & Operations Research 32(8), 1953–1981 (2005)
6. Qu, R., Xu, Y., Kendall, G.: A variable neighborhood search algorithm for delay-constrained least-cost multicast routing. In: Stützle, T. (ed.) LION 3. LNCS, vol. 5851, pp. 15–29. Springer, Heidelberg (2009)

7. Sun, J., Fang, W., Wu, X., Xie, Z., Xu, W.: QoS multicast routing using a quantum-behaved particle swarm optimization algorithm. Engineering Applications of Artificial Intelligence 24(2011), 123–131 (2011)

8. Qu, R., Xu, Y., Castro, J.P., Landa-Silva, D.: Particle swarm optimization for the Steiner tree in graph and delay-constrained multicast routing problems. Journal of Heuristics 19(2), 317–342 (2013)

9. Crichigno, J., Barán, B.: Multiobjective Multicast Routing Algorithm. In: de Souza, J.N., Dini, P., Lorenz, P. (eds.) ICT 2004. LNCS, vol. 3124, pp. 1029–1034. Springer, Heidelberg (2004)

10. Li, C.-B., Cao, C.-X., Li, Y.-G., Yu, Y.-B.: Hybrid of genetic algorithm and particle swarm optimization for multicast QoS routing. In: Proceedings of IEEE International Conference on Control and Automation, pp. 2355–2359 (2007)

11. Xu, Y., Qu, R.: Solving multi-objective multicast routing problems by evolutionary multi-objective simulated annealing algorithms with variable neighbourhoods. Journal of the Operational Research Society 62, 313–325 (2011)

12. Xu, Y., Qu, R., Li, R.: A simulated annealing based genetic local search algorithm for multi-objective multicast routing problems. Annuals Operational Research 206, 527–555 (2013)

13. Lalwani, S., Sianhal, S., Kumar, R., Gupta, N.: A comprehensive survey: Applications of multi-objective particle swarm optimization (MOPSO) algorithm. Transactions on Combinatories 2, 39–101 (2013)

14. Kennedy, J., Eberhart, R.C.: A discrete binary version of the particle swarm algorithm. In: Proceedings of the World Multiconference on Systemics, Cybernetics and Informatics, Piscataway, NJ, pp. 4104–4109 (1997)

15. Tasgetiren, M.F., Liang, Y.C., Sevkli, M., Gencyilmaz, G.: A particle swarm optimization algorithm for makespan and total flowtime minimization in the permutation flowshop sequencing problem. European Journal of Operational Research 177, 1930–1947 (2007)

16. Anghinolfi, D., Paolucci, M.: A new discrete particle swarm optimization approach for the single-machine total weighted tardiness scheduling problem with sequence-dependent setup times. European Journal of Operational Research 193, 73–85 (2009)

17. Moreno-Perez, J.A., Castro-Gutierrez, J.P., Martinez-Garcia, F.J., Melian, B., Moreno-Vega, J.M., Ramos, J.: Discrete particle swarm optimization for the p-median problem. In: Proceedings of the 7th Metaheuristics International Conference, Montreal, Canada (2007)

18. Consoli, S., Moreno-Perez, J.A., Darby-Dowman, K., Mladenovic, N.: Discrete particle swarm optimization for the minimum labelling Steiner tree problem. In: Krasnogor, N., Nicosia, G., Pavone, M., Pelta, D. (eds.) NICSO 2007. SCI, vol. 129, pp. 313–322. Springer, Heidelberg (2008)

19. Castro, J.P., Landa-Silva, D., Moreno Perez, J.A.: Exploring feasible and infeasible regions in the vehicle routing problem with time windows using a multi-objective particle swarm optimization approach. In: NICSO (2008)

20. Sun, J., Fang, W., Wu, X., Xie, Z., Xu, W.: QoS multicast routing using a quantum-behaved particle swarm optimization algorithm. Engineering Applications of Artificial Intelligence 24, 123–131 (2011)

21. Jain, S., Sharma, J.D.: Tree structured encoding based multi-objective multicast routing algorithm. International Journal of Physical Sciences 7, 1622–1632 (2012)

22. Wang, S., Lei, X., Huang, X.: Multi-objective optimization of reservoir flood dispatch based on MOPSO algorithm. In: Proceedings of the 8th International Conference on Natural Computation, pp. 827–832 (2013)

23. Konak, A., Coit, D.W., Smith, A.E.: Multi-objective optimization using genetic algorithms: A tutorial. Reliability Engineering and System Safety 91, 992–1007 (2006)
24. Zhang, Q., Li, H.: MOEA/D: A multiobjective evolutionary algorithm based on decomposition. IEEE Trans. on Evolutionary Computation 11, 712–731 (2007)
25. Eppstein, D.: Finding the k shortest paths. SIAM J. Computing (1998)
26. Betsekas, D., Gallager, R.: Data networks, 2nd edn. Prentice-Hall, Englewood Cliffs (1992)
27. Deb, K., Goldberg, D.E.: An investigation of niche and species formation in genetic function optimization. In: Third International Conference on Genetic Algorithms, pp. 42–50 (1989)
28. Chang, Y.C.: Multi-Objective Optimal SVC Installation for Power System Loading Margin Improvement. IEEE Transactions on Power Systems 27, 984–992 (2012)
29. Kapur, J., Kesavan, H.: Entropy Optimization Principles with Applications. Academic Press, San Diego (1992)

A *Physarum*-Inspired Multi-Agent System to Solve Maze

Yuxin Liu[1], Chao Gao[1], Yuheng Wu[1], Li Tao[1], Yuxiao Lu[1], and Zili Zhang[1,2,*]

[1] School of Computer and Information Science,
Southwest University, Chongqing 400715, China
[2] School of Information Technology, Deakin University, VIC 3217, Australia
zhangzl@swu.edu.cn

Abstract. *Physarum Polycephalum* is a primitive unicellular organism. Its foraging behavior demonstrates a unique feature to form a shortest path among food sources, which can be used to solve a maze. This paper proposes a *Physarum*-inspired multi-agent system to reveal the evolution of *Physarum* transportation networks. Two types of agents – one type for search and the other for convergence – are used in the proposed model, and three transition rules are identified to simulate the foraging behavior of *Physarum*. Based on the experiments conducted, the proposed multi-agent system can solve the two possible routes of maze, and exhibits the reconfiguration ability when cutting down one route. This indicates that the proposed system is a new way to reveal the intelligence of *Physarum* during the evolution process of its transportation networks.

Keywords: *Physarum Polycephalum*, Multi-Agent System, Maze.

1 Introduction

Maze is an interesting game, as well as a complex searching problem, which can puzzle some high-level organism sometimes. However, based on the latest biological studies, researchers find an amazing result: a primitive single-cell organism, called *Physarum Polycephalum*, has the ability to solve it [1]. *Physarum Polycephalum*, commonly known as a multi-headed slime mould, can reshape itself in the plasmodium stage of its complicated life-cycle. The plasmodium of *Physarum Polycephalum* reserves the shortest protoplasmic tube connecting the entrance and the exit in the maze. The amazing observation inspires an innovative spark based on biological and physical theories to uncover the key of intelligence hidden in *Physarum* over the last few years [2–6].

Jones [5, 7] presents a low-level particle-based agent model to reveal the evolution of plasmodium networks. For enhancing the self-organization of Jones' model, Wu et al. [8] have presented a multi-agent system (MAS) with self-adaptive population. Their system can automatically adjust the population of agents, maintain its homeostasis and keep the stable of the macro formation. For

---

* Corresponding author.

Y. Tan et al. (Eds.): ICSI 2014, Part I, LNCS 8794, pp. 424–430, 2014.

further developing Wu's MAS, this paper derives two types of agents to simulate the search and convergence behaviors of *Physarum* in foraging and utilizes agent's type transition to pass information globally. After the improvement, the new model, denoted as the PMAS, has the self-organization ability and can solve a maze.

The rest of this paper is organized as follows. Section 2 presents the basis of the PMAS, especially the behaviors of agents. Section 3 validates the route-finding ability of the PMAS when solving a maze. Section 4 concludes this paper.

2   The *Physarum*-Inspired Multi-Agent System

This section introduces the basic idea of the PMAS from three aspects: environment, the structure and behaviors of agents.

2.1   Environment Setting

The environment of the PMAS is a 2D scenario discretized with $n \times m$ grids, which corresponding to the substrate in biological experiments. The plasmodium of *Physarum* and food sources are modelled as agents and data points, respectively. Each grid only holds one agent. One time step is defined as that each agent performs once asynchronously.

An agent secretes $depT$ of trail when it moves into a new grid. A data point deposits CN of chemo-nutrient in its grids at each time step. Both chemo-attractants (i.e., trail and chemo-nutrient) diffuse independently in the scenario by means of a simple average filter as shown in Eq. (1), where $V_{(i,j)}^t$ represents the value of chemo-attractants in grid (i, j) at time step t. For trail information, V means trail value (denoted as TV), and for chemo-nutrient information, V means chemo-nutrient value (denoted as NV). What's more, two damping factors, $dampT$ and $dampN$, are used to describe the volatilization of trail and chemo-nutrient, respectively. Taking trail information as an example, at each time step, the amount of trail in all grids updates simultaneously through that filter and descreases to $1 - dampT$ of the previous value.

$$V_{(i,j)}^t = \frac{\sum_{k=1}^{m} \sum_{l=1}^{m} V_{(i+k-m+1, j+l-m+1)}^{t-1}}{m^2} \tag{1}$$

2.2   Agent Architecture

In corresponding with the foraging behaviors of *Physarum*, two types of agents, one type for search, named as agent_T1 and the other for convergence, named as agent_T2, are proposed. They have the same architecture and just one difference in basic behaviors. As shown in Fig. 1, an agent is composed of three components, i.e., the main body, the left sensor and the right sensor. The body determines the position of an agent. The sensor is used to sample the chemo-attractants in the grid where the sensor locates.

The "Forward" of an agent indicates the direction that it would move along. The "Sensor Arm Length" is used to represent the sense distance of an agent. The "Sensor Angle" is fixed to 45°. Each sensor is armed with a "Trail Sampling" module and a "Chemo-nutrient Sampling" module, which are used to measure the trail value (TV) and chemo-nutrient value (NV), respectively. The total sampled chemo-attractants value (SV) is calculated by a linear weighted method $SV = WT \times TV + WN \times NV$, which WT and WN are the weights of trail and chemo-nutrient, respectively. The "Synthesis Comparator" module of the main body controls the forward direction of an agent by comparing SV between the left sensor and right sensor. For an agent_T1, it selects the direction with lower SV. But an agent_T2 selects that with greater one. The "Motion Counter" is a module to record the motion of an agent. It is crucial for the self-organization of the quantity of agents in the system.

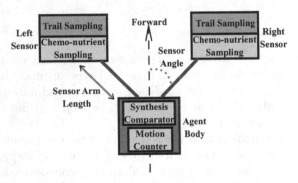

Fig. 1. The architecture and morphology of an agent

2.3   Behavior Rules

The behaviors of an agent can be divided into two categories: the basic behaviors of each type of agent and the interaction between two types of agents.

2.3.1   Basic Behaviors of Each Type of Agent

The basic behaviors of an agent mainly include movement, reproduction and elimination. Initially, the integer in the "Motion Counter" of an agent, denoted as MC, is zero. At each time step, an agent attempts to move forward to a neighbor grid. If that grid is occupied by other agent, then the agent stays at the current grid and changes its forward direction randomly. At the same time, MC is decreased one. If that grid is empty, then the agent moves to the grid, deposits $depT$ of trail in the grid and rotates the new direction according to "Synthesis Comparator". Meanwhile, MC is increased one.

The parameters RT and ET are set to trigger the reproduction and the elimination of an agent, respectively. When an agent moves successfully and MC is greater than RT, the agent clones a child agent at its previous grid (i.e., reproduction behavior). If MC is less than ET, the elimination behavior is triggered, i.e., the agent disappears from the environment.

2.3.2 Transition Rules Between Two Types of Agents

There are three transition rules for establishing the cooperation between two types of agents. Specifically, Rule 1 supports for the system search of the scenario. Rule 2 indicates a local information that a new data point is found. Rule 3 implements the message passing in the system by type transition, which globalizes the local information and makes the chaotic system converge to an organized pattern.

Rule 1. *The population evolution starts with an agent_T1.*

Rule 2. *An agent_T1 transmits to an agent_T2, when the agent_T1 enters into the grids of a new data point.*

Rule 3. *An agent_T1 transmits to an agent_T2, if the agent_T1 has an agent_T2 neighbor.*

What's more, Algorithm 1 presents details about the behaviors of an agent.

Algorithm 1. Framework of the behaviors of an agent

1. **if** $MC < ET$ **then**
2.   eliminate itself;
3.   return;
4. **end if**
5. **if** the forward grid is empty **then**
6.   **if** $MC > RT$ **then**
7.   clone a child agent at the current grid;
8.   **end if**
9.   move to the forward grid;
10. $MC + +$;
11.   rotate a new direction
12. **else**
13.   reset the forward direction randomly;
14. $MC - -$;
15. **end if**
16. **if** the agent is an agent_T1 **then**
17.   **if** the agent enters into the grid of a new data point ‖ the agent has an agent_T2 neighbor **then**
18.   transform to an agent_T2
19.   **end if**
20. **end if**

---

3 Numerical Experiments

A maze scenario in this section is used to explore the ability of the system on maze solving and path reconfiguration. The basic parameters and their values are listed in Table 1. As shown in Fig. 2, the black parts stand for walls of the maze, and the white areas are passages. The widths of walls and passages are both 20 grids. Two data points, which represent the entrance and exit of the maze, are arranged at the southwest and northeast, respectively.

Table 1. Basic parameters and their values used in this paper

Parameter	Explanation	Value
Scenario Size	The planar area for network formation	200×200 grids
Sensor Arm Length	The distance between sensor and its body	4 grids
$depT$	The amount of trail deposited by an agent	5
$dampT$	Diffusion damping factor of trail	0.1
WT	The weight of trail value	0.4
m	The size of mean filter for trail	3
	The size of mean filter for chemo-nutrient	5
CN	The amount of chemo-nutrient in the girds of data points	10
$dampN$	Diffusion damping factor of chemo-nutrient	0.2
WN	The weight of chemo-nutrient value	0.6
RT	A parameter that is used to trigger the reproduction behavior	5
ET	A parameter that is used to trigger the elimination behavior	-5

(a) t=51   (b) t=266   (c) t=356   (d) t=489

(e) t=1110   (f) t=2352   (g) t=2600   (h) t=3196

Fig. 2. (Color online) Maze solving of the system. (a)-(f) show the process that how the two routes of the maze are constructed. (g)and (h) capture the reconfiguration of path after the maze been changed.

For startup the system, an agent_T1 is generated in the entrance data point. A mass of agents_T1 appears and covers almost the entire passages, as shown in Fig. 2(a) and Fig. 2(b). When the data point in the exit is found, the agent's type transition begins along the passage that agents_T1 propagating before (Fig. 2(c) and Fig. 2(d)). With the last agent_T1 transforming to agent_T2, agents_T2 distributed on the blind sides shrink to the right path gradually (Fig. 2(e)). At last, agents_T2 compose two stable paths connecting the entrance and the exit, which are the exact two routes in this maze, as shown in Fig. 2(f). So far, it is revealed that the system as a maze solver can find all routes of the maze.

For further exploring the robustness of the system, the structure of the maze in t=2600 is reset. The part of the passage marked as the gray block in Fig. 2(g) is transformed into a wall, which results in the cutting down of that branch. However, this sudden change to the environment and the population does not collapse the system. The system keeps running and the other path is maintained naturally. Eventually, the system organizes the only route in the new maze at t=3196, as shown in Fig. 2(h).

4 Conclusions

In this paper, we propose a nature-inspired multi-agent system, dawing on the foraging behavior of *Physarum*, to solve a maze. The proposed system, denoted as the PMAS, imports two types of agents (i.e., agent_T1 for search and agent_T2 for convergence) and three transition rules between agents to simulate the foraging behavior of *Physarum* in network evolution. The PMAS shows the same intelligence as the plasmodium of *Physarum* in foraging, such as self-reshape, path finding and reconfiguration. In particular, experimental results validate that the PMAS can solve a maze by finding all possible routes and manifest reconfiguration features when altering the maze dynamically. What's more, our system would provide a novel perspective to understand the intelligence of *Physarum* during its transportation network evolution process.

Acknowledgement. This work was supported by the National Science and Technology Support Program (No. 2012BAD35B08), the Specialized Research Fund for the Doctoral Program of Higher Education (No. 20120182120016), Natural Science Foundation of Chongqing (Nos. cstc2012jjA40013, cstc2013jcyjA400-22), and the Fundamental Research Funds for the Central Universities (Nos. XDJK2012B016, XDJK2012C018, XDJK2013D017).

References

1. Nakagaki, T., Yamada, H., Tóth, A.: Maze-Solving by an Amoeboid Organism. Nature 407(6803), 470 (2000)
2. Tero, A., Kobayashi, R., Nakagaki, T.: A Mathematical Model for Adaptive Transport Network in Path Finding by True Slime Mold. Journal of Theoretical Biology 244(4), 553–564 (2007)
3. Adamatzky, A.: Physarum Machines: Encapsulating Reaction Diffusion to Compute Spanning Tree. Naturwissenschaften 94(12), 975–980 (2007)
4. Gunji, Y.P., Shirakawa, T., Niizato, T., Haruna, T.: Minimal Model of a Cell Connecting Amoebic Motion and Adaptive Transport Networks. Journal of Theoretical Biology 253(4), 659–667 (2008)
5. Jones, J.: The Emergence and Dynamical Evolution of Complex Transport Networks from Simple Low-Level Behaviours. International Journal of Unconventional Computing 6(2), 125–144 (2010)

6. Liu, Y., Zhang, Z., Gao, C., Wu, Y., Qian, T.: A Physarum Network Evolution Model Based on IBTM. In: Tan, Y., Shi, Y., Mo, H. (eds.) ICSI 2013, Part II. LNCS, vol. 7929, pp. 19–26. Springer, Heidelberg (2013)
7. Jones, J.: Characteristics of Pattern Formation and Evolution in Approximations of Physarum Transport Networks. Artificial Life 16(2), 127–153 (2010)
8. Wu, Y., Zhang, Z., Deng, Y., Zhou, H., Qian, T.: An Enhanced Multi-Agent System with Evolution Mechanism to Approximate Physarum Transport Networks. In: Thielscher, M., Zhang, D. (eds.) AI 2012. LNCS, vol. 7691, pp. 27–38. Springer, Heidelberg (2012)

Consensus of Single-Integrator Multi-Agent Systems at a Preset Time

Cong Liu, Qiang Zhou, and Yabin Liu

School of Automation Science and Electrical Engineering,
Beihang University (BUAA), Beijing 100191, P.R. China
liucong_09@126.com

Abstract. This paper studies the preset time consensus problem of sing-integrator multi-agent systems for reaching the desired state in both undirected and directed communication networks. Two linear consensus protocols with time-varying gain are proposed. In fixed undirected networks, if the undirected topology is connected, the proposed protocol can achieve the consensus at the preset time even if only a portion of agents can obtain the desired state. In fixed directed networks, if the directed topology has a directed spanning tree, the proposed protocol can solve the consensus problem at a given preset time. Finally, numerical simulation results are presented to demonstrate the effectiveness of the theoretical results.

Keywords: consensus, single-integrator, multi-agent systems, preset time, desired state.

1   Introduction

Coordinate control of multi-agent systems has drawn much attention due to its wide applications such as formation control, flocking, rendezvous, attitude alignment of clusters of satellites in the last decade. As the foundation of coordinate control, consensus problems are aimed to design appropriate protocols or algorithms based on local information such that agents reach an agreement on some quantities of interest. There are many results about consensus problems of multi-agent systems in the existing publications. In [1], a theoretical explanation is provided for the alignment behavior observed in the Vicsek model [2]. This is a pioneering work on consensus. Then, a general framework for the consensus problem of networks of integrators is proposed in [3]. Since then, the consensus problem has been extensively studied. In [4-6], consensus of single-integrators is considered. The consensus problem of multi-agent systems with second order dynamics is studied in [7-10]. High order linear dynamic multi-agent systems have also been paid much attention in [11-14].

However, in the above mentioned literatures, all the multi-agent systems only can achieve asymptotic consensus. In practice, finite-time consensus is sometimes more desirable, especially when high precise performance and strict convergence time is required. So, finite-time consensus problems of multi-agent systems are considered in

Y. Tan et al. (Eds.): ICSI 2014, Part I, LNCS 8794, pp. 431–441, 2014.
© Springer International Publishing Switzerland 2014

[15-20]. In the most existing finite-time consensus results, the proposed protocols are generally discontinue and nonlinear and only the upper bound of convergence time is given. In [21], a preset time dependent time-varying but linear consensus protocol is designed and agents in the system reach the same state at a preset time.

In this paper, based on the results of [21], we consider the desired state consensus problem of multi-agent systems with single-integrator dynamics at a preset time for both the undirected and directed communication network cases. With the proposed control protocol, under the same communication condition as asymptotic consensus, the desired state can be achieved at a preset time.

The rest of this paper is organized as follows. Some useful concepts and results about graph theory are reviewed in Section 2. Section 3 states the problem to be investigated. In Section 4, main results are presented. The simulation results are given in Section 5. Conclusions are drawn in Section 6.

2   Preliminaries

The following notations will be used throughout this paper. The sets of real numbers and complex numbers are denoted by R and C , respectively. 1 denotes the appropriate dimension column vector with all entries equal to one. I denotes the appropriate dimension identity matrix. $\text{Re}(\sigma_i)$ denotes the real part of $\sigma_i \in C$. The Euclidean norm is denoted by $\|\cdot\|$. A diagonal matrix is represented by $diag\{b_1, \cdots, b_n\}$ with b_i being the i th main diagonal element. An upper triangular matrix is denoted by $triag\{a_1, \cdots, a_n\}$ with a_i being the i th main diagonal element. The Kronecker product is denoted by $\otimes$.

The multi-agent system can always be modeled by graph. The agents in the multi-agent system are regarded as vertices of graph. The communication links among multi-agents are regarded as edges of graph. Thus, graph theory is an important tool to investigate the distributed coordinate control problem of multi-agent systems. In this section, some concepts and properties about algebraic graph theory are presented. For more details, please refer to [22].

Let $G(A) = (V, E, A)$ be an n th order weighted directed graph with a vertex set $V = \{v_i : i = 1, 2, \cdots, n\}$, an edge set $E = \{e_{ij} = (v_j, v_i) | v_i, v_j \in V\}$ and a weighted adjacency matrix $A = [a_{ij}] \in R^{n \times n}$. An edge e_{ij} of G, where v_j is called the parent vertex of v_i and v_i is the child vertex of v_j, denotes that v_j is a neighbor of v_i. The neighbor set of agent v_i is denoted by $N_i = \{v_j : e_{ij} \in E\}$. The element a_{ij} of weighted adjacency matrix A associated with the edge e_{ij} is nonnegative, i.e., $a_{ij} > 0 \Leftrightarrow e_{ij} \in E$ and $a_{ij} = 0$ otherwise. Moreover, we assume $a_{ii} = 0$ for $i = 1, 2, \cdots, n$. A directed path in the directed graph is a finite ordered sequence of vertices $v_{i_1}, v_{i_2}, \cdots, v_{i_j}$ such that $(v_{i_l}, v_{i_{l+1}}) \in E$ for $l = 1, 2, \cdots, j-1$. A directed tree is a directed graph in which every vertex has exactly one parent vertex except for one vertex called root vertex which has a directed path to every other vertex. A directed spanning tree of G is a directed tree which consists of all vertices and some edges of G. If $a_{ij}=a_{ji}$, the graph G is a undirected graph. If there exists a path between any two distinct vertices of the undirected graph, we say that the undirected graph is connected. The Laplacian matrix $L=[l_{ij}]\in R^{n\times n}$ of G is defined as $L=\Delta-A$, where $\Delta=diag\{\Delta_{11},\Delta_{22},\cdots,\Delta_{nn}\}$ and $\Delta_{ii}=\sum_{j=1,j\neq i}^{n}a_{ij}$ for $i=1,2,\cdots,n$. Apparently, for the undirected graph, the weighted adjacency matrix A and Laplacian matrix L are symmetric matrices.

Lemma 1. [23] If the undirected graph G is connected and D is any nonnegative diagonal matrix with at least one of the main diagonal entries being positive, the $H=L+D$ is symmetric positive definite, where L is the Laplacian matrix of G.

Lemma 2. [24] For a directed graph G, zero is an eigenvalue of L with $\mathbf{1}$ as a right eigenvector and all nonzero eigenvalues have positive real parts. Furthermore, zero is a simple eigenvalue of L if and only if G has a directed spanning tree.

3   Problem Formulation

Consider a multi-agent system consisting of n agents labeled 1 through n. The i th agent is described by

$$\dot{x}_i(t)=u_i(t),\tag{1}$$

where $x_i(t)\in R^N$ is the state and $u_i(t)\in R^N$ is control input. Let $x_c\in R^N$ be the desired state.

Definition 1. Given any finite time $t_f>0$ and the desired state x_c, the multi-agent system (1) is said to achieve the preset time consensus, if for any initial states, the solution of (1) satisfies

$$\lim_{t\to t_f^-}\|x_i(t)-x_c\|=0.\tag{2}$$

The object of this paper is to design the consensus protocol (control input) $u_i(t)$ to achieve the preset time consensus of Definition 1.

4   Main Results

In this section, preset time consensus problems of Definition 1 under the undirected topology and directed topologies are considered, respectively.

4.1   Consensus under Undirected Topology

Suppose that the undirected graph G is connected and at least one agent attains the desired state x_c. Denote by $D = diag\{d_1, \cdots, d_n\}$ the relation between the desired state and agents. When the i th agent can get the desired state, $d_i > 0$ and $d_i = 0$ otherwise.

In order to achieve the object, the following consensus protocol is proposed:

$$u_i(t) = \frac{c}{t_f - t}\left[\sum_{j \in N_i} a_{ij}(x_j(t) - x_i(t)) - d_i(x_i(t) - x_c)\right], i = 1, 2, \cdots, n , \qquad (3)$$

where c is a positive constant scalar. The disagreement error is defined as $e_i(t) = x_i(t) - x_c$. Denote $e(t) = \left[e_1^T(t), e_2^T(t), \cdots, e_n^T(t)\right]^T$, then the multi-agent system (1) with the protocol (3) can be written in the compact form:

$$\dot{e}(t) = -\frac{c}{t_f - t}(H \otimes I)e(t) , \qquad (4)$$

where $H = L + D$.

Remark 1. When designing the control input $u_i(t)$, only the relative state $x_j(t) - x_i(t)$ between the i th agent and its neighboring agents can be used which makes the consensus protocol is distributed. In order to driver all the agents to arrive the desired state, a portion of agents should be pinned by using the desired state as the prior knowledge. Therefore, the term $x_i(t) - x_c$ is introduced and d_i determines whether the i th agent is pinned. In the most existing finite-time consensus results, only the upper bound of convergence time is given and the convergence time is related to the initial states. However, in the consensus protocol (3), the control gain is adjusted in the light of the preset time and thus, the convergence time is independent of initial states of agents.

Theorem 1. Suppose that the undirected graph G is connected. Given any finite-time $t_f > 0$ and the desired state x_c , the consensus protocol (3) solves the desired state consensus problem of system (1) as $t \to t_f^-$.

Proof: D is a nonnegative diagonal matrix with at least one of the main diagonal entries being positive and L is the Laplacian matrix of G . It follows from Lemma 1 that H is a positive definite matrix. Thus, there exists a nonsingular matrix $P \in R^{n \times n}$ such that $P^{-1}HP = diag\{\lambda_1, \lambda_2, \cdots, \lambda_n\}$, where $\lambda_i > 0 \in R, i = 1, 2, \cdots, n$ are eigenvalues of H . Let $\xi(t) = (P^{-1} \otimes I)e(t)$, we have

$$\dot{\xi}(t) = -\frac{c}{t_f - t}(diag\{\lambda_1, \lambda_2, \cdots, \lambda_n\} \otimes I)\xi(t) . \qquad (5)$$

From (5), we have

$$\dot{\xi}_i(t) = \frac{-c\lambda_i}{t_f - t}\xi_i(t), i = 1, 2, \cdots, n, t \in [0, t_f) ,$$

(6)

which implies that

$$\xi_i(t) = \left(\frac{t_f - t}{t_f}\right)^{c\lambda_i}\xi_i(0), t \in [0, t_f) .$$

(7)

Since c and λ_i are positive, it follows that $\xi_i(t) \to 0$, as $t \to t_f^-$. Then we have

$$\lim_{t \to t_f^-} e(t) = (P \otimes I)\lim_{t \to t_f^-} \xi(t) = 0 ,$$

(8)

which implies $x_i(t) \to x_c$, as $t \to t_f^-$. Theorem 1 holds.

Remark 2.   In the proof of Theorem 1, from (7) we have

$$\dot{\xi}_i(t) = -c\lambda_i \frac{\left(t_f - t\right)^{c\lambda_i - 1}}{t_f^{c\lambda_i}}\xi_i(0) .$$

(9)

If we select c such that

$$c\lambda_i > 1, i = 1, 2, \cdots, n ,$$

(10)

then $\dot{\xi}_i(t)$ is bounded for all $t \in [0, t_f)$. It follows that $u_i(t)$ is bounded for all $t \in [0, t_f)$.

Remark 3.   In the consensus protocol (3), some parameters have to be determined. t_f and x_c can be chose by the need of practice application. The value of c influences the convergence rate. Intuitively, under the condition satisfying (10), the lager c makes the convergence faster due to it means that agents can get the larger control input in this situation. It is verified in Section 5. However, in the practice application, the input saturation should be considered when determining c.

4.2   Consensus under Directed Topology

Now consider the case with directed topology. Suppose that the directed graph G has a directed spanning tree and all agents knows the desired state x_c. In order to achieve the object, the following consensus protocol is proposed:

$$u_i(t) = \frac{c}{t_f - t}\left[\sum_{j \in N_i} a_{ij}(x_j(t) - x_i(t)) - (x_i(t) - x_c)\right], i = 1, 2, \cdots, n ,$$

(11)

where c is a positive constant scalar. The disagreement error is defined as $e_i(t) = x_i(t) - x_c$. Denote $e(t) = \left[e_1^T(t), e_2^T(t), \cdots, e_n^T(t) \right]^T$, then the multi-agent system (1) with the protocol (11) can be written in the compact form:

$$\dot{e}(t) = -\frac{c}{t_f - t}(H \otimes I)e(t) , \qquad (12)$$

where $H = L + I$.

Remark 4. When designing the control input $u_i(t)$, only the relative state $x_j(t) - x_i(t)$ between the i th agent and its neighboring agents can be used which makes the consensus protocol is distributed. In order to driver all the agents to arrive the desired state, agents should be pinned by using the desired state as the prior knowledge. Therefore, the term $x_i(t) - x_c$ is introduced to assure agents are pinned. In the most existing finite-time consensus results, only the upper bound of convergence time is given and the convergence time is related to the initial states. However, in the consensus protocol (11), the control gain is adjusted in the light of the preset time and thus, the convergence time is independent of initial states of agents.

Lemma 3. [21] Consider the following time-varying linear differential equation

$$\dot{\theta}(t) = \dot{f}(t)M\,\theta(t) , \qquad (13)$$

where $\theta(t) \in C^n$, $f(t) \in C$ is differentiable, and $M \in C^{n \times n}$ is a constant matrix. Then, the solution of equation (13) is given by

$$\theta(t) = \exp\left[(f(t) - f(0))M\right]\theta(0) . \qquad (14)$$

Theorem 2. Suppose that the directed graph G has a directed spanning tree. Given any finite-time $t_f > 0$ and the desired state x_c , the consensus protocol (11) solves the desired state consensus problem of system (1) as $t \to t_f^-$.

Proof: The directed graph G has a directed spanning tree, so it follows from Lemma 2 there exists a nonsingular matrix $Q \in C^{n \times n}$ such that $Q^{-1}LQ = triag\{0, \lambda_2, \cdots, \lambda_n\}$, where $\lambda_i \in C, i = 2, 3, \cdots, n$ are eigenvalues of L and $Re(\lambda_i) > 0$. Let $\xi(t) = (Q^{-1} \otimes I)e(t)$, and then from (12) we have

$$\dot{\xi}(t) = -\frac{c}{t_f - t}(T \otimes I)\xi(t) , \qquad (15)$$

where $T = triag\{\sigma_1 = 1, \sigma_2 = \lambda_2 + 1, \cdots, \sigma_n = \lambda_n + 1\}$.

By Lemma 3, based on (15), we have

$$\xi(t) = \left[\exp\left(c\ln\frac{t_f - t}{t_f} T \right) \otimes I \right] \xi(0), t \in [0, t_f) \ . \tag{16}$$

Since c and $\mathrm{Re}(\sigma_i)$ are positive, it is easy to verify that $\exp\left(c\ln\dfrac{t_f - t}{t_f} T \right) \to 0$, as

$t \to t_f^-$. It follows that $\xi(t) \to 0$, as $t \to t_f^-$. Then we have

$$\lim_{t \to t_f^-} e(t) = (Q \otimes I) \lim_{t \to t_f^-} \xi(t) = 0 \ , \tag{17}$$

which implies $x_i(t) \to x_c$, as $t \to t_f^-$. Theorem 2 holds.

Remark 5. If we select c such that

$$c\,\mathrm{Re}(\sigma_i) > 1, i = 1, 2, \cdots, n \ , \tag{18}$$

then $u_i(t)$ is always bounded for all $t \in [0, t_f)$.

Remark 6. In the consensus protocol (11), some parameters have to be determined. t_f and x_c can be chose by the need of practice application. The value of c influences the convergence rate. Intuitively, under the condition satisfying (18), the lager c makes the convergence faster due to it means that agents can get the larger control input in this situation. It is verified in Section 5. However, in the practice application, the input saturation should be considered when determining c.

5   Simulation

In this section, two illustrative examples are provided to verify the effectiveness of the proposed consensus protocols. Without loss of generality, choose $N = 1$ here.

Consider a multi-agent system consisting of five agents under the undirected graph. Suppose the interaction topology of agents is described by Fig. 1. The weight of every communication link is set to 1. Only agent v_1 can get the desired state and then $d_1 = 1$. By calculating, the eigenvalues of $H = L + D$ are $\lambda_1 = 0.1401, \lambda_2 = 0.8803$, $\lambda_3 = 2.4229, \lambda_4 = 2.8341, \lambda_5 = 4.7226$. Choose the value of c such that (10) holds and let $x_c = 3$. Under the control protocol (3), the simulation result is shown in Fig. 2‒4. We can see the states of all the agents reach the desired state as $t \to t_f^-$ for any initial states. Fig. 3 implies the larger c will make the convergence faster.

Fig. 1. The interaction topology of agents: the undirected topology (*left*) and the directed topology (*right*)

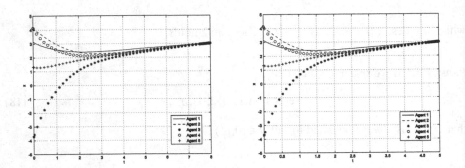

Fig. 2. The simulation result of system (1) under protocol (3) with the same initial states and different parameters: $c = 8, t_f = 8$ (*left*) and $c = 8, t_f = 5$ (*right*)

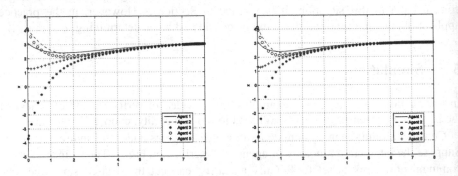

Fig. 3. The simulation result of system (1) under protocol (3) with the same initial states and different parameters: $c = 12, t_f = 8$ (*left*) and $c = 20, t_f = 8$ (*right*)

Next consider a multi-agent system consisting of five agents under the directed graph. Suppose the interaction topology of agents is described by Fig. 1. The weight of every communication link is set to 1. All the agent can get the desired state. By calculating, the eigenvalues of $H = L + I$ are $\sigma_1 = 1, \sigma_{2,3} = 2 \pm 1j, \sigma_4 = \sigma_5 = 3$. Choose c such that (18) holds and let $x_c = 1$. Under the control protocol (11), the simulation result is shown in Fig. 5~7. We can see the states of all the agents reach the desired state as $t \to t_f^-$ for any initial states. Fig. 6 implies the larger c will make the convergence faster.

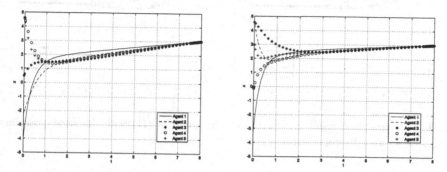

Fig. 4. The simulation result of system (1) under protocol (3) with the same parameters $c = 8, t_f = 8$ and different initial states

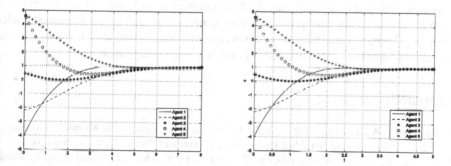

Fig. 5. The simulation result of system (1) with protocol (11) with the same initial states and different parameters: $c = 2, t_f = 8$ (*left*) and $c = 2, t_f = 5$ (*right*)

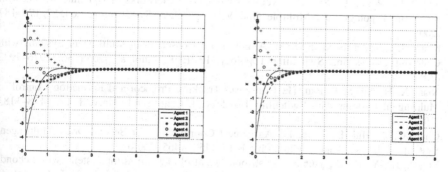

Fig. 6. The simulation result of system (1) with protocol (11) with the same initial states and different parameters: $c = 6, t_f = 8$ (*left*) and $c = 10, t_f = 8$ (*right*)

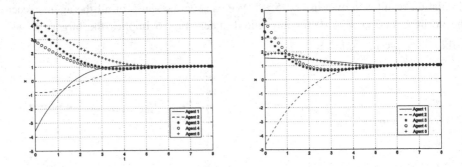

Fig. 7. The simulation result of system (1) with protocol (11) with the same parameters $c = 2, t_f = 8$ and different initial states

6   Conclusion

This paper studies the preset time consensus problem of sing-integrator multi-agent systems reaching the desired state in both undirected and directed communication network. Linear consensus protocols with time-varying gains are presented under which sing-integrator multi-agent systems can reach the desired state if the undirected network is connected or the directed network has a directed spanning tree.

References

1. Jadbabaie, A., Lin, J., Morse, A.S.: Coordination of Groups of Mobile Autonomous Agents Using Nearest Neighbor Rules. IEEE T. Automat. Contr. 48, 988–1001 (2003)
2. Vicsek, T., Czirók, A., Ben-Jacob, E., Cohen, I., Shochet, O.: Novel Type of Phase Transition in a System of Self-Driven Particles. Phys. Rev. Lett. 75, 122–1229 (1995)
3. Olfati-Saber, R., Murray, R.M.: Consensus Problems in Networks of Agents with Switching Topology and Time-Delays. IEEE T. Automat. Contr. 49, 1520–1533 (2004)
4. Sun, Y.G., Wang, L., Xie, G.: Average Consensus in Networks of Dynamic Agents with Switching Topologies and Multiple Time-Varying Delays. Syst. Control Lett. 57, 175–183 (2008)
5. Munz, U., Papachristodoulou, A., Allgower, F.: Consensus in Multi-agent Systems with Coupling Delays and Switching Topology. IEEE T. Automat. Contr. 56, 2976–2982 (2011)
6. Nian, X.-H., Su, S.-J., Pan, H.: Consensus Tracking Protocol and Formation Control of Multi-agent Systems with Switching Topology. J. Cent. South. Univ. T. 18, 1178–1183 (2011)
7. Chun-Xi, Z., Hui, L., Peng, L.: Agreement Coordination for Second-Order Multi-agent Systems with Disturbances. Chin. Phy. B 17, 4458–4465 (2008)
8. Lin, P., Jia, Y., Du, J., Yuan, S.: Distributed Control of Multi-agent Systems with Second-Order Agent Dynamics and Delay-Dependent Communications. Asian J. Control 10, 254–259 (2008)

9. Yu, W., Chen, G., Cao, M.: Some Necessary and Sufficient Conditions for Second-Order Consensus in Multi-agent Dynamical Systems. Automatica 46, 1089–1095 (2010)
10. Yang, H., Zhang, Z., Zhang, S.: Consensus of Second-Order Multi-agent Systems with Exogenous Disturbances. Int. J. Robust Nonlin. 21, 945–956 (2011)
11. Seo, J.H., Shim, H., Back, J.: Consensus of High-Order Linear Systems Using Dynamic Output Feedback Compensator: Low Gain Approach. Automatica 45, 2659–2664 (2009)
12. Jiang, F., Wang, L., Xie, G.: Consensus of High-Order Dynamic Multi-agent Systems with Switching Topology and Time-Varying Delays. Journal of Control Theory and Applications 8, 52–60 (2010)
13. Tan, Y., Yue-Hui, J., Wei, W., Ying-Jing, S.: Consensus of High-Order Continuous-Time Multi-agent Systems with Time-Delays and Switching Topologies. Chin. Phy. B 20, 20511–20516 (2011)
14. Zhang, Q., Niu, Y., Wang, L., Shen, L., Zhu, H.: Average Consensus Seeking of High-Order Continuous-Time Multi-agent Systems with Multiple Time-Varying Communication Delays. Int. J. Control Autom. 9, 1209–1218 (2011)
15. Cortés, J.: Finite-Time Convergent Gradient Flows with Applications to Network Consensus. Automatica 42, 1993–2000 (2006)
16. Chen, G., Lewis, F.L., Xie, L.: Finite-Time Distributed Consensus via Binary Control Protocols. Automatica 47, 1962–1968 (2011)
17. Li, S., Du, H., Lin, X.: Finite-Time Consensus Algorithm for Multi-agent Systems with Double-Integrator Dynamics. Automatica 47, 1706–1712 (2011)
18. Lu, X., Chen, S., Lü, J.: Finite-Time Tracking for Double-Integrator Multi-agent Systems with Bounded Control Input. IET Control Theory. A. 7, 1562–1573 (2013)
19. Zhao, Y., Duan, Z., Wen, G., Zhang, Y.: Distributed Finite-Time Tracking Control for Multi-agent Systems: An Observer-Based Approach. Syst. Control Lett. 62, 22–28 (2013)
20. Zuo, Z., Tie, L.: A new class of finite-time nonlinear consensus protocols for multi-agent systems. Int. J. Control 87, 363–370 (2014)
21. Cai, Y., Xie, G.M., Liu, H.Y.: Reaching Consensus at a Preset Time: Single-Integrator Dynamics Case. In: Proceedings of the 31st Chinese Control Conference, pp. 6220–6225. IEEE Computer Society, Washington, DC (2012)
22. Biggs, N.: Algebraic Graph Theory. Cambridge University Press, Cambridge (1993)
23. Hong, Y., Hu, J., Gao, L.: Tracking Control for Multi-agent Consensus with an Active Leader and Variable Topology. Automatica 42, 1177–1182 (2006)
24. Ren, W., Beard, R.W.: Consensus Seeking in Multi-agent Systems under Dynamically Changing Interaction Topologies. IEEE T. Automat. Contr., 655–661 (2005)

Representation of the Environment and Dynamic Perception in Agent-Based Software Evolution

Qingshan Li, Hua Chu, Lihang Zhang, and Liang Diao

Software Engineering Institute, Xidian University,
Xi'an 710071, P.R. China
qshli@mail.xidian.edu.cn

Abstract. As the Internet become mainstream software system environment, software systems shift from closed, static and controllable to open, dynamic and difficult to control. The changes in the environment are unpredictable; it is major challenge for software system research to ensure that the software systems can deal with dynamic environment and change themselves appropriately. In this paper, according to the Multi-Agent environment, we divide environmental perception mechanism into three parts: by defining the environment, the composition problems in Multi-Agent Systems environment are solved; by designing method by which dynamic environmental data is generated and changes, we propose a dynamic environmental perception model based on the "publish / subscribe" model; by customized rules, the system can change itself in the environment according to the appropriate action to achieve the entire software adaptive process. Finally, we present examples to verify the feasibility and effectiveness of the theory.

Keywords: Software Evolution, Multi-Agent System, Dynamic Environment.

1 Introduction

Software evolution is the process that software continues to change itself and reach people's requirement [1]. Adaptive dynamic evolution mean that during the process of dynamic evolution, the software can sense the environment and changes in demand according to predefined strategy, as well as change its structure or behavior automatically so that it can adapt itself better to the external environment and the changed needs [2]. But in the field of adaptive dynamic evolution software systems, the existing work focuses on the manner adopted by the system when the environment changes [3]. By this way, the system can adjust itself to the environment. However, the studies on expressing themselves in the environment and detecting the changes of environment are inadequate [4]. Agent environment has been considered an important research, many studies are not available as an important factor of the environment into the MAS models and tools, or are the responsibility of the environment for a weakened.

The main contribution of this paper is to address the problem of multi-Agent system environment which supports explicit representation and effective perception mechanisms and methodologies aimed at improving the development of Multi-Agent

Y. Tan et al. (Eds.): ICSI 2014, Part I, LNCS 8794, pp. 442–449, 2014.

Systems, Multi-Agent System to enhance the ability to adapt to complex environments, improve system maintainability [6].

2   Environment Abstraction and Representation

2.1   Environment Definition and Classification

In MAS, usually the software environment is divided into static environment and dynamic environment [7]. Static environment refers to the objective physical world, such as CPU and memory. Dynamic environment refers to the system environment, such as the Agent's state and behavior. Wedivide dynamic environment of the software into two categories: internal environment and external environment.

Definition 2.1the internal environment is expressed as a tripletassociated withAgentInEVR = <AgentSet, AgentCaps, LifeMessage>. AgentSetrepresents a collection of Agent in system. AgentCaps represents the current Agent's ability. LifeMessage represents the Agent lifecycle state.

Definition 2.2External environment is represented as a tupleOutEVR = <AID, Attribute, Type, Value>. AID representsthe unique identifier of Agent running in simulation platform. Attribute representsAgent attribute name. Type represents type of attribute. Value represents the value of the attributes.

2.2   Environmental Perception Framework

This paper presents a dynamic environment perception framework as shown in Fig. 1.

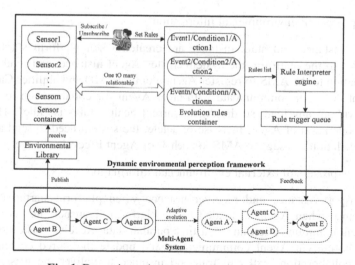

Fig. 1. Dynamic environment perception frameworks

Framework overall similar to the style of pipeline structure, mainly consist of sensing module and the evolution rules setting module. Users can subscribe and unsubscribe environmental information by sensors. Users can develop the "Event - Condition - Action" evolution rules at any time and send it to the evolution rule container. An evolution rule contains a reference to one or more sensors. In the MAC operation, ongoing environmental information publishes by the environmental information database to save and update the latest real-time environmental information. Sensor container screens of concerned environmental information, and update the corresponding sensor records. Meanwhile, the Rule Interpreter engine constantly on the evolution rules customized loop through a list, according to the reference sensor data and conditional logic to determine the evolution rule of whether all the conditions met, if it satisfies the reference then set the evolution rule into the rule trigger queue. The rule trigger queue is similar to the closed-loop system, gives a new task integration relationship and processed by MAS. The evolution control engine parses the action and form new integrated relationship. The task of the current system is still running old tasks until the new distribution of tasks successfully and then Multi-Agent collaboration operating logic will switch to the new collaborative relationship. The whole process dynamic switches task without involved the users to realize the adaptive evolution.

3   Environment Perception Mechanism

3.1   Dynamic Environmental Perception Mechanism

This article divides mechanism of Adaptive Agent perception environment into four stages: publish, subscribe, receive and unsubscribe.

- Publish process of environmental information

First, AMS instance and state instance are created when platform startup. When Agent loaded to the platform, the system creates Agent instance, and sends registration information to the AMS to register Agent identifies AID and abilities Cap. At this point Agent does not communicate with other Agent for collaboration to deal with tasks and without any load, so the Agent current health status is "idle". Then when there is an operation of Agent increase or delete, the system need to send a registration or cancellation message to AMS, on behalf of Agent lifecycle change.

- Publishing process of external environmental information

Create transport instance and receiver instance when platform startup and create Agent instance when Agent is loaded to the platform. Transport instance is called during Agent communication and real-time publish data. Receiver instance intercept and store the current Agent communication and update the stored contents. When the Agent collaboration, will call transport instance to deliver the message which content using the string type. Then when transport instance is called, meanwhile call receiver instance to update the content of the communication and implement the release process of the external environment information. Receive instance get AID of sent messages Agent and identifies message the type of information to the external environment.

- Subscribe/unsubscribe process of Environmental information

Create SensorContainer instance when platform startup and initialize PublicSensor lists and InstanceSensor list to empty. Suppose the user needs to subscribe to the internal environment information that addition or deletion of Agent changes, then add the common sensors, create PublicSensor instance, initializing its name and Agent logo and initialize the value of the life state to null. Finally put PublicSensor instance to SensorContainer list of PublicSensor.

- Reception process of environmental information

When platform startup, it creates environment Information Base Environment Library instance which constantly receives environment information stored in StateReceiver and MsgReceiver. If the identification of environmental information in the Agent and the Agent subscribed sensors identify are same, then the sensor updates the content of the environmental information.

3.2   Rule Triggering Process Based on the "Event/Condition/Action"

In a dynamic environment perception framework, we propose the ECA evolution rule based on the "event/condition/action". If the match succeeds which means evolution rule conditions are met, then get the corresponding evolution action, shown in Fig. 2.

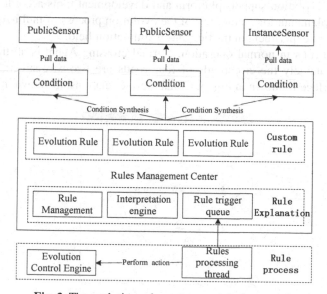

Fig. 2. The evolution rule custom module framework

Evolution rule custom module is used to provide to the user for the evolution rules of customization. It can screen environment information from environmental library and store in the user's sensor, and do a logic judgment to the data. It can work with integration rule and evolution control engine to achieve the environment data feedback process.

Evolution rule custom module is divided into three parts: custom rule, rule explanation and rule process.

Associated with the perception module, is responsible for collecting the condition and action information customized by the user, while saves the evolution rule as file storage in the form of XML.

Rule explanation associates with the rule processing threads at run time, including three parts. Rule management module reads the stored evolution rule during initialization and can also read and store the new rule at runtime. , When an event occurs in the system, through explaining the conditions for each rule and if the match is successful then the rule interpretation engine puts the rule into the trigger queue.

Rules process thread periodic check rule trigger queue, if there is a triggered rule, then evolution control engine call its actions to switch the integrate rule.

4   Experiment

In order to verify the proposed environment model and environment perception mechanism, this chapter designs an Agent-based software adaptive integration evolution experiment by testing specific case to verify the feasibility and effectiveness of dynamic environment perception mechanism. This article develops Agent-based software adaptive evolution support platform and development tools according to design principles, implementation techniques of the evolution process of multi-Agent system model and do an experiment in traffic control simulation field.

The system runs in normal operation. Normal crossing Agent is abstract of unrestricted road, namely the crossroads of four roads are two-way street. The system operation interface is shown in Fig. 3. Collaborative relationship shown in Fig. 4.

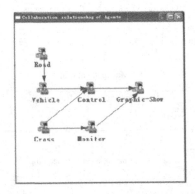

Fig. 3. Collaborative relationships during normal operation

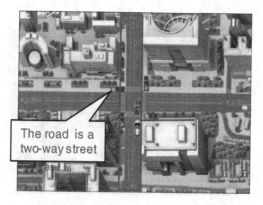

Fig. 4. Normal system operations

When user needs to implement controls on the roads, some sections can only be one-way traffic or impassable, and the traffic lights being out of action in order to ensure the adoption of special vehicles. The evolution rules trigger interface shown in Fig. 5. Thus, the new integrated logic is shown in Fig. 6. Fig. 7shows the road becoming a one-way street and the traffic lights in a paralyzed state.

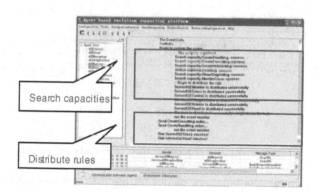

Fig. 5. Evolution rules trigger interface

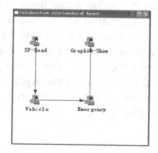

Fig. 6. Collaborative relationship after system evolution

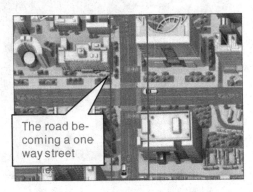

Fig. 7. The changed system operations

In this paper, we give a performance test of switching time overhead in traditional artificial modeand automatic mode. The experiment process is that using the same integrated logic execution evolution, in the same simulation condition and loaded in the same number of Agents, we record the time of switching integration script by using the artificial mode and automatic mode, as shown in Fig. 8.

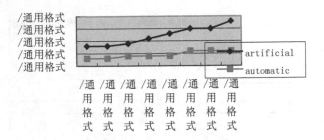

Fig. 8. Switching processing time overhead comparison

From the figure two evolutionary modes testing time can be seen in the same conditions, the time overhead required for automatic mode is smaller than that for the artificial mode, mainly analyzed in two ways. For public event, artificial mode time overhead mainly consuming decision-making and operations in artificial parts, including the selection of the integration script, load and start the operation of such commands as well as on the response of the system. For instance event, it generally cannot adaptive evolution by manually switching. Because the user is often difficult to determine whether by visual observation of the conduct of the script to switch, for example, if the condition for the evolution is the Agent currently position, as the user may determine the error cause visual integration scripts switched earlier or later, resulting in unpredictable consequences.

5   Conclusion

Currently a number of software systems operate with changing demands and changing environmental conditions. To effectively make the software system can self-aware and adapt to the complex environment is a major challenge for computer software technology. With the proposed Multi-Agent system environment representation and dynamic perception, we can describe the system environment of software system according to Agent features and characteristics of the area; we can also design a dynamic sensing mechanism with which the system is able to perceive the environment at runtime and can take appropriate actions to meet users' needs and environment changes according to its own knowledge. This improves the adaptability and intelligence of the system. But there still exist some shortcomings. The proposed environment representation is not comprehensive enough and it doesn't have the versatility. We will continue to improve the environment representation, parsing evolution rules and study the Agent learning and reasoning mechanisms.

Acknowledgments. This work is supported by the Projects (61173026, 61373045, 61202039) supported by the National Natural Science Foundation of China; Projects (BDY221411, K5051223008, K5051223002) supported by the Fundamental Research Funds for the Central Universities of China; Project (513***103E) supported by the Pre-Research Project of the "Twelfth Five-Year-Plan" of China; Project (2012AA02A603) supported by the National High Technology Research and Development Program of China.

References

1. Andersson, J., Lemos, R., Malek, S.: Reflecting on self-adaptive software systems. In. The ICSE 2009 Workshop on Software Engineering for Adaptive and Self-managing Systems, pp. 38–47. IEEE Press, Vancouver (2009)
2. Yoder, W., Johnson, R.: The adaptive object-model architectural style. In: The Working IEEE/IFIP Conf on Software Architecture, pp. 25–31. IEEE Press, New York (2002)
3. Huang, G., Mei, H., Yang, F.: Based on reflective middleware software runtime software architecture. Science in China 34(2), 1233–1253 (2005)
4. Shen, X.: The Research and Design on Mechanism of Specifying and Sensing the Environment of Self-adaptive Agents. PhD thesis, National Defense Science and Technology University (2008)
5. Weyns, D., Omicini, A., Odell, J.: Environment as a first-class abstraction in multiagent systems. Autonomous Agents and Multiagent Systems 14(1), 5–30 (2007)
6. Kwon, S., Choi, J.: An agent-based adaptive monitoring system. In: Shi, Z.-Z., Sadananda, R. (eds.) PRIMA 2006. LNCS (LNAI), vol. 4088, pp. 672–677. Springer, Heidelberg (2006)
7. Smarsly, K., Law, H., Hartmann, D.: Multiagent-based collaborative framework for a self-managing structural health monitoring system. Journal of Computing in Civil Engineering 26(1), 76–89 (2011)

Cooperative Parallel Multi Swarm Model
for Clustering in Gene Expression Profiling

Zakaria Benmounah, Souham Meshoul, and Mohamed Batouche

Computer Science Department, College of NTIC, Constantine University 2,
25000 Constantine, Algeria
{zakaria.benmounah,souham.meshoul,mohamed.batouche}
@univ-constantine2.dz

Abstract. Clustering of gene expression profiles is a mandatory task in cancer classification. Querying the expression of thousands of genes simultaneously imposes the use of powerful clustering techniques. Swarm based methods have shown their ability to perform data clustering. However, they may be faced to premature convergence problem and may be time consuming when large data sets need to be processed. Nowadays, the availability and widespread of parallel processing resources make possible the use of cooperative parallel methods. Within this context, we propose in this paper an archipelago based model that allows to reap advantage from the dynamics and the intrinsic parallelism of three swarm based methods namely PSO, ABC and ACO. Cooperation is achieved by sharing information through migration inside and between archipelagoes. The proposed cooperative parallel model for clustering gene expression profiles has been implemented on multicore computers and applied to several data sets. Experimental results show that it competes and even outperforms existing methods.

Keywords: Clustering, Swarm Intelligence, Parallel Metaheuristics, Gene Expression Profiling.

1   Introduction

Gene expression profiling (GEP) is a popular technique in molecular biology used to study the expression of very large numbers of genes simultaneously. More particularly it aims at accurately classifying tumors as the majority of cells within a tumor share a common profile of gene expression. Managing the mass of microarray data is one of GEP challenges. GEP can be performed for class discovery or class prediction [1]. The need is to identify characteristics of expression profiles related to non-predefined subset in the first case or to predefined subset in the second case.

In order to study gene expression profiling datasets, one often used computational method is clustering. Clustering is the process of grouping genes on the basis of their profiles. This classification of genes aims at interpreting and extracting possible inferences from microarrays and other throughput bioinformatics data sets to identify new

Y. Tan et al. (Eds.): ICSI 2014, Part I, LNCS 8794, pp. 450–459, 2014.
© Springer International Publishing Switzerland 2014

subsets of tumors. To this end, many methods covering gene expression profiling clustering have been proposed in the past decades [2]. No clustering method can adequately handle all types of clusters structures and properties due to the variety of data-sets regarding overlapping, size, shape and density [3].

One alternative is the use of techniques that combine multiple clustering algorithms at once such as clustering ensemble. This later evolves over two steps namely a generation step and a consensus step. While the generation step consists in generating multiple clustering solutions using the same data sets, the consensus step constructs new clustering solutions using the results obtained by the generation step. These two tasks are in fact NP-hard problems that require effective methods to be solved [4]. Swarm intelligence (SI) algorithms have been investigated to gauge their potential to solve clustering problem [5, 6]. Although metaheuristics allow solving NP-hard problems in reasonable time they still lead to unsatisfactory convergence if not improved. This is essentially due to the possibility to get stuck into local optimum.

The use of parallelism to design new nature inspired algorithms deals with complex problems such as clustering. Parallel algorithms target not only speeding up the search but enhancing the quality of obtained solutions as well. The use of parallel computing is fostered by the rapid development of technology in designing processors (e.g. multicore processors). Nowadays, parallel computing resources have become increasingly available and the cost/performance ratio is constantly decreasing.

In this paper, we propose a multi-swarm approach to deal with clustering problem. The key feature consists in organizing swarms into archipelagos where each archipelago acts as a generalized island model. Swarms evolve in parallel. Cooperation between swarms is performed at two levels. Inside each archipelago through intra-archipelago migration and between archipelagos through inter-archipelago migration. Three metheuristics are considered namely Particle Swarm Optimization (PSO), Ant Colony Optimization (ACO) and Artificial Bee Colony (ABC).

The remainder is organized as follows. In section 2, we present the background and related works. Section 3 is devoted to the description of the proposed approach for gene expression profiling. In section 4, we present the experimental results obtained by using available data sets. Finally, conclusions and future work are drawn.

2   Background and Related Work

2.1   Background

Generalized Island Model (GIM). GIM is a parallel model which can be applied to a large class of optimization problems. Based on metaheuristics cooperation, it allows an effective parallel execution of multiple algorithms across multiple islands. This cooperation is maintained by the exchange of solutions between these islands. The exchange operator (migration operator) aims at improving the overall performance of the different algorithms used [7].

Clustering. The purpose of clustering approach is to find natural groups in a given data set. This objective is realized by strengthening the similarity and the dissimilarity between genes within the same cluster and in different clusters respectively.

Let D denotes the data set, a clustering of D consists in finding a partition of D denoted by $\tau(D)$ such that $\tau(D) = \{c_1, c_2, ..., c_K\}$ where each c_i is a cluster and K is the number of clusters. Clusters should be nonempty, disjoint and their union leads to D.

Clustering Ensemble. Is an approach that suggests clustering data over two steps namely a generation step and a consensus step. Generation step $\pi(D)$ consists in generating multiple clustering solutions. The consensus step $\tau(D)^*$ consists in constructing the final clustering solution using the information provided by the first step that is $\pi(D) = \{\tau(D)_1, \tau(D)_2, ..., \tau(D)_n\}$ and $\tau(D)^* = \{c_1^*, c_2^*, ..., c_K^*\}$ [4].

2.2   Related Work

The collective behavior of decentralized and self-organized natural systems such as colonies of ants, bees and swarm of birds among other has inspired many researchers and lead to what is known as swarm intelligence systems. Modern metaheuristics are nature inspired like: Ant colony optimization (ACO) proposed by Dorigo et al. [8], Particles warm optimization (PSO) proposed by Kennedy and Eberhart [9], Artificial bee colony (ABC) proposed by Karaboga and Basturk in [10]. On the other hand, data clustering can be easily cast as a global optimization problem, thereby making the application of SI tools becomes more obvious and appropriate. In [11], a constrained ACO (C-ACO) is presented on the basis of handling arbitrary shaped clusters and outliers present in the data. Adaptive ACO was proposed in [12] to determine the optimal number of clusters and also to improve the convergence rate. It imitates the ants' behavior of grouping dead bodies based on their size, so as to keep the nest clean. In fact, the first algorithm using this concept is proposed by LF (Lumer and Faieta) algorithm [13]. In [14], several ant colonies are used in parallel to find clustering solutions that are sent to the queen (each colony has one queen). At the final stage, the results at queens level are combined using a hypergraph model. From another side, PSO based Clustering was first introduced by Omran et al. in [15]. Their results showed that PSO based clustering outperform k-means and FCM (fuzzy c-means). Authors in [16] have introduced the quantum-behaved PSO for clustering analysis as a QPSO application to gene expression profiling. Also, to determine the unknown number of clusters, a recent clustering algorithm was developed by Cura et al. [17]. Clustering has been tackled as well using ABC metaheuristic. Karaboga is the first algorithm proposed [18]. In [19], the authors proposed ABCK algorithm, which tries to find the Centroid initialization to be given as an input for the K-means algorithm.

Algorithms based on these nature inspired metaheuristics generally suffer from premature convergence. Many tentative methods to avoid such issue were proposed in the literature. In [20], Ghosh et al have proposed a variant of ACO, known as APC (aggregation pheromone density- based clustering) algorithm. This latter attempts to solve the issue by the pheromone matrix updating policy. Cohen and Castro in [21] have proposed a particle swarm clustering algorithm in which the particle's velocity update is influenced by the particle's previous position along with cognitive, social and self-organizing terms. These latter are helpful to guide the particle towards better solutions avoiding local optima. In [22], an ABC algorithm proposes the introduction of a new group of bees called the scout bees, which creates a random search to help the employed bees finding the food source. All the methods reported so far remain limited in terms of quality of results and induce an extra cost regarding time complexity.

3   Proposed Cooperative Parallel Multi-swarm Model for Clustering

The key idea behind the proposed work is to organize swarms into archipelagos as shown on figure 1. A fully connected and bidirectional topology is used. First a generation step is performed followed by a consensus step. The use of cluster ensemble can be justified by the fact that weak clustering solutions when combined using a consensus clustering technique will give rise to high quality clustering solutions [23].

3.1   Generation Step

Each archipelago is in fact a generalized island model that encompasses three swarms that evolve according to ACO, PSO and ABC dynamics. Each of these algorithms explores the space of centroids as search space. The same solution encoding is adopted for the three algorithms that is a vector of centroids.

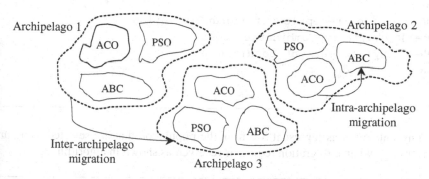

Fig. 1. Proposed model

Inside each archipelago, the clustering algorithms evolve in parallel; they generate clustering solutions and cooperate using an intra-archipelago migration. This latter is triggered by each algorithm whenever it gets into stagnation that is no further improvement of its global best solution. By another side, whenever all islands within an archipelago get into stagnation simultaneously, an inter archipelago migration is performed. After migration, a stagnant island updates its population by introducing the migrant solutions using a recombination policy that consists in replacing worst solutions by the migrant ones to form the new population. Each island evolves according to algorithm 1 shown below.

Algorithm 1. Island$_{i,j}$ (the jth island in the ith archipelago)

```
begin
 input
 P :population
 S :selection strategy
 R :recombination policy
 Aⱼ :clustering based SI algorithm (ACO, PSO or ABC)
 Bestᵢ,ⱼ :Best solution found in the Islandᵢ,ⱼ
 k ∈ { 1,2,...,mᵢ}-j, mᵢ represents the number of islands in the iᵗʰ
 archipelago
 r ∈ { 1,2,...,n}-i, n represents the number of archipelagos
 initialize P :initialize population
 while non_stop_criterion
  P'←Ai(P) :get the new population P'
  if isStagnant (P') :stagnation test
   initiate intra-Archipelago migration
   send('start migration') to Islandᵢ,ₖ
   M'←receive :receive migrant solutions
   if Bestᵢ,ⱼ=Bestᵢ,ₖ :one dominant solution
    initiate inter-Archipelago migration
    send('start migration') to Islandᵣ,ₖ
   end
   M'←receive :reception of migrant solution
   P'← R(P', M') :recombination policy
   P←P' :update current population
  end
 end
end.
```

The migrant solutions represent 5% from the population that are selected using an Elitism strategy when a migration request is received as shown in algorithm 2.

Algorithm 2. Island$_{i,j}$ at receiving of ('start migration')

```
begin
 M ← S(P) :select a subset M from local population
 send(M,j) :send selected solutions
end.
```

3.2   Consensus Step

Clustering solutions obtained after the generation step are given as input for the consensus step. The task of this latter is to combine them with the aim to obtain a final better solution. Therefore, there is no constraint regarding the way partitions must be obtained. In our model we used GA based consensus clustering (HCE) proposed in [24]. The method uses the search capability of genetic algorithms to obtain the consensus clustering. Since convergence is not guaranteed, we propose using three instances of HCE in parallel and introducing solution migration to allow cooperation between them as shown on figure 2. The outline of each HCE is given in algorithm 3.

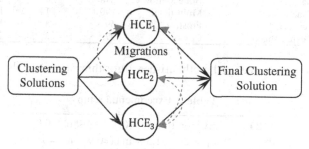

Fig. 2. Consensus step

Algorithm 3. HCE$_1$

```
begin
  input:
  P  :initial population
  S  :selection strategy
  R  :recombination policy
  Ui  :number of iteration in HCE
  initalize P :initialize population
  while non_stop_criterion
    P'← HCE(P, Ui) :P' represents the new population
    M ← S(P') :select M chromosomes from P'
    send M :send to all other islands
    M'←Receive :reception of M' chromosomes from neighboring
    islands
    P''← R(P', M') :combine M' with P' to generate new  population
    P ← P''
  end
end.
```

---

4   Experimental Results and Discussion

The proposed model has been implemented in C++ using a cluster of 8 workstations (HP Z600) where each workstation has an Intel Xeon with 8 cores at 2.4 GHz and 16

Go memory. Message Passing Interface (MPI) has been used for the inter processor information exchange. In order to assess the performance of the proposed method, real data sets borrowed from [25] have been used. These sets represent different complexities regarding data size and distribution. As described in Table 1.

Table 1. Description of used gene expression datasets

Dataset	Tissue	Dimension	Num. classes	Selected # of genes
Breast Cancer	Breast	49	2	1198
Endometrial cancer	Endometrium	42	4	1771
Leukemia	Bone Marrow	248	6	2526
Multi-tissue	Multi-tissue	190	14	1363
Prostate cancer	Prostate	69	3	1625
Serrated carcinomas	Colon	37	2	2202

Parameters of the three used metheuristics algorithms are set as shown in Table 2.

Table 2. Experimental setup

ACO	Evaporation rate=0.1, Threshold=0.9
PSO	$c_1 = c_2 = 1.2$, wmax=0.9, wmin=0.4
ABC	Upper bounce=5, limit=10

4.1   Evaluating Clustering Result Quality

To assess the quality of our results, we used cluster cohesion as an internal performance measure. It measures the compactness of clusters that is how closely related are objects in a cluster. Cohesion is measured by the within cluster sum of squares:
$Cohesion = \frac{1}{N}\sum_{i=1}^{K}\sum_{x\in C_i}\|x - z_i\|^2$, Where N is the number of genes in the datasets, K the number of cluster and z_i the center of cluster C_i.

The proposed approach has been executed 20 times during which values of cohesion have been gathered. Each of the used swarm algorithms has been processed in the same manner separately. In table 2, we show the obtained best values in each case. It is observed from these results that the proposed cooperative method achieved the best results in all cases when compared to each algorithm alone. Furthermore, we illustrated the distributions of the obtained results over the 20 runs in terms of boxplots as shown on figure 3. We see that in terms of median values, the proposed model outperforms the three algorithms. Except for Multi-tissue data set, our algorithm exhibits more stability and robustness. In order to study how significant is the difference between algorithms a Friedman test followed by a multi-comparison test have been performed. The p-values given by the Friedman test are shown on table 3 and multi-comparison graphs are shown on figure 3. The p-value indicate significant difference between the proposed model and the other algorithms in all cases at significance level α=0.05 while giving the best results as shown by the boxplots. The multi-comparison graphs clearly show that our algorithm is significantly different from the other algorithms. This gain shows the advantage of the use of cooperation and parallelism between metaheuristics.

Table 3. Best cohesion values obtained over 20 runs and Friedman's test p-values

Data set	ACO	PSO	ABC	Proposed	Friedman's Test p-value
Breast Cancer	14.52	17.89	20.08	**12.05**	4.8921e-11
Endometrial cancer	6.10	6.06	6.16	**2.09**	1.9925e-07
Leukemia	44.08	42.09	57.02	**34.56**	8.5126e-11
Multi-tissue	3.06	2.43	2.09	**1.03**	1.1631e-05
Prostate cancer	24.07	20.37	25.53	**15.62**	4.8921e-11
Serrated carcinomas	28.10	26.73	29.33	**24.32**	1.3177e-08

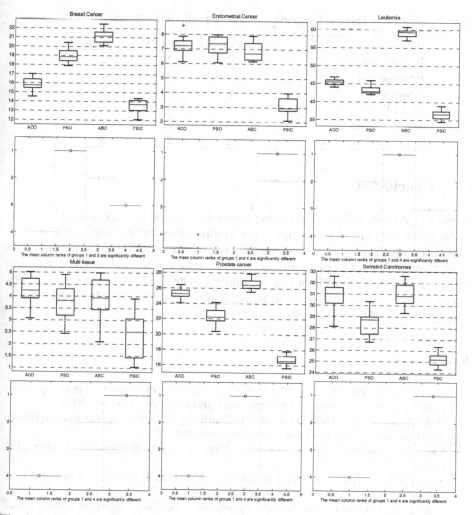

Fig. 3. Boxplots showing distribution of results over 20 runs for all data sets and the corresponding multi-comparison graph in which 1: ACO, 2: PSO, 3: ABC and 4: PSIC (PSIC: Parallel Swarm Intelligence in Clustering, it refers to the proposed model)

Another set of experiments has been conducted to evaluate the ability of our model to find the right number of cluster. We compared our results with the expected optimal number of classes related to each data set and also with DiCLEANS (Divisive Clustering Ensemble with Automatic Cluster Number) algorithm reported in the literature [23]. The experimental results show that our approach competes with and even outperforms DiCLEANS in some cases as shown in table 4. However, when the number of clusters grows like in Multi-tissue and Leukemia data sets both algorithms fail. In our case this can be explained by the fact that clustering algorithms in all islands have been run with fixed number of clusters.

Table 4. Comparison of obtained number of clusters

Data	True Cluster	DICLEANS	Proposed
Breast Cancer	2	2	2
Endometrial cancer	4	4	4
Leukemia	6	3	4
Multi-tissue	14	5	6
Prostate cancer	3	2	3
Serrated carcinomas	2	2	2

5   Conclusion

In this paper we described a model that allows cooperation between several heterogeneous swarms organized into archipelagos to solve clustering of gene expression profiles. Each archipelago is a GIM. A fully connected and bidirectional topology is used. Communication is performed in asynchronous manner and an elitist selection strategy is used to decide about migrant solutions. Obtained results are very competitive. As ongoing work, we intend to further improve results using MapReduce model and dynamic determination of cluster numbers.

References

1. Tarca, A.L., Roberto, R., Sorin, D.: Analysis of microarray experiments of gene expression profiling. American Journal of Obstetrics and Gynecology 195, 373–388 (2006)
2. Harun, P., Burak, E., Andy, D.P., Cetin,Y.: Clustering of high throughput gene expression data. Computers & Operations Research 39, 3046–3061 (2012)
3. Rasha, K., Mohamed, S.K.: Cooperative clustering. Pattern Recognition 43, 2315–2329 (2010)
4. Sandro, V.P., José, R.S.: A Survey of Clustering Ensemble Algorithms. International Journal of Pattern Recognition and Artificial Intelligence 25, 337–372 (2011)
5. Senthilnath, J., Omkar, S.N., Mani, V.: Clustering using firefly algorithm – Performance study. Swarm and Evolutionary Computation 1(3), 164–171 (2011)
6. Suresh, S., Sundararajan, N., Saratchandran, P.: A sequential multi-category classifier using radial basis function networks. Neurocomputing 71, 1345–1358 (2008)

7. Dario, I., Marek, R., Francesco, B.: The Generalized Island Model. Parallel Architectures & Bioinspired Algorithms 415, 151–169 (2012)
8. Dorigo, M.: Optimization, learning and natural algorithms. Ph.D. thesis Politecnico di Milano (1992)
9. Kennedy, J., Eberhart, R.: Particle swarm optimization. IEEE International Conference on Neural Network 4, 1942–1948 (1995)
10. Basturk, B., Karaboga, D.: Anartificial bee colony (ABC) algorithm for numeric function optimization. In: IEEE Swarm Intelligence Symposium USA (2006)
11. Chu, S.-C., Roddick, J., Su, C.-J., Pan, J.-S.: Constrained ant colony optimization for data clustering. In: Zhang, C., Guesgen, H.W., Yeap, W.-K. (eds.) PRICAI 2004. LNCS (LNAI), vol. 3157, pp. 534–543. Springer, Heidelberg (2004)
12. Ingaramo, D., Leguizamon, A., Errecalde, G., Adaptive, M.: clustering with artificial ants. J. Comput. Sci. Technol. 4, 264–271 (2005)
13. Lumer, E., Faieta, B.: Diversity and adaptation in populations of clustering ants. In: Third International Conference on Simulation of Adaptive Behavior, pp. 501–508 (1994)
14. Yang, Y., Kamel, M.: An aggregated clustering approach using multi-ant colonies algorithms. Pattern Recognit. 39, 1278–1289 (2006)
15. Omran, M., Salman, A., Engelbrecht, A.: Image Classification using Particle Swarm Optimization. Simulated Evolution and Learning 1, 370–374 (2002)
16. Li, X.: A new intelligent optimization artificial fish swarm algorithm (Doctor Thesis). Zhejiang University of Zhejiang, China (2003)
17. Cura, T.: A particle swarm optimization approach to clustering. Expert System Application 39, 1582–1588 (2012)
18. Karaboga, D., Ozturk, C.: A novel clustering approach: artificial Bee Colony (ABC) algorithm. Application Soft Computing 11, 652–657 (2011)
19. Giuliano, A., Mohammad, R.F.: Clustering Analysis with Combination of Artificial Bee Colony Algorithm and K-means Technique. International Journal of Computer Theory and Engineering 6(2), 146–150 (2014)
20. Ghosh, S., Kothari, M., Halder, A., Ghosh, A.: Use of aggregation pheromone density for imagesegmentation. Pattern Recognition 30, 939–949 (2009)
21. Cohen, S, C, M., de Castro, L, N.: Data Clustering with Particle Swarms. IEEE Congress on Evolutionary Computation, 1792–1798 (2006)
22. Zhang, C., Ouyang, D., Ning, J.: An artificial bee colony approach for clustering. Expert System Application 37, 4761–4767 (2010)
23. Selim, M., Emin, A.: DICLEANS: Divisive Clustering Ensemble with Automatic Cluster Number. IEEE/ACM Tran. Computational Biology and Bioinformatics 9, 408–420 (2012)
24. Yoon, H.-S., Lee, S.-H., Cho, S.-B., Kim, J.H.: A Novel Framework for Discovering Robust Cluster Results. In: Todorovski, L., Lavrač, N., Jantke, K.P. (eds.) DS 2006. LNCS (LNAI), vol. 4265, pp. 373–377. Springer, Heidelberg (2006)
25. Souto, M., Costa, I., de Araujo, D., Ludermir, T., Schliep, A.: Clustering Cancer Gene Expression Data: A Comparative Study. BMC Bioinformatics 9, 497 (2008)

Self-aggregation and Eccentricity Analysis: New Tools to Enhance Clustering Performance via Swarm Intelligence

Jiangshao Gu and Kunmei Wen

School of Computer Science and Technology,
Huazhong University of Science and Technology, Wuhan, China
{jsgu,kmwen}@hust.edu.cn

Abstract. In view of the intrinsic drawbacks of traditional clustering methods, e.g. the sensitivity to initialization and the risk of falling into local optima, we introduce two new tools to enhance clustering performance via Swarm Intelligence (SI), i.e. Self-Aggregation (SA) and Eccentricity Analysis (EA), which are based on Firefly Algorithm (FA) in this paper. In order to confirm the effectiveness of the techniques, an improved k-means++ method is given as an instance. Large experiments illustrate that our algorithm performs better on both accuracy and robustness than the existing ones.

Keywords: Cluster Analysis, Eccentricity Analysis, Firefly Algorithm, Self-Aggregation, Swarm Intelligence.

1   Introduction

Cluster analysis is a vital task of unsupervised learning, which aims at grouping a set of objects in the way that the objects in the same cluster share more common features than those in other clusters. Over the years, it has been used in many fields, e.g. machine learning, image analysis, pattern recognition, etc [1, 2, 3]. There are four major categories in the approaches, i.e. partitional clustering, density-based clustering, hierarchical clustering, and grid-based clustering.

Swarm Intelligence (SI) is a class of methods to solve optimization problems, inspired by the collective behavior in nature such as bird flocking and fish schooling. A series of clustering methods hybrid with SI have been proposed in recent years [4, 5, 6, 7, 8], whose advantages have been studied and confirmed to a certain extent. In the following pages, we will go on to compare their characters and provide some analyses together with our algorithms.

In this paper, we will talk about two new thinkings on using SI to improve the clustering methods, i.e. Self-Aggregation (SA) and Eccentricity Analysis (EA). We would like to point out that the ideas are substantially different from the traditional approaches, which transform the clustering into optimization problems. In fact, our research focuses on more essential ways for SI to work on clustering, which mean

Y. Tan et al. (Eds.): ICSI 2014, Part I, LNCS 8794, pp. 460–469, 2014.

using some kind of swarm behavior to make the expected results arise spontaneously. In practice, we choose Firefly Algorithm (FA) [9, 10] as the frame for some reasons listed later. Furthermore, we design a new k-means++ method hybrid with these techniques and demonstrate their superiority by experiments.

The remainder of this paper is organized as follows. Section 2 and 3 briefly look back at k-means, k-means++, and FA methods. Section 4 introduces the motivations and implementations of the proposed algorithms in detail. The experiment setup and comparison analyses are displayed in Section 5. The contributions and future work are concluded in Section 6.

2 The K-means/K-means++ Method

The k-means method [11] is one of the most popular clustering approaches. Given a set of N objects, where each object is a d-dimensional real vector X_i, the target of k-means is to partition the objects into k sets S_i, so as to minimize the *Sum Squared Errors (SSE)*,

$$SSE = \sum_{j=1}^{k} \sum_{i=1}^{|S_j|} \left\| X_i - C_j \right\|^2$$

(1)

where C_i is the mean of the objects belonging to S_i. The k-means uses an iterative refinement approach as follows.

Algorithm 1. The k-means method.
1. Arbitrarily choose k centers in the domain.
2. Assign each object to the nearest center, where the Euclidean distance is most often used.
3. Recalculate the means to be the centers of new clusters.
4. Loop to 2 until terminal condition is met, e.g. the assignment no longer changes.

The k-means method is a simple and fast algorithm which is very appropriate for spherical clusters. However, its performance badly depends on the initial distribution. To address the uncertainties, Arthur et al. [12] propose a specific way of choosing centers based on the idea to make them as far as possible away from each other. The whole algorithm is called k-means++.

Algorithm 2. The k-means++ method.
1. Take an object uniformly at random as a center.
2. Choose another center with probability proportional to the distance, from the object to the closest center we have already chosen.
3. Loop to 2 until we have obtained k centers in total.
4. Proceed as with the standard k-means algorithm.

---

3   Firefly Algorithm

In summer, fireflies will produce short and rhythmic flashes to attract mating partners or potential prey. Through communication, the swarm will gather in the vicinity of the most suitable places, which behaviour could be referred to in solving optimization problems. There are two important issues in FA, i.e. the variation of light intensity I_i and formulation of the attractiveness $\beta_{i,j}$. Based on some assumptions and analyses, they are recommended to be defined as follows for maximization goals.

$$I_i = f(X_i) \tag{2}$$

$$\beta_{i,j} = \beta_0 e^{-\gamma_{i,j}^2} \tag{3}$$

where β_0 is the attractiveness at $r_{i,j} = 0$, and γ is the fixed light absorption coefficient. The $r_{i,j}$ gives the Euclidean distance between any two fireflies. When a firefly moves towards another, its location will be updated in

$$X_i = X_i + \beta_{i,j}(X_j - X_i) + \alpha(rand - \frac{1}{2}) \tag{4}$$

where the second term is due to the attraction while the third term is randomization with scale factor α, and *rand* is a random number distributed uniformly in the range [0, 1.0], i.e. $rand \sim U(0, 1.0)$. Now we can summarize FA as pseudo-code:

Algorithm 3. Firefly algorithm.

1. For each firefly:
    1.1. Initialize its position randomly in the space.
    1.2. Evaluate fitness value and update the current optimum.
2. For each moving firefly:
    2.1. For each targeted firefly:
        2.1.1. Check if the brightness of targeted one is greater using Eq.(2).
        2.1.2. If so, move towards the targeted firefly using Eq.(3) and Eq.(4).
    2.2. Evaluate fitness value and update the current optimum.
3. Loop to 2 until terminal condition is met, e.g. iteration limit is reached.

Compared with other SI-based algorithms, there are some distinct strengths of FA such as few parameters and strong capacity to capture extremum areas. Its inherent characteristics of aggregation and differentiation also meet our expectations well for clustering. What's more, we would like to point out that there is another SI-based algorithm called Glowworm Swarm Optimization (GSO) [13, 14], which is also inspired by the luminescent insects. Although the two methods share similar observations, the implementations of them are different or even opposite.

4   Proposed Algorithms

In this section, we will explain in detail how to employ FA to realize Self-Aggregation (SA) and Eccentricity Analysis (EA), and then introduce a hybrid k-mean++ algorithm to outline the universal improving strategy.

4.1   Approach for Self-aggregation

In the previous studies, there are two representative clustering methods called K-FA [6] and GSOCA+KM [7]. Both of them adopt SI-based algorithms for pretreatment, and then use k-means to obtain final results. It is observed that their processing ways stand for two different ideas:

1. For K-FA, it transforms clustering into a classic optimization problem defined by Eq.(1) and utilizes FA to acquire an optimal solution, which could be taken as the seeds of k-means.
2. For GSOCA+KM, it regards the data as individuals in a glowworm swarm and makes them reorganize spontaneously by evolutions, to strengthen the differentiation of clusters and make those in the same cluster aggregating closer.

According to our research, we find the latter can not only give better performance, but also enhance the theoretical upper limit. Now we provide some suggestive interpretations.

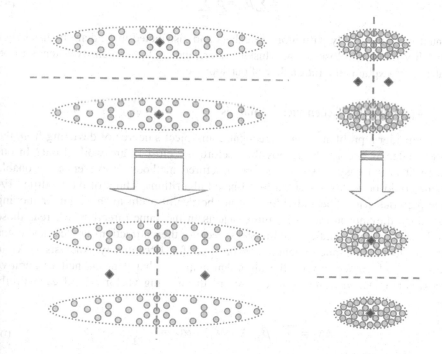

Fig. 1. Two clustering processes from unstable states to stable, where the circles refer to objects and the squares refer to cluster centers. The left is executed on original data and reorganized data are adopted for the right.

It is interesting to see that the unprocessed data set in the left case leads to a false partition according to our common understanding. What is the root cause of this? Note that the *SSE* (similar to potential energy in physics) of the false partition is yet smaller. Therefore the reason is that the seemingly wrong partition is just a "right" one in the prototype of k-means, i.e. an object should be assigned to the nearest cluster center. This rule is not correct in most actual cases, and fortunately it could be amended through Self-Aggregation (SA).

Compared with GSO, there are at least three facts indicating the superiority of FA on realizing SA as follows.

1. With the removal of luciferin and other simplifications, the number of parameters in FA is far less than GSO's.
2. In FA, the attractiveness attenuates with distance rather than being simply restricted by the threshold radius, leading to more consistent performance.
3. The step size in FA is adaptive and surely better than the constant in GSO, e.g. for avoiding the oscillatory phenomenon.

Considering the above, we propose our algorithm of SA based on FA, by the redefinition of light intensity I_i as

$$I_i = \sum_{j=1}^{N} \beta_{i,j} = \beta_0 \sum_{j=1}^{N} e^{-\gamma r_{i,j}^2} \tag{5}$$

Intuitively, the firefly with more neighbors and distributed more densely, has larger probability of being close to the cluster center, and its light intensity I_i tends to be greater. This explains the rationality of our approach.

4.2   Approach for Eccentricity Analysis

Let us consider a problem, how to recognize an object's degree of deviating from the cluster center as accurately as possible, before the clusters are worked out? In our opinion, it is not easy to solve through structured methods. However, a reasonable solution could be neatly provided by SI-based algorithms. Think of the modified FA in the previous part. The individuals at the border of a cluster have larger moving steps, since they are attracted by more objects in the same direction, whereas those near the center share smaller moving steps, because their attractions are less and come from all sides. This phenomenon can be used for Eccentricity Analysis (EA) in reverse, for which we rewrite the algorithm again so that it needs neither iterative process nor actual movement, just to record the moving vector ΔX_i of each firefly using

$$\Delta X_i = \sum_{I_i < I_j} \left(\beta_{i,j}(X_j - X_i) + \alpha(rand - \frac{1}{2}) \right) \tag{6}$$

Then we can obtain the *Eccentricity* of each firefly (*ECC_i*) in

$$ECC_i = \left| \{ X_j \mid norm\ of\ \Delta X_j < norm\ of\ \Delta X_i, j \in [1, N] \} \right| \tag{7}$$

whose intuitive sense is the number of fireflies moving a shorter distance than the specific one, and on the rise from each center to the boundary. Formally, we have grounds for the correctness of Eq.(8) in a statistical sense.

$$\forall i, \forall j, ECC_i < ECC_j \rightarrow \|X_i - C_{X_i}\| < \|X_j - C_{X_j}\| \tag{8}$$

Given ECC_i, we can know the approximate distribution of the swarm, which could be utilized in many ways, e.g. designing particular regulations for different individuals in an optimization or clustering procedure.

4.3   Hybrid K-means++ Algorithm

Given the two helpful instruments, there is a natural strategy to improve an existing clustering method, i.e. 1) do SA for reorganization; 2) do EA to label ECC_i; 3) run clustering method taking the heuristic information into use. Now we introduce a k-means++ improved in this way, by interpreting the 3) step as follows.

Since the k-means++ attempts to generate the initial centers as far as possible away from each other, it probably leads to a layout where the chosen objects are distributed on the edge of the clusters, which is not the most ideal situation. Nevertheless, if we make the centers picked out from the objects with the lowest ECC_i at P percent, the clustering seeds will come close to the accurate centers to a certain extent. This modification can improve the robustness of the algorithm, as well as accelerate the convergence. We take the improved k-means++ for comparative trials in the next section.

5   Experiments and Results

In this section, we first make a brief description of the experimental setup including benchmark data sets and parameter settings, then make a comparison with some competitors, i.e. k-means, k-means++, K-FA, and GSOCA+KM to prove the advantages of our proposals.

The experimental platform was a laptop computer with Intel (R) Core (TM) i7-3630QM CPU @ 2.4 GHz (8 CPUs), containing 8.0GB of RAM, running Microsoft Windows 7 Ultimate 64-bit. The simulation programs were developed using Dev-C++ 4.9.9.2.

5.1   Experimental Setup

The experiments were conducted on six real-world data sets widely used in the literature, i.e. Iris, Wine, Glass, Balance, Seeds, and WDBC (Wisconsin Diagnostic Breast Cancer) downloaded from the UCI database repository. The detailed features of them are listed as follows.

Before clustering, we normalized the data to make them proportionally distributed in the unit space, by executing in each dimension

$$X_i = \frac{X_i - X^{\min}}{X^{\max} - X^{\min}} \tag{9}$$

Table 1. Summary of the data sets

Data Set	Records	Features	Clusters	Area
Iris	150	4	3	Life
Wine	178	13	3	Physical
Glass	214	9	6	Physical
Balance	625	4	3	Social
Seeds	210	7	3	Life
WDBC	569	30	2	Life

where X^{min} and X^{max} are the minimum and maximum coordinates of the given data set respectively.

Each time for pretreatment, we ran K-FA for 50 iterations with swarm size 50, GSOCA+KM for 10 iterations, and our algorithm for 5 iterations keeping the run-times close. Then k-means/k-means++ was iterated for 100 rounds. All the tests were repeated for 10 times independently with configures the same. It should be noted that only a few iterations of SA are enough to achieve obvious improvements.

For K-FA and GSOCA+KM, we set the parameters in Table 2 and 3 with reference to [6, 7] and our experience.

Table 2. Parameter settings in K-FA method

Parameter	γ	α	β_0
Value	1.0	0.7	1.0

Table 3. Parameter settings in GSOCA+KM method

Parameter	l_0	ρ	γ	s	β	r_s	n_t	r
Value	0	0.4	1.0	0.02	0.02	0.2	10	0.1

For our algorithm, we believe that the following settings provide satisfactory performance most often, according to plenty of tuning practice.

Table 4. Parameter settings in hybrid k-means++ proposed

Parameter	γ	α	β_0	P
Value	100.0	0.005	0.5	50.0

5.2 Performance Comparison

In order to measure the effect of clustering, the classification accuracy is usually taken into use. However, we don't want to waste run-time on the correspondence between the real and artificial clusters, so we construct the *Associative Error Rate* (*AER*) given by

$$AER = \frac{\sum_{i=1}^{N}\sum_{j=1}^{N}(IDC_i^{reality} = IDC_j^{reality}) \oplus (IDC_i^{result} = IDC_j^{result})}{N^2} \qquad (10)$$

where the $IDC_i^{reality}$ is the identification number of the cluster that the i-th object belongs to in reality, while the IDC_i^{result} is the similar term given by the algorithms. The result of logical operation is 0 or 1. Obviously AER varies in the range [0, 1.0] and the smaller is the better, especially 0 if the division is exactly the same as the truth.

Table 5. Performance comparison in *associative error rate*

Algorithm	Comparison Index	Benchmark Data Sets					
		Iris	Wine	Glass	Balance	Seeds	WDBC
	Best	12.50%	6.50%	33.10%	39.50%	13.00%	13.40%
k-means	Mean	23.50%	9.40%	45.10%	41.00%	13.20%	16.70%
	Std	15.30%	7.60%	14.50%	1.20%	0.20%	10.00%
	Best	12.50%	5.80%	32.00%	37.70%	13.00%	13.40%
k-means++	Mean	14.30%	7.00%	33.50%	41.50%	13.30%	13.40%
	Std	4.50%	0.70%	1.30%	2.20%	0.30%	0.00%
	Best	12.50%	6.00%	32.70%	37.70%	13.00%	13.40%
K-FA	Mean	25.50%	6.90%	44.30%	41.90%	14.40%	13.40%
	Std	20.90%	0.40%	15.10%	2.70%	3.90%	0.00%
	Best	9.40%	6.50%	33.60%	37.90%	10.40%	13.40%
GSOCA+KM	Mean	18.40%	9.60%	47.40%	40.60%	16.40%	16.70%
	Std	16.40%	7.20%	17.30%	1.50%	16.80%	10.00%
k-means++	Best	4.20%	5.30%	32.50%	34.10%	9.90%	12.50%
with	Mean	10.40%	7.80%	35.20%	42.40%	12.40%	13.80%
SA & EA	Std	7.60%	1.90%	1.90%	4.10%	1.30%	0.70%

The detailed comparison indexes and experimental records are enumerated in Table 5. We can easily find that k-means++ and K-FA indeed enhanced the performance of standard k-means on Wine, Glass, and Balance, however, they played an insignificant role on the rest of data sets, since the theoretical limits had been reached as we discuss. The GSOCA+KM was successful in upgrading the best results on Iris and Seeds, but failed on WDBC which might be partially attributed to the massive parameters increasing the complexity of tuning.

It is demonstrated that our algorithm outperformed the competitors consistently in all the instance, which not only broke the theoretical limits on Iris, Seeds, and WDBC, but also enhanced the results on the other data sets especially on Balance. In addition, it is fairly reliable for the stability embodied in good mean performance.

Table 6. Performance comparison in *sum squared errors*

Category of Algorithms	Benchmark Data Sets					
	Iris	Wine	Glass	Balance	Seeds	WDBC
Methods without SA	7.00E+00	4.90E+01	1.84E+01	2.17E+02	2.20E+01	2.16E+02
Methods with SA	7.79E+00	4.90E+01	1.99E+01	2.20E+02	2.22E+01	2.16E+02

In Table 6, we list the *SSE* of the best results generated by the methods without SA, i.e. k-means, k-means++, and K-FA, as well as the opponents, i.e. GSOCA+KM and our hybrid algorithm. It is proved once again that the better partitions in the sense of *SSE* are not in fact as good as those reorganized first, for which our pretreatment tools are highly recommended.

6   Conclusion and Future Work

This paper first introduces the ideas and realizations of Self-Aggregation (SA) and Eccentricity Analysis (EA) based on Firefly Algorithm (FA), then provides a variant of k-means++ hybrid with these techniques, whose superiority is verified by experiments. The work indicates a new direction to enhance clustering performance via Swarm Intelligence (SI).

Future work mainly focuses on making general optimizations in the case of high-dimensional and large-sized conditions, and finding a more efficient or adaptive way to determine the several parameters.

Acknowledgments. This work is supported by National Natural Science Foundation of China under Grant No. 61300222, central university basic research special funding (Innovation Fund of Huazhong University of Science and Technology under Grant No. 2013QN120).

References

1. Anderberg, M.R.: Cluster analysis for applications (No. OAS-TR-73-9). Office of the Assistant for Study Support Kirtland Afb N Mex (1973)
2. Ketchen, D.J., Shook, C.L.: The application of cluster analysis in strategic management research: An analysis and critique. Strategic Management Journal 17(6), 441–458 (1996)
3. Punj, G., Stewart, D.W.: Cluster analysis in marketing research: Review and suggestions for application. Journal of Marketing Research, 134–148 (1983)
4. Abshouri, A.A., Bakhtiary, A.: A New Clustering Method Based on Firefly and KHM. Journal of Communication and Computer 9(4), 387–391 (2012)
5. Aljarah, I., Ludwig, S.A.: A new clustering approach based on glowworm swarm optimization. In: 2013 IEEE Congress on Evolutionary Computation (CEC), pp. 2642–2649. IEEE (2013)

6. Hassanzadeh, T., Meybodi, M.R.: A new hybrid approach for data clustering using firefly algorithm and K-means. In: 2012 16th CSI International Symposium on Artificial Intelligence and Signal Processing (AISP), pp. 007-011. IEEE (2012)
7. Huang, Z., Zhou, Y.: Using glowworm swarm optimization algorithm for clustering analysis. Journal of Convergence Information Technology 6(2), 78–85 (2011)
8. Senthilnath, J., Omkar, S.N., Mani, V.: Clustering using firefly algorithm: Performance study. Swarm and Evolutionary Computation 1(3), 164–171 (2011)
9. Yang, X.-S.: Firefly algorithms for multimodal optimization. In: Watanabe, O., Zeugmann, T. (eds.) SAGA 2009. LNCS, vol. 5792, pp. 169–178. Springer, Heidelberg (2009)
10. Yang, X.S.: Nature-inspired metaheuristic algorithms. Luniver Press (2010)
11. MacQueen, J.: Some methods for classification and analysis of multivariate observations. In: The Fifth Berkeley Symposium on Mathematical Statistics and Probability, vol. 1(281-297), p. 14 (1967)
12. Arthur, D., Vassilvitskii, S.: k-means++: The advantages of careful seeding. In: The Eighteenth Annual ACM-SIAM Symposium on Discrete Algorithms, pp. 1027–1035. Society for Industrial and Applied Mathematics (2007)
13. Krishnanand, K.N.: Glowworm swarm optimization: A multimodal function optimization paradigm with applications to multiple signal source localization tasks (2009)
14. Krishnanand, K.N., Ghose, D.: Glowworm swarm optimisation: A new method for optimising multi-modal functions. International Journal of Computational Intelligence Studies 1(1), 93–119 (2009)

DNA Computation Based Clustering Algorithm

Zhenhua Kang[1,2], Xiyu Liu[1], and Jie Xue[1]

[1] Shandong normal university, Jinan, Shandong
[2] Shandong institute of business and technology, Yantai, Shandong
{kangzhh1007,sdxyliu,xiaozhuzhu1113}@163.com

Abstract. Using DNA computation to solve clustering problem is a new approach in this field. In the process of problem solving, we use DNA strands to assign vertices and edges, constructing the shortest Hamilton path and cutting branches whose length is longer than the threshold we gavegetting the initial clustering result. For improving the quality, we do the iterative calculation, getting clusters for every produced cluster, we deal all of the process with DNA computation in test tubes, reducing the time complexity obviously by DNAs high parallelism. In this paper, we give the process and analysis of our algorithm, illustrating the feasibility of the method.

Keywords: DNA computation, clustering, shortest Hamilton path, parallelism.

1 Introduction

In 1994 ,Adleman [1] computed the seven vertices of Hamiltonian path problem with DNA molecules in test tube, which shows a great power in combinational problems [2] [3] by DNA computing. Then DNA computing is widely used in various fields. So far, it solves many practical problems successfully. Compared with traditional computing, DNA computing is more suitable for solving complex combinatorial optimization problem. DNA computing belongs to the biochemical reaction in essence, which loads information in DNA chain. Through the relevant operation and under the environment of a particular reaction, it output the solution set. DNA computing has three advantages: (1)huge parallelism (2)high speed (3)large storage, 1 bit information can be stored in $1 \ nm^3$ [4].

Clustering plays an important and indispensable role in data mining. Although several methods are available in these areas [5], these algorithms exhibit polynomial or exponential complexity when the number of clusters is unknown and the data set is huge, which make problems more challenging. There are many types of clustering, such as hierarchical clustering, Density-based clustering, Subspace clustering, etc. However, when the number of clusters is unknown and the data set become huge, these algorithms exhibit polynomial or exponential complexity, which make the problem being more challenging [6].

William Rowan Hamilton was Astronomer Royal of Ireland in the mid-19th century, the problem that has come to bear his name. A Hamiltonian path is

Y. Tan et al. (Eds.): ICSI 2014, Part I, LNCS 8794, pp. 470–478, 2014.

a path that visits each vertex exactly once, given starting and ending vertex beforehand. The Hamiltonian path problem is to decide for any given graph with specified start and end vertices whether a Hamiltonian path exists or not. So it is a decision problem.

There have been many algorithms to deal with the Hamiltonian path problem, just like greedy algorithm, dynamic planning algorithm, divide and conquer algorithm. However, it seemed that there were any efficient methods in solving it. In the early 1970s, it was shown to be NP complete. Until 1994, Adleman used DNA computation to find the Hamilton path in a seven vertex directed graph, which proved to be a much powerful algorithm.

DNA computing has been used in many fields, but there has not many researches in clustering. Bakar and Watada presented some ideas to use DNA computing to solve clustering problems [5]. They proposed a new DNA approach to solve clustering problem based on k-means and fuzzy c-means algorithm. Kim and Watada (2009) gave the similar method for heterogeneous coordinate data. Zhang Hongyan, Liu xiyu [7] presented another research on clustering based on the idea of using DNA computing to find Hamilton circuit.

Motivated by researches above, we combine DNA computing with clustering through Hamilton path. At first, we find the shortest Hamilton path, then, we use the thought of mimimum spanning tree, cut edges beyond the thresholds by the help of grids. All the process conduct in test tubes The DNA-Hamilton based algorithm provides an alternative for traditional computing.

2 Preliminaries

DNA computing is essentially a biochemical reaction, The information included in the DNA chain, under certain circumstances, and through some related operations, we can get the solutions. A strand of DNA is encoded with four nucleotides, A(Adenine), G(Guanine), T(Thymine), C(Cytosine). Each strand, according to chemical convention, has a 5' and a 3' end; hence, any single strand has a natural orientation. According to the famous Watson-Crick complementary which is deduced by James D. Watson and Francis H. C. Crick in 1953, A bonds with T and G bonds with C. The pairs are (A, T) and (G, C).

The manipulations about DNA computation are shown as follows: Gel electrophoresis used for the separation of DNA molecules. Electrophoresis is a procedure which enables the sorting of DNA molecules based on size and charge. Using an electric field, DNA molecules can be made to move through a gel made of agar. The DNA molecules being divided are dispensed into a well in the gel material. The gel is placed in an electrophoresis chamber, which is connected to a power source. When the electric current is applied, the negatively charged DNA molecules move toward the anode, and we know that the larger molecules move more slowly than the smaller molecules through the gel. So, the smaller molecules go to the positive electrode preferentially. Then, the different sized molecules form distinct bands on the gel, from this bands, we can know the length of different DNA molecules.

DNA denaturation means the melting of double stranded DNA to generate two single strands. The main work is to break the hydrogen bonds between the bases in the duplex. We often heat the DNA to a temperature above its melting point in the laboratory to melt the double stranded DNA. If we cool the tube again, the denatured DNA will bump into each other and stick tightly, and becomes double stranded DNA. This process is named DNA annealing. Polymerase Chain Reaction can produce large amounts of a specific DNA fragment from small amounts of a complex template. Recombinant DNA techniques create molecular clones by conferring on a specific sequence the ability to replicate by inserting it into a vector and introducing the vector into a host cell. Because of Polymerase Chain Reactions high replication ability, it has had an enormous impact in both basic and diagnostic aspects of molecular biology.

DNA sequencing is the method to determine the order of the nucleotide bases in a molecule of DNA. Some important methods of DNA sequencing are as follows: Maxam - Gilbert sequencing (this method is based on chemical modification of DNA and subsequent cleavage at specific bases), and then Frederick Sanger invented a new method named Chain-termination methods, because of its high efficient and lower amounts of radioactivity than the method of Maxam and Gilbert, it rapidly became peoples choice. There are more ways about DNA sequencing, we do not introduce here.

3   DNA Computing Design for Clustering Algorithms

The basic idea of this algorithm is using grids to divide dataset firstly. Then, it extracts the mapping data to form graph. Using DNA computation does the clustering at last.

3.1   Framework of Algorithm

Input D;
  Output C;
  Step 1 meshing data set;
  Step 2 take each center of the grid which contains data as vertex of the generated graph, and connect the center.
  Step 3 calculating distance between vertices and expressed as matrix A. generating the directed weight complete graph.
  Step 4 using DNA sequences to encode vertex and edge, reacting fully in a test tube, getting the shortest Hamilton path.
  Step 5 deleting edged not meeting with the threshold Φ (Φ is the function of x, y), forming the preliminary clusters.
  Step 6 doing step 2 - step 4 of data points in each cluster, finding the shortest Hamilton path weights and ε, if $\epsilon < \xi$ (upper and lower bounds of the factor for the weight, and the new grid of x, y). The algorithm stops. Otherwise, it goes to step 7.

Step 7 for $\epsilon > \xi$ cluster, repeated step 1 - step 6, until all the cluster are satisfied with $\epsilon < \xi$.

Step 8 Return clustering result C.

3.2 Process of Algorithms

According to the characteristics of data sets, corresponding grid is given, data are divided. In step 3, the distance between vertices is

$$D = \sqrt{(x_2 - x_1)^2 + (y_2 - y_1)^2} \tag{1}$$

Graph $G = (V, E)$ is generated. In step 4, each vertex design for nmr, the DNA strand of edge is the complement strands of its precursor vertex + weight information coding + complement of subsequent vertex. Each vertex is set as an independent cluster and encode by DNA sequence. Through a series of biological response, the shortest Hamilton path is obtained. The process is as follows [4]: (a) assuming for any graph, the beginning is v_1 and ending is v_n, (b) generating all arbitrary path in graph, (c) keeping paths with starting v_1 and ending v_n, (d) keeping paths containing all vertex at least once, (e) choosing the shortest path, which is the initial clustering result.

3.3 Process of Biology Reaction

(1) input(T):

putting single-stranded DNA molecules representing each vertex and edge into test tube T. doing full annealing and connection reaction, amplifying random product by using polymerase chain reaction.

(2) prefix(T, v_1):

extracting all DNA molecules containing vertex v_1 from test tube T

(3) postfix(T, v_n):

extracting all DNA molecules containing vertex v_n from test tube T

(4) amplify(T):amplifying DNA strands with v_1 and v_n

(5) length-separate(T$m * n$):

extracting all paths with n vertices (each vertex encoding length of m)

(6) for i from 2 to $n - 1$ do

(7) {

(8) substring-separate(Tv_i):extracting DNA stands with all vertexes recursively.

(9) }

(10) end for

(11) return detect(T):

using gel electrophoresis technique to find the shortest path, and reading out DNA sequence through DNA sequencing technology.

In step 5, we use DNA molecular probe to isolate the DNA sequence whose length is greater than the threshold, so that we can remove edges who are not satisfied with the conditions. In step 6, Hamilton shortest path of each cluster was conducted in test tubes at the same time. The consumed time is equal to find one Hamilton path, which reduces the time complexity of the algorithm greatly.

3.4   Analysis of Time Complexity

In Step 4, the process of finding the shortest Hamilton path depends on great parallelism of DNA computing. All shortest Hamilton paths are found at the same time. Thus, time complexity of step 1 - step 7 is $O(n)$. Finding the Hamilton shortest path for every cluster executes in parallelism too. The time complexity is $O(n)$. As a result, the total time consumption is $O(kn)$ (k is the conducting number of step 1 - step 7). Compared with other clustering algorithm (as shown in table below), our algorithm consumes less time.

Table 1. Time complexity of different algorithms

Algorithms	Time complexity
K-means	$O(nkl)$
K-median	$O(n^{(i+\varepsilon)})$
Hierarchical agglomerative algorithms	$O(n^2 logn)$
MST	$O(n^2 logn)$
DBSCAN	$O(nlogn)$

4   Example

Take each center of the grid who contains data, and connect edges who are linked to these centers($v_1(2.5, 12.5)$, $v_2(2.5, 17.5)$, $v_3(17.5, 7.5)$, $v_4(27.5, 12.5)$, $v_5(27.5, 22.5)$) as Fig.2.

According to 1, calculate distances between vertexes. Then, we obtain the directed weight complement graph as Fig 3.

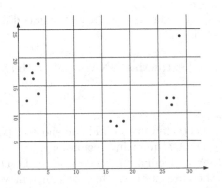

Fig. 1. Example

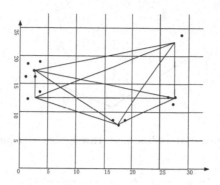

Fig. 2. Clustering result of the example

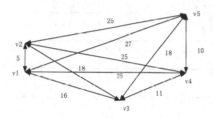

Fig. 3. Directed weight complement graph

Each vertex is designed as 20 nm. DNA strand of edges are complementary of its precursor vertex + weight coding + complementary of its subsequent vertex. Coding length of weights are equal to matrix A. Vertices v_1 is *AT ATC GCGGG TTCAA CGTGC*, v_2 is *GCAGT TGACA TGCAG GATCG*, edge e_{12} is shown below.

TATAGCGCCCAAGTTGCACG GTCGT CGTCAACTGTACGTCCTAGCT

Thus, there is:

ATATCGCGGGTTCAACGTGC   GCAGTTGACATGCAGGATCG
TATAGCGCCCAAGTTGCACG GTCGT CGTCAACTGTACGTCCTAGCT

By annealing and connection reaction, all double-stranded DNA molecules of arbitrary paths are got. Then paths are amplified by polymerase chain reaction, only path with v_1 (as a starting) and v_5 are amplified. Shortest paths are isolated by gel. We use affinity purification technology for purification. Finally the shortest Hamilton path is read out through sequencing technology, as shown in Fig. 4.

Set the threshold as Φ, delete edges with $D > 10$, initial clustering result is shown in Fig. 5.

So far, we get the preliminary clustering result. However, it only ensures the quality between clusters, failing to consider the relationship within clusters after clustering. This paper not only focuses on the distance between clusters, but also ensures the quality in cluster.

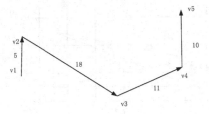

Fig. 4. The shortest Hamilton path through sequencing technology

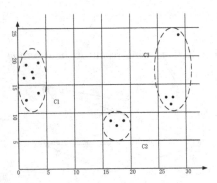

Fig. 5. Initial clustering result deleted edges with $D > 10$, threshold as Φ

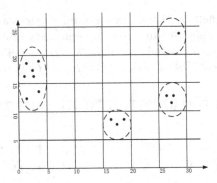

Fig. 6. Final clustering result

According to this algorithm, it finds shortest Hamilton path for every cluster and calculate sum of weights, then, get w(C1), w(C2), w(C3). Because of $w(C3) > \xi$. Edges do not meet with threshold are deleted. New groups are obtained. The process will continue until all clusters meet the requirements. The algorithm halts. Final clustering result is shown in Fig. 6.

This paper uses data set as Fig. 7 to show the effectiveness of DNA computation based clustering algorithm. This experiment executes in Windows xp,

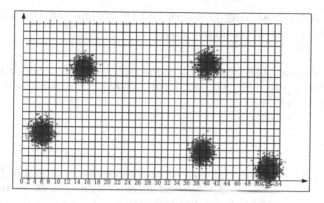

Fig. 7. The data set this paper used

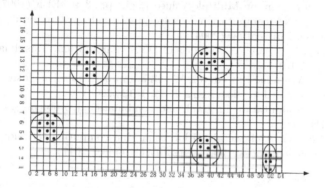

Fig. 8. The clustering result of DNA computation based clustering algorithm

MATLAB 7.0. Data set is two dimension with $0 \leq x \leq 54$ and $0 \leq y \leq 17$ .The grid is divided into 53*16. Clustering result is shown in Fig. 8.

The experimental results show that the algorithm can obtain solutions, the data points are divided into 4 clusters, similarity within cluster is high, between clusters is low, and the algorithm has obvious advantages on time complexity and clustering quality.

5   Conclusion

One of the main advantage of DNA computing is solving multiple problems in parallel. Combing DNA computing grid, and Hamilton shortest path to solve clustering prove to be feasible. Every step is to look for local optimal clustering. The algorithm also illustrates the feasibility of DNA computing for clustering and powerful parallelism of DNA computing. But we still have a lot of work to do.We just proved that it is feasible in theory. There are many problems to be solved about biotechnology. In the future, we will continue to study with DNA computing to solve clustering problems and try to find out the more effective method to solve multi-dimensional data.

Acknowlegements. This research is supported by the Natural Science Foundation of China(No.61170038), the Natural Science Foundation of Shandong Province(No.ZR2011FM001).

References

1. Xue, J., Liu, X.: Applying dna computation to clustering in graph. In: 2011 2nd International Conference on Artificial Intelligence, Management Science and Electronic Commerce (AIMSEC), pp. 986–989 (2011)
2. Rooß, D.: Recent developments in dna-computing. In: Proceedings of the 27th International Symposium on Multiple-Valued Logic, pp. 3–9. IEEE (1997)
3. Yin, Z.: Using DNA computation to solve graph and combinatorial optimization problem. Science Publication (2004)
4. Hao, F., Wang, S., Qiang, X.: Models of DNA computation. Tsinghua University Press, Beijing (2010)
5. Bakar, R.B.A., Watada, J., Pedrycz, W.: Dna approach to solve clustering problem based on a mutual order. Biosystems 91(1), 1–12 (2008)
6. Yin, Z.: DNA computation in graph and combinational problems. Science Press, Beijing (2004)
7. Zhang, H., Liu, X.: A clique algorithm using dna computing techniques based on closed-circle dna sequences. Biosystems 105(1), 73–82 (2011)

Clustering Using Improved Cuckoo Search Algorithm

Jie Zhao[1], Xiujuan Lei[1,2], Zhenqiang Wu[1], and Ying Tan[2]

[1] School of Computer Science, Shaanxi Normal University, Xi'an, 710062, China
[2] School of Electronics Engineering and Computer Science, Peking University,
Beijing, 100871, China
xjlei168@163.com

Abstract. Cuckoo search (CS) is one of the new swarm intelligence optimization algorithms inspired by the obligate brood parasitic behavior of cuckoo, which used the idea of Lévy flights. But the convergence and stability of the algorithm is not ideal due to the heavy-tail property of Lévy flights. Therefore an improved cuckoo search (ICS) algorithm for clustering is proposed, in which the movement and randomization of the cuckoo is modified. The simulation results of ICS clustering method on UCI benchmark data sets compared with other different clustering algorithms show that the new algorithm is feasible and efficient in data clustering, and the stability and convergence speed both get improved obviously.

Keywords: Clustering, cuckoo search, Lévy flights, swarm intelligence optimization algorithm.

1 Introduction

Clustering is the process of separating similar objects or multi-dimensional data vectors into a number of clusters or groups. It is an unsupervised problem. Clustering techniques have been used successfully in data analysis, image analysis, data mining and other fields of science and engineering [1].

Many algorithms have been developed for clustering. The traditional clustering methods can be classified into four categories: partitioning methods, hierarchical methods, density-based methods and grid-based methods [2].

Swarm intelligence optimization algorithm such as genetic algorithms (GA) [3], ant colony optimization [4], particle swarm optimization (PSO) [5, 6], artificial bee colony (ABC) [7, 8], bacteria foraging optimization algorithm (BFO) [9], firefly algorithm (FA) [10] has been widely used in the clustering in recent years. Cuckoo Search (CS) algorithm is a new intelligence optimization algorithm which has been successfully applied to the global optimization problem [11], economic dispatch [12], clustering [13-16] and other fields [17]. However, the cuckoo search clustering algorithm has several drawbacks such as slow convergence speed and vibration of the convergence.

Y. Tan et al. (Eds.): ICSI 2014, Part I, LNCS 8794, pp. 479–488, 2014.
© Springer International Publishing Switzerland 2014

In this paper, we propose an improved cuckoo search (ICS) algorithm for clustering, in which the movement of cuckoo and random disturbance was modified to find optimal cluster center. The algorithm was tested on four UCI benchmark datasets, and its performance was compared respectively with K-means, PSO, GA, FA and CS clustering algorithm. The simulation results illustrated that this algorithm not only own higher convergence performance but also can find out the optimal solution than the other algorithms.

2   Cuckoo Search Algorithm

Cuckoo search algorithm is a novel metaheuristic optimization algorithm developed by Xin-she Yang and Suash Deb in 2009 [18], which is based on the obligate brood parasitic behaviour of some cuckoo species in combination with the Lévy flights behavior of some birds and fruit flies.

2.1   Cuckoo Brood Parasitic Behaviour

Cuckoos are fascinating birds not only because of the beautiful sounds they can make, but also because of their aggressive reproduction strategy they share [19]. Quite a number of species engage the obligate brood parasitism by laying their eggs in the nests of other host birds, which may be different species. They may remove others' eggs to increase the hatching probability of their own eggs [20]. If a host bird discovers that the eggs are not their own' eggs, they will either throw these alien eggs away or simply abandon its nest and build a new nest elsewhere.

Studies also indicated that the cuckoo eggs hatch slightly earlier than their host eggs. Once the first cuckoo chick is hatched, the first instinct action it will take is to evict the host eggs by blindly propelling the eggs out of the host, which increases the cuckoo chick's share of food provided by its host bird. In addition, a cuckoo chick can also mimic the call of host chicks to gain access to more feeding opportunity.

2.2   Lévy Flights

In nature, animals search for food in a random or quasi-random manner. Various studies have shown that the flight behavior of many animals and insects demonstrates the typical characteristics of Lévy flights [21]. Lévy flights comprise sequences of randomly orientated straight-line movements. Frequently occurring but relatively short straight-line movement randomly alternate with more occasionally occurring longer movements, which in turn are punctuated by even rarer, even longer movements, and so on with this pattern repeated at all scales. As a consequence, the straight-line movements have no characteristic scales, and Lévy flights are said to be scale-free, the distribution of straight-line movement lengths have a power-law tail [22]. Fig. 1 shows the path of Lévy flights of 60 steps starting from (0, 0).

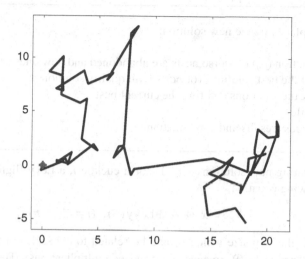

Fig. 1. Lévy flights in consecutive 60 steps starting at the origin (0, 0) which marked with " * "

2.3   Cuckoo Search Algorithm

Here we simply describe the cuckoo search algorithm as follows which contain three idealized rules [19]:

(1) Each cuckoo lays one egg at a time, and dumps its egg in a randomly chosen nest;

(2) The best nest with high quality of eggs will carry over to the next generations;

(3) The number of available host nests is fixed, and then the egg laid by a cuckoo is discovered by the host bird with a probability $p_a \in [0,1]$. In this case, the host bird can either throw the eggs away or abandon the nest, and build a completely new nest. For simplicity, this rule can be approximated by the fraction p_a of the n nests are replaced by new nests (with new random solutions).

For a maximization problem, the quality of fitness of a solution can be proportional to the objective function. Other forms of fitness can be defined in a similar way to the fitness function in genetic algorithm and other optimization algorithms [23].

Based on these three rules, the basic steps of the cuckoo search can be summarized as the pseudo code as Table 1 [19]:

Table 1. Pseudo code of the cuckoo search

```
begin
    Objective function f (x), x = (x₁, ..., x_d)ᵀ
    Generate initial population of n host nests xᵢ (i = 1,2, ..., n)
    while (t < MaxGeneration) or (stop criterion)
        Get a cuckoo randomly by Lévy flights
            evaluate its quality/fitness Fᵢ
        Choose a nest among n (say, j) randomly
```

```
    if (F_i > F_j),
        replace j by the new solution;
    end
    A fraction (p_a) of worse nests are abandoned and new ones are built;
    Keep the best solutions (or nests with quality solutions);
    Rank the solutions and find the current best
end while
Postprocess results and visualization
end
```

When generating new solutions x $(t+1)$ for a cuckoo i, a Lévy flight is performed using the following equation:

$$x_i^{(t+1)} = x_i^{(t)} + \alpha \oplus \text{Lévy}(\beta), \quad (i = 1,2,...,n) \ . \tag{1}$$

where $\alpha > 0$ is the step size which should be related to the scales of the problem of interests. The product $\oplus$ means entry-wise multiplications. The Lévy flight essentially provides a random walk while the random step length is drawn from a Lévy distribution which has an infinite variance with an infinite mean [19].

$$\text{Lévy}(\beta) \sim u = t^{-1-\beta}, \quad (0 < \beta \le 2) \ . \tag{2}$$

Mantegna puts forward a most efficient and yet straightforward ways to calculate Lévy distribution [18, 24].

$$\text{Lévy}(\beta) \sim \frac{u}{|v|^{1/\beta}} \ . \tag{3}$$

$$u \sim N(0, \sigma_u^2) \ , \quad v \sim N(0, \sigma_v^2) \ . \tag{4}$$

$$\sigma_u = \left\{ \frac{\Gamma(1 + \beta) \sin(\pi\beta / 2)}{\Gamma[(1 + \beta) / 2]\beta 2^{(\beta-1)/2}} \right\}^{1/\beta} \ , \quad \sigma_v = 1 \ . \tag{5}$$

In CS Algorithm, the worst nest is abandoned with a probability p_a and a new nest is built with random walks [19].

$$x_{worst}^{(t+1)} = x_{worst}^{(t)} + \alpha * rand() \ . \tag{6}$$

3   Clustering Using ICS Algorithm

3.1   The Clustering Criterion

Clustering is the process of grouping a set of data objects into multiple groups or clusters so that objects within a cluster have high similarity, but are very dissimilar to objects in other clusters. Dissimilarities and similarities are assessed based on the attribute values describing the objects and often involve distance measures. The most popular distance measure is Euclidean distance [1].

Let $i = (x_{i1}, x_{i2}, ..., x_{ip})$ and $j = (x_{j1}, x_{j2}, ..., x_{jp})$ be two objects described by p numeric attributes, the Euclidean distance between object i and j is define as:

$$d(i, j) = \sqrt{(x_{i1} - x_{j1})^2 + (x_{i2} - x_{j2})^2 + \cdots + (x_{ip} - x_{jp})^2} \ . \tag{7}$$

For a given N objects the clustering problem is to minimize the sum of squared Euclidean distances between each object and allocate each object to one of k cluster centers [1]. The main goal of the clustering method is to find the centers of the clusters by minimizing the objective function. The clustering objective function is the sum of Euclidean distances of the objects to their centers as given in Eq. (8) [2]

$$J_c = \sum_{k=1}^{m} \sum_{X_i \in C_k} d(X_i, Z_k) \ . \tag{8}$$

where m denotes the number of clusters, C_k denotes the kth cluster, $d(X_i, Z_k)$ denotes the Euclidean distance between object X_i and cluster center Z_k.

3.2   Clustering Using ICS Algorithm

Some researchers [13-16] have been designed CS for clustering. The Lévy flight is more efficient because the step length is heavy-tailed and any large step is possible, which makes the whole search space to be covered [14]. However, Lévy flight often leads to slow convergence rate and vibration when clustering using CS. We propose ICS algorithm for data clustering, in the algorithm, each egg in a nest represents a cluster center, the cuckoo searches for a new nest in line with Eq. (9)

$$x_i^{(t+1)} = x_i^{(t)} + \alpha \oplus \text{Lévy} (\beta) + w * rand * (Z_{best} - x_i^{(t)}), \quad (i = 1, 2, ..., n) \ . \tag{9}$$

where $rand$ is random number, w denotes disturbance constant, Z_{best} denotes the cluster center of the best cluster.

The ICS algorithm for data clustering is as the following steps:

Step 1: Generate initial population of n host nests randomly, the host nests position denote the cluster centers. Initialize the iterations *iter*, maximum iteration *maxiter* and cluster number *nc*;

Step 2: Clustering and calculate the clustering objective function *Fold* using Eq. (8) to find the best nest *bestnest*;

Step 3: Generate n-1 new nests using Eq. (9) except the *bestnest*, clustering and calculate the clustering objective function *Fnew*,

Step 4: Compare *Fnew* with *Fold*, if *Fnew* < *Fold*, replace the old nests by the new ones;

Step 5: A fraction (p_a) of worst nests are abandoned and new ones are built using Eq. (6);

Step 6: Find the best solution, set *iter* = *iter* + 1;

Step 7: If *iter* < = *maxiter*, goto *Step* 3, otherwise output the clustering result.

4   Implementation and Results

In order to test the accuracy and the efficiency of ICS data clustering algorithm, experiments have been performed on four datasets including Iris, Glass, Wine and Sonar selected from standard set UCI [25]. All algorithms are implemented in Matlab R2011b and executed on a Pentium dual-core processor 3.10GHz PC with 4G RAM. The parameter values used in our algorithm are $n=15$, $p_a=0.25$, $\alpha=0.01$ and $w=0.06$.

4.1   Data Set Description

The four clustering data sets Iris, Glass, Wine and Sonar are well-known and popular-used benchmark datasets. Table 2 shows the characteristics of the datasets.

Table 2. Summary of the characteristics of the considered data sets

Name of data set	Number of classes	Number of features	Size of data set (size of classes)
Iris	3	4	150(50,50,50)
Glass	6	9	214(29,76,70,17,13,9)
Wine	3	13	178(59,71,48)
Sonar	2	60	208(111,97)

4.2   Analysis of Algorithm Convergence

To evaluate the convergence performance, we have compared the ICS algorithm with traditional K-means, PSO and CS clustering algorithm on Iris data set.

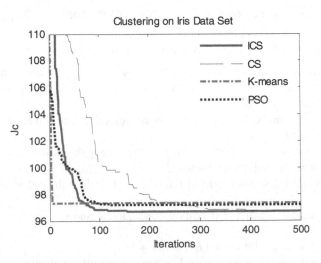

Fig. 2. Convergence curve of clustering on Iris data set

Fig. 2 illustrates that the ICS clustering algorithm has achieved the best convergence performance in the terms of the clustering objective function. K-means algorithm is easily to fall into local optimum due to the premature convergence [26]. The disadvantage of CS clustering algorithm is the slow convergence rate and vibration of the convergence; the convergence rate is insufficient when it searches global optimum.

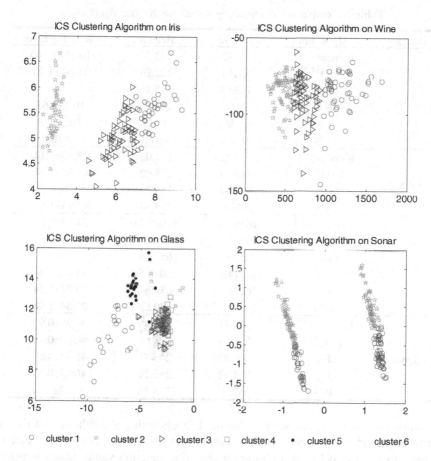

○ cluster 1   ☆ cluster 2   ▷ cluster 3   □ cluster 4   • cluster 5   cluster 6

Fig. 3. Results of ICS clustering algorithms on Iris, Wine, Glass and Sonar data sets

The results of ICS clustering algorithms on Iris, Wine, Glass and Sonar data sets is given in Fig. 3 which can make it visualized clearly. Principal component analysis was utilized to reduce the dimensionality of the data set. It can be seen from Fig. 3 clearly that ICS clustering algorithm possess better effect on Iris data set and Wine data set. The Glass data set has six classes and the Sonar data set has sixty features, so the higher data complexity leads to the larger clustering error.

4.3   Clustering Results

The best clustering objective function, mean clustering objective function and clustering error for K-means, PSO, GA, FA, CS and the proposed algorithm of ICS on different data set including Iris, Glass, Wine and Sonar are shown in Table 3. Experiments were repeated 30 times.

Table 3. Comparison of clustering results via the four algorithms

Data set	Algorithm	Best Jc	Mean Jc	Clustering Error
Iris	K-means [27]	97.32	102.57	16.05±10.10
	PSO [27]	97.10	102.26	10.64±4.50
	GA[7]	113.98	125.19	—
	FA	99.06	103.18	10.3±3.61
	CS	96.89	97.67	10.2±1.1
	ICS	96.66	96.68	9.6±0.6
Glass	K-means [27]	213.42	241.03	48.30±3.14
	PSO [27]	230.54	258.02	48.72±1.34
	CS	212.74	215.22	52.34±2.3
	ICS	210.95	213.84	43.93±1.89
Wine	K-means [27]	16555.68	17662.73	34.38±6.08
	PSO [27]	16307.16	16320.67	28.74±0.39
	GA[7]	16530.53	16530.53	—
	FA	16714.00	18070.59	31.46±3.45
	CS	16298.79	16309.24	29.21±1.34
	ICS	16295.67	16302.40	27.64±1.08
Sonar	K-means [27]	234.77	235.06	44.95±0.97
	PSO [27]	271.83	276.68	46.60±0.42
	FA	239.75	245.71	45.34±4.67
	CS	271.52	282.70	46.63±0.53
	ICS	232.20	238.58	44.23±0.24

As shown in Table 3, it is obvious that the ICS clustering algorithm could find the optimal clustering objective function value, and the mean clustering objective function value close to the best clustering objective function value, which illustrates the new algorithm has good stability. The results are exactly same as the phenomenon showed in Fig. 3.

5   Conclusion and Discussion

The ICS algorithm to solve clustering problems has been developed in this paper. To evaluate the performance of the ICS, it is compared with K-means, PSO and CS clustering algorithms on four well known UCI data sets. The experimental results indicated that the ICS clustering algorithm has best convergence performance, stability and better clustering effect. In order to improve the obtained results, we plan to apply the proposed approach into other clustering areas as our future work.

Acknowledgement. This paper is supported by the National Natural Science Foundation of China (61100164, 61173190), Scientific Research Start-up Foundation for Returned Scholars, Ministry of Education of China ([2012]1707) and the Fundamental Research Funds for the Central Universities, Shaanxi Normal University (GK201402035, GK201302025).

References

1. Han, J.W., Kamber, M., Pei, J.: Data Mining: Concepts and Techniques. Morgan Kaufmann Publishers (2011)
2. Lei, X.J.: Swarm Intelligent Optimization Algorithms and their Applications. Science Press (2012)
3. Maulik, U., Bandyopadhyay, S.: Genetic Algorithm-based Clustering Technique. Pattern Recognition 33, 1455–1465 (2000)
4. Kao, Y., Cheng, K.: An ACO-based clustering algorithm. In: 5th International Workshop on Ant Colony Optimization and Swarm Intelligence, pp. 340–347 (2006)
5. Van Der Merwe, D.W., Engelbrecht, A.P.: Data Clustering Using Particle Swarm Optimization. In: Congress on Evolutionary Computation (CEC 2003), pp. 215–220 (2003)
6. Zhang, Q., Lei, X.J., Huang, X., Zhang, A.D.: An Improved Projection Pursuit Clustering Model and its Application Based on Quantum-behaved PSO. In: 2010 Sixth International Conference on Natural Computation (ICNC 2010), vol. 5, pp. 2581–2585 (2010)
7. Zhang, C.S., Ouyang, D.T., Ning, J.X.: An Artificial Bee Colony Approach for Clustering. Expert Systems with Applications 37, 4761–4767 (2010)
8. Lei, X.J., Tian, J.F., Ge, L., Zhang, A.D.: The Clustering Model and Algorithm of PPI Network Based on Propagating Mechanism of Artificial Bee Colony. Information Sciences 247, 21–39 (2013)
9. Lei, X.J., Wu, S., Ge, L., Zhang, A.D.: Clustering and Overlapping Modules Detection in PPI Network Based on IBFO. Proteomics 13, 278–290 (2013)
10. Senthilnath, J., Omkar, S.N., Mani, V.: Clustering Using Firefly Algorithm: Performance study. Swarm and Evolutionary Computation 1, 164–171 (2011)
11. Ghodrati, A., Lotfi, S.: A Hybrid CS/PSO Algorithm for Global Optimization. In: Pan, J.-S., Chen, S.-M., Nguyen, N.T. (eds.) ACIIDS 2012, Part III. LNCS, vol. 7198, pp. 89–98. Springer, Heidelberg (2012)
12. Basu, M., Chowdhury, A.: Cuckoo Search Algorithm for Economic Dispatch. Energy 60, 99–108 (2013)
13. Saida, I.B., Nadjet, K., Omar, B.: A New Algorithm for Data Clustering Based on Cuckoo Search Optimization. Genetic and Evolutionary Computing 238, 55–64 (2014)
14. Senthilnath, J., Das, V., Omkar, S.N., Mani, V.: Clustering Using Lévy Flight Cuckoo Search. In: Bansal, J.C., Singh, P., Deep, K., Pant, M., Nagar, A. (eds.) Proceedings of Seventh International Conference on Bio-Inspired Computing: Theories and Applications, (BIC-TA 2012). AISC, vol. 202, pp. 65–75. Springer, Heidelberg (2013)
15. Goel, S., Sharma, A., Bedi, P.: Cuckoo Search Clustering Algorithm: A Novel Strategy of Biomimicry. In: World Congress on Information and Communication Technologies, pp. 916–926 (2011)

16. Manikandan, P., Selvarajan, S.: Data Clustering Using Cuckoo Search Algorithm (CSA). In: Babu, B.V., Nagar, A., Deep, K., Pant, M., Bansal, J.C., Ray, K., Gupta, U. (eds.) Proceedings of the Second International Conference on Soft Computing for Problem Solving (SocProS 2012). AISC, vol. 236, pp. 1275–1283. Springer, Heidelberg (2014)

17. Bulatović, R.R., Đorđević, S.R., Đorđević, V.S.: Cuckoo Search Algorithm: A Metaheuristic Approach to Solving the Problem of Optimum Synthesis of a Six-bar Double Dwell Linkage. Mechanism and Machine Theory 61, 1–13 (2013)

18. Yang, X.S.: Nature-Inspired Metaheuristic Algorithms, 2nd edn. Luniver Press (2010)

19. Yang, X.S., Deb, S.: Cuckoo Search via Lévy Flights. In: World Congress on Nature & Biologically Inspired Computind, pp. 210–214. IEEE Publications, USA (2009)

20. Payne, R.B., Sorenson, M.D., Klitz, K.: The Cuckoos. Oxford University Press (2005)

21. Valian, E., Mohanna, S., Tavakoli, S.: Improved Cuckoo Search Algorithm for Feedforward Neural Network Training. International Journal of Artificial Intelligence & Applications 2, 36–43 (2011)

22. Reynolds, A.M., Rhodes, C.J.: The Lévy Flight Paradigm: Random Search Patterns and Mechanisms. Concepts & Synthesis 90, 877–887 (2009)

23. Gandomi, A.H., Yang, X.S., Alavi, A.H.: Cuckoo Search Algorithm: a Metaheuristic Approach to Solve Structural Optimization Problems. Engineering with Computers 29, 17–35 (2013)

24. Mantegna, R.N.: Fast, Accurate Algorithm for Numerical Simulation of Lévy Stable Stochastic Processes. Physical Review E 49, 4677–4689 (1994)

25. UCI Machine Learning Repository, http://archive.ics.uci.edu/ml/

26. Kanungo, T., Mount, D.M., Netanyahu, N.S., Piatko, C.D., Silverman, R., Wu, A.Y.: An Efficient k-Means Clustering Algorithm. In: Analysis and Implementation. IEEE Transactions on Pattern Analysis and Machine Intelligence, pp. 881–892. IEEE Press, New York (2002)

27. Hassanzadeh, T., Meybodi, M.R.: A New Hybrid Approach for Data Clustering using Firefly Algorithm and K-means. In: CSI International Symposium on Artificial Intelligence and Signal Processing, pp. 7–11. IEEE Press, New York (2012)

Sample Index Based Encoding for Clustering Using Evolutionary Computation

Xiang Yang[1,2] and Ying Tan[2]

[1] AVIC Leihua Electronic Technology Research Institute, Wuxi 214063, China
[2] Key Laboratory of Machine Perception (Ministry of Education), Peking University,
Department of Machine Intelligence, School of EECS, Peking University,
Beijing 100871, China
{yangxiang,ytan}@pku.edu.cn

Abstract. Clustering is a commonly used unsupervised machine learning method, which automatically organized data into different clusters according to their similarities. In this paper, we carried out a throughout research on evolutionary computation based clustering. This paper proposed a sample index based encoding method, which significantly reduces the search space of evolutionary computation so the clustering algorithm converged quickly. Evolutionary computation has a good global search ability while the traditional clustering method k-means has a better capability at local search. In order to combine the strengths of both, this paper researched on the effect of initializing k-means by evolutionary computation algorithms. Experiments were conducted on five commonly used evolutionary computation algorithms. Experimental results show that the sample index based encoding method and evolutionary computation initialized k-means both perform well and demonstrate great potential.

Keywords: Clustering, evolutionary computation, sample index based encoding, initializing k-means.

1   Introduction

Clustering refers to partitioning data into different clusters according to their similarities. It is an unsupervised machine learning method that does not require the pre-specified categories information. It directly mining unlabeled data according to their inherent structure, and automatically partitions them into several clusters based on the intrinsic properties of the data. The goal of clustering is to get the greatest similarity between samples of the same cluster and the smallest similarity between samples of the different clusters. Clustering has been widely used in many fields. For example, in the field of text processing, document clustering can effectively organize different documents together based on subject, which can help people to filter the documents and find what they needed quickly.

There are many traditional clustering algorithms, such as k-means [1], hirarchical clustering [2] and expectation maximization (EM) [3]. The target of k-means algorithm is to get the smallest sum of distances within the same classes.

Y. Tan et al. (Eds.): ICSI 2014, Part I, LNCS 8794, pp. 489–498, 2014.
© Springer International Publishing Switzerland 2014

It will get a satisfying result through iterative updating the center of each cluster and updating the cluster each sample belongs to. However, using such a clustering approach always has the risk of falling into local optimum.

Evolutionary computation is a kind of optimization algorithm by simulating the natural biological process of evolution. Evolutionary computation algorithms generally maintain a population of solutions, and gradually improve the quality of the solution through the evolution of the population. There are some common evolutionary computation algorithms, like genetic algorithms [4], evolutionary strategies [5], differential evolution [6], particle swarm optimization algorithm [7] and firework algorithms [8]. Evolutionary computation has incomparable superiority compared to traditional optimization algorithms. It does not need to calculate the gradient of the objective function for achieving the optimization, and it has strong robustness, not easy to fall into local optimum.

In recent years, evolutionary computation is used by many researchers to improve the quality of clustering. Ujjwal Maulik et al. used the genetic algorithm to optimize the inter-clustering distance [9,10], Das, S. et al. and Paterlini, S. et al. applied differential evolution to the clustering problem [11,12], while Van der Merwe, DW et al. and Chen, ChingCYi et al. studied the particle swarm optimization based clustering [13,14].

In this paper, we first reviewed the basic principle and process of using evolutionary computation to cluster, and then we proposed a sample index based encoding method, this method can effectively reduce the search space of evolutionary computation algorithms, which will make the clustering algorithm converge quickly. Finally we focused on the study of initializing k-means by evolutionary computation algorithms.

This paper is organized as follows: Chapter 1 is introduction. Chapter 2 gives a brief introduction to k-means and the clustering based on evolutionary computation. Chapter 3 presents a sample index based encoding method. Chapter 4 demonstrates the using of evolutionary computation to initialize k-means. The experimental results are given in Chapter 5. Chapter 6 makes the summaries.

2   Formulated Description of Clustering

Given a data set D, whose amount of the samples is N and the dimension of each sample is d. $D = x_1, x_2, ..., x_N$ where x_i represents a d-dimensional vector, $i = 1, 2, ...N$. Clustering algorithms require these N samples be partitioned into K clusters. Many clustering algorithm use the centroids of the clusters to determine the cluster attribution. Assuming the centroid of the i-th cluster is c_i, $i = 1, 2, ..., K$. For the sample x_j, clustering algorithm will calculate the distance between x_j and all the centers of K clusters, and partition x_j into the k-th cluster if the distance between x_j and the centroid of the k-th category is the smallest. The process can be represented mathematically by the following formula:

$$k = \arg\min_i \|x_j - c_i\|^2 \tag{1}$$

where $\|x_j - c_i\|^2$ represents the Euclidean distance between the two vectors.

The goal of clustering is to make the samples inside the same cluster have the greatest similarity. This goal can be achieved by minimizing the within-cluster sum of squares (WCSS). The definition of WCSS is as follows [9]:

$$J = \sum_{i=1}^{K} \sum_{x_j \in C_i} \|x_j - c_i\|^2 \tag{2}$$

where C_i represents a collection of samples in class i, the WCSS represents the sum of all samples distances to their corresponding clusters centroid.

K-means applies an iterative way to update the centroid of the cluster to obtain a promising clustering result. It first randomly initialize a centroid for each cluster, then each samples cluster label is determined according to Formula 1, then the average of all samples within the cluster i is set as the new centroid of the i-th cluster, as shown in the following equation [1]:

$$c_i = \frac{\sum_{x_j \in C_i} x_j}{|C_i|} \tag{3}$$

After obtaining the new centroid for each cluster, k-means process the clustering according to Formula 1 again, and then get new cluster centroids. This process is iterated until the termination condition is satisfied.

The clustering result of k-means is susceptible to the initial cluster centroid. If the selection of initial cluster centroid is not good, k-means is easy to fall into local optimum. While the evolutionary computation has excellent ability of global optimization. Therefore using evolutionary computation to cluster data can get a better quality of clustering.

When using evolutionary computation to clustering, we have to encode the way of clustering into individuals. The individual is represented by a multi-dimensional vector and the way of encoding has significant impact on the clustering results. There two common ways used for encoding: cluster centroid based encoding [9] and sample category based encoding [15].

When using cluster centroid based encoding, all clusters centroids will be joined into a single vector, as an individual in evolutionary computation, which is represented as $< c_1, c_2, , c_K >$.

Assuming a data set has 1000 samples in total, and they can be partitioned into 20 clusters, each dimension of the data is 10, the clustering problem will be encoded as a 200-dimensional vector. The vector is represented by the conjunction of 20 10-dimensional vectors, the 20 vectors stand for the 20 centroids for all the 20 clusters.

Given the centroid of each category, each sample is partitioned to the corresponding cluster according to Formula 1, and then use the formula Formula 2 to get the WCSS, which will then be used as an individuals fitness function.

Sample category based encoding method encodes each sample's cluster directly. Each individuals dimension equals to the number of samples, the individual is expressed as $< L_1, L_2, , L_N >$, where L_j stands for sample x_js cluster label, $j = 1, 2, ..., N$, L_j ranges between 1 to K as an integer. If the j-th sample belonging to the category i, $i = 1, 2, ..., K$, then the individual value of the j-th dimension L_j is i.

After given each samples category, the center of each class is calculated by using the Formula 3, then the WCSS will be used as an individuals fitness function.

3   Sample Index Based Encoding

In order to further improve the quality of clustering, we propose to use the sample index based encoding method. Cluster centroid based encoding using a point in d-dimensional space to represent a cluster centroid, so each individuals dimension after encoding is $K * d$. Sample index based encoding using the sample from the sample set as a cluster centroid; we just have to record the samples index in the sample set. Therefore the individuals dimension after encoding is K. The individual is represented as $< I_1, I_2, , I_K >$.

Under such encoding method, the i-th clusters centroid is the I_i-th sample in the sample set x_{I_i}. After each clusters centroid is determined, each samples cluster is calculated by using Formula 1, then the WCSS will be used as an individuals fitness function.

Assuming a data set has 1000 samples in total, and they can be partitioned into 20 clusters, each dimension of the data is 10, the clustering problem will be encoded as a 20-dimensional vector. If the 10-th dimension of the vector is 426, the centroid of the 10-th cluster is the 426-th sample in the sample set.

The individuals dimension of cluster centroid based encoding is $K * d$, the individuals dimension of sample category based coding is N, while the individuals dimension of sample index based encoding is K. Generally K is much less than $K * d$ and N. Therefore the dimension of the proposed encoding method is much lower than the other two ones. The relationship between the search space of evolutionary computation and the individuals dimension is exponential, so the low-dimensional encoding means you can significantly reduce the search space. After the reduction, evolutionary computation algorithms can easily find the optimal solution and get a better result.

General speaking, before using evolutionary computation to search for the optimal solution, we need to specify the upper and lower bounds for each dimension in the search space. The upper and lower bounds of the cluster centroid based encoding is determined by the range of training data. For example the range of c_is x-th dimension should be equal to the range of the x-th dimension of all the data. So actually the process of evolutionary computation is to search a solution within a hypercube. But in general the data are not evenly distributed in the hypercube, the data may be concentrated in certain areas, and most areas in the search space don't have any data. The evolutionary computation algorithm will inevitably enter areas without data to search. This will waste a lot of time.

When using sample index based encoding, since each centroid comes from the sample set, the selection of centroids is more close to the real distribution of the data. So evolutionary computation algorithm will not enter the areas without data, but it will only search within areas of data. This mechanism ensures that the evolutionary algorithms search time is spent where it makes sense, thus evolutionary algorithm can quickly converge to the optimal solution, resulting in better clustering results.

4   K-means Initialized by Evolutionary Computation

Evolutionary computation is well known for its excellent global search capability. Evolutionary computation algorithms adopt the stochastic strategy to avoid trapping into the local optimum. Individuals have certain probability to jump out of current searching areas, which enables the evolutionary computation algorithms to explore the global optimum in the whole search space. But such algorithms may not perform well when searching the local area to further improve the current best solution. The stochastic way of local search cannot finely guide the current best solution to the actual best solution near it.

K-means minimizes the WCSS by iterating the following procedures: updating the cluster label of each sample according to the centroids of clusters, and then updating the centroids of clusters according the cluster label of each sample. The updating of centroids of clusters at successive iterations takes place in the local area; the centroids at the next iteration are not far from the centroids at the previous iteration. Therefore k-means has strong local search capability. But such searching strategy cannot explore the whole search space sufficiently, leading to a poor global search capability. If the initial centroids are not well chosen, k-means will trapping into the local optimum.

In order to exploit the synergy of global search ability and local search capability, the combination of evolutionary computation with k-means have been studied. For example Ahmadyfard, A. et al. combined particle swarm optimization algorithm and k-means algorithm to to get a better clustering algorithm [16].

In this section we combine the proposed sample index based encoding method with k-means to study their synergy. First we use an evolutionary computation algorithm to get K centroids. Evolutionary computation is able to obtain quite good centroids over the whole search space due to its excellent global search capability. But these centroids need to be further tuned in the local area to improve the clustering performance. We use k-means to tune these centroids by taking these centroids as the initial centroids of k-means. k-means will exploit the local area to search for better centroids in an iterative way. k-means initialized by evolutionary computation combines the strength of evolutionary computation and k-means, leading to both excellent global search capability and excellent local search capability.

The procedure of k-means initialized by evolutionary computation are shown in Algorithm 1.

Algorithm 1. K-means initialized by evolutionary computation.

1: Randomly initialize a population of individuals, each individuals dimension is K.
2: Calculate the fitness function for each individual in the population. First, we parse the centroid for each cluster from the sample index based encoding method. Then each samples cluster is determined according to the clusters centroids, and the WCSS will be used as an individuals fitness function.
3: Apply the evolutionary operations (such as selection, crossover, mutation, etc.) of evolutionary algorithms to get the next generation of the population from the current population.
4: If the termination condition of evolutionary algorithm meets, get the optimal solution and go to Step 5, otherwise go to Step 2.
5: Figure out the centroid of each cluster from the optimal individual obtained by evolutionary computation.
6: Each samples cluster is determined according to its closest centroid.
7: Calculate the mean vector of all the samples in each cluster. Then use the mean vector as the new cluster centroid.
8: If the termination condition of k-means is satisfied, go to Step 9, otherwise go to Step 6.
9: Output clustering result.

---

5 Experiments

5.1 Experimental Setup

In this paper, we use six evolutionary computation algorithms for clustering, which are differential evolution (DE) [6], the conventional fireworks algorithm (FWA) [8], enhanced fireworks algorithm (EFWA) [17], evolutionary strategies (ES) [5], genetic algorithms (GA) [4] and particle swarm optimization algorithm (PSO) [7].

For fireworks algorithm and enhanced fireworks algorithm, we use the default parameters from their original papers. We use the java library jMetal [18] to implement the other four algorithms, and the default parameters of jMetal is used for these algorithms. The maximum number of evaluations of all algorithms are set to 25000.

These algorithms will use a lot of random numbers at running time, so the results obtained by the algorithm will be different when run repeatedly. In order to obtain a stable measurement evolutionary computation based clustering, each experiment were run 20 times, and the average of the results will be used as the final result.

Experiments are conducted on eight commonly used document clustering data sets. Stacked auto-encoder is used to extract document feature [19]. At first the tf-idf feature [20] is extracted for the most frequent 2000 words. Then a stacked auto-encoder with the structure of $2000 - 500 - 250 - 125 - 10$ is used to extract abstract document feature. After such feature extraction each document is represented as a 10-dimensional vector. The name of each dataset, the number of samples, the dimension and the number of categories are shown in Table 1.

Table 1. Detailed information of the eight datasets

Dataset	Number of Samples	Dimension	Number of Categories
re0	1504	10	13
re1	1657	10	25
wap	1560	10	20
tr31	927	10	7
tr45	690	10	10
fbis	2463	10	17
la1	3204	10	6
la2	3075	10	6

Clustering algorithms need to set the number of clusters K in advance. These data in the dataset have the original category, we set the number of clusters equal to the number of original categories.

5.2 Comparison among Different Encoding Methods

In this section we will compare cluster centroid based encoding, sample category based encoding and sample index based encoding. The average WCSS over all of the 8 data sets is shown in Table 2.

Table 2. Within-cluster sum of squares of cluster centroid based encoding, sample category based encoding and sample index based encoding

Evolutionary Algorithms	cluster centroid	sample category	sample index
DE	935.08	1582.72	828.44
EFWA	1183.19	1645.08	814.41
ES	920.33	1425.40	800.64
FWA	1023.67	1634.44	819.05
GA	940.97	1524.56	811.12
PSO	1216.97	1639.51	881.40

We can see from the above results that the proposed sample index based encoding performs better than the two existing ways of encoding after optimization by all of the six algorithms. This shows that the proposed encoding method can effectively reflect the essence of the document sets structure, which made the optimization easily to be done. Thus the proposed sample index based encoding method has a great potential in the future development of clustering.

From Table 2, we can get the performance of different evolutionary computation algorithms. Evolutionary strategy (ES) performed best among all of the optimization algorithms, which was followed by the genetic algorithm (GA), enhanced fireworks algorithm (EFWA), fireworks algorithm (FWA) and differential evolution (DE), the worst one is the particle swarm optimization (PSO). Therefore the evolutionary computation algorithm used also has an import impact on the clustering performance.

We also compare the clustering results between evolutionary computation based clustering and k-means. The average WCSS of k-means over the 8 data sets is 811.32. As shown in Table 2, evolutionary strategy achieves better result than the k-means, so it can be used as a new and effective clustering methods. However, several other optimization algorithms effect is not significant. While the usage of evolutionary computation to initialize k-means can effectively improve the quality of clustering, experimental results are shown in the next section.

5.3   K-means Initialized by Evolutionary Computation

Fig. 1 gives the average WCSS over eight datasets of three clustering algorithms. The three clustering algorithms include clustering using evolutionary computation directly, k-means and k-means initialized by evolutionary computation. The average WCSS of clustering using PSO directly is 881.4, while other clustering algorithms are all below 835. We cut this extreme value in Fig. 1 to get a suitable figure; only the part below 835 is shown in Fig. 1.

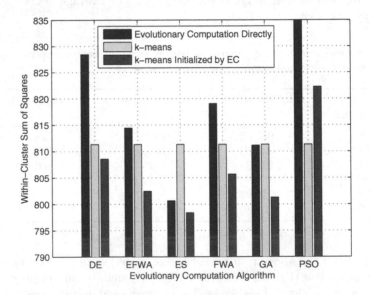

Fig. 1. Within-cluster sum of squares of clustering using evolutionary computation directly, k-means and k-means initialized by evolutionary computation

From the figure we can see that k-means initialized by differential evolution, enhanced fireworks algorithm, evolution strategy, fireworks algorithm and genetic algorithm is superior to original k-means and the direct evolutionary computation algorithms. Therefore the initialization strategy of k-means is able to improve the performance of clustering evidently.

It is easy to fall into local optimum for traditional k-means initialized at random. While k-means initialized by evolutionary computation will locate the initial centroids near the optimal centroids. In such case k-means who has excellent local searching ability will find the optimal centroids easily. The clustering performance of particle swarm optimization is slightly poor. This is because particle swarm optimization doesn't converge well for clustering.

6 Conclusions

This paper introduces the basic principle and procedures of k-means and clustering using evolutionary computation. A novel encoding method based on sample index is proposed in this paper. We combine k-means and evolutionary computation by initializing k-means by evolutionary computation to enhance the clustering performance. At last this paper gives the experimental results of evolutionary computation based clustering over six common evolutionary computation algorithms.

The proposed sample index based encoding method significantly outperforms cluster centroid based encoding and sample category based encoding. The encoding method based on sample index is able to restrict the centroids within the training data. The evolutionary computation algorithms will concentrate on meaningful search space to get a better solution. Whats more, the search space of this encoding method is much smaller than the other two encoding methods due to its lower dimension, therefore evolutionary computation algorithms will find the optimal centroids more easily.

K-means initialized by evolutionary computation is superior to the original k-means and using evolutionary computation directly. Evolutionary computation is good at global search, while k-means is good at local search. Initializing k-means by evolutionary computation is able to combine the two advantages and improve the clustering performance.

Acknowledgments. This work was supported by National Natural Science Foundation of China (NSFC) with Grant No. 61375119, No. 61170057 and No. 60875080.

References

1. Hartigan, J.A., Wong, M.A.: Algorithm as 136: A k-means clustering algorithm. Applied Statistics, 100–108 (1979)
2. Johnson, S.C.: Hierarchical clustering schemes. Psychometrika 32(3), 241–254 (1967)
3. Moon, T.K.: The expectation-maximization algorithm. IEEE Signal Processing Magazine 13(6), 47–60 (1996)

4. Goldberg, D.E., et al.: Genetic algorithms in search, optimization, and machine learning, vol. 412. Addison-wesley, Reading (1989)
5. Liem, K.F.: Evolutionary strategies and morphological innovations: cichlid pharyngeal jaws. Systematic Biology 22(4), 425–441 (1973)
6. Storn, R., Price, K.: Differential evolution–a simple and efficient heuristic for global optimization over continuous spaces. Journal of Global Optimization 11(4), 341–359 (1997)
7. Kennedy, J., Eberhart, R., et al.: Particle swarm optimization. In: Proceedings of IEEE International Conference on Neural Networks, Perth, Australia, vol. 4, pp. 1942–1948 (1995)
8. Tan, Y., Zhu, Y.: Fireworks algorithm for optimization. In: Tan, Y., Shi, Y., Tan, K.C. (eds.) ICSI 2010, Part I. LNCS, vol. 6145, pp. 355–364. Springer, Heidelberg (2010)
9. Maulik, U., Bandyopadhyay, S.: Genetic algorithm-based clustering technique. Pattern Recognition 33(9), 1455–1465 (2000)
10. Bandyopadhyay, S., Maulik, U.: An evolutionary technique based on k-means algorithm for optimal clustering in rn. Information Sciences 146(1), 221–237 (2002)
11. Das, S., Abraham, A., Konar, A.: Automatic clustering using an improved differential evolution algorithm. IEEE Transactions on Systems, Man and Cybernetics, Part A: Systems and Humans 38(1), 218–237 (2008)
12. Paterlini, S., Krink, T.: Differential evolution and particle swarm optimisation in partitional clustering. Computational Statistics & Data Analysis 50(5), 1220–1247 (2006)
13. Van der Merwe, D., Engelbrecht, A.P.: Data clustering using particle swarm optimization. In: The 2003 Congress on Evolutionary Computation, CEC 2003, vol. 1, pp. 215–220. IEEE (2003)
14. Chen, C.Y., Ye, F.: Particle swarm optimization algorithm and its application to clustering analysis. In: 2004 IEEE International Conference on Networking, Sensing and Control, vol. 2, pp. 789–794. IEEE (2004)
15. Forsati, R., Mahdavi, M., Shamsfard, M., Reza Meybodi, M.: Efficient stochastic algorithms for document clustering. Information Sciences 220, 269–291 (2013)
16. Ahmadyfard, A., Modares, H.: Combining pso and k-means to enhance data clustering. In: International Symposium on Telecommunications, IST 2008, pp. 688–691. IEEE (2008)
17. Zheng, S., Janecek, A., Tan, Y.: Enhanced fireworks algorithm. In: 2013 IEEE Congress on Evolutionary Computation (CEC), pp. 2069–2077. IEEE (2013)
18. Durillo, J.J., Nebro, A.J., Alba, E.: The jmetal framework for multi-objective optimization: Design and architecture. In: 2010 IEEE Congress on Evolutionary Computation (CEC), pp. 1–8. IEEE (2010)
19. Hinton, G.E., Salakhutdinov, R.R.: Reducing the dimensionality of data with neural networks. Science 313(5786), 504–507 (2006)
20. Salton, G., McGill, M.J.: Introduction to Modern Information Retrieval. McGraw-Hill, Inc., New York (1986)

Data Mining Tools Design with Co-operation of Biology Related Algorithms[*]

Shakhnaz Akhmedova and Eugene Semenkin

Siberian Aerospace University, Krasnoyarsky rabochy avenue 31,
660014, Krasnoyarsk, Russia
shahnaz@inbox.ru, eugenesemenkin@yandex.ru

Abstract. Artificial neural network (ANN) and support vector machine (SVM) based classifier design by a meta-heuristic called Co-Operation of Biology Related Algorithms (COBRA) is presented. For the ANN's structure selection the modification of COBRA that solves unconstrained optimization problems with binary variables is used. The ANN's weight coefficients are adjusted with the original version of COBRA. For the SVM-based classifier design the original version of COBRA and its modification for solving constrained optimization problems are used. Three text categorization problems from the DEFT'07 competition were solved with these techniques. Experiments showed that all variants of COBRA demonstrate high performance and reliability in spite of the complexity of the solved optimization problems. ANN-based and SVM-based classifiers developed in this way outperform many alternative methods on the mentioned benchmark classification problems. The workability of the proposed meta-heuristic optimization algorithms was confirmed.

Keywords: Bio-inspired optimization algorithms, neural networks, support vector machines, text categorization.

1 Introduction

Data Mining is the computational process of discovering patterns in large data sets involving methods at the intersection of artificial intelligence, machine learning, statistics, and database systems. The basic idea of the data mining process is to extract information from a data set and transform it into an understandable structure for further use.

Data mining involves different classes of tasks and classification is one of them. Classification is the problem of identifying which of a set of categories a new observation belongs to on the basis of a training set of data that contains observations whose category is known. In this work we consider text categorization tasks as a representative of data mining problems. There are various approaches which can be used for solving this kind of problems and among the most famous are Artificial Neural

Research is fulfilled with the support of the Ministry of Education and Science of Russian Federation within State assignment project 140/14.

Networks (ANNs) and Support Vector Machines (SVMs). In machine learning, the SVMs and ANNs are supervised learning models; the basic SVM and ANN take a set of input data and predict which of the possible classes will form the output.

The ANN models have three components: the input data layer, the hidden layer(s) and the output layer. Each of these layers contains nodes and these nodes are connected to nodes at adjacent layer(s). Also there is an activation function on each node. So the ANN's structure contains a number of hidden layers, a number of nodes (neurons) on each layer, and a type of activation function on each node. Nodes in the network are interconnected and each connection has a weight coefficient; the number of these coefficients depends on the problem solved (number of inputs) and the number of hidden layers and nodes. Thus, networks with a more or less complex structure usually have many weight coefficients which should be adjusted with some optimization algorithm. In this study we use the collective bionic meta-heuristic called Co-operation of Biology Related Algorithms (COBRA) [1] for the ANNs' weight coefficients adjustment and its binary modification (COBRA-b) for the ANNs' structure design.

The SVM model is a representation of the examples (input data) as points in a space, mapped so that examples of the separate classes are divided by a clear gap that should be as wide as possible. New examples are then mapped into the same space and are predicted to belong to a category depending on which side of the gap they fall on [2]. In this study we have chosen the polynomial kernel function with three parameters for solving text categorization problems, so COBRA and its modification for constrained optimization (COBRA-c) were used for SVM-classifiers design.

The rest of the paper is organized as follows. In Section 2 the developed meta-heuristic and its modifications are presented. Section 3 describes the ANN-based and SVM-based classifier design by the mentioned optimization tools. In Section 4 the term relevance estimation for text categorization problems [3] is explained. Section 5 demonstrates the workability of the optimization of meta-heuristics within the ANN-based and SVM-based classifier design for three text categorization problems from the DEFT'07 competition. The conclusion contains the discussion of results and considerations for further research directions.

2 Co-Operation of Biology Related Algorithms (COBRA)

Existing meta-heuristic algorithms, such as Particle Swarm Optimization or the Firefly Algorithm, started the demonstration of their power dealing with tough optimization problems and even NP-hard problems. Five well-known and similar nature-inspired algorithms such as Particle Swarm Optimization (PSO), Wolf Pack Search (WPS), the Firefly Algorithm (FFA), the Cuckoo Search Algorithm (CSA) and the Bat Algorithm (BA) were used for the development of a new meta-heuristic.

The results of our investigation into the effectiveness of these optimization methods [1] showed that we cannot say which approach is the most appropriate for the given optimization problem and the given dimension (number of variables). The best results were obtained by different methods for different problems and for different dimensions; in some cases the best algorithm differs even for the same test problem if the dimension varies. Each strategy has its advantages and disadvantages. So it brought researchers to the idea of formulating a new meta-heuristic approach that combines the major advantages of the algorithms listed above. The proposed approach was called Co-Operation of Biology Related Algorithms (COBRA) [1]. Its basic idea consists of generating five populations (one population for each algorithm) which are then executed in parallel, cooperating with each other (so-called island model). The parameters of component optimization algorithms were set according to the original recommendations of algorithms' authors. Up-to-date adjusting techniques of these algorithms were not used in order to check the co-operation idea itself.

The proposed algorithm is a self-tuning meta-heuristic. That is why there is no necessity to choose the population size for each algorithm. The number of individuals in each algorithm's population can increase or decrease depending on whether the fitness value was improving on the current stage or not. If the fitness value has not improved during a given number of generations, then the size of all populations increases. And vice versa; if the fitness value has constantly improved, then the size of all populations decreases. Besides, each population can "grow" by accepting individuals removed from other populations. A population "grows" only if its average fitness is better than the average fitness of all other populations. Thereby we can determine the "winner algorithm" on each iteration/generation. The result of this kind of competition allows us to present the biggest resource (population size) to the most appropriate (in the current generation) algorithm. This property can be very useful in the case of a hard optimization problem when, as it is known, there is no single best algorithm on all stages of the optimization process execution.

Besides, all populations communicate with each other: they exchange individuals in such a way that a part of the worst individuals of each population is replaced by the best individuals of other populations. It brings up-to-date information of the best achievements to all component algorithms and prevents their preliminary convergence to their own local optimum which improves the group performance of all algorithms.

The performance of the proposed algorithm was evaluated on the set of benchmark problems (28 unconstrained real-parameter optimization problems with 10, 30 and 50 variables) from the CEC'2013 competition. Experiments showed that COBRA works successfully and is reliable on this benchmark and demonstrates competitive behaviour. Results also showed that COBRA outperforms its component algorithms when the dimension grows and more complicated problems are solved [1].

2.1   Binary Modification of COBRA

All the algorithms listed above (PSO, WPS, FFA, CSA and BA) were originally developed for continuous valued spaces. However many applied problems are defined in discrete valued spaces where the domain of the variables is finite. For this purpose the binary modification of COBRA, COBRA-b, was developed.

COBRA was adapted to search in binary spaces by applying a sigmoid transformation to the velocity component (PSO, BA) and coordinates (FFA, CSA, WPS) to squash them into a range [0, 1] and force the component values of the positions of the particles to be 0's or 1's.

The basic idea of this adaptation was taken from [4]; firstly it was used for the PSO algorithm. It's known that in PSO each particle has a velocity, so the binarization of individuals is conducted by the use of the calculation value of the sigmoid function which is also given in [4]:

$$s(v) = 1/(1+exp(-v)). \tag{1}$$

After that a random number from the range [0, 1] is generated and the corresponding component value of the particle's position is 1 if this number is smaller than s(v) and 0 otherwise.

In BA each bat also has a velocity, which is why we can apply exactly the same procedure for the binarization of this algorithm. But in WPS, FFA and CSA individuals have no velocities. For this reason, the sigmoid transformation is applied to position components of individuals and then a random number is compared with the obtained value. The performance of COBRA-b evaluated on benchmark optimization problems from [5]. Experiments showed that COBRA-b works successfully and reliably enough but slower than the original version of COBRA for the same problems with a smaller success rate obtained [6].

2.2   COBRA Modification for Constrained Optimization Problems

The next step in our study was about the development and investigation of COBRA-c, i.e., COBRA's modification that can be used for solving constrained real-parameter optimization problems.

For these purposes three constraint handling methods were used: widely known dynamic penalties, Deb's rule [7] and the technique that was described in [8]. The method proposed in [8] was implemented for the PSO-component of COBRA; at the same time other components were modified by implementing Deb's rule followed by calculating function values using dynamic penalties. The performance of the proposed algorithm was evaluated on the set of 18 scalable benchmark functions provided for the CEC 2010 competition [9], when the dimension of decision variables is set to 10 and 30. COBRA-c was compared with algorithms that took part in the competition CEC 2010 and was superior to 3-4 of the 14 methods from this competition. Besides, COBRA-c outperforms all its component algorithms.

3   Data Mining Tools Design with COBRA

3.1   Artificial Neural Networks Design

The neural networks' structure design and the tuning of its weight coefficients are considered as the solving of two unconstrained optimization problems: the first one with binary variables and the second one with real-valued variables. The type of variables depends on the representation of the ANN's structure and coefficients.

We set the maximum number of hidden layers to equal 5 and the maximum number of neurons in each hidden layer equal to 5. Each node is represented by a binary string of length 4. If the string consists of zeros ("0000") then this node does not exist in ANN. So, the whole structure of the neural network is represented by a binary string of length 100 (25x4), and each 20 variables represent one hidden layer. The number of input layers depends on the problem in hand. ANN has one output neuron.

We use 15 known activation functions for nodes: the sigmoidal function, the linear function, the hyperbolic tangent function and others. For determining which activation function will be used on a given node, the integer that corresponds to its binary string is calculated.

Thus we use the optimization method for problems with binary variables (COBRA-b) for finding the best structure and the optimization method for problems with real-valued variables (COBRA) for the adjustment of every structure weight coefficients.

3.2   Support Vector Machine Classifiers Design

In the most common formulation, support vector machines are classification mechanisms, which, given a training set

$$X^l = \{(x_1, y_1),...,(x_l, y_l)\}, x_i \in R^m, y_i \in \{-1;1\}, \tag{2}$$

assuming l examples with m real attributes, learn a hyper-plane:

$$<w, x> +b = 0, \tag{3}$$

where $<...>$ – dot product, which separates examples labelled as -1 from the ones labelled as +1. So using this hyper-plane, a new instance x is classified using the following classifier:

$$f(x) = \begin{cases} 1, (< w, x > +b) \geq 1 \\ -1, (< w, x > +b) \leq -1 \end{cases}. \tag{4}$$

However, the given data set is not always linearly separable, and in this case SVM (as a linear classifier) does not provide satisfying classification results. One way to solve this problem is to map the data onto a higher dimension space and then to use a linear classifier in that space. The general idea is to map the original feature space to some higher-dimensional feature space where the training set is linearly separable. SVM provides an easy and efficient way of doing this mapping to a higher dimension space, which is referred to as the kernel trick [2].

In this study the polynomial kernel is used. So, let $K(x, x') = (\alpha < x, x' > +\beta)^d$, where α, β, d are parameters of the kernel function K. Then the classifier is:

$$f(x) = \begin{cases} 1, ((K(w, x) + \beta)^d + b) \geq 1 \\ -1, ((K(w, x) + \beta)^d + b) \leq -1 \end{cases}. \tag{5}$$

It means that the following constrained optimization problem should be solved:

$$\|w\|^2 \to \min, \tag{6}$$

$$y_i ((\alpha < w, x_i > +\beta)^d + b) \geq 1, i = \overline{1, l}. \tag{7}$$

Thus for solving a classification problem, the kernel function's parameters α, β, d, a vector w and a shift factor b should be determined, i.e. the constrained optimization problem with continuous variables must be solved.

4   Term Relevance Estimating

It is well known that the way that documents are represented influences on the performance of the text classification algorithms. Generally, documents are not classified as sequences of symbols. They are usually transformed into vector representation because most machine learning algorithms are designed for vector space models. The document mapping into the feature space remains a complex non trivial task. Many researchers develop new algorithms for text preprocessing. In this work, the technique described in [3] was used.

The basic idea is that every word that appears in the text has to contribute some value to a certain class. So, a real number term relevance is assigned for each word; and this number depends on the frequency of the word occurrence. The term relevance is calculated using a modified formula of fuzzy rules relevance estimation for the fuzzy classifier. The membership function has been replaced by word frequency in the current class.

Let L be the number of classes; n_i is the number of instances of the i-th class; N_{ji} is the number of j-th word occurrence in all instances of the i-th class; $T_{ji} = \dfrac{N_{ji}}{n_i}$ is the relative frequency of j-th word occurrence in the i-th class. $R_j = \max_i T_{ji}, S_j = \arg(\max_i T_{ji})$ is the number of class which we assign to j-th word. The term relevance, C_j, is calculated in the following way:

$$C_j = \frac{1}{\sum\limits_{i=1}^{L} T_{ji}} (R_j - \frac{1}{L-1} \sum\limits_{\substack{i=1 \\ i \neq S_j}}^{L} T_{ji}). \tag{8}$$

So each instance is represented by a vector of $L+1$ numbers, where the first number is a class identifier, and the other numbers are the sum of C_j values of all words that occurred in this instance (according to their S_j).

5   Experimental Results

The DEFT07 ("Défi Fouille de Texte") Evaluation Package [10] has been used for the application of algorithms and the comparison of results. For the testing of the proposed approach three corpora were used: "A voir à lire", "Video games" and "Debates in Parliament".

Table 1. Test corpora

Corpus	Description	Marking scale
A voir à lire	3,000 commentaries about books, films and shows	0:unfavorable, 1:neutral, 2:favorable
Video games	4,000 commentaries about video games	0:unfavorable, 1:neutral, 2:favorable
Debates in Parliament	28,800 interventions by Represen-tatives in the French Assembly	0:against the proposed law, 1:for it

The F-score value with β=1 was used for evaluating the obtained results:

$$F - score = \frac{(\beta^2 + 1) * precision * recall}{\beta^2 * (precision + recall)}, \qquad (9)$$

The classification "precision" for each class is calculated as the number of correctly classified instances for a given class divided by the number of all instances of the algorithm assigned for this class. "Recall" is the number of correctly classified instances for a given class divided by the number of instances that should have been in this class.

The results for all text categorization problems are presented in Table 2 (there are also results obtained for the best submission of other researchers for each corpus). It should be mentioned that our algorithms exhibit good stability demonstrating the same best results in the majority of runs.

Table 2. Comparison of results obtained by different research teams

Researchers	A voir à lire (rank1)	Video Games (rank2)	Debates (rank3)	Rank
J.-M. Torres-Moreno (LIA)	0.603 (2)	0.784 (1)	0.720 (1)	1
G. Denhiere (EPHE et U. Wurzburg)	0.599 (3)	0.699 (5)	0.681 (6)	4
S. Maurel (CELI France)	0.519 (8)	0.706 (4)	0.697 (4)	5
M. Vernier (GREYC)	0.577 (5)	0.761 (3)	0.673 (7)	6
E. Crestan (Yahoo ! Inc.)	0.529 (7)	0.673 (8)	0.703 (3)	7
M. Plantie (LGI2P et LIRMM)	0.472 (10)	0.783 (2)	0.671 (9)	8
A.-P. Trinh (LIP6)	0.542 (6)	0.659 (9)	0.676 (8)	9
M. Genereux (NLTG)	0.464 (11)	0.626 (10)	0.569 (12)	11
E. Charton (LIA)	0.504 (9)	0.619 (11)	0.616 (10)	10
A. Acosta (Lattice)	0.392 (12)	0.536 (12)	0.582 (11)	12
SVM+COBRA	0.619 (1)	0.696 (6)	0.692 (5)	2
ANN+COBRA	0.585 (4)	0.692 (7)	0.7032 (2)	3

Results for each corpus were ranked and then the total rank was evaluated by the following formula:

$$Rank = \frac{rank1 + rank2 + rank3}{3} \qquad (14)$$

6   Conclusion

In this paper we have described a new meta-heuristic, called Co-Operation of Biology Related Algorithms, and introduced its modification for solving unconstrained optimization problems with binary variables (COBRA-b) and constrained optimization problems with real-valued variables (COBRA-c). We illustrated the performance estimation of the proposed algorithms on sets of test functions.

Then we used the described optimization methods for the automated design of ANN-based and SVM-based classifiers. These approaches were applied to three text categorization problems which were taken from the DEFT'07 competition. For this purpose an alternative formula for word relevance estimation was used.

Solving these problems is equivalent to solving big and hard optimization problems where objective functions have many variables and are given in the form of a computational program. The suggested algorithms successfully solved all problems designing classifiers with a competitive performance, which allows us to consider the study results as confirmation of the algorithm's reliability, workability and usefulness in solving real world optimization problems.

Having these appropriate tools for data mining we consider the following directions of approach development: the design of other types of neural network models, the design of SVMs with alternative kinds of kernel, the application to the design of fuzzy systems, and the improvement of COBRAs optimization performance.

References

1. Akhmedova, S., Semenkin, E.: Co-Operation of Biology Related Algorithms. In: IEEE Congress on Evolutionary Computation (CEC 2013), Cancún, México, pp. 2207–2214 (2013)
2. Boser, B., Guyon, I., Vapnik, V.: A Training Algorithm for Optimal Margin Classifiers. In: Haussler, D. (ed.) 5th Annual ACM Workshop on COLT, Pittsburgh, pp. 144–152 (1992)
3. Gasanova, T., Sergienko, R., Minker, W., Semenkin, E., Zhukov, E.: A Semi-supervised Approach for Natural Language Call Routing. In: SIGDIAL 2013 Conference, pp. 344–348 (2013)
4. Kennedy, J., Eberhart, R.: A Discrete Binary Version of the Particle Swarm Algorithm. In: World Multiconference on Systemics, Cybernetics and Informatics, pp. 4104–4109, Piscataway, NJ (1997)
5. Molga, M.: Smutnicki, C.: Test Functions for Optimization Need (2005)
6. Akhmedova, S., Semenkin, E.: New Optimization Metaheuristic Based on Co-Operation of Biology Related Algorithms. Vestnik. Bulletine of Siberian State Aerospace University 4(50), 92–99 (2013)
7. Deb, K.: An Efficient Constraint Handling Method for Genetic Algorithms. Computer Methods in Applied Mechanics and Engineering 186(2-4), 311–338 (2000)
8. Liang, J.J., Shang, Z., Li, Z.: Coevolutionary Comprehensive Learning Particle Swarm Optimizer. In: Congress on Evolutionary Computation (CEC 2010), pp. 1505–1512 (2010)
9. Mallipeddi, R., Suganthan, P.N.: Problem Definitions and Evaluation Criteria for the CEC 2010 Competition on Constrained Real-Parameter Optimization. Technical report, Nanyang Technological University, Singapore (2009)
10. Actes de l'atelier DEFT 2007, Plate-forme AFIA 2007, Grenoble, Juillet (2007), http://deft07.limsi.fr/actes.php

Author Index